Coulson and Richardson's
Chemical Engineering

Coulson & Richardson's Chemical Engineering Series

Chemical Engineering, Volume 1A, Seventh edition
Fluid Flow: Fundamentals and Applications
Raj Chhabra and V. Shankar

Chemical Engineering, Volume 1B, Seventh edition
Heat and Mass Transfer: Fundamentals and Applications
Raj Chhabra and V. Shankar

Chemical Engineering, Volume 2A, Sixth edition
Particulate Systems and Particle Technology
Raj Chhabra and Basa M. Gurappa

Chemical Engineering, Volume 2B, Sixth edition
Separation Processes
A. K. Ray

Chemical Engineering, Volume 3A, Fourth edition
Chemical and Biochemical Reactors and Reaction Engineering
R. Ravi, R. Vinu and S. N. Gummadi

Chemical Engineering, Volume 3B, Fourth edition
Process Control
Sohrab Rohani

Chemical Engineering
Solutions to the Problems in Volume 1
J. R. Backhurst, J. H. Harker and J. F. Richardson

Chemical Engineering
Solutions to the Problems in Volumes 2 and 3
J. R. Backhurst, J. H. Harker and J. F. Richardson

Chemical Engineering, Volume 6, Third edition
Chemical Engineering Design
R. K. Sinnott

Coulson and Richardson's Chemical Engineering

Volume 1B: Heat and Mass Transfer: Fundamentals and Applications

Seventh Edition

Raj Chhabra
V. Shankar

Butterworth-Heinemann
An imprint of Elsevier

Butterworth-Heinemann is an imprint of Elsevier
The Boulevard, Langford Lane, Kidlington, Oxford OX5 1GB, United Kingdom
50 Hampshire Street, 5th Floor, Cambridge, MA 02139, United States

Notices

Knowledge and best practice in this field are constantly changing. As new research and experience broaden our understanding,
changes in research methods, professional practices, or medical treatment may become necessary.

Practitioners and researchers must always rely on their own experience and knowledge in evaluating and using any information,
methods, compounds, or experiments described herein. In using such information or methods they should be mindful of their
own safety and the safety of others, including parties for whom they have a professional responsibility.

To the fullest extent of the law, neither the Publisher nor the authors, contributors, or editors, assume any liability for any injury and/
or damage to persons or property as a matter of products liability, negligence or otherwise, or from any use or operation of
any methods, products, instructions, or ideas contained in the material herein.

Library of Congress Cataloging-in-Publication Data
A catalog record for this book is available from the Library of Congress

British Library Cataloguing-in-Publication Data
A catalogue record for this book is available from the British Library

ISBN: 978-0-08-102550-5

For information on all Butterworth-Heinemann publications
visit our website at https://www.elsevier.com/books-and-journals

www.elsevier.com • www.bookaid.org

Publisher: Joe Hayton
Acquisition Editor: Anita Koch
Editorial Project Manager: Ashlie Jackman
Production Project Manager: Sruthi Satheesh
Cover Designer: Victoria Pearson

Typeset by SPi Global, India

Contents

PART 1 HEAT TRANSFER

PART 2 MASS TRANSFER

PART 3 MOMENTUM, HEAT AND MASS TRANSFER

About Professor Coulson

John Coulson, who died on 6 January 1990 at the age of 79, came from a family with close involvement with education. Both he and his twin brother Charles (renowned physicist and mathematician), who predeceased him, became professors. John did his undergraduate studies at Cambridge and then moved to Imperial College where he took the postgraduate course in chemical engineering—the normal way to qualify at that time—and then carried out research on the flow of fluids through packed beds. He then became an assistant lecturer at Imperial College and, after wartime service in the Royal Ordnance Factories, returned as lecturer and was subsequently promoted to a readership. At Imperial College he initially had to run the final year of the undergraduate course almost single-handed, a very demanding assignment. During this period, he collaborated with Sir Frederick (Ned) Warner to write a model design exercise for the IChemE home paper on 'The Manufacture of Nitrotoluene'. He published research papers on heat transfer and evaporation, on distillation, and on liquid extraction and co-authored this textbook of chemical engineering. He did valiant work for the Institution of Chemical Engineers that awarded him its Davis medal in 1973 and was also a member of the advisory board for what was then a new Pergamon journal, Chemical Engineering Science.

In 1954, he was appointed to the newly established chair at Newcastle upon Tyne, where chemical engineering became a separate department and independent of mechanical engineering of which it was formerly part and remained there until his retirement in 1975. He took a period of secondment to Heriot Watt University where, following the splitting of the joint Department of Chemical Engineering with Edinburgh, he acted as adviser and de facto head of department. The Scottish university awarded him an honorary DSc in 1973.

John's first wife Dora sadly died in 1961; they had two sons, Anthony and Simon. He remarried in 1965 and is survived by Christine.

John F. Richardson

About Professor Richardson

Professor **John Francis Richardson**, Jack to all who knew him, was born at Palmers Green, North London, on 29 July 1920 and attended the Dame Alice Owen's School in Islington. Subsequently, after studying chemical engineering at Imperial College, he embarked on research into the suppression of burning liquids and of fires. This early work contributed much to our understanding of the extinguishing properties of foams, carbon dioxide, and halogenated hydrocarbons, and he spent much time during the war years on large-scale fire control experiments in Manchester and at the Llandarcy refinery in South Wales. At the end of the war, Jack returned to Imperial College as a lecturer where he focussed on research in the broad area of multiphase fluid mechanics, especially sedimentation and fluidisation, two-phase flow of a gas and a liquid in pipes. This laid the foundation for the design of industrial processes like catalytic crackers and led to a long-lasting collaboration with the Nuclear Research Laboratories at Harwell. This work also led to the publication of the famous paper, now common knowledge, the so-called Richardson-Zaki equation, which was selected as the Week's citation classic (*Current Contents*, 12 February 1979)!

After a brief spell with Boake Roberts in East London, where he worked on the development of novel processes for flavours and fragrances, he was appointed the professor of chemical engineering at the then University College of Swansea (now University of Swansea) in 1960. He remained there until his retirement in 1987 and thereafter continued as an emeritus professor until his death on 4 January 2011.

Throughout his career, his major thrust was on the wellbeing of the discipline of chemical engineering. In the early years of his teaching duties at Imperial College, he and his colleague John Coulson recognised the lack of satisfactory textbooks available in the field of chemical engineering. They set about rectifying the situation, and this is how the now well-known Coulson-Richardson series of books on chemical engineering was born. The fact that this series of books (six volumes) is as relevant today as it was at the time of their first appearance is a testimony to the foresight of John Coulson and Jack Richardson.

Throughout his entire career spanning almost 40 years, Jack contributed significantly to all facets of professional life, teaching, research in multiphase fluid mechanics, and service to the Institution of Chemical Engineers (IChemE, the United Kingdom). His professional work and long-standing public service was well recognised. Jack was the president of IChemE during the period 1975–76; he was named a fellow of the Royal Academy of Engineering in 1978 and was awarded an OBE in 1981.

In his spare time, Jack and his wife Joan were keen dancers, having been founder members of the Society of International Folk Dancing, and they also shared a love of hill walking.

Raj Chhabra

Preface to Seventh Edition

The sixth edition of this title appeared in 1999 at the dawn of the new millennium (with reprints in 2000 and 2003). There is a gap of almost two decades between the sixth edition and the seventh edition in your hands now. The fact that this title has never been out of print is a testimony to its timelessness and 'evergreen' character, both in terms of content and style. The sole reason for this unusually long gap between the sixth and seventh editions is obviously the fact that Jack Richardson passed away in 2011 and, therefore, the publishers needed to establish whether it was a worthwhile project to continue with. The question was easily answered in the affirmative by numerous independent formal reviews and by the continuous feedback from students, teachers, and working professionals from all over the world. Having established that there was a definite need for this title, the next step was to identify individuals who would have the inclination to carry forward the legacy of Coulson and Richardson. Indeed, we feel privileged to have been entrusted with this onerous task.

The basic philosophy and the objectives of this edition remain the same as articulated so very well by the previous authors of the sixth edition. In essence, this volume continues to concentrate on the fundamentals of heat and mass transfer, as applied to a wide-ranging industrial settings. Most of the concepts have been illustrated by including examples of practical applications in the areas of estimating heat losses by conduction, convection and radiation. Similarly, consideration is given to the process design of heat exchangers with and without phase change, to the analysis of cooling towers, etc. The entire volume has been reviewed keeping in mind the feedback received from the readers and the reviewers. Wherever needed, both contents and presentation have been improved by reorganising the existing material for easier understanding, or new material has been added to provide updated and reliable information. Apart from the general revision of all the chapters, the specific changes made in this edition are summarised below:

(i) In Chapter 1, a short section on transient conduction in a semi-infinite medium has been added (Examples 1.7 and 1.8).
(ii) A new Section 1.37 on multi-dimensional steady state conduction has been introduced. This deals with numerical as well as the shape factor approaches to solve such problems.

(iii) In the context of radiation, analogy with electrical networks has now been emphasized explicitly and illustrated via solved examples (Example 1.32).

(iv) A new section on Taylor–Aris dispersion has been added in the context of mass transfer (Section 2.8).

(v) A new chapter (Chapter 6) on microscale transport phenomena has been added to this edition.

Most of these changes are based on the first author's extensive conversations and discussion with Jack Richardson over a period of 30 years.

We are grateful to the many individuals who have facilitated the publication of the seventh edition. Over the past 2 years, it has been a wonderful experience working with the staff at Butterworth–Heinemann. Each one of them has been extremely helpful, and some of these individuals deserve a mention here. First and foremost, we are grateful to Fiona Geraghty for commissioning the new edition. She not only patiently answered my endless queries but also came to my rescue on several occasions. Similarly, Maria Convoy and Ashlie Jackman went much beyond their call of duty to see this project through. Finally, Mohana Natarajan assembled the numerous fragments in different forms and formats—ranging from handwritten notes to latex files—into the finished product in your hands. We end this preface with an appeal to our readers to please let us know as and when you spot errors or inconsistencies so that these can be rectified at the earliest opportunity.

Raj Chhabra
V. Shankar
Kanpur, September 2017

Preface to Sixth Edition

It is somewhat sobering to realise that the sixth edition of Volume 1 appears 45 years after the publication of the first edition in 1954. Over the intervening period, there have been considerable advances in both the underlying theory and the practical applications of chemical engineering, all of which are reflected in parallel developments in undergraduate courses. In successive editions, we have attempted to adapt the scope and depth of treatment in the text to meet the changes in the needs of both students and practitioners of the subject.

Volume 1 continues to concentrate on the basic processes of momentum transfer (as in fluid flow), heat transfer, and mass transfer, and it also includes examples of practical applications of these topics in areas of commercial interest such as the pumping of fluids, the design of shell and tube heat exchangers, and the operation and performance of cooling towers. In response to the many requests from the readers (and the occasional note of encouragement from our reviewers), additional examples and their solutions have now been included in the main text. The principal areas of application, particularly of the theories of mass transfer across a phase boundary, form the core material of Volume 2; however, whilst in Volume 6, material presented in other volumes is utilised in the practical design of process plant.

The more important additions and modifications that have been introduced into this sixth edition of Volume 1 are the following:

Dimensionless analysis. The idea and advantages of treating length as a vector quantity and of distinguishing between the separate role of mass in representing a quantity of matter as opposed to its inertia are introduced.

Fluid flow. The treatment of the behaviour of non-Newtonian fluids is extended, and the methods used for pumping and metering of such fluids are updated.

Heat transfer. A more detailed discussion of the problem of unsteady-state heat transfer by conduction where bodies of various shapes are heated or cooled is offered together with a more complete treatment of heat transfer by radiation and a reorientation of the introduction to the design of shell and tube heat exchangers.

Mass transfer. The section on mass transfer accompanied by chemical reaction has been considerably expanded, and it is hoped that this will provide a good basis for the understanding of the operation of both homogeneous and heterogeneous catalytic reactions.

As ever, we are grateful for a great deal of help in the preparation of this new edition from a number of people. In particular, we should like to thank Dr. D.G. Peacock for the great enthusiasm and dedication he has shown in the production of the index, a task he has undertaken for us over many years. We would also mention especially Dr. R.P. Chhabra of the Indian Institute of Technology at Kanpur for his contribution on unsteady-state heat transfer by conduction, those commercial organisations that have so generously contributed new figures and diagrams of equipment, our publishers who cope with our perhaps overwhelming number of suggestions and alterations with a never-failing patience, and, most of all, our readers who with great kindness make so many extremely useful and helpful suggestions all of which are incorporated wherever practicable. With their continued help and support, the signs are that this present work will continue to be of real value as we move into the new millennium.

Swansea, 1999
Newcastle upon Tyne, 1999

John F. Richardson
John R. Backhurst
John H. Harker

Preface to Fifth Edition

This textbook has been the subject of continual updating since it was first published in 1954. There have been numerous revised impressions, and the opportunity has been taken on the occasion of each reprinting to make corrections and revisions, many of them in response to readers who have kindly pointed out errors or who have suggested modifications and additions. When the summation of the changes has reached a sufficiently high level, a new edition has been produced. We have now reached this point again, and the fifth edition incorporates all the alterations in the 1993 revision of the fourth edition, together with new material, particularly on simultaneous mass transfer and chemical reaction for unsteady-state processes.

There have been changes in the publisher too. Since the appearance of the fourth edition in 1990, Pergamon Press has become part of Elsevier Science, and now, following a reorganisation in the Reed Elsevier group of companies, the responsibility for publishing the Chemical Engineering series has passed to Butterworth-Heinemann, another Reed Elsevier company.

We are grateful to our readers for their interest and very much hope they will continue to make suggestions for the improvement of the series.

John F. Richardson

Preface to Fourth Edition

The first edition of Volume 1 was published in 1954, and Volume 2 appeared a year later. In the intervening 35 years or so, there have been far-reaching developments in *Chemical Engineering*, and the whole approach to the subject has undergone a number of fundamental changes. The question therefore arises as to whether it is feasible to update a textbook written to meet the needs of the final-year students of an undergraduate course in the 1950s so that it can continue to fulfil a useful purpose in the last decade of the century. Perhaps, it would have been better if a new textbook had been written by an entirely new set of authors. Although at one stage this had seemed likely through the sponsorship of the Institution of Chemical Engineers, there is now no sign of any such replacement book appearing in the United Kingdom.

In producing the fourth edition, it has been necessary to consider whether to start again with a clean sheet of paper—an impossibly daunting task—or whether to retain the original basic structure with relatively small modifications. In following the latter course, the authors were guided by the results of a questionnaire sent to a wide range of university ('Do not tamper overmuch with the devil we know, in spite of all his faults!' and polytechnic) departments throughout the English-speaking world. The clear message that came back was

It was in 1971 that Volume 3 was added to the series, essentially to make good some of the more glaring omissions in the earlier volumes. Volume 3 contains a series of seven specialist chapters written by members of the staff of the Chemical Engineering Department at the University College of Swansea, with Dr. D.G. Peacock of the School of Pharmacy, London, as a joint editor. In 1977–79, as well as contributing significantly to the new editions of Volumes 1 and 2, two colleagues at the University of Newcastle upon Tyne, Dr. J.R. Backhurst and the Revd Dr. J.H. Harker, prepared Volumes 4 and 5, the solutions to the problems in Volumes 1 and 2, respectively. The final major development was the publication of Volume 6 on chemical engineering design by Mr. R.K. Sinnott in 1983. With the preparation of a fourth edition, the opportunity has presented itself for a degree of rationalisation, without introducing major changes to the structure. This has led to the following format:

Volume 1	Fluid Flow, Heat Transfer, and Mass Transfer
Volume 2	Particle Technology and Separation Processes
Volume 3	Chemical and Biochemical Reactor Engineering and Control
Volume 4/5	Solutions to the Problems in Volumes 1, 2, and 3
Volume 6	Chemical Engineering Design

The details of this new arrangement are as follows:

Volume 1 has acquired an abbreviated treatment of non-Newtonian flow, formerly in Volume 3.

Liquid Mixing appears as a new chapter in Volume 1, which incorporates the relevant material formerly in Volumes 2 and 3.

Separate chapters now appear in Volume 1 on compressible flow and on multiphase flow, the latter absorbing material previously scattered between Volumes 1 and 2.

New chapters are added to Volume 2 to cover four separation processes of increasing importance—adsorption (from Volume 3), ion exchange, chromatographic separations, and membrane separations.

Volume 3 is now devoted to various aspects of reaction engineering and control, material that is considerably expanded.

Some aspects of design, previously in the earlier volumes, are now transferred to a more appropriate home in Volume 6.

As far as Volume 1 is concerned, the opportunity has been taken to update existing material. The major changes in fluid flow include the incorporation of non-Newtonian flow, an extensive revision of compressible flow and the new chapters on multiphase flow and liquid mixing. Material for this last chapter has been contributed by Dr. R.P. Chhabra of the Indian Institute of Technology at Kanpur. There has also been a substantial revision of the presentation of material on mass transfer and momentum, heat, and mass transfer. To the Appendix have been added the tables of Laplace transform and error functions, which were formerly in Volume 3, and throughout this new edition, all the diagrams have been redrawn. Some further problems have been added at the end.

Sadly, John Coulson was not able to contribute as he had done previously, and his death in January 1990 leaves us with a gap that is difficult to fill. John Backhurst and John Harker, who made a substantial contribution to the preparation of the third edition in 1977, have taken an increased share of the burden of revising the book and contributing new material and have taken a special responsibility for those sections that originated from John Coulson, in addition to the special task of updating the illustrations. Without their continued support and willing cooperation, there would have been no fourth edition.

Finally, we would all like to thank our many readers who have made such helpful suggestions in the past and have pointed out errors, many of which the authors would never have spotted. It is hoped that readers will continue to act in this way as unseen authors.

John F. Richardson
June 1990

Preface to Third Edition

The introduction of the SI system of units by the United Kingdom and many other countries has itself necessitated the revision of this engineering text. This clear implementation of a single system of units will be welcomed not only by those already in the engineering profession but also by those who are about to join. The system that is based on the cgs and mks systems using length (L), mass (M), and time (T) as the three basic dimensions, as is the practice in the physical sciences, has the very great advantage that it removes any possible confusion between mass and force that arises in the engineering system from the common use of the term *pound* for both quantities. We have therefore presented the text, problems, and examples in the SI system but have arranged the tables of physical data in the Appendix to include both SI and other systems wherever possible. This we regard as important because so many of the physical data have been published in cgs units. For similar reasons, engineering units have been retained as an alternative where appropriate.

In addition to the change to the SI system of units, we have taken the opportunity to update and to clarify the text. A new section on the flow of two-phase gas–liquid mixtures has been added to reflect the increased interest in the gas and petroleum industries and in its application to the boiling of liquids in vertical tubes.

The chapter on mass transfer, the subject that is so central and specific to chemical engineering, has been considerably extended and modernised. Here, we have thought it important in presenting some of the theoretical work to stress its tentative nature and to show that, although some of the theories may often lack a full scientific basis, they provide the basis of a workable technique for solving problems. In the discussion on fluid flow, reference has been made to American methods, and the emphasis on flow measurement has been slanted more to the use of instruments as part of a control system. We have emphasised the importance of pipe-flow networks, which represent a substantial cost item in modern large-scale enterprises.

This text covers the physical basis of the three major transfer operations of fluid flow, heat transfer, and mass transfer. We feel that it is necessary to provide a thorough grounding in these operations before introducing techniques that have been developed to give workable solutions in the most convenient manner for practical application. At the same time, we have directed the attention of the reader to such invaluable design codes as TEMA and the British Standards for heat exchanger design and to other manuals for pipe-flow systems.

It is important for designers always to have in their minds the need for reliability and safety; this is likely to follow from an understanding of the basic principles involved, many of which are brought out in the text.

We would like to thank our many friends from several countries who have written with suggestions, and it is our hope that this edition will help in furthering growth and interest in the profession. We should also like to thank a number of industrialists who have made available much useful information for incorporation in this edition; this help is acknowledged at the appropriate point. Our particular thanks are due to B. Waldie for his contribution to the high temperature aspects of heat transfer and to the Kellogg International Corporation and Humphreys and Glasgow Limited for their help. In conclusion, we would like to thank J.R. Backhurst and J.H. Harker for their editorial work and for recalculating the problems in SI units and converting the charts and tables.

Since the publication of the second edition of this volume, Volume 3 of *Chemical Engineering* has been published in order to give a more complete coverage of those areas of chemical engineering that are of importance in both universities and industry in the 1970s.

John M. Coulson
John F. Richardson
January 1976

Preface to Second Edition

In presenting this second edition, we should like to thank our many friends from various parts of the world who have so kindly made suggestions for clarifying parts of the text and for additions that they have felt to be important. During the last eight years, there have been changes in the general approach to chemical engineering in the universities with a shift in emphasis towards the physical mechanisms of transport processes and with a greater interest in unsteady-state conditions. We have taken this opportunity to strengthen those sections dealing with the mechanisms of processes, particularly in Chapter 7 on mass transfer and in the chapters on fluid mechanics where we have laid greater emphasis on the use of momentum exchange. Many chemical engineers are primarily concerned with the practical design of plant, and we have tried to include a little more material of use in this field in Chapter 6 on heat transfer. An introductory section on dimensional analysis has been added, but it has been possible to do no more than outline the possibilities opened up by the use of this technique. Small changes will be found throughout the text, and we have tried to meet many readers' requests by adding some more worked examples and a further selection of problems for the student. The selection of material and its arrangement are becoming more difficult and must be to a great extent a matter of personal choice, but we hope that this new edition will provide a sound basis for the study of the fundamentals of the subject and will perhaps be of some value to practising engineers.

John M. Coulson
John F. Richardson

Preface to First Edition

The idea of treating the various processes of the chemical industry as a series of unit operations was first brought out as a basis for a new technology by Walker, Lewis, and McAdams in their book in 1923. Before this, the engineering of chemical plants had been regarded as individual to an industry, and there was little common ground between one industry and another. Since the early 1920s, chemical engineering as a separate subject has been introduced into the universities of both America and England and has expanded considerably in recent years so that there are now a number of university courses in both countries. During the past 20 years, the subject matter has been extensively increased by various researches described in a number of technical journals to which frequent reference is made in the present work.

Despite the increased attention given to the subject, there are few general books, although there have been a number of specialised books on certain sections such as distillation and heat transfer. It is the purpose of the present work to present to the student an account of the fundamentals of the subject. The physical basis of the mechanisms of many of the chemical engineering operations forms a major feature of chemical engineering technology. Before tackling the individual operations, it is important to stress the general mechanisms that are found in so many of the operations. We have therefore divided the subject matter into two volumes, the first of which contains an account of these fundamentals—diffusion, fluid flow, and heat transfer. In Volume 2, we shall show how these theoretical foundations are applied in the design of individual units such as distillation columns, filters, crystallisers, and evaporators.

Volume 1 is divided into four sections—fluid flow, heat transfer, mass transfer and humidification. Since the chemical engineer must handle fluids of all kinds, including compressible gases at high pressures, we believe that it is a good plan to consider the problem from a thermodynamic aspect and to derive general equations for flow that can be used in a wide range of circumstances. We have paid special attention to showing how the boundary layer is developed over plane surfaces and in pipes, since it is so important in controlling heat and mass transfer. At the same time, we have included a chapter on pumping since chemical engineering is an essentially practical subject, and the normal engineering texts do not cover the problem as experienced in the chemical and petroleum industries.

The chapter on heat transfer contains an account of the generally accepted techniques for calculation of film transfer coefficients for a wide range of conditions and includes a section on the general construction of tubular exchangers that form a major feature of many works. The possibilities of the newer plate type units are indicated.

In Section 3, the chapter on mass transfer introduces the mechanism of diffusion, and this is followed by an account of the common relationships between heat, mass, and momentum transfer and the elementary boundary layer theory. The final section includes the practical problem of humidification where both heat and mass transfer are taking place simultaneously.

It will be seen that in all chapters, there are sections in small print. In a subject such as this, which ranges from very theoretical and idealised systems to the practical problems with empirical or experimentally determined relations, there is much to be said for omitting the more theoretical features in a first reading, and in fact, this is frequently done in the more practical courses. For this reason, the more difficult theoretical sections have been put in small print, and the whole of Chapter 9 may be omitted by those who are more concerned with the practical utility of the subject.

In many of the derivations, we have given the mathematical analysis in more detail than is customary. It is our experience that the mathematical treatment should be given in full and that the student should then apply similar analysis to a variety of problems.

We have introduced into each chapter a number of worked examples that we believe are essential to a proper understanding of the methods of treatment given in the text. It is very desirable for a student to understand a worked example before tackling fresh practical problems himself. Chemical engineering problems require a numerical answer, and it is essential to become familiar with the different techniques so that the answer is obtained by systematic methods rather than by intuition.

In preparing this text, we have been guided by courses of lectures that we have given over a period of years and have presented an account of the subject with the major emphasis on the theoretical side. With a subject that has grown so rapidly and that extends from the physical sciences to practical techniques, the choice of material must be a matter of personal selection. It is, however, more important to give the principles than the practice, which is best acquired in the factory. We hope that the text may also prove useful to those in industry who, whilst perhaps successfully employing empirical relationships, feel that they would like to find the extent to which the fundamentals are of help.

We should like to take this opportunity of thanking a number of friends who have helped by their criticism and suggestions. We are particularly indebted to Mr F.E. Warner, to M. Guter, to D.J. Rasbash, and to L.L. Katan. We are also indebted to a number of companies who have kindly permitted us to use illustrations of their equipment. We have given a number of references to technical journals, and we are grateful to the publishers for permission to use illustrations from their works. In particular, we would thank the Institution of Chemical Engineers, the American Institute of Chemical Engineers, the American Chemical Society, the Oxford University Press, and the McGraw-Hill Book Company.

South Kensington,
London S.W.7
1953

Acknowledgements

The authors and publishers acknowledge the kind assistance of the following organisations in providing illustrative material:

Fig. 1.96A, Brown Fintube Co
Fig. 1.96B, G. A. Harvey and Co Ltd
Fig. 1.99A, Alfa Laval Ltd, Brentford, Middlesex
Fig. 1.99B, APV Ltd
Fig. 1.100, Ashmore, Benson, Pease and Co Ltd
Figs 3.2 and 3.3, Professor F. N. M. Brown, University of Notre Dame
Fig. 5.10A, B, and D, Casella London Ltd, Bedford
Fig. 5.10C, Protimeter plc, Marlow, the United Kingdom
Figs 5.12 and 5.13, Visco Ltd, Croydon, Surrey
Fig. 5.14, Davenport Engineering

Introduction

Welcome to the next generation of Coulson–Richardson series of books on *chemical engineering*. I would like to convey to you all my feelings about this project that have evolved over the past 30 years and are based on numerous conversations with Jack Richardson himself (1981 onwards until his death in 2011) and with some of the other contributors to previous editions including Tony Wardle, Ray Sinnott, Bill Wilkinson, and John Smith. So what follows here is the essence of these interactions combined with what the independent (solicited and unsolicited) reviewers had to say about this series of books on several occasions.

The Coulson–Richardson series of books has served the academia, students, and working professionals extremely well since their first publication more than 50 years ago. This is a testimony to their robustness and, to some extent, their timelessness. I have often heard much praise, from different parts of the world, for these volumes both for their informal and user-friendly yet authoritative style and for their extensive coverage. Therefore, there is a strong case for continuing with its present style and pedagogical approach.

On the other hand, advances in our discipline in terms of new applications (for instance, energy, bio, microfluidics, nanoscale engineering, smart materials, new control strategies, and reactor configurations) are occurring so rapidly and in such a significant manner that it will be naive, even detrimental, to ignore them. Therefore, while we have tried to retain the basic structure of this series, the contents have been thoroughly revised. Wherever the need was felt, the material has been updated, revised, and expanded as deemed appropriate. Therefore, the reader whether a student or a researcher or a working professional should feel confident that what is in the book is the most up-to-date, accurate, and reliable piece of information on the topic he/she is interested in.

Evidently, this is a massive undertaking that cannot be managed by a single individual. Therefore, we now have a team of volume editors responsible for each volume having the individual chapters being written by experts in some cases. I am most grateful to all of them for having joined us in the endeavour. Furthermore, based on extensive deliberations and feedback from a large number of individuals, some structural changes were deemed to be appropriate as detailed here. In this edition, Volumes 1 to 3 have been split into two volumes each as follows:

Volume 1A: Fluid Flow: Fundamentals and Applications.

Volume 1B: Heat and Mass Transfer: Fundamentals and Applications.

Volume 2A: Particulate Systems and Particle Technology.

Volume 2B: Separation Processes.

Volume 3A: Chemical and Biochemical Reactors and Reaction Engineering.

Volume 3B: Process Control.

Undoubtedly, the success of a project of such a vast scope and magnitude hinges on the cooperation and assistance of many individuals. In this regard, we have been extremely fortunate in working with some of the outstanding individuals at Butterworth–Heinemann, a few of whom deserve to be singled out: Jonathan Simpson, Fiona Geraghty, Maria Convey, and Ashlie Jackman who have taken personal interest in this project and have come to our rescue whenever needed, going much beyond the call of duty.

Finally, this series has had a glorious past, but I sincerely hope that its future will be even brighter by presenting the best possible books to the global chemical engineering community for the next 50 years, if not for longer. I sincerely hope that the new edition of this series will meet (if not exceed) your expectations! Lastly, a request to the readers, please continue to do the good work by letting me know if, no not if, but when you spot a mistake so that these can be corrected at the first opportunity.

Raj Chhabra
Editor-in-chief
Kanpur, September 2017

PART 1

Heat Transfer

CHAPTER 1

Heat Transfer

1.1 Introduction

In the majority of chemical processes heat is either given out or absorbed, and fluids must often be either heated or cooled in a wide range of plant, such as furnaces, evaporators, distillation units, dryers, and reaction vessels where one of the major problems is that of transferring heat at the desired rate. In addition, it may be necessary to prevent the loss of heat from a hot vessel or pipe system. Additional examples are found in the context of food storage and preservation, cooling of electronic components, thermal efficiency of buildings, energy production and storage devices, biomedical related applications, etc. The control of the flow of heat at the desired rate forms one of the most important areas of chemical engineering. Provided that a temperature difference exists between two parts of a system, heat transfer will take place in one or more of three different ways.

Conduction. In a solid or in a stagnant fluid medium, the flow of heat by conduction is the result of the transfer of vibrational energy from one molecule to another, and in fluids it occurs in addition as a result of the transfer of kinetic energy. Heat transfer by conduction may also arise from the movement of free electrons, a process, which is particularly important with metals and accounts for their high thermal conductivities.

Convection. Heat transfer by convection arises from the bulk flow and mixing of elements of fluid. If this mixing occurs as a result of density differences as, for example, when a pool of liquid is heated from below, the process is known as *natural* (also known as *free*) *convection.* If the mixing results from eddy movement in the fluid, for example when a fluid flows through a pipe heated on the outside, it is called *forced convection.* It is important to note that convection requires mixing of fluid elements, and is not governed by temperature difference alone, as is the case in conduction and radiation. The density of each fluid varies with temperature to some extent, and therefore, natural convection, howsoever small, is always present in all practical applications. For instance, on a calm day (or little wind), steam-carrying pipes lose heat to surroundings by natural convection. This contribution, however, diminishes as the wind velocity gradually increases.

Radiation. All materials radiate thermal energy in the form of electromagnetic waves. When this radiation falls on a second body it may be partially reflected, transmitted, or

Coulson and Richardson's Chemical Engineering. https://doi.org/10.1016/B978-0-08-102550-5.00001-8

absorbed. It is only the fraction that is absorbed that appears as heat in the body. Thus, radiation heat transfer differs from conduction and convection in a fundamental way.

1.2 Basic Considerations

1.2.1 Individual and Overall Coefficients of Heat Transfer

In many of the applications of heat transfer in process plants, one or more of the mechanisms of heat transfer may be involved. In the majority of heat exchangers heat passes through a series of different intervening layers before reaching the second fluid (Fig. 1.1). These layers may be of different thicknesses and of different thermal conductivities. The problem of transferring heat to crude oil in the primary furnace before it enters the first distillation column may be considered as an example. The heat from the flames passes by radiation and convection to the pipes in the furnace, by conduction through the pipe walls, and by forced convection from the inside of the pipe to the oil. Here all three modes of transfer are involved. After prolonged usage, solid deposits may form on both the inner and outer walls of the pipes, and these will then contribute additional thermal resistance to the transfer of heat. The simplest form of equation, which represents this heat transfer operation may be written as:

$$Q = UA\Delta T \tag{1.1}$$

where Q is the heat transferred per unit time, A the area available for the flow of heat, ΔT the difference in temperature between the flame and the boiling oil, and U is known as the overall heat transfer coefficient (W/m^2 K in SI units). Eq. (1.1) is nothing more than the familiar Newton's law of cooling.

At first sight, Eq. (1.1) implies that the relationship between Q and ΔT is linear. Whereas this is approximately so over limited ranges of temperature difference for which U is nearly constant, in practice U may well be influenced both by the temperature difference and by the absolute value of the temperatures.

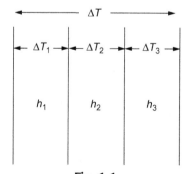

Fig. 1.1
Heat transfer through a composite wall.

If it is required to know the area needed for the transfer of heat at a specified rate, the temperature difference ΔT, and the value of the overall heat-transfer coefficient must be known. Thus the calculation of the value of U is a key requirement in any design problem in which heating or cooling is involved. A large part of the study of heat transfer is therefore devoted to the evaluation of this coefficient for a given situation.

The value of the coefficient will depend on the mechanism by which heat is transferred, will also depend on the fluid dynamics of both the heated and the cooled fluids, on the properties of the materials through which the heat must pass, and on the geometry of the fluid paths. In solids, heat is normally transferred by conduction; some materials such as metals have a high thermal conductivity, whilst others such as ceramics have a low conductivity. Transparent solids like glass also transmit radiant energy particularly in the visible part of the spectrum.

Liquids also transmit heat readily by conduction, though circulating currents are frequently set up and the resulting convective transfer may be considerably greater than the transfer by conduction. Many liquids also transmit radiant energy. Gases are poor conductors of heat and circulating currents are difficult to suppress; convection is therefore much more important than conduction in a gas. Radiant energy is transmitted with only limited absorption in gases and, of course, without any absorption *in vacuo*. Radiation is the only mode of heat transfer, which does not require the presence of an intervening medium.

If the heat is being transmitted through a number of media in series, the overall heat transfer coefficient may be broken down into individual coefficients h each relating to a single medium. This is as shown in Fig. 1.1. It is assumed that there is good contact between each pair of elements so that the temperature is the same at the two sides of each junction.

If heat is being transferred through three media, each of area A, and individual coefficients for each of the media are h_1, h_2, and h_3, and the corresponding temperature changes are ΔT_1, ΔT_2, and ΔT_3 then, provided that there is no accumulation of heat in the media (that is, the system is at a steady state), the heat transfer rate Q will be the same through each medium. Three equations, analogous to Eq. (1.1) can therefore be written as:

$$\left.\begin{array}{l} Q = h_1 A \Delta T_1 \\ Q = h_2 A \Delta T_2 \\ Q = h_3 A \Delta T_3 \end{array}\right\} \tag{1.2}$$

Rearranging:

$$\Delta T_1 = \frac{Q}{A}\frac{1}{h_1}$$

$$\Delta T_2 = \frac{Q}{A}\frac{1}{h_2}$$

$$\Delta T_3 = \frac{Q}{A}\frac{1}{h_3}$$

Adding:

$$\Delta T_1 + \Delta T_2 + \Delta T_3 = \frac{Q}{A}\left(\frac{1}{h_1} + \frac{1}{h_2} + \frac{1}{h_3}\right) \qquad (1.3)$$

Noting that $(\Delta T_1 + \Delta T_2 + \Delta T_3) =$ total temperature difference ΔT: then:

$$\Delta T = \frac{Q}{A}\left(\frac{1}{h_1} + \frac{1}{h_2} + \frac{1}{h_3}\right) \qquad (1.4)$$

From Eq. (1.1):

$$\Delta T = \frac{Q}{A}\frac{1}{U} \qquad (1.5)$$

Comparing Eqs. (1.4), (1.5):

$$\frac{1}{U} = \frac{1}{h_1} + \frac{1}{h_2} + \frac{1}{h_3} \qquad (1.6)$$

The reciprocals of the heat transfer coefficients are resistances, and Eq. (1.6) therefore illustrates that the resistances are additive. Thus, one can make an analogy with electrical circuits where the Ohm's law applies, i.e. current is proportional to potential difference and inversely proportional to the electrical resistance, i.e.

$$\text{Current} = \frac{\text{Potential difference}}{\text{Resistance}} \qquad (1.7)$$

Now comparing Eq. (1.6) with Eq. (1.7), Q is the current, ΔT acts as the potential difference, and $(1/hA)$ is the thermal resistance; the corresponding electric circuit is (Fig. 1.2):

In some cases, particularly for the radial flow of heat through a thick pipe wall or cylinder, the area for heat transfer is a function of position. Thus the area for transfer applicable to each of the three media could differ and may be A_1, A_2, and A_3. Eq. (1.3) then becomes:

$$\Delta T_1 + \Delta T_2 + \Delta T_3 = Q\left(\frac{1}{h_1 A_1} + \frac{1}{h_2 A_2} + \frac{1}{h_3 A_3}\right) \qquad (1.8)$$

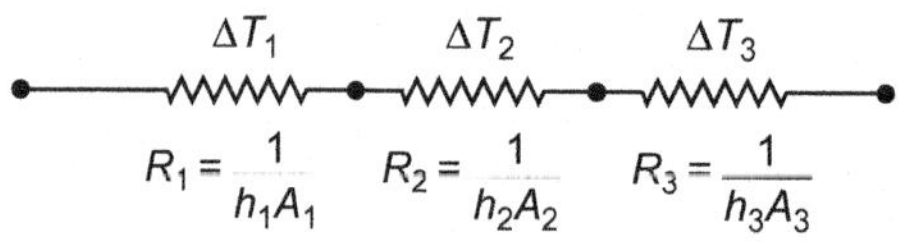

Fig. 1.2
Electrical circuit analogy for Fig. 1.1.

Eq. (1.8) must then be written in terms of one of the area terms A_1, A_2, and A_3, or sometimes in terms of a mean area. Since Q and ΔT must be independent of the particular area considered, the value of U will vary according to which area is used as the basis. Thus Eq. (1.8) may be written, for example as:

$$Q = U_1 A_1 \Delta T \quad \text{or} \quad \Delta T = \frac{Q}{U_1 A_1}$$

This will then give U_1 as:

$$\frac{1}{U_1} = \frac{1}{h_1} + \frac{A_1}{A_2}\left(\frac{1}{h_2}\right) + \frac{A_1}{A_3}\left(\frac{1}{h_3}\right) \tag{1.9}$$

In this analysis, it is assumed that the heat flowing per unit time through each of the media is the same, i.e. the system is in a steady state.

Now that the overall coefficient U has been broken down into its component parts, each of the individual coefficients h_1, h_2, and h_3 must be evaluated. This can be done from a knowledge of the nature of the heat transfer process in each of the media. A study will therefore be made on how these individual coefficients can be calculated for conduction, convection, and radiation, in the ensuing sections in this chapter.

1.2.2 Mean Temperature Difference

Where heat is being transferred from one fluid to a second fluid through the wall of a vessel and the temperature is the same throughout the bulk of each of the fluids, there is no difficulty in specifying the overall temperature difference ΔT. Frequently, however, each fluid is flowing through a heat exchanger such as a pipe or a series of pipes in parallel, and its temperature changes as it flows, and consequently the temperature difference is continuously changing along the length. If the two fluids are flowing in the same direction (*cocurrent flow*), the temperatures of the two streams progressively approach one another as shown in Fig. 1.3. In these circumstances the outlet temperature of the heating fluid must always be higher

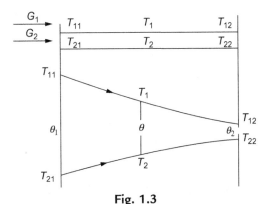

Fig. 1.3
Mean temperature difference for cocurrent flow.

than that of the cooling fluid. If the fluids are flowing in opposite directions (*countercurrent flow*), the temperature difference will show less variation throughout the heat exchanger as shown in Fig. 1.4. In this case it is possible for the cooling liquid to leave at a higher temperature than the heating liquid, and one of the great advantages of countercurrent flow is that it is possible to extract a higher proportion of the heat content of the heating fluid. The calculation of the appropriate value of the temperature difference for cocurrent and for countercurrent flow is now considered. It is assumed that the overall heat transfer coefficient U remains constant throughout the heat exchanger. This essentially implies that the thermophysical properties of fluids (thermal conductivity, density, viscosity, heat capacity) are nearly independent of the temperature over the range of interest.

It is necessary to find the average value of the temperature difference θ_m to be used in the general equation:

$$Q = UA\theta_m \quad (Eq.\ 1.1)$$

Fig. 1.4 shows the temperature conditions for the fluids flowing in opposite directions, a condition known as the countercurrent flow.

The outside stream (of specific heat C_{p1}) and mass flow rate G_1 falls in temperature from T_{11} to T_{12}.

The inside stream (of specific heat C_{p2}) and mass flow rate G_2 rises in temperature from T_{21} to T_{22}.

Over a small element of area dA where the local temperatures of the streams are T_1 and T_2, the local temperature difference is:

$$\theta = T_1 - T_2$$

$$\therefore \quad d\theta = dT_1 - dT_2$$

$$\text{Heat given out by the hot stream} = dQ = -G_1 C_{p1} dT_1$$

$$\text{Heat taken up by the cold stream} = dQ = G_2 C_{p2} dT_2$$

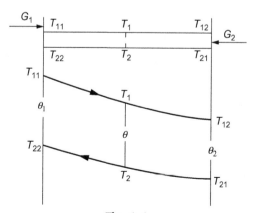

Fig. 1.4
Mean temperature difference for countercurrent flow.

$$\therefore \quad d\theta = dT_1 - dT_2 = -\frac{dQ}{G_1 C_{p1}} - \frac{dQ}{G_2 C_{p2}} = -dQ\left(\frac{G_1 C_{p1} + G_2 C_{p2}}{G_1 C_{p1} \times G_2 C_{p2}}\right) = -\psi dQ \,(\text{say})$$

$$\therefore \quad \theta_1 - \theta_2 = \psi Q$$

Over this element:

$$U dA\theta = dQ$$

$$\therefore U dA\theta = -\frac{d\theta}{\psi}$$

If U may be taken as constant:

$$-\psi U \int_0^A dA = \int_{\theta_1}^{\theta_2} \frac{d\theta}{\theta}$$

$$\therefore \quad -\psi UA = -\ln\frac{\theta_1}{\theta_2}$$

From the definition of θ_m, $Q = UA\theta_m$.

$$\therefore \quad \theta_1 - \theta_2 = \psi Q = \psi UA\theta_m = \ln\frac{\theta_1}{\theta_2}(\theta_m)$$

and:

$$\theta_m = \frac{\theta_1 - \theta_2}{\ln(\theta_1/\theta_2)} \tag{1.10}$$

where θ_m is known as the *logarithmic mean temperature difference* or simply as LMTD.

Underwood[1] proposed the following approximation for the logarithmic mean temperature difference:

$$(\theta_m)^{1/3} = \frac{1}{2}\left(\theta_1^{1/3} + \theta_2^{1/3}\right) \tag{1.11}$$

and, for example, when $\theta_1 = 1$ K and $\theta_2 = 100$ K, θ_m is 22.4 K compared with the true logarithmic mean of 21.5 K. When $\theta_1 = 10$ K and $\theta_2 = 100$ K, both the approximation and the logarithmic mean values coincide at 39 K. Of course, for the special case of $\theta_1 = \theta_2$, all three coincide, i.e. $\theta_1 = \theta_2 = \theta_m$.

If the two fluids flow in the same direction on each side of a tube, cocurrent flow takes place and the general shape of the temperature profile along the tube is as shown in Fig. 1.3. A similar analysis will show that this gives the same expression for θ_m, the logarithmic mean temperature difference. For the same terminal temperatures it is important to note that the value of θ_m for countercurrent flow is appreciably greater than the value for cocurrent flow. This is seen from the temperature profiles, where with cocurrent flow the cold fluid cannot be heated to a higher temperature than the exit temperature of the hot fluid as illustrated in Example 1.1.

Example 1.1

A heat exchanger is required to cool 20 kg/s of water from 360 to 340 K by means of 25 kg/s water entering at 300 K. If the overall coefficient of heat transfer is constant at 2 kW/m² K, calculate the surface area required in (a) a countercurrent concentric tube exchanger, and (b) a cocurrent flow concentric tube exchanger.

Solution

Heat load: $Q = 20 \times 4.18(360 - 340) = 1672\,kW$

The cooling water outlet temperature is given by:

$$1672 = 25 \times 4.18(T_{22} - 300) \quad \text{or} \quad T_{22} = 316\,K$$

(a) *Counterflow*

$$\theta_1 = 360 - 316 = 44\,K; \quad \theta_2 = 340 - 300 = 40\,K$$

In Eq. (1.10):

$$\theta_m = \frac{44 - 40}{\ln(44/40)} = 41.9\,K$$

Heat transfer area:

$$A = \frac{Q}{U\theta_m}$$

$$= \frac{1672}{2 \times 41.9}$$

$$= \underline{19.95\,m^2}$$

(b) *Cocurrent flow*

$$\theta_1 = 360 - 300 = 60\,K; \quad \theta_2 = 340 - 316 = 24\,K$$

In Eq. (1.10):

$$\theta_m = \frac{60 - 24}{\ln(60/24)} = 39.3\,K$$

Heat transfer area:

$$A = \frac{1672}{2 \times 39.3}$$

$$= \underline{21.27\,m^2}$$

It may be noted that using Underwood's approximation (Eq. 1.11), the calculated values for the mean temperature driving forces are 41.9 and 39.3 K for counter- and cocurrent flow respectively, which agree exactly with the logarithmic mean values calculated above.

1.3 Heat Transfer by Conduction

1.3.1 Conduction Through a Plane Wall

This important mechanism of heat transfer is now considered in more detail for the flow of heat , through a plane wall of thickness x as shown in Fig. 1.5.

The rate of heat flow Q over the area A and a small distance dx may be written as:

$$Q = -kA\left(\frac{dT}{dx}\right) \tag{1.12}$$

which is often known as the *Fourier's equation*, where the negative sign indicates that the temperature gradient is in the opposite direction to the flow of heat (that is, heat flows downhill from high temperature to low temperature) and k is the thermal conductivity of the material. Integrating for a wall of thickness x with boundary temperatures T_1 and T_2, as shown in Fig. 1.5:

$$Q = \frac{kA(T_1 - T_2)}{x} \tag{1.13}$$

Implicit in Eq. (1.13) is the assumption of constant thermal conductivity. Eq. (1.13) can also be arranged in the form of Eq. (1.7) and the thermal resistance in this case is given by (x/kA).

Thermal conductivity is a function of temperature and experimental data may often be expressed by a linear relationship of the form:

$$k = k_0(1 + k'T) \tag{1.14}$$

where k is the thermal conductivity at the temperature T and k_0 and k' are constants for a specific material. Combining Eqs. (1.12), (1.14):

$$-kdT = -k_0(1 + k'T)dT = \frac{Qdx}{A}$$

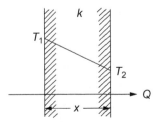

Fig. 1.5
Conduction of heat through a plane wall.

Integrating between the temperature limits T_1 and T_2 over the interval x_1 to x_2,

$$-\int_{T_1}^{T_2} k\, dT = (T_1 - T_2)k_0 \left\{ 1 + k' \left(\frac{T_1 + T_2}{2} \right) \right\} = Q \int_{x_1}^{x_2} \frac{dx}{A} \qquad (1.15)$$

where k is a linear function of T, the following equation may therefore be used:

$$k_a(T_1 - T_2) = Q \int_{x_1}^{x_2} \frac{dx}{A} \qquad (1.16)$$

where k_a is the arithmetic mean of k_1 and k_2 at T_1 and T_2 respectively or the thermal conductivity at the arithmetic mean of T_1 and T_2.

Where k is a nonlinear function of T, some mean value, k_m will apply, where:

$$k_m = \frac{1}{T_2 - T_1} \int_{T_1}^{T_2} k\, dT \qquad (1.17)$$

Note that none of these equations (Eqs. 1.16 or 1.17) can be rearranged in the form of Eq. (1.7). Thus, the expression $R = x/kA$ is only valid when the thermal conductivity is constant.

From Table 1.1 it will be seen that metals have very high thermal conductivities, nonmetallic solids lower values, nonmetallic liquids low values, and gases very low values. It is important to note that amongst metals, stainless steel has a low value, that water has a very high value for liquids (due to partial ionisation), and that hydrogen has a high value for gases (due to the high mobility of the molecules). With gases, k decreases with increase in molecular mass and increases with the temperature. In addition, for gases the dimensionless *Prandtl group* $C_p\mu/k$, which is approximately constant (where C_p is the specific heat at constant pressure and μ is the viscosity), can be used to evaluate k at high temperatures where it is difficult to determine a value experimentally because of the convection effects. In contrast, k does not vary significantly with pressure, except where this is reduced to a value so low that the mean free path of the molecules becomes comparable with the dimensions of the flow passages; further reduction of pressure then causes k to decrease.

Typical values of Prandtl numbers for a range of materials are as follows:

Air	0.71	*n*-Butanol	50
Oxygen	0.63	Light oil	600
Ammonia (gas)	1.38	Glycerol	1000
Water	5–10	Polymer melts	10,000
		Mercury	0.02

The low conductivity of heat insulating materials, such as cork, glass wool, asbestos, foams, and so on, is largely accounted for by their high proportion of air space. The flow of heat

Table 1.1 Thermal conductivities of selected materials

Solids—Metals	Temp (K)	k (Btu/ h ft² °F/ft)	k (W/ m² K)	Liquids	Temp (K)	k (Btu/ h ft² °F/ft)	k (W/ m² K)
Aluminium	573	133	230	Acetic acid 50%	293	0.20	0.35
Cadmium	291	54	94	Acetone	303	0.10	0.17
Copper	373	218	377	Aniline	273–293	0.10	0.17
Iron (wrought)	291	35	61	Benzene	303	0.09	0.16
Iron (cast)	326	27.6	48	Calcium chloride brine 30%	303	0.32	0.55
Lead	373	19	33	Ethyl alcohol 80%	293	0.137	0.24
Nickel	373	33	57	Glycerol 60%	293	0.22	0.38
Silver	373	238	412	Glycerol 40%	293	0.26	0.45
Steel 1% C	291	26	45	n-Heptane	303	0.08	0.14
Tantalum	291	32	55	Mercury	301	4.83	8.36
				Sulphuric acid 90%	303	0.21	0.36
Admiralty metal	303	65	113	Sulphuric acid 60%	303	0.25	0.43
Bronze	–	109	189	Water	303	0.356	0.62
Stainless Steel	293	9.2	16	Water	333	0.381	0.66
Solids—Nonmetals				Gases			
Asbestos sheet	323	0.096	0.17	Hydrogen	273	0.10	0.17
Asbestos	273	0.09	0.16	Carbon dioxide	273	0.0085	0.015
Asbestos	373	0.11	0.19	Air	273	0.014	0.024
Asbestos	473	0.12	0.21	Air	373	0.018	0.031
Bricks (alumina)	703	1.8	3.1	Methane	273	0.017	0.029
Bricks (building)	293	0.4	0.69	Water vapour	373	0.0145	0.025
Magnesite	473	2.2	3.8	Nitrogen	273	0.0138	0.024
Cotton wool	303	0.029	0.050	Ethylene	273	0.0097	0.017
Glass	303	0.63	1.09	Oxygen	273	0.0141	0.024
Mica	323	0.25	0.43	Ethane	273	0.0106	0.018
Rubber (hard)	273	0.087	0.15				
Sawdust	293	0.03	0.052				
Cork	303	0.025	0.043				
Glass wool	–	0.024	0.041				
85% Magnesia	–	0.04	0.070				
Graphite	273	87	151				

through such materials is governed mainly by the resistance of the air spaces, which should be sufficiently small for convection currents to be suppressed.

It is convenient to rearrange Eq. (1.13) to give:

$$Q = \frac{(T_1 - T_2)A}{(x/k)} \qquad (1.18)$$

where x/k is known as the *thermal resistance* per unit area and k/x is the *transfer coefficient*.

Example 1.2

Estimate the heat loss per square metre of surface through a brick wall 0.5 m thick when the inner surface is at 400 K and the outside surface is at 300 K. The thermal conductivity of the brick may be taken as 0.7 W/m K.

Solution

From Eq. (1.13):

$$Q = \frac{0.7 \times 1 \times (400 - 300)}{0.5}$$
$$= 140 \, \text{W/m}^2$$

1.3.2 Thermal Resistances in Series

It has been noted earlier that thermal resistances may be added together for the case of heat transfer through a composite section formed from different media in series.

Fig. 1.6 shows a composite wall made up of three materials with thermal conductivities k_1, k_2, and k_3, with thicknesses as shown and with the temperatures T_1, T_2, T_3, and T_4 at the faces. Applying Eq. (1.13) to each section in turn, and noting that the same quantity of heat Q (steady state) must pass through each area A:

$$T_1 - T_2 = \frac{x_1}{k_1 A} Q, \quad T_2 - T_3 = \frac{x_2}{k_2 A} Q \quad \text{and} \quad T_3 - T_4 = \frac{x_3}{k_3 A} Q$$

On addition:

$$(T_1 - T_4) = \left(\frac{x_1}{k_1 A} + \frac{x_2}{k_2 A} + \frac{x_3}{k_3 A} \right) Q \tag{1.19}$$

or:

$$Q = \frac{T_1 - T_4}{\Sigma(x_1/k_1 A)}$$

$$= \frac{\text{Total driving force}}{\text{Total (thermal resistance/area)}} \tag{1.20}$$

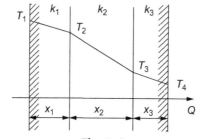

Fig. 1.6

Conduction of heat through a composite wall.

Example 1.3

A furnace is constructed with 0.20 m of firebrick, 0.10 m of insulating brick, and 0.20 m of ordinary building brick. The inside temperature is 1200 K and the outside temperature is 330 K. If the thermal conductivities are as shown in Fig. 1.7, estimate the heat loss per unit area and the temperature at the junction of the firebrick and the insulating brick.

Solution
From Eq. (1.20):

$$Q = (1200 - 330) / \left[\left(\frac{0.20}{1.4 \times 1} \right) + \left(\frac{0.10}{0.21 \times 1} \right) + \left(\frac{0.20}{0.7 \times 1} \right) \right]$$

$$= \frac{870}{(0.143 + 0.476 + 0.286)} = \frac{870}{0.905}$$

$$= \underline{\underline{961\, W/m^2}}$$

The ratio (Temperature drop over firebrick)/(Total temperature drop) $= (0.143/0.905)$

$$\therefore \quad \text{Temperature drop over firebrick} = \left(\frac{870 \times 0.143}{0.905} \right) = 137\, K$$

Hence the temperature at the firebrick-insulating brick interface (junction A),
$T_A = (1200 - 137) = \underline{1063\, K}$

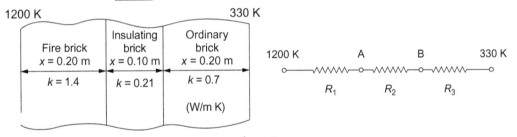

Fig. 1.7
Schematic for Example 1.3 and equivalent electrical circuit.

Example 1.4

In Example 1.3, the inner surface of the firebrick is in contact with hot gases at a temperature of 1700 K and the corresponding heat transfer coefficient is 150 W/m² K. Similarly, the exterior of the furnace is exposed to atmosphere (ambient temperature of 300 K) and it loses heat by natural convection with heat transfer coefficient of 6 W/m² K. Estimate the rate of heat loss per unit area to atmosphere and the intermediate junction temperatures.

Solution
The equivalent electrical circuit now has two additional resistances as:

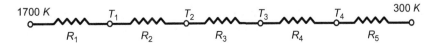

Now evaluating the individual resistances:

$$R_1 = \frac{1}{hA} = \frac{1}{150 \times 1} = 6.67 \times 10^{-3}\,K/W$$

The values of R_2, R_3, and R_4 are the same as in example 1.3 and R_5 is calculated similar to R_1 as:

$$R_2 = \frac{0.20}{1.4 \times 1} = 0.143\,K/W$$

$$R_3 = \frac{0.10}{0.21 \times 1} = 0.476\,K/W$$

$$R_4 = \frac{0.20}{0.7 \times 1} = 0.286\,K/W$$

$$R_5 = \frac{1}{hA} = \frac{1}{6 \times 1} = 0.167\,K/W$$

Total thermal resistance:

$$\sum R = 6.67 \times 10^{-3} + 0.143 + 0.476 + 0.286 + 0.167$$

Or

$$\sum R = 1.079\,K/W$$

The rate of heat loss per unit area:

$$Q = \frac{\Delta T}{\sum R} = \frac{1700 - 300}{1.079} = 1297.5\,W$$

Since the resistances are in series, the rate of heat flow in each element is the same and one can now calculate the intermediate temperature as:

$$\frac{1700 - T_1}{6.6 \times 10^{-3}} = 1297.5, \ \ \text{i.e. } T_1 = 1700 - 8.7 = 1691.3\,K$$

Similarly,

$$\frac{1691.3 - T_2}{0.143} = 1297.5, \ \ \text{i.e. } T_2 = 1505.8\,K$$

$$\frac{1505.8 - T_3}{0.476} = 1297.5, \ \ \text{i.e. } T_3 = 888.2\,K$$

and

$$\frac{888.2 - T_4}{0.286} = 1297.5 = \frac{T_4 - 300}{0.167}$$

$$T_4 = 517\,K$$

Therefore, the intermediate temperatures are:

$$T_1 = 1691.3\,K, \ \ T_2 = 1505.8\,K, \ \ T_3 = 888.2\,K \ \text{ and } \ T_4 = 517\,K$$

Example 1.5

It is desired to reduce the rate of heat loss from the plane wall of a furnace to the atmosphere by tripling the thickness of the wall of the furnace. The initial temperature of the inside furnace wall is 900 K and that of the ambient air is 300 K and these two values do not change even after modifications. The outer surface of the wall in the initial design is at 400 K. Estimate the % reduction in the rate of heat loss per square metre to the atmosphere.

Solution

Assume that the heat transfer coefficient from the exterior to the atmosphere and the thermal conductivity of the material of construction of the furnace wall are constant. There are two resistances in series-conduction through the furnace wall and convection from the wall to the atmosphere. So the electrical circuits for the two designs are:

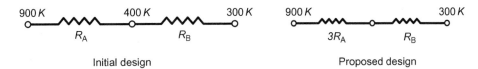

Initial design	Proposed design

In the initial design:

$$Q_i = \frac{900 - 300}{R_A + R_B} = \frac{600}{R_A + R_B} \, \text{W/m}^2 \, \text{K}$$

Also,

$$\frac{900 - 400}{R_A} = \frac{400 - 300}{R_B}$$

$$\therefore \frac{R_A}{R_B} = \frac{500}{100} = 5$$

i.e. $R_A = 5R_B$

$$\therefore Q_i = \frac{600}{6R_B} \, \text{W/m}^2 \, \text{K}$$

In the new design:

$$Q_{ii} = \frac{900 - 300}{3R_A + R_B} = \frac{600}{16R_B} \, \text{W/m}^2 \, \text{K}$$

$$\% \text{reduction} = \left(\frac{Q_i - Q_{ii}}{Q_i} \right) \times 100$$

$$\therefore \% \text{reduction} = \left(1 - \frac{600/16R_B}{600/6R_B} \right) \times 100$$

$$= \left(1 - \frac{6}{16} \right) \times 100 = 62.5\%$$

% reduction in the rate of heat loss to the atmosphere $= 62.5\%$

1.3.3 Conduction Through a Thick-Walled Tube

The conditions for heat flow through a thick-walled tube when the temperatures on the inside and outside are held constant are shown in Fig. 1.8. Here the area for heat flow is proportional to the radius and hence the temperature gradient is inversely proportional to the radius.

The heat flow at any radius r is given by:

$$Q = -k2\pi rl\frac{\mathrm{d}T}{\mathrm{d}r} \tag{1.21}$$

where l is the length of tube.

Integrating Eq. (1.21) between the limits r_1 and r_2:

$$Q\int_{r_1}^{r_2}\frac{\mathrm{d}r}{r} = 2\pi lk\int_{T_1}^{T_2}\mathrm{d}T$$

or:

$$Q = \frac{2\pi lk(T_1 - T_2)}{\ln(r_2/r_1)} \tag{1.22}$$

This equation may be put into the form of Eq. (1.13) to give:

$$Q = \frac{k(2\pi r_m l)(T_1 - T_2)}{r_2 - r_1} \tag{1.23}$$

where $r_m = (r_2 - r_1)/\ln(r_2/r_1)$, is known as the *logarithmic mean radius*. For thin-walled tubes, the arithmetic mean radius r_a may be used, giving:

$$Q = \frac{k(2\pi r_a l)(T_1 - T_2)}{r_2 - r_1} \tag{1.24}$$

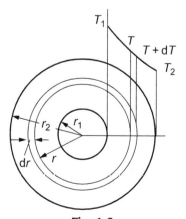

Fig. 1.8
Conduction through thick-walled tube or spherical shell.

By comparing Eqs. (1.7), (1.22), it is readily seen that in this case, the thermal resistance is given by $\left(\frac{1}{2\pi kl}\right) \ln \frac{r_2}{r_1}$.

1.3.4 Conduction Through a Spherical Shell and to a Particle

For one dimensional steady heat conduction through a spherical shell, the heat flow at radius r is given by:

$$Q = -k4\pi r^2 \frac{dT}{dr} \tag{1.25}$$

Integrating between the limits r_1 and r_2:

$$Q \int_{r_1}^{r_2} \frac{dr}{r^2} = -4\pi k \int_{T_1}^{T_2} dT$$

$$\therefore \quad Q = \frac{4\pi k(T_1 - T_2)}{(1/r_1) - (1/r_2)} \tag{1.26}$$

In this case, the thermal resistance is given by $\left(\dfrac{1}{4\pi k}\right)\left(\dfrac{1}{r_1} - \dfrac{1}{r_2}\right)$

An important application of heat transfer to a sphere is that of conduction through a stationary fluid (of thermal conductivity k) surrounding a spherical particle or droplet of radius r as encountered for example in fluidised beds, rotary kilns, spray dryers, and plasma devices. If the temperature difference $(T_1 - T_2)$ is spread over a very large distance such that $r_2 = \infty$ and T_1 is the temperature of the surface of the drop, then:

$$\frac{Qr}{(4\pi r^2)(T_1 - T_2)k} = 1$$

or:

$$\frac{hd}{k} = Nu' = 2 \tag{1.27}$$

where $Q/4\pi r^2(T_1 - T_2) = h$ is the heat transfer coefficient, d is the diameter of the particle or droplet, and hd/k is a dimensionless group known as the *Nusselt number (Nu')* for the particle. The more general use of the Nusselt number, with particular reference to heat transfer by convection, is discussed in Section 1.4. This value of 2 for the Nusselt number is the theoretical minimum for heat transfer through a *continuous* medium. It is greater if the temperature difference is applied over a finite distance, when Eq. (1.26) must be used. When there is a relative motion between the particle and the fluid, the heat transfer rate will be further increased, as discussed in Section 1.4.6.

In this approach, heat transfer to a spherical particle by conduction through the surrounding fluid has been the prime consideration. In many practical situations, the flow of heat from the surface to the internal parts of the particle is of importance. For example, if the particle is a poor conductor then the rate at which the particulate material reaches some desired average temperature may be limited by conduction inside the particle rather than by conduction to the outside surface of the particle. This problem involves unsteady state transfer of heat, which is considered in Section 1.3.5.

Equations may be developed to predict the rate of change of diameter d of evaporating droplets. If the latent heat of vapourisation is provided by heat conducted through a hotter stagnant gas to the droplet surface, and heat transfer is the rate controlling step, it is shown by Spalding[2] that the surface area, i.e., d^2 decreases linearly with time. A closely related and important practical problem is the prediction of the residence time required in a combustion chamber to ensure virtually complete burning of the oil droplets or coal particles. Complete combustion is desirable to obtain maximum utilisation of energy and to minimise pollution of the atmosphere by partially burned oil droplets. Here a droplet is surrounded by a flame and heat conducted back from the flame to the droplet surface provides the heat to vapourise the oil and sustain the surrounding flame. Again, the surface area, i.e., d^2 decreases approximately linearly with time though the derivation of the equation is more complex due to mass transfer effects, steep temperature gradients,[3] and circulation in the drop.[4] Reference must be made to specialised books for further details.[5,6]

1.3.5 Unsteady State Conduction

Basic considerations

In the problems, which have been considered so far, it has been assumed that the conditions at any point in the system remain constant with respect to time. The case of heat transfer by conduction in a medium in which the temperature is changing with time is now considered. This problem is of importance in the calculation of the temperature distribution in a body which is being heated or cooled. If, in an element of dimensions dx by dy by dz (Fig. 1.9), the

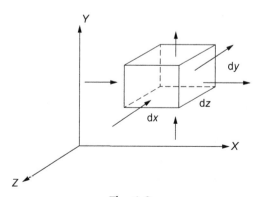

Fig. 1.9
Three-dimensional element for heat conduction.

temperature at the point (x, y, z) is θ and at the point $(x + dx, y + dy, z + dz)$ is $(\theta + d\theta)$, then assuming that the thermal conductivity k is constant and that no heat is generated in the medium, the rate of conduction of heat through the element is:

$$= -k\,dy\,dz\left(\frac{\partial\theta}{\partial x}\right)_{yz} \quad \text{in the } x-\text{direction}$$

$$= -k\,dz\,dx\left(\frac{\partial\theta}{\partial y}\right)_{zx} \quad \text{in the } y-\text{direction}$$

$$= -k\,dx\,dy\left(\frac{\partial\theta}{\partial z}\right)_{xy} \quad \text{in the } z-\text{direction}$$

The rate of change of heat content of the element is equal to *minus* the rate of increase of heat flow from (x, y, z) to $(x + dx, y + dy, z + dz)$. Thus the rate of change of the heat content of the element is:

$$= k\,dy\,dz\left(\frac{\partial^2\theta}{\partial x^2}\right)_{yz} dx + k\,dz\,dx\left(\frac{\partial^2\theta}{\partial y^2}\right)_{zx} dy + k\,dx\,dy\left(\frac{\partial^2\theta}{\partial z^2}\right)_{xy} dz$$

$$= k\,dx\,dy\,dz\left[\left(\frac{\partial^2\theta}{\partial x^2}\right)_{yz} + \left(\frac{\partial^2\theta}{\partial y^2}\right)_{zx} + \left(\frac{\partial^2\theta}{\partial z^2}\right)_{xy}\right]$$

(1.28)

The rate of increase of heat content is also equal, however, to the product of the heat capacity of the element and the rate of rise of temperature.

Thus:

$$k\,dx\,dy\,dz\left[\left(\frac{\partial^2\theta}{\partial x^2}\right)_{yz} + \left(\frac{\partial^2\theta}{\partial y^2}\right)_{zx} + \left(\frac{\partial^2\theta}{\partial z^2}\right)_{xy}\right] = C_p\rho\,dx\,dy\,dz\frac{\partial\theta}{\partial t}$$

or:

$$\frac{\partial\theta}{\partial t} = \frac{k}{C_p\rho}\left[\left(\frac{\partial^2\theta}{\partial x^2}\right)_{yz} + \left(\frac{\partial^2\theta}{\partial y^2}\right)_{zx} + \left(\frac{\partial^2\theta}{\partial z^2}\right)_{xy}\right]$$

$$= D_H\left[\left(\frac{\partial^2\theta}{\partial x^2}\right)_{yz} + \left(\frac{\partial^2\theta}{\partial y^2}\right)_{zx} + \left(\frac{\partial^2\theta}{\partial z^2}\right)_{xy}\right]$$

(1.29)

where $D_H = k/C_p\rho$ is known as the *thermal diffusivity*, which plays the same role in heat transfer as the momentum diffusivity (μ/ρ) in momentum transfer. In fact, the familiar Prandtl number is simply the ratio of the momentum and thermal diffusivities. Implicit in Eq. (1.29) is the assumption that k is independent of temperature and the medium is isotropic

This partial differential equation is most conveniently solved by the use of the Laplace transform of temperature with respect to time. As an illustration of the method of solution, the problem of the unidirectional flow of heat in a continuous medium will be considered. The basic differential equation for the X-direction is:

$$\frac{\partial \theta}{\partial t} = D_H \frac{\partial^2 \theta}{\partial x^2} \tag{1.30}$$

This equation cannot be integrated directly since the temperature θ is expressed as a function of two independent variables, distance x and time t. The method of solution involves transforming the equation so that the Laplace transform of θ with respect to time is used in place of θ. The equation then involves only the Laplace transform $\bar{\theta}$ and the distance x. The Laplace transform of θ is defined by the relation:

$$\bar{\theta} = \int_0^\infty \theta e^{-pt} dt \tag{1.31}$$

where p is a parameter.

Thus $\bar{\theta}$ is obtained by operating on θ with respect to t with x constant.

Then:

$$\frac{\partial^2 \bar{\theta}}{\partial x^2} = \overline{\frac{\partial^2 \theta}{\partial x^2}} \tag{1.32a}$$

and:

$$\begin{aligned}
\overline{\frac{\partial \theta}{\partial t}} &= \int_0^\infty \frac{\partial \theta}{\partial t} e^{-pt} dt \\
&= [\theta e^{-pt}]_0^\infty + p \int_0^\infty e^{-pt} \theta \, dt \\
&= -\theta_{t=0} + p\bar{\theta}
\end{aligned} \tag{1.32b}$$

Then, taking the Laplace transforms of each side of Eq. (1.30):

$$\overline{\frac{\partial \theta}{\partial t}} = D_H \overline{\frac{\partial^2 \theta}{\partial x^2}}$$

or:

$$p\bar{\theta} - \theta_{t=0} = D_H \frac{\partial^2 \bar{\theta}}{\partial x^2} \quad \text{(from Eqs 1.32a, b)}$$

and:

$$\frac{\partial^2 \bar{\theta}}{\partial x^2} - \frac{p}{D_H} \bar{\theta} = -\frac{\theta_{t=0}}{D_H}$$

If the temperature everywhere is constant initially, $\theta_{t=0}$ is a constant and the equation may be integrated as a normal second-order differential equation since p is not a function of x.

Thus:

$$\bar{\theta} = B_1 e^{\sqrt{(p/D_H)}x} + B_2 e^{-\sqrt{(p/D_H)}x} + \theta_{t=0} p^{-1} \tag{1.33}$$

and therefore:

$$\frac{\partial \bar{\theta}}{\partial x} = B_1 \sqrt{\frac{p}{D_H}} e^{\sqrt{(p/D_H)}x} - B_2 \sqrt{\frac{p}{D_H}} e^{-\sqrt{(p/D_H)}x} \tag{1.34}$$

The temperature θ, corresponding to the transform $\bar{\theta}$, may now be found by reference to tables of the Laplace transform. It is first necessary, however, to evaluate the constants B_1 and B_2 using the boundary conditions for the particular problem since these constants will in general involve the parameter p which was introduced in the transformation.

Considering the particular problem of the unidirectional flow of heat through a body with plane parallel faces a distance l apart, the heat flow is normal to these faces and the temperature of the body is initially constant throughout. The temperature scale will be so chosen that this uniform initial temperature is zero. At time, $t=0$, one face (at $x=0$) will be brought into contact with a source at a constant temperature θ' and the other face (at $x=l$) will be assumed to be perfectly insulated thermally.

The initial and boundary conditions are therefore:

$$
\begin{array}{lll}
t=0, & \theta = 0 & 0 \le x \le l \\
t>0, & \theta = \theta' & \text{when } x=0 \\
t>0, & \dfrac{\partial \theta}{\partial x} = 0 & \text{when } x=l
\end{array}
$$

Thus:

$$\bar{\theta}_{x=0} = \int_0^\infty \theta' e^{-pt} dt = \frac{\theta'}{p}$$

and:

$$\left(\frac{\partial \bar{\theta}}{\partial x} \right)_{x=l} = 0$$

Substitution of these boundary conditions in Eqs. (1.33), (1.34) gives:

$$B_1 + B_2 = \frac{\theta'}{p}$$

and:

$$B_1 e^{\sqrt{(p/D_H)}l} - B_2 e^{-\sqrt{(p/D_H)}l} = 0 \tag{1.35}$$

Hence:

$$B_1 = \frac{(\theta'/p)e^{-\sqrt{(p/D_H)}l}}{e^{\sqrt{(p/D_H)}l} + e^{-\sqrt{(p/D_H)}l}}$$

and:

$$B_2 = \frac{(\theta'/p)e^{\sqrt{(p/D_H)}l}}{e^{\sqrt{(p/D_H)}l} + e^{-\sqrt{(p/D_H)}l}}$$

Then:

$$\begin{aligned}
\bar{\theta} &= \frac{e^{(l-x)\sqrt{(p/D_H)}} + e^{-(l-x)\sqrt{(p/D_H)}}}{e^{\sqrt{(p/D_H)}l} + e^{-\sqrt{(p/D_H)}l}} \frac{\theta'}{p} \\
&= \frac{\theta'}{p} \left(e^{(l-x)\sqrt{(p/D_H)}} + e^{-(l-x)\sqrt{(p/D_H)}} \right) \left(1 + e^{-2\sqrt{(p/D_H)}l} \right)^{-1} \left(e^{-\sqrt{(p/D_H)}l} \right) \\
&= \frac{\theta'}{p} \left(e^{-x\sqrt{(p/D_H)}} + e^{-(2l-x)\sqrt{(p/D_H)}} \right) \left(1 - e^{-2l\sqrt{(p/D_H)}} + \cdots + (-1)^N e^{-2Nl\sqrt{(p/D_H)}} + \cdots \right) \\
&= \sum_{N=0}^{N=\infty} \frac{\theta'}{p} (-1)^N \left(e^{-(2lN+x)\sqrt{(p/D_H)}} + e^{-\{2(N+1)l-x\}\sqrt{(p/D_H)}} \right)
\end{aligned} \tag{1.36}$$

The temperature θ is then obtained from the tables of inverse Laplace transforms in the Appendix (Table 12, No. 83) and is given by:

$$\theta = \sum_{N=0}^{N=\infty} (-1)^N \theta' \left(\text{erfc} \frac{2lN + x}{2\sqrt{D_H t}} + \text{erfc} \frac{2(N+1)l - x}{2\sqrt{D_H t}} \right) \tag{1.37}$$

where:

$$\text{erfc}\, x = \frac{2}{\sqrt{\pi}} \int_x^\infty e^{-\xi^2} d\xi$$

Values of $\text{erfc}\, x (= 1 - \text{erf}\, x)$ are given in the Appendix (Table 13) and in specialist sources.[7]

Eq. (1.37) may be written in the form:

$$\frac{\theta}{\theta'} = \sum_{N=0}^{N=\infty} (-1)^N \left\{ \text{erfc}\left[Fo_l^{-1/2} \left(N + \frac{1}{2}\frac{x}{l} \right) \right] + \text{erfc}\left[Fo_l^{-1/2} \left((N+1) - \frac{1}{2}\frac{x}{l} \right) \right] \right\} \tag{1.38}$$

where $Fo_l = (D_H t/l^2)$ and is known as the *Fourier number*.

Thus:

$$\frac{\theta}{\theta'} = f\left(Fo_l, \frac{x}{l}\right) \tag{1.39}$$

The numerical solution to this problem is then obtained by inserting the appropriate values for the physical properties of the system and using as many terms in the series as are necessary for the degree of accuracy required. In most cases, the above series converges quite rapidly.

This method of solution to problems of unsteady flow is particularly useful because it is applicable when there are discontinuities in the physical properties of the material.[8] The boundary conditions, however, become a little more complicated, but the problem is intrinsically no more difficult.

A general method of estimating the temperature distribution in a body of any shape involves replacing the heat flow problem by the analogous electrical situation and measuring the electrical potentials at various points. The heat capacity per unit volume ρC_p is represented by an electrical capacitance, and the thermal conductivity k by an electrical conductivity. This method can be used to take account of variations in the thermal properties over the body.

Example 1.6

Calculate the time taken for the distant face of a brick wall, of thermal diffusivity $D_H = 4.2 \times 10^{-7}$ m^2/s and thickness $l = 0.45$ m, to rise from 295 to 375 K, if the whole wall is initially at a constant temperature of 295 K and the near face is suddenly raised to 900 K and maintained at this temperature. Assume that all the flow of heat is perpendicular to the faces of the wall and that the distant face is perfectly insulated.

Solution
The temperature at any distance x from the near face at time t is given by:

$$\theta = \sum_{N=0}^{N=\infty} (-1)^N \theta' \left\{ \text{erfc}\left[\frac{2lN + x}{2\sqrt{D_H t}}\right] + \text{erfc}\left[\frac{2(N+1)l - x}{2\sqrt{D_H t}}\right] \right\} \quad \text{(Eq. 1.37)}$$

The temperature at the distant face ($x = l$) is therefore given by:

$$\theta = 2 \sum_{N=0}^{N=\infty} (-1)^N \theta' \text{erfc}\left[\frac{(2N+1)l}{2\sqrt{D_H t}}\right]$$

Choosing the temperature scale so that the initial temperature is everywhere zero, then:

$$\frac{\theta}{2\theta'} = \frac{375 - 295}{2(900 - 295)} = 0.066$$

$$D_H = 4.2 \times 10^{-7} \, \text{m}^2/\text{s} \quad \therefore \quad \sqrt{D_H} = 6.5 \times 10^{-4}$$

$$l = 0.45 \, \text{m}$$

Thus:

$$0.066 = \sum_{N=0}^{N=\infty} (-1)^N \text{erfc} \left[\frac{0.45(2N+1)}{2 \times 6.5 \times 10^{-4} t^{0.5}} \right]$$

$$0.066 = \sum_{N=0}^{N=\infty} (-1)^N \text{erfc} \left[\frac{346(2N+1)}{t^{0.5}} \right]$$

$$= \text{erfc}(346 t^{-0.5}) - \text{erfc}(1038 t^{-0.5}) + \text{erfc}(1730 t^{-0.5}) - \cdots$$

An approximate solution is obtained by taking the first term only, to give:

$$346 t^{-0.5} = 1.30$$

from which

$$t = 70,840 \, \text{s}$$
$$= \underline{70.8 \, \text{ks}} \text{ or } \underline{19.7 \, \text{h}}$$

Example 1.7

This example illustrates the solution of the one-dimensional unsteady conduction Eq. (1.30) for a semiinfinite body. Initially, the entire semiinfinite body is at a uniform temperature T_o as shown below. At $t=0$, the temperature of the face of the body at $x=0$ is suddenly raised to and maintained at $T=T_s$. Calculate the temperature in the body as a function of time and x-coordinate.

Solution
The schematics of this situation are shown in Fig. 1.10:

Since it is the case of one-directional conduction, the temperature in the slab is governed by Eq. (1.30):

$$\frac{\partial \theta}{\partial t} = D_H \frac{\partial^2 \theta}{\partial x^2} \quad \text{(Eq. 1.30)}$$

The initial and boundary conditions for this situation are:

$$\text{For } t=0, \quad T=T_o \quad 0 \le x \le \infty$$

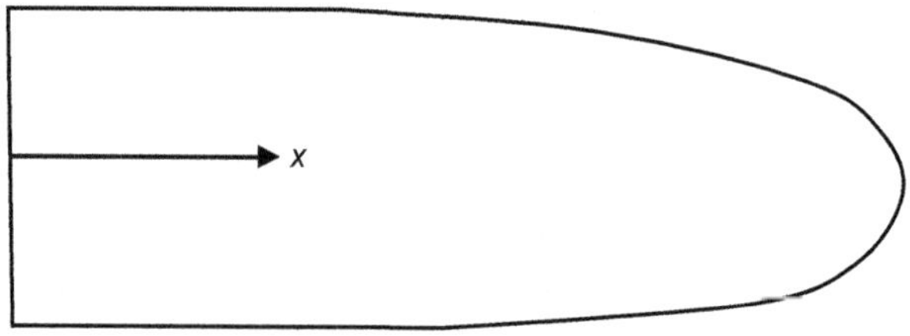

Fig. 1.10
Unsteady conduction in a semiinfinite medium.

$$\text{For } t > 0, \quad T = T_s \text{ at } x = 0$$

$$T = T_0 \text{ at } x \to \infty$$

The last condition implies that it will take infinite time for the temperature change at $x = 0$ to reach $x = \infty$.

In this case, the dimensionless temperature θ:

$$\theta = \frac{T - T_0}{T_s - T_0}$$

and the corresponding boundary conditions now become:

$$t = 0, \theta = 0 \quad 0 \leq x \leq \infty$$

$$t > 0, \ \theta = 1 \text{ at } x = 0$$

$$\theta = 0 \text{ at } x \to \infty$$

One can now introduce a similarity variable ξ defined as:

$$\xi = \frac{x}{\sqrt{4D_H t}}$$

such that $\theta(\xi)$. Now, using the chain rule of differentiation:

$$\frac{\partial \theta}{\partial t} = \frac{d\theta}{d\xi} \cdot \frac{\partial \xi}{\partial t} \quad \text{and} \quad \frac{\partial \theta}{\partial x} = \frac{d\theta}{d\xi} \cdot \frac{\partial \xi}{\partial x}$$

the original partial differential equation and the boundary conditions transform as:

$$\frac{d^2\theta}{d\xi^2} + 2\theta\xi = 0$$

$$\theta = 0 \text{ when } \xi \to \infty$$
$$\theta = 1 \text{ when } \xi = 0$$

The solution of the new differential equation subject to the above-noted boundary conditions is in the form of the standard error function:

$$\theta = 1 - \text{erf}(\xi)$$

Note that $\text{erf}(0) = 0$ and $\text{erf}(\infty) = 1$. In fact, $\text{erf}(1.8) = 0.99$, i.e. $\theta = 0.01$ at $\xi = 1.8$. This means that at $\xi = 1.8$, the temperature difference has dropped to 1% of the maximum available $\Delta T = T_s - T_0$. This idea is used to introduce the notions of the 'depth of penetration' and the 'time of penetration' as follows:

$\xi = \frac{x_p}{\sqrt{4D_H t_p}} = 1.8$, i.e. $x_p \propto \sqrt{D_H}$. In other words, the depth of penetration x_p will be greater for materials with high thermal diffusivity, i.e. the effect of temperature change at $x = 0$ will be felt up to larger values of x in materials with high D_H. Similarly, $t_p \propto 1/D_H$, i.e. lower the value of D_H, longer it will take to experience the change in temperature.

This analysis is used to estimate the depth below the ground to install buried pipelines so as to prevent freezing of water in cold climates, as illustrated in Example 1.8.

Example 1.8

While laying underground water pipelines, one is generally concerned with the possibility of freezing of water (thereby bursting of pipe) during cold temperature conditions. One can approximately calculate the safe depth for such a pipe below the ground surface by assuming that the ground surface temperature to remain constant over a prolonged period and by assuming mean properties of the soil. Calculate the minimum depth of the water main to avoid freezing if the soil temperature, initially at a uniform temperature of 15°C changes to −20°C and stays at this value for 45 days. The thermal diffusivity of soil is 1.4×10^{-7} m²/s.

Solution
This situation is sketched in Fig. 1.11:

One can assume the ground to extend to infinity and therefore, it is the case of the transient conduction in a semiinfinite body with $D_H = 1.4 \times 10^{-7}$ m²/s. Thus, one can use the following equation derived in Example 1.7:

$$\frac{T - T_o}{T_s - T_o} = 1 - \text{erf}\, \frac{x}{\sqrt{4D_H t}}$$

Here, $T_o = 15°C$, $T_s = -20°C$, $T = 0°C$

$$t = 45 \times 24 \times 3600\,\text{s}$$

Thus, the depth to attain the freezing point of water is obtained as:

$$\frac{0 - 15}{-20 - 15} = 1 - \text{erf}\, \frac{x_m}{\sqrt{4D_H t}}$$

$$\text{i.e. } \text{erf}\, \frac{x_m}{\sqrt{4D_H t}} = 0.571$$

$$\frac{x_m}{\sqrt{4D_H t}} = 0.56 \quad \text{(From error function tables)}$$

$$\therefore \ x_m = 0.56 \times \sqrt{4 \times 1.4 \times 10^{-7} \times 45 \times 24 \times 3600}$$

$$\text{or } x_m = 0.83\,\text{m}$$

Therefore, it will be safe to install the water pipe at a depth greater than 0.83 m.

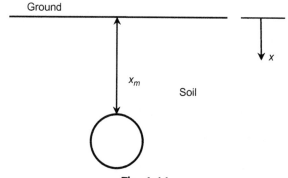

Ground

x_m

Soil

x

Fig. 1.11
Underground water pipe.

Schmidt's method

Numerical methods have been developed by replacing the differential equation by a finite difference equation. Thus in a problem of unidirectional flow of heat:

$$\frac{\partial \theta}{\partial t} \approx \frac{\theta_{x(t+\Delta t)} - \theta_{x(t-\Delta t)}}{2\Delta t} \approx \frac{\theta_{x(t+\Delta t)} - \theta_{xt}}{\Delta t}$$

$$\frac{\partial^2 \theta}{\partial x^2} \approx \frac{\left(\dfrac{\theta_{(x+\Delta x)t} - \theta_{xt}}{\Delta x} - \dfrac{\theta_{xt} - \theta_{(x-\Delta x)t}}{\Delta x}\right)}{\Delta x}$$

$$= \frac{\theta_{(x+\Delta x)t} + \theta_{(x-\Delta x)t} - 2\theta_{xt}}{(\Delta x)^2}$$

where θ_{xt} is the value of θ at time t and distance x from the surface, and the other values of θ are at intervals Δx and Δt as shown in Fig. 1.12.

Substituting these values in Eq. (1.30):

$$\theta_{x(t+\Delta t)} - \theta_{x(t-\Delta t)} = D_H \frac{2\Delta t}{(\Delta x)^2} \left(\theta_{(x+\Delta x)t} + \theta_{(x-\Delta x)t} - 2\theta_{xt}\right) \tag{1.40}$$

and:

$$\theta_{x(t+\Delta t)} - \theta_{xt} = D_H \frac{\Delta t}{(\Delta x)^2} \left(\theta_{(x+\Delta x)t} + \theta_{(x-\Delta x)t} - 2\theta_{xt}\right) \tag{1.41}$$

Thus, if the temperature distribution at time t, is known, the corresponding distribution at time $t + \Delta t$ can be calculated by the application of Eq. (1.41) over the whole extent of the body

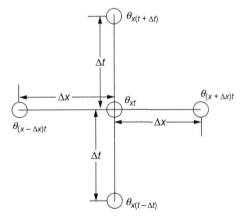

Fig. 1.12
Variation of temperature with time and distance.

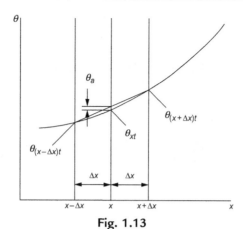

Fig. 1.13
Schematic representation of Schmidt's method.

in question. The intervals Δx and Δt are so chosen that the required degree of accuracy is obtained.

A graphical procedure has been proposed by Schmidt.[9] If the temperature distribution at time t is represented by the curve shown in Fig. 1.13 and the points representing the temperatures at $x - \Delta x$ and $x + \Delta x$ are joined by a straight line, then the temperature θ_a is given by:

$$\theta_a = \frac{\theta_{(x+\Delta x)t} + \theta_{(x-\Delta x)t}}{2} - \theta_{xt}$$

$$= \frac{(\Delta x)^2}{2D_H \Delta t}\left(\theta_{x(t+\Delta t)} - \theta_{xt}\right) \quad \text{(from Eq. 1.41)}$$

(1.42)

Thus, θ_a represents the change in θ_{xt} after a time interval Δt, such that:

$$\Delta t = \frac{(\Delta x)^2}{2D_H}$$

(1.43)

If this simple construction is carried out over the whole body, the temperature distribution after time Δt is obtained. The temperature distribution after an interval $2\Delta t$ is then obtained by repeating this procedure.

The most general method of tackling the problem is the use of the *finite-element* technique[10] to determine the temperature distribution at any time by using the finite difference equation in the form of Eq. (1.40).

Example 1.9

Solve Example 1.6 using Schmidt's method.

Solution

The development of the temperature profile is shown in Fig. 1.14. At time $t=0$ the temperature is constant at 295 K throughout and the temperature of the hot face is raised to 900 K. The problem will be solved by taking relatively large intervals for Δx.

Choosing $\Delta x = 50$ mm, the construction shown in Fig. 1.14 is carried out starting at the hot face.

Points corresponding to temperature after a time interval Δt are marked 1, after a time interval $2\Delta t$ by 2, and so on. Because the second face is perfectly insulated, the temperature gradient must be zero at this point. Thus, in obtaining temperatures at $x=450$ mm it is assumed that the temperature at $x=500$ mm will be the same as at $x=400$ mm, that is, horizontal lines are drawn in the diagram. It is seen that the temperature is less than 375 K after time $23\Delta t$ and greater than 375 K after time $25\Delta t$.

Thus:

$$t \approx 24\Delta t$$

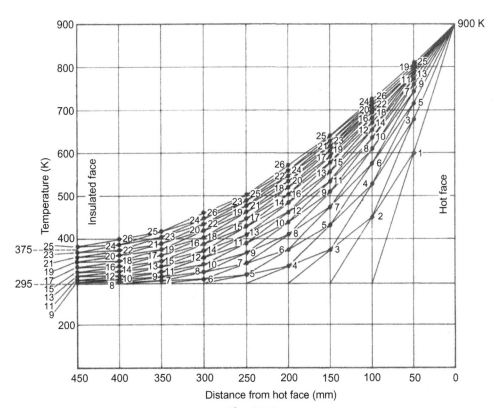

Fig. 1.14

Development of temperature profile.

From Eq. (1.43):

$$\Delta t = 0.05^2 / (2 \times 4.2 \times 10^{-7}) = 2976\,\text{s}$$

Thus time required

$$= 24 \times 2976 = 71,400\,\text{s}$$

or:

$$\underline{71.4\,\text{ks} = 19.8\,\text{h}}$$

This value is quite close to that obtained by calculation, even using the coarse increments in Δx.

Heating and cooling of solids and particles

The exact mathematical solution of problems involving unsteady thermal conduction may be very difficult, and sometimes impossible, especially where bodies of irregular shapes are concerned, and other methods are therefore required.

When a body of characteristic linear dimension L, initially at a uniform temperature θ_0, is exposed suddenly to surroundings at a temperature θ', the temperature distribution at any time t is found from dimensional analysis to be:

$$\frac{\theta' - \theta}{\theta' - \theta_0} = f\left(\frac{hL}{k_p}, D_H \frac{t}{L^2}, \frac{x}{L}\right) \tag{1.44}$$

where D_H is the thermal diffusivity $(k_p / C_p \rho)$ of the solid, x is distance within the solid body (measured from the line of symmetry), and h is the heat transfer coefficient in the fluid at the surface of the body.

Analytical solutions of Eq. (1.44) in the form of infinite series are available for some simple regular shapes of particles, such as rectangular slabs, long cylinders, and spheres, for conditions where there is heat transfer by conduction or convection to or from the surrounding fluid. These solutions tend to be quite complex, even for simple shapes. The heat transfer process may be characterised by the value of the *Biot number Bi* where:

$$Bi = \frac{hL}{k_p} = \frac{L/k_p}{1/h} \tag{1.45}$$

where h is the external heat transfer coefficient, L is a characteristic dimension, such as radius in the case of a sphere or long cylinder, or half the thickness in the case of a slab, and k_p is the thermal conductivity of the particle.

The Biot number is essentially the ratio of the resistance to heat transfer within the particle to that within the external fluid. At first sight, it appears to be similar in form to the Nusselt Number Nu' where:

$$Nu' = \frac{hd}{k} = \frac{2hr_0}{k} \tag{1.46}$$

However, the Nusselt number refers to a single fluid phase, whereas the Biot number is related to the properties of both the fluid and the solid phases.

Three cases are now considered:

(1) Very large Biot numbers, $Bi \to \infty$
(2) Very low Biot numbers, $Bi \to 0$
(3) Intermediate values of the Biot number.

(1) *Bi very large.* The resistance to heat transfer in the fluid is then low compared with that in the solid with the temperature of the surface of the particle being approximately equal to the bulk temperature of the fluid, and the heat transfer rate is independent of the Biot number. Eq. (1.44) then simplifies to:

$$\frac{\theta' - \theta}{\theta' - \theta_0} = f\left(D_H \frac{t}{L^2}, \frac{x}{L}\right) = f\left(Fo_L, \frac{x}{L}\right) \tag{1.47}$$

where $Fo_L \left(= D_H \dfrac{t}{L^2}\right)$ is known as the *Fourier* number, using L in this case to denote the characteristic length, and x is distance from the centre of the particle. Curves connecting these groups have been plotted by a number of researchers for bodies of various shapes, although the method is limited to those shapes which have been studied experimentally.

In Fig. 1.15, taken from Carslaw and Jaeger,[7] the value of $(\theta' - \theta_c)/(\theta' - \theta_0)$ is plotted to give the temperature θ_c at the centre of bodies of various shapes, initially at a uniform temperature θ_0, at a time t after the surfaces have been suddenly altered and maintained at a constant temperature θ'.

In this case (x/L) is constant at 0 and the results are shown as a function of the particular value of the Fourier number $Fo_L (= D_H t / L^2)$.

(2) *Bi very small,* (say, <0.1). Here the main resistance to heat transfer lies within the fluid; this occurs when the thermal conductivity of the particle in very high and/or when the particle is very small. Under these conditions, the temperature within the particle is uniform and a 'lumped capacity' analysis may be performed. Thus, if a solid body of volume V and initial temperature θ_0 is suddenly immersed in a volume of fluid large enough for its temperature θ' to remain effectively constant, the rate of heat transfer from the body may be expressed as:

$$-\rho C_p V \frac{d\theta}{dt} = hA_e(\theta - \theta') \tag{1.48}$$

where A_e is the external surface area of the solid body.

Then:

$$\int_{\theta_0}^{\theta} \frac{d\theta}{\theta - \theta'} = -\int_{0}^{t} \frac{hA_e}{\rho C_p V} dt$$

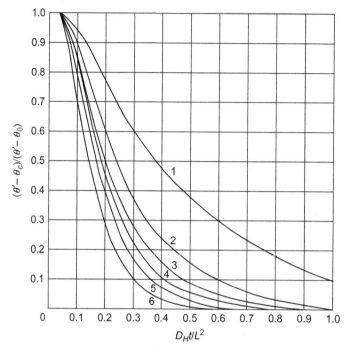

Fig. 1.15

Cooling curves for bodies of various shapes: 1, slab (2L=thickness); 2, square bar (2L=side); 3, long cylinder (L=radius); 4, cube (2L=length of side); 5, cylinder (L=radius, length=2L); 6, sphere (L=radius).

i.e.:

$$\frac{\theta - \theta'}{\theta_0 - \theta'} = e^{-t/\tau} \tag{1.49}$$

where $\tau = \dfrac{\rho C_p V}{hA_e}$ is known as the *response time constant*.

It will be noted that the relevant characteristic dimension in the Biot number is defined as the ratio of the volume to the external surface area of the particle (V/A_e), and the higher the value of V/A_e, the slower will be the response time. With the characteristic dimension defined in this way, this analysis is valid for particles of any shape at values of the Biot number less than 0.1

Example 1.10

A 25 mm diameter copper sphere and a 25 mm copper cube are both heated in a furnace to 650°C (923 K). They are then annealed in air at 95°C (368 K). If the external heat transfer coefficient h is 75 W/m^2 K in both cases, what is the temperature of the sphere and of the cube at the end of 5 min?

The physical properties at the mean temperature for copper are:

$$\rho = 8950 \,\text{kg/m}^3 \quad C_p = 0.38 \,\text{kJ/kgK} \quad k_p = 385 \,\text{W/mK}$$

Solution

$$V/A_e \text{ for the sphere} \quad = \frac{\frac{\pi}{6}d^3}{\pi d^2} = \frac{d}{6} = \frac{25 \times 10^{-3}}{6} = 4.17 \times 10^{-3}\,\text{m}$$

$$V/A_e \text{ for the cube} \quad = \frac{l^3}{6l^2} = \frac{l}{6} = \frac{25 \times 10^{-3}}{6} = 4.17 \times 10^{-3}\,\text{m}$$

$$\therefore \quad Bi = \frac{h(V/A_e)}{k} = \frac{75 \times 4.17 \times 10^{-3}}{385} = 8.1 \times 10^{-4} \ll 0.1$$

The use of a lumped capacity method is therefore justified.

$$\tau = \frac{\rho C_p V}{h A_e} = \frac{8950 \times 380}{75} \times \frac{25 \times 10^{-3}}{6} = 189\,\text{s}$$

Then using Eq. (1.49):

$$\frac{\theta - 368}{923 - 368} = \exp\left(-\frac{5 \times 60}{189}\right)$$

and:

$$\theta = 368 + 0.2045(923 - 368) = 481\,\text{K} = 208\,°\text{C}$$

Since the sphere and the cube have the same value of V/A_e, after 5 min they will both attain a temperature of 208°C.

(3) *Intermediate values of Bi.* In this case the resistances to heat transfer within the solid body and the fluid are of comparable magnitude. Neither will the temperature within the solid be uniform (case 1), nor will the surface temperature be equal to that in the bulk of the fluid (case 2).

Analytical solutions in the form of infinite series can be obtained for some regular shapes (thin plates, spheres, and long cylinders (length $\gg$ radius)), and numerical solutions using *finite element methods*[10] have been obtained for bodies of other shapes, both regular and irregular. Some of the results have been presented by Heisler[11] in the form of charts, examples of which are shown in Figs 1.16–1.18 for thin slabs, long cylinders, and spheres, respectively. It may be noted that in this case the characteristic length L is the half-thickness of the slab and the external radius r_o of the cylinder and sphere.

Figs 1.16–1.18 enable the temperature θ_c at the centre of the solid (centre-plane, centre-line, or centre-point) to be obtained as a function of the Fourier number, and hence of time, with the reciprocal of the Biot number (Bi^{-1}) as parameter.

Temperatures at off-centre locations within the solid body can then be obtained from a further series of charts given by Heisler (Figs 1.19–1.21) which link the desired temperature to the centre-temperature as a function of Biot number, with location within the particle as parameter

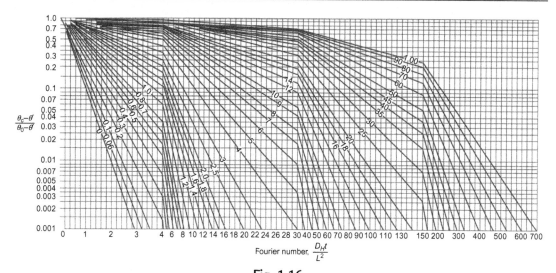

Fig. 1.16

Mid-plane temperature for an infinite plate of thickness 2L, for various values of parameters $k_p/hL (=Bi^{-1})$.

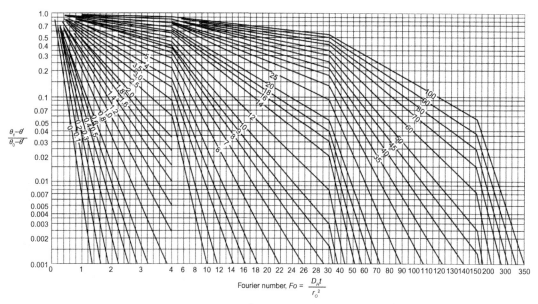

Fig. 1.17

Axis temperature for an infinite cylinder of radius r_o, for various parameters k_p/hr_o $(=Bi^{-1})$.

(that is the distance x from the centre plane in the slab or radius in the cylinder or sphere). Additional charts are given by Heisler for the quantity of heat transferred from the particle in a given time in terms of the initial heat content of the particle, e.g. see reference 12.

Figs 1.19–1.21 clearly show that, as the Biot number approaches zero, the temperature becomes uniform within the solid, and the lumped capacity method may be used for calculating the

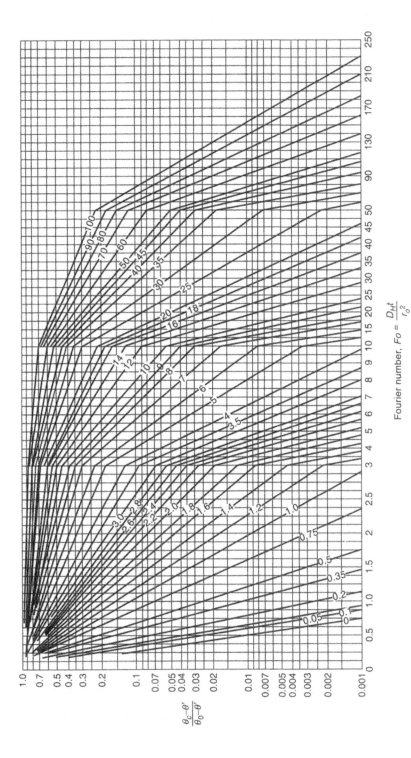

Fourier number, $Fo = \dfrac{D_H t}{r_o^2}$

Fig. 1.18

Centre-temperature for a sphere of radius r_o, for various values of parameters k_p/hr_o ($= Bi^{-1}$).

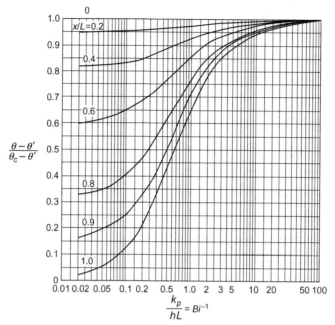

Fig. 1.19

Temperature as a function of mid-plane temperature in an infinite plate of thickness 2L.

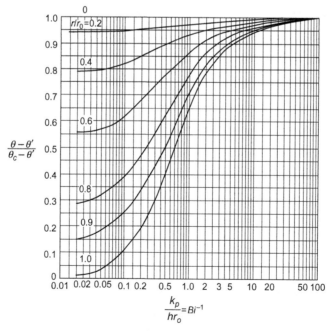

Fig. 1.20

Temperature as a function of axis temperature in an infinite cylinder of radius r_o.

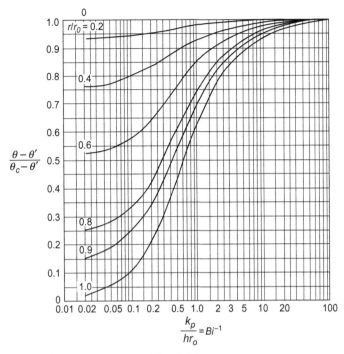

Fig. 1.21

Temperature as a function of centre-temperature for a sphere of radius r_0.

unsteady-state heating of the particles, as discussed in section (2). The charts are applicable for Fourier numbers greater than about 0.2.

Example 1.11

A large thermoplastic sheet, 10 mm thick, at an initial temperature of 20°C (293 K), is to be heated in an oven in order to enable it to be moulded. The oven temperature is maintained at 100°C (373 K), the maximum temperature to which the plastic may be subjected, and it is necessary to ensure that the temperature throughout the sheet reaches a minimum of 80°C (353 K). Calculate the minimum length of time for which the sheet must be heated.

Thermal conductivity k_p of the plastic $= 2.5$ W/m^2 K

Thermal diffusivity $D_H = 2 \times 10^{-7}$ m^2/s

External heat transfer coefficient $h = 100$ W/m^2 K

Solution

Throughout the heating process, the temperature within the sheet will be a minimum at the centre-plane ($x = 0$) and therefore the required time is that for the centre to reach 80°C (353 K).

$$\text{For this process, the Biot number } Bi = \frac{hL}{k_p} = \frac{100 \times 5 \times 10^{-3}}{2.5} = 0.2 \text{ and } Bi^{-1} = 5$$

Since L, the half-thickness of the plate is 5 mm. Here $\theta' = 373$ K and $\theta_0 = 293$ K.

$$\text{The limiting value of } \frac{\theta' - \theta_\iota}{\theta' - \theta_0} = \frac{373 - 353}{373 - 293} = 0.25$$

From Fig. 1.19, the Fourier number $\frac{D_H t}{L^2} \approx 7.7$

Thus:

$$t = \frac{7.7 \times (5 \times 10^{-3})^2}{2 \times 10^{-7}} = 960\,\text{s or } \underline{\underline{16\,\text{min}}}$$

Heating and melting of fine particles

There are many situations in which particles are heated or cooled by a surrounding gas and these may be classified according to the degree of movement of the particle as follows:

(i)   *Static beds*

Although most beds of particles involve relatively large particle diameters, such as in pebble bed units used for the transfer of heat from flue gases to the incoming air for example, smaller particles, such as sand, are used in beds and again, these are mainly used for heat recovery. One such application is the heating and cooling of buildings in hotter climes where the cool nocturnal air is used to cool a bed of particles which is then used to cool the incoming air during the heat of the day as it enters a building. In this way, an almost constant temperature may be achieved in a given enclosed environment in spite of the widely fluctuating ambient condition. A similar system has been used in less tropical areas where it is necessary to maintain a constant temperature in an environment in which heat is generated, such as a telephone exchange, for example. Such systems have the merit of very low capital and modest operating costs and, in most cases, the resistance to heat transfer by conduction within the solids is not dissimilar to the resistance in the gas film surrounding the particles.

(ii)   *Partial movement of particles*

The most obvious example of a process in which particles undergo only limited movement is the fluidised bed, which is discussed in some detail in Volume 2A. Applications here involve, not only heating and cooling, but also drying as in the case of grain dryers for example, and on occasions, chemical reaction as, for example, with fluidised-bed combustion. In such cases, conditions in the bed may, to all intents and purposes, be regarded as steady-state, with unsteady-state conduction taking place only in the entering 'process stream' which by and large, is only a small proportion of the total bed mass in the bed.

(iii)   *Falling particles*

Particles fall by gravity through either static or moving gas streams in rotary dryers, for example, but they also fall through heating or cooling gases in specially designed columns. Examples here include the cooling of sand after it has been dried—again recovering heat in the process—salt cooling and also the spray drying of materials such as detergents which are sprayed as a concentrated solution of the material at the top of the tower and emerge as a dry powder. A similar situation occurs in fertiliser production where solid particles or granules are obtained from a spray of the molten material by counterflow against a cooling gas stream. Convection to such materials is discussed in Section 1.4.6.

One important problem involving unsteady state conduction of heat to particles is in the melting of powders in plasma spraying[13] where Biot numbers can range from 0.005 to 5. In this case, there is initially a very high relative velocity between the fluid and the powder. The plasmas referred to here are partially ionised gases with temperatures of around 10,000 K formed by electric discharges such as arcs. There is an increasing industrial use of the technique of plasma spraying in which powders are injected into a high-velocity plasma jet so that they are both melted and projected at velocities of several hundred metres per second onto a surface. The molten particles with diameters typically of the order 10–100 µm impinge to form an integral layer on the surface. Applications include the building up of worn shafts of pumps, for example, and the deposition of erosion-resistant ceramic layers on centrifugal pump impellers and other equipment prone to erosion damage. When a powder particle first enters the plasma jet, the relative velocity may be hundreds of metres per second and heat transfer to the particle is enhanced by convection, as discussed in Section 1.4.6. Often, and more particularly for smaller particles, the particle is quickly accelerated to essentially the same velocity as the plasma jet[2] and conduction becomes the main mechanism of heat transfer from plasma to particle. From a design point of view, neglecting the convective contribution will ease calculations and give a more conservative and safer estimate of the size of the largest particle which can be melted before it strikes the surface. In the absence of complications due to noncontinuum conditions discussed later, the value of $Nu' = hd/k$ is therefore often taken as 2, as in Eq. (1.27).

One complication, which arises in the application of this equation to powder heating in high temperature plasmas lies in the dependence of k, the thermal conductivity of the gas or plasma surrounding the particle, on temperature. For example, the temperature of the particle surface may be 1000 K, whilst that of the plasma away from the particle may be about 10,000 K or even higher. The thermal conductivity of argon increases by a factor of about 20 over this range of temperature and that of nitrogen gas passes through a pronounced peak at about 7100 K due to dissociation-recombination effects. Thus, the temperature at which the thermal conductivity k is evaluated will have a pronounced effect on the value of the external heat transfer coefficient. A mean value of k would seem appropriate where:

$$(k)_{\text{mean}} = \frac{1}{T_2 - T_1} \int_{T_1}^{T_2} k \, dT \quad \text{(Eq. 1.17)}$$

Some researchers have correlated experimental data in terms of k at the arithmetic mean temperature, and some at the temperature of the bulk plasma. Experimental validation of the true effective thermal conductivity is difficult because of the high temperatures, small particle sizes, and variations in velocity and temperature in plasma jets.

In view of the high temperatures involved in plasma devices and the dependence of radiation heat transfer on T^4, as discussed in Section 1.5, it is surprising at first sight that conduction is more significant than radiation in heating particles in plasma spraying. The explanation lies in the small values of d and relatively high values of k for the gas, both of which contribute to high values of h for any given value of Nu'. Also the emissivities of most gases are, as seen later in Section 1.5, rather low.

In situations where the surrounding fluid behaves as a noncontinuum fluid, for example at very high temperatures and/or at low pressures, it is possible for Nu' to be less than 2. A gas begins to exhibit noncontinuum behaviour when the mean free path between collisions of gas molecules or atoms with each other is greater than about 1/100 of the characteristic size of the surface considered. The molecules or atoms are then sufficiently far apart on average for the gas to begin to lose the character of a homogeneous or continuum fluid, which is normally assumed in the majority of heat transfer or fluid dynamics problems. For example, with a particle of diameter 25 μm as encountered in, for example, oil-burner sprays, pulverised coal flames, and in plasma spraying in air at room temperature and atmospheric pressure, the mean free path of gas molecules is about 0.06 μm and the air then behaves as a continuum fluid. If however, the temperature were say 1800 K, as in a flame, then the mean free path would be about 0.33 μm, which is greater than 1/100 of the particle diameter. Noncontinuum effects, leading to values of Nu' lower than 2 would then be likely according to theory.[14,15] The exact value of Nu' depends on the surface accommodation coefficient. This is a difficult parameter to measure for the examples considered here, and hence experimental confirmation of the theory is difficult. At the still higher temperatures that exist in thermal plasma devices, noncontinuum effects should be more pronounced and there is limited evidence that values of Nu' below 1 are obtained.[13] In general, noncontinuum effects, leading in particular to values of Nu' less than 2, would be more likely at high temperatures, low pressures, and small particle sizes. Thus, there is an interest in these effects in the aerospace industry when considering, for example, the behaviour of small particles present in rocket engine exhausts.

1.3.6 Conduction With Internal Heat Source

If an electric current flows through a wire, the heat generated internally will result in a temperature distribution between the central axis and the surface of the wire. This type of problem will also arise in chemical or nuclear reactors where heat is generated internally. It is necessary to determine the temperature distribution in such a system and the maximum temperature which will occur.

If the temperature at the surface of the wire is T_o and the rate of heat generation per unit volume is Q_G, then considering unit length of a cylindrical element of radius r, the heat generated must be transmitted in an outward direction by conduction so that the energy balance becomes:

$$-k2\pi r\frac{\mathrm{d}T}{\mathrm{d}r}=\pi r^2 Q_G$$

Hence:

$$\frac{\mathrm{d}T}{\mathrm{d}r}=-\frac{Q_G r}{2k} \tag{1.50}$$

Integrating:

$$T=-\frac{Q_G r^2}{4k}+C$$

$T=T_o$ when $r=r_o$ the radius of wire and hence:

$$T=T_0+Q_G\frac{r_0^2-r^2}{4k}$$

or:

$$T-T_0=\frac{Q_G r_0^2}{4k}\left(1-\frac{r^2}{r_0^2}\right) \tag{1.51}$$

This gives a parabolic distribution of temperature and the maximum temperature will occur at the axis of the wire where $(T-T_0)=Q_G r_0^2/4k$. The arithmetic mean temperature difference being, $(T-T_0)_{av}=Q_G r_0^2/8k$.

Since $Q_G\pi r_o^2$ is the rate of heat release per unit length of the wire then, putting T_1 as the temperature at the centre:

$$T_1-T_0=\frac{\text{rate of heat release per unit length}}{4\pi k} \tag{1.52}$$

Example 1.12

A fuel channel in a natural uranium reactor is 5 m long and has a heat release of 0.25 MW. If the thermal conductivity of the uranium is 33 W/m K, what is the temperature difference between the surface and the centre of the uranium element, assuming that the heat release is uniform along the rod?

Solution
Heat release rate per unit length

$$\text{Heat release rate}=0.25\times10^6\,\text{W}$$

$$=\frac{0.25\times10^6}{5}=5\times10^4\,\text{W/m}$$

Thus, from Eq. (1.52):

$$T_1 - T_0 = \frac{5 \times 10^4}{4\pi \times 33}$$
$$= \underline{121 \, \text{K}}$$

It should be noted that the temperature difference is independent of the diameter of the fuel rod for a cylindrical geometry, and that the heat released per unit volume has been considered as being uniform.

In practice the assumption of the uniform heat release per unit length of the rod is not valid since the neutron flux, and hence the heat generation rate varies along its length. In the simplest case where the neutron flux may be taken as zero at the ends of the fuel element, the heat flux may be represented by a sinusoidal function, and the conditions become as shown in Fig. 1.22.

Since the heat generated is proportional to the neutron flux, the heat dQ developed per unit time in a differential element of the fuel rod of length dx may be written as:

$$dQ = C \sin\left(\frac{\pi x}{L}\right) dx$$

The total heat generated by the rod Q is then given by:

$$Q = C \int_0^L \sin\left(\frac{\pi x}{L}\right) dx = \frac{2CL}{\pi}$$

Thus, $C = \pi Q/2L$. The heat release per unit length at any point is then given by:

$$\frac{dQ}{dx} = \frac{\pi Q}{2L} \sin\left(\frac{\pi x}{L}\right)$$

Substituting into Eq. (1.52) gives:

$$T_1 - T_0 = \frac{\left(\frac{\pi Q}{2L}\right) \sin\left(\frac{\pi x}{L}\right)}{4\pi k} \tag{1.53}$$

It may be noted that when $x = 0$ or $x = L$, then $T_1 - T_0$ is zero as would be expected since the neutron flux was taken as zero at these positions.

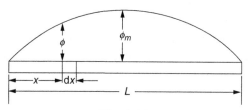

Fig. 1.22
Variation of neutron flux along a length of fuel rod.

Example 1.13

A refrigerant carrying pipe (outside diameter of 20 mm) is to be covered with two layers of insulation each of 15 mm thickness. The average thermal conductivity of one material is one third of the other material. In order to minimise the rate of heat flow from the outside into the pipe, which of the two materials should be put next to the pipe? What is the difference in the rate of heat flow in these two cases? Assume that the temperature of the outer surface of pipe and that of the outer surface of the insulation are held fixed in both cases.

Solution

Let material *A* be of thermal conductivity *k* and material *B* of thermal conductivity 3*k*. The two possibilities are:

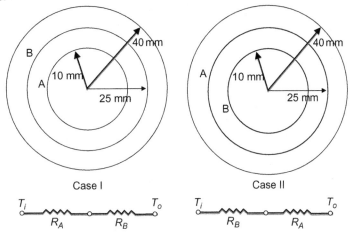

The total thermal resistances (per unit length of pipe) are calculated as:

Case I:

$$R_I = \frac{1}{2\pi\kappa_A} \ln\frac{25}{10} + \frac{1}{2\pi\kappa_B} \ln\frac{40}{25} = \frac{1}{2\pi\kappa}\left(\ln\frac{25}{10} + \frac{1}{3}\ln\frac{40}{25}\right)$$

$$R_I = \frac{1.073}{2\pi\kappa}$$

Case II:

$$R_{II} = \frac{1}{2\pi\kappa_B} \ln\frac{25}{10} + \frac{1}{2\pi\kappa_A} \ln\frac{40}{25}$$

$$R_{II} = \frac{1}{2\pi\kappa}\left(\frac{1}{3}\ln\frac{25}{10} + \ln\frac{40}{25}\right) = \frac{0.7754}{2\pi\kappa}$$

Since the overall temperature difference is $(T_i - T_0)$ in both cases, the ratio of the rate of heat transfer:

$$\frac{Q_I}{Q_{II}} = \frac{R_{II}}{R_I} = \frac{(0.7754/2\pi\kappa)}{(1.073/2\pi\kappa)} = \frac{0.7754}{1.073} = 0.723$$

Thus, the heat transfer is reduced by $(1 - 0.723) \times 100 = 27.7\%$ when material *A* is used adjacent to the pipe wall than when *B* is next to the pipe wall.

1.3.7 Multidimensional Steady Conduction

The discussion so far has been restricted to one dimensional conduction. This is a good approximation when the temperature gradient in one direction is much larger than that in the other two directions. For instance, by symmetry, heat transfer occurs in the radial direction for a spherical object. In long cylindrical pipes and wires, conduction predominantly occurs in the radial direction only. In the case of a plane surface, like the wall of a building, the wall thickness is typically much smaller than the other two dimensions. However, there are situations where heat conduction occurs in two- or even three-dimensions. For instance, in a long bar of square cross-section, conduction will occur in two directions whereas in a cube (all three sides are same), conduction will be three dimensional. The numerical methods are generally employed to solve 2-D and 3-D conduction problems with additional effects arising from complex geometry, or internal heat generation, or variable thermal conductivity, etc.[7,10,12] In this section, we consider the simplest possible case of the two-dimensional steady conduction in a plane geometry to illustrate the nature of two-dimensional conduction and to introduce a graphical method to obtain approximate results, without having to resort to numerical solutions.

Let us consider a long bar of cross-section $H \times W$, as shown in Fig. 1.23, whose three sides are maintained at a temperature T_0 and the top side is exposed to a temperature which varies with x (as shown in Fig. 1.23).

Since H and W are comparable, while the length of the bar $L \gg H$ and $L \gg W$, this implies that $\dfrac{\partial T}{\partial z} \ll \dfrac{\partial T}{\partial x}$ and $\dfrac{\partial T}{\partial z} \ll \dfrac{\partial T}{\partial y}$ thereby suggesting no heat transfer occurs along the z-direction. By writing the energy balance on a small element $dx \times dy$, one can obtain the following differential equation:

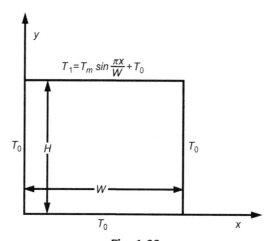

Fig. 1.23
Two-dimensional steady conduction.

$$\frac{\partial^2 T}{\partial x^2} + \frac{\partial^2 T}{\partial y^2} = 0 \tag{1.54}$$

Implicit in the derivation of Eq. (1.54) are the assumptions of the constant thermal conductivity of the material, no internal source and steady state. Thus, the temperature field $T(x, y)$ is given by the solution of Eq. (1.54) subject to the following boundary conditions:

$$
\begin{aligned}
&\text{At } x = 0, \quad 0 \le y \le H, \quad T = T_0 \\
&\text{At } x = W, \quad 0 \le y \le H, \quad T = T_0 \\
&\text{At } y = 0, \quad 0 \le x \le W, \quad T = T_0 \\
&\text{At } y = H, \quad 0 \le x \le W, \quad T_1 = T_m \sin \frac{\pi x}{W} + T_0
\end{aligned}
$$

Since the rate of heat transfer is governed by the temperature differences, we can rescale the temperature as $\theta = T - T_0$, Eq. (1.54) and the boundary conditions can now be rewritten as:

$$\frac{\partial^2 \theta}{\partial x^2} + \frac{\partial^2 \theta}{\partial y^2} = 0 \tag{1.55}$$

$$\text{At } x = 0, \quad 0 \le y \le H, \quad \theta = 0 \tag{1.56i}$$

$$\text{At } x = W, \quad 0 \le y \le H, \quad \theta = 0 \tag{1.56ii}$$

$$\text{At } y = 0, \quad 0 \le x \le W, \quad \theta = 0 \tag{1.56iii}$$

$$\text{At } y = H, \quad 0 \le x \le W, \quad \theta_1 = T_m \sin \frac{\pi x}{W} \tag{1.56iv}$$

Eq. (1.55) can be easily solved by the standard separation of variables approach as:

$$\theta = F_1(x) \cdot F_2(y) \tag{1.57}$$

This form of solution enables Eq. (1.55) to be split into two ordinary differential equations:

$$-\frac{1}{F_1}\frac{d^2 F_1}{dx^2} = \frac{1}{F_2}\frac{d^2 F_2}{dy^2} = \lambda_1^2$$

where λ_1^2 is a positive constant and it leads to:

$$\frac{d^2 F_1}{dx^2} + \lambda_1^2 F_1 = 0 \tag{1.58i}$$

$$\frac{d^2 F_2}{dy^2} - \lambda_1^2 F_2 = 0 \tag{1.58ii}$$

The solution to these two differential equations is now obtained as:

$$F_1 = C_1 \cos \lambda_1 x + C_2 \sin \lambda_1 x$$

$$F_2 = C_3 e^{-\lambda_1 y} + C_4 e^{\lambda_1 y}$$

and hence

$$\theta = F_1 \cdot F_2 = (C_1 \cos \lambda_1 x + C_2 \sin \lambda_1 x)\left(C_3 e^{-\lambda_1 y} + C_4 e^{\lambda_1 y}\right) \tag{1.59}$$

Using the boundary condition given in Eq. (1.56iii):

$$0 = (C_1 \cos \lambda_1 x + C_2 \sin \lambda_1 x)(C_3 + C_4)$$

which implies $C_3 = -C_4$ and substituting it in Eq. (1.59):

$$\theta = C_3 (C_1 \cos \lambda_1 x + C_2 \sin \lambda_1 x)\left(e^{-\lambda_1 y} - e^{\lambda_1 y}\right) \tag{1.60}$$

Now using Eq. (1.56i):

$$0 = C_3 (C_1 + C_2 \times 0)\left(e^{-\lambda_1 y} - e^{\lambda_1 y}\right)$$

which leads to $C_1 = 0$ and with this Eq. (1.60) becomes:

$$\theta = -C_3 C_2 \sin \lambda_1 x \left(e^{\lambda_1 y} - e^{-\lambda_1 y}\right) \tag{1.61}$$

Now using Eq. (1.56ii):

$$0 = -C_3 C_2 \sin \lambda_1 W \left(e^{\lambda_1 y} - e^{-\lambda_1 y}\right)$$

Neither C_2 nor C_3 can be equal to 0 and the only possibility is $\sin \lambda_1 W = 0$

$$\therefore \ \lambda_1 W = n\pi$$

where n is an integer and hence $\lambda_1 = \dfrac{n\pi}{W}$.

Thus $n = 1, 2, 3, \ldots$ will all satisfy this condition and since we are dealing with a linear equation, the sum of all such solutions corresponding to $n = 1, 2, \ldots$ will also be an admissible solution. One can thus rewrite Eq. (1.61) as:

$$\theta = T - T_0 = \sum_{n=1}^{\infty} C_n \sin \frac{n\pi x}{W} \sinh \frac{n\pi y}{W} \tag{1.62}$$

Finally using condition given in Eq. (1.56iv):

$$T_m \sin \frac{\pi x}{W} = \sum_{n=1}^{\infty} C_n \sin \frac{n\pi x}{W} \sinh \frac{n\pi H}{W}$$

Upon expansion:

$$T_m \sin \frac{\pi x}{W} = C_1 \sin \frac{\pi x}{W} \sinh \frac{\pi H}{W} + C_2 \sin \frac{2\pi x}{W} \sinh \frac{2\pi H}{W} + \cdots$$

Evidently, C_2 and subsequent coefficients must be zero, i.e. $C_n = 0$ for $n > 1$:

$$\therefore \ C_1 = \frac{T_m}{\sin \dfrac{\pi H}{W}}$$

and thus, the final solution is obtained by substituting this value of C_1 in Eq. (1.62), with all other C's being zero:

$$\theta = T - T_0 = T_m \frac{\sin \frac{\pi x}{W} \sinh \frac{\pi y}{W}}{\sinh \frac{\pi H}{W}}$$

or

$$T = T_0 + \left(\frac{T_m}{\sinh \frac{\pi H}{W}} \right) \sin \frac{\pi x}{W} \sinh \frac{\pi y}{W} \tag{1.63a}$$

Having obtained the temperature field $T(x, y)$, via Eq. (1.63a), one can obtain the heat flux in x- and y-directions as $-k\frac{\partial T}{\partial x}$ and $-k\frac{\partial T}{\partial y}$ respectively, or one can now plot the graphs of constant temperature contours known as isotherms. Note that the heat flow lines (given by $-k\nabla T$) will always be perpendicular to the isotherms, as is seen in Example 1.14.

Example 1.14

Reconsider the case shown in Fig. 1.23 when the top surface is subjected to a constant temperature T_1 which is different from T_0. How will the final answer change under these conditions?

Plot the isotherms and heat flow lines for $H = 50$ mm, $W = 60$ mm, $T_0 = 100°C$, and $T_1 = 150°C$ for a material of thermal conductivity of 40 W/m^2 K. Also, calculate the temperature at the points given by $(0.25W, 0.25H)$, $(0.5W, 0.5H)$, and $(0.75W, 0.75H)$.

Solution

In this case, evidently $T_m = 0$ and therefore, Eq. (1.56iv) changes as:

$$y = H, \quad 0 \leq x \leq W, \quad \theta = T_1 - T_0 = \theta_1 \text{ (say)}$$

It is immediately obvious that substituting $T_m = 0$ into Eq. (1.63) leads to the absurd result of $T = T_0$ everywhere! Since the other three boundary conditions, Eqs. (1.56i)–(1.56iii), remain unchanged, we can begin from Eq. (1.62):

$$\theta = \sum_{n=1}^{\infty} C_n \sin \frac{n\pi x}{W} \sinh \frac{n\pi y}{W} \qquad \text{(From Eq. 1.62)}$$

Now using the new condition of $y = H, \quad 0 \leq x \leq W, \quad \theta = \theta_1$:

$$\theta_1 = \sum_{n=1}^{\infty} C_n \sin \frac{n\pi x}{W} \sinh \frac{n\pi H}{W} \tag{1.63b}$$

This is a Fourier series and the constant C_n can be evaluated by expanding θ_1 in the form of a Fourier series over the interval $0 \leq x \leq W$ as:

$$\theta_1 = \theta_1 \frac{2}{\pi} \sum_{n=1}^{\infty} \frac{(-1)^{n+1} + 1}{n} \sin \frac{n\pi x}{W}$$

Comparing the last two equations:

$$C_n = \frac{2}{\pi}\theta_1 \frac{(-1)^{n+1}+1}{n} \cdot \frac{1}{\sinh\dfrac{n\pi H}{W}}$$

Alternatively, one can exploit the orthogonal properties of the sine function. Multiply both sides of Eq. (1.63b) by $\sin\frac{m\pi x}{W}$ and integrating it over the interval $x=0$ to $x=W$:

$$\int_0^W \theta_1 \sin\frac{m\pi x}{W}dx = \int_0^W \sin\frac{m\pi x}{W}\left\{\sum_{n=1}^{\infty} C_n \sin\frac{n\pi x}{W}\sinh\frac{n\pi H}{W}\right\}dx \tag{1.64}$$

where m is an integer and using the well-known result

$$\int_0^\pi (\sin mx \cdot \sin nx)\, dx = 0 \quad \text{when } m \neq n$$

$$\neq 0 \quad \text{when } m = n$$

This implies that all terms on the right hand side of Eq. (1.64) will be zero except when $m = n$. This also leads to the same value of C_n as obtained above. Thus, the final solution is given by:

$$\theta = T - T_0 = \frac{2}{\pi}(T_1 - T_0)\sum_{n=1}^{\infty}\left\{\frac{(-1)^{n+1}+1}{n}\right\}\sin\frac{n\pi x}{W} \cdot \frac{\sinh\dfrac{n\pi y}{W}}{\sinh\dfrac{n\pi H}{W}} \tag{1.65}$$

This is a rapidly converging series and one only needs to consider the first few terms. For instance, for the numerical values given here:

$H = 50$ mm; $W = 60$ mm; $k = 40$ W/m^2 K, $T_0 = 100°$C, and $T_1 = 150°$C

$$\therefore \quad \theta_1 = T_1 - T_0 = 50°C$$

$\therefore$ $\theta(0.5W, 0.5H)$ is obtained by substituting $x = 0.5W$ and $y = 0.5H$ in Eq. (1.65):

$$\theta(0.5W, 0.5H) = T - 100 = \frac{2}{\pi}(50)\sum_{n=1}^{\infty}\left\{\frac{(-1)^{n+1}+1}{n}\right\}\sin\frac{n\pi}{2} \cdot \frac{\sinh\dfrac{5n\pi}{12}}{\sinh\dfrac{5n\pi}{6}}$$

The results for temperature at different locations are summarised in the table below:

		Value of θ	
n	θ (0.5W, 0.5H)	θ (0.25W, 0.25H)	θ (0.75W, 0.75H)
1	16.02	4.64	23.06
2	16.02	4.64	23.06
3	15.61	4.68	25.16
4	15.61	4.68	25.16
5	15.62	4.68	24.82

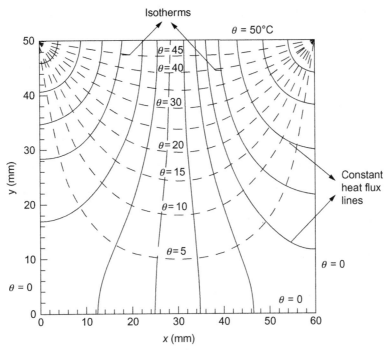

Fig. 1.24

Isotherms (*broken lines*) and constant heat flux lines (*solid*).

6	15.62	–	24.82
7	15.62	–	24.76
8	–	–	24.76
9	–	–	24.77

Evidently, the results converge very quickly. Fig. 1.24 shows the corresponding isotherm and constant heat flux lines for this example.

Graphical method

This is an approximate method to solve two dimensional steady conduction problems. It simply hinges on the fact that isotherms and heat flow lines are perpendicular to each other. It is best illustrated through an example. We would like to calculate the rate of heat transfer through the brick wall whose inner and outer surfaces are maintained at different temperatures (Fig. 1.25A). For the sake of simplicity, let us consider it to be a square.

The step-by-step procedure is as follows[12]:

(1) Identify lines of symmetry: These lines are determined by geometrical and thermal considerations. Thus, for instance, there is a perfect symmetry about the *x*- and *y*-axis in

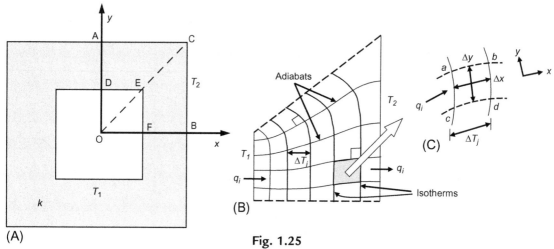

Fig. 1.25
(A) Steady state conduction in a two-dimensional square thick wall. (B) Flux plot. (C) Typical curvilinear square.

Fig. 1.25A. It is thus sufficient to obtain the solution only in one-quarter, say OACB, in this case. However, the diagonal OC is also a line of symmetry from geometric considerations and hence our solution domain is further reduced to only 1/8 of the total domain in this case, i.e. OCB. By virtue of symmetry, these lines behave like adiabatic lines (surfaces), i.e. no heat transfer is possible perpendicular to these lines, akin to no fluid flow across a streamline.

(2)   Identify lines of known temperature: These are known as isotherms. In Fig. 1.25A, A-C-B and D-E-F are isotherms because the temperature is constant along these lines. Obviously, isotherms will always be perpendicular to adiabat lines.

(3)   Now, the heat flow lines should be drawn to map the domain in terms of the curvilinear squares, as shown in Fig. 1.25B. Naturally, the isotherms and heat flow lines must always be normal to each other, and also as far as possible, the four sides of each square must be approximately of the same length. This is not always easy or possible to achieve, as can be seen in Fig. 1.25C. However, this condition can be satisfied in the following approximate manner (Fig. 1.25C):

$$\Delta x \approx \frac{ab + cd}{2} \approx \Delta y = \frac{ac + bd}{2}$$

Naturally, smaller the values of Δx and Δy, more accurate are the results. Once the graphical construction has been made, it is rather straight forward to obtain the rate of heat transfer by simply applying the Fourier's law of conduction as follows:

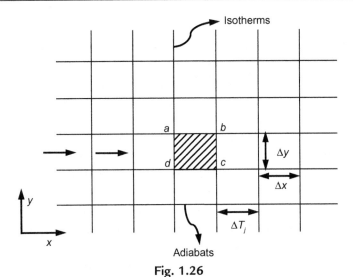

Fig. 1.26
Schematics of isotherms and constant heat flow lines.

Let the temperature difference between the two successive isotherms be $\triangle T_j$, as shown in Fig. 1.26. Normal to the isotherms are the adiabats and the heat flow in between two adiabats is q_k. So for M heat flow lanes, the total heat flow Q:

$$Q = \sum_{k=1}^{M} q_k = Mq_k \tag{1.66}$$

i.e. if the heat flow through each line is q_k. Using Fourier's law for the square abcd shown in Fig. 1.26:

$$q_k = k(\Delta y \cdot l)\frac{\Delta T_j}{\Delta x} \tag{1.67}$$

where k is the thermal conductivity of material and l is the dimension normal to the xy plane. When ΔT_j is same for each pair of neighbouring isotherms:

$$\Delta T_{\text{overall}} = N\Delta T_j \tag{1.68}$$

where N is the number of temperature intervals. Now combining these three equations and assuming $\Delta x \approx \Delta y$:

$$Q = \frac{M}{N}lk\Delta T_{\text{overall}} \tag{1.69}$$

Thus, the value of (M/N) can be obtained from the isotherm-heat flow passages construction depending upon the number of temperature intervals (N). For a given value of N, the value of M is automatically fixed in order to satisfy the condition outlined in step 3 of the procedure.

Thus, M need not be an integer. Thus, for a given configuration, the precision of the value of (M/N) increases with the increasing values of M and N. Eq. (1.69) affords the possibility of introducing a conduction shape factor S as:

$$Q = Sk\Delta T_{overall} \tag{1.70}$$

$$\text{where } S = \frac{M}{N} \tag{1.71}$$

Table 1.2 summarises the value of S for a range of configurations. More detailed listings are available in the literature.[7,16] This approach can also be extended to three-dimensional conduction.

Table 1.2 Conduction shape factors[7,16]

System	Schematic	Shape Factor, S	Limitations
Isothermal cylinder of radius r buried in semiinfinite medium having isothermal surface		$\dfrac{2\pi l}{\cosh^{-1}(D/r)}$ $\dfrac{2\pi l}{\ln(D/r)}$	$l \gg r$ $l \gg r$ $D > 3r$
Isothermal sphere of radius r buried in semiinfinite medium having isothermal surface		$\dfrac{4\pi r}{1 - (r/2D)}$	–
Conduction between two isothermal cylinders of length l buried in infinite medium		$\dfrac{2\pi l}{\cosh^{-1}\left(\dfrac{D^2 - r_1^2 - r_2^2}{2r_1 r_2}\right)}$	$l \gg r$ $l \gg D$
Row of horizontal cylinders of length l in semiinfinite medium with isothermal surface		$\dfrac{2\pi l}{\ln\left[\left(\dfrac{W}{\pi r}\right)\sinh(2\pi D/W)\right]}$	$D > 2r$

Table 1.2 **Conduction shape factors'—cont'd**

System	Schematic	Shape Factor, S	Limitations
Buried cube in infinite medium, l on a side		$8.24l$	–
Isothermal cylinder of radius r placed in semiinfinite medium as shown		$\dfrac{2\pi l}{\ln(2l/r)}$	$l \gg 2r$
Isothermal rectangular parallelepiped buried in semiinfinite medium having isothermal surface		$1.685L\left[\log\left(1+\dfrac{b}{a}\right)\right]^{-0.59}\left(\dfrac{b}{c}\right)^{-0.078}$	Ref. 7
Plane wall		A/L	One-dimensional heat flow
Thin horizontal disc buried in semiinfinite medium with isothermal surface		$4r$ $8r$ $\dfrac{8r}{1-(2/\pi)\tan^{-1}(r/2D)}$	$D=0$ $D \gg 2r$ $(D/2r)>1$ $\tan^{-1}(r/2D)$ in radians
Hemisphere buried in semiinfinite medium $\Delta T = T_s - T_\infty$		$2\pi r$	–

Continued

Table 1.2   Conduction shape factors'—cont'd

System	Schematic	Shape Factor, S	Limitations
Isothermal sphere buried in semiinfinite medium with insulated surface	Insulated; D; T_∞; r	$\dfrac{4\pi r}{1 + (r/2D)}$	—
Two isothermal spheres buried in infinite medium	r_1; r_2; D	$\dfrac{4\pi r_2}{\dfrac{r_2}{r_1}\left[1 - \dfrac{(r_1/D)^4}{1 - (r_2/D)^2}\right] - \dfrac{2r_2}{D}}$	$D > 5r_{\max}$
Thin rectangular plate of length L, buried in semiinfinite medium having isothermal surface	Isothermal; D; L; W	$\dfrac{\pi W}{\ln(4W/L)}$	$D = 0$ $W > L$
		$\dfrac{2\pi W}{\ln(4W/L)}$	$D \gg W$ $W > L$
		$\dfrac{2\pi W}{\ln(2\pi D/L)}$	$W \gg L$ $D > 2W$
Eccentric cylinders of length l	r_2; r_1; D	$\dfrac{2\pi l}{\cosh^{-1}\left(\dfrac{r_1^2 + r_2^2 - D^2}{2r_1 r_2}\right)}$	$L \gg r_2$
Cylinder centred in a square of length l	l; W; r; W	$\dfrac{2\pi l}{\ln(0.54W/r)}$	$L \gg W$
Horizontal cylinder of length l centred in infinite plate	Isothermal; D; r; D; Isothermal	$\dfrac{2\pi l}{\ln(4D/r)}$	

For buried objects, $\Delta T = T_s - T_\infty$, where T_∞ is the same as the isothermal surface temperature for semiinfinite media.

Example 1.15

A 20 m long square block ($k = 150$ W/m K) of cross-section 5 m × 5 m has a hole of elliptic cross-section (major axis = 2 m, minor axis = 1 m) cut through it along its length. The inner surface of the elliptic opening is at a temperature of 100°C whereas the outer surface of the square block is at 20°C. Using the graphical method, obtain the shape factor and estimate the rate of heat transfer per unit length of the block (Fig. 1.27).

Solution

In this case, it is sufficient to use one quarter of the overall geometry due to symmetry considerations as shown below. The construction is shown by dividing the overall $\triangle T = 100 - 20 = 80°C$ into 8 equal intervals. Next, the corresponding heat flux lines are drawn ensuring that these are orthogonal to isotherms and that curvilinear squares are formed as prescribed in step 3 of the graphical method. The values of M and N are summarised in the table below where it is clearly seen that the value of S attains a limiting value of 1.225. This value compares rather well with the value of 1.229 based on the numerical solution of the conduction equation for this geometry.

The rate of heat transfer per unit length:

$$Q = Sk\Delta T$$
$$= 1.225 \times 150 \times (100 - 20) = 14.7\,\text{kW/m}$$

$\triangle T_j$ (°C)	M	N	$S = M/N$
20	5.2	4	1.3
10	10	8	1.25
5	19.7	16	1.231
2	49	40	1.225

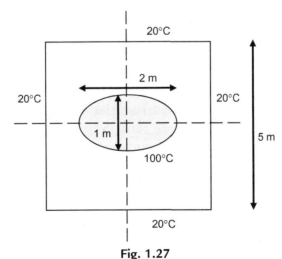

Fig. 1.27
Schematics for Example 1.15.

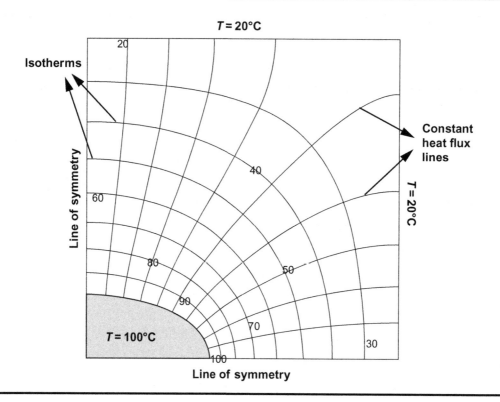

Numerical method

Undoubtedly, a large number of analytical solutions for steady and unsteady, one- and multidimensional conduction problems have been obtained over the years, but most of these are limited to the simple geometrical shapes. There are numerous practical situations where the complex geometry and/or the boundary conditions preclude the possibility of analytical solutions and one must thus resort to the use of finite difference techniques. The basic ideas of this approach are presented here as applied to steady conduction problems.

The main virtue of an analytical solution lies in the fact that the temperature (and hence the heat flux) is known at each point. In contrast, numerical solutions provide the value of temperature at selected points in the domain. Thus, the starting point of this approach is to create a grid (divide the domain into smaller regions) as shown in Fig. 1.28A for a two-dimensional shape. The so-called nodal (reference) point and the collection of such points is called the computational mesh or grid. Each node is thus identified by a double index notation (m, n), as seen in Fig. 1.28B where m and n denote the x- and y-coordinates respectively of the node. The numerical solution of the conduction equation thus gives values of temperature at discrete points like (m, n), $(m+1, n+1)$, etc. and naturally by varying the values of $\triangle x$ and $\triangle y$, the number of discrete points can be increased or decreased in a given application.

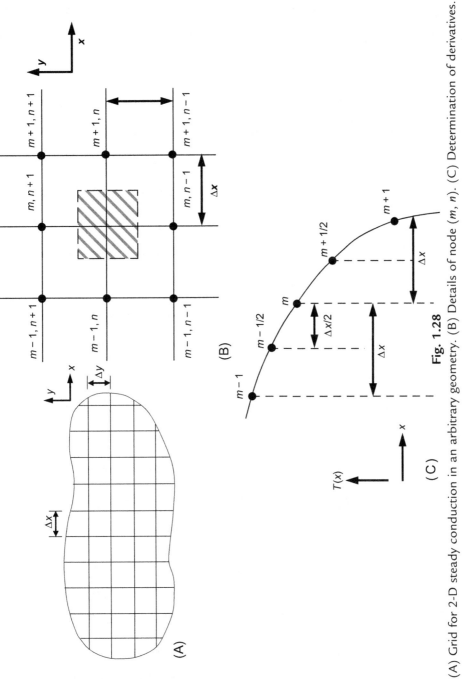

Fig. 1.28

(A) Grid for 2-D steady conduction in an arbitrary geometry. (B) Details of node (m, n). (C) Determination of derivatives.

From a physical point, the value of temperature $T_{m,n}$ corresponding to the node (m, n) is a measure of the average temperature of the hashed region, shown in Fig. 1.28B.

The conduction Eq. (1.54) is now applied to each node by writing the derivatives appearing in Eq. (1.54) in terms of the temperatures of the neighbouring points. This can be achieved via the use of Taylor series expansion to write as:

$$T_{x+\Delta x} = T_x + \frac{\partial T}{\partial x} \cdot \Delta x + \text{higher order terms}$$

Neglecting the higher order terms, we can obtain:

$$\frac{\partial T}{\partial x} \approx \frac{T_{x+\Delta x} - T_x}{\Delta x} \tag{1.72}$$

Now referring to Fig. 1.28C and applying the above result:

$$\left.\frac{\partial T}{\partial x}\right|_{m+1/2,n} = \frac{T_{m+1,n} - T_{m,n}}{\Delta x} \tag{1.73a}$$

$$\left.\frac{\partial T}{\partial x}\right|_{m-1/2,n} = \frac{T_{m,n} - T_{m-1,n}}{\Delta x} \tag{1.73b}$$

One can now use these values to evaluate the second order derivatives $\left.\frac{\partial^2 T}{\partial x^2}\right|_{m,n}$ terms as follows:

$$\left.\frac{\partial^2 T}{\partial x^2}\right|_{m,n} \approx \frac{\left.\frac{\partial T}{\partial x}\right|_{m+1/2,n} - \left.\frac{\partial T}{\partial x}\right|_{m-1/2,n}}{\Delta x} \tag{1.74}$$

Substituting from Eqs. (1.73a), (1.73b):

$$\left.\frac{\partial^2 T}{\partial x^2}\right|_{m,n} \approx \frac{T_{m+1,n} - 2T_{m,n} + T_{m-1,n}}{(\Delta x)^2} \tag{1.75}$$

Note that a similar result can also be obtained by retaining the second order term in Eq. (1.54).

By the same reasoning:

$$\left.\frac{\partial^2 T}{\partial y^2}\right|_{m,n} \approx \frac{T_{m,n+1} - 2T_{m,n} + T_{m,n-1}}{(\Delta y)^2} \tag{1.76}$$

Now substituting from Eqs. (1.75), (1.76) into Eq. (1.54) for node (m, n):

$$\frac{T_{m+1,n} - 2T_{m,n} + T_{m-1,n}}{(\Delta x)^2} + \frac{T_{m,n+1} - 2T_{m,n} + T_{m,n-1}}{(\Delta y)^2} = 0 \tag{1.77}$$

Using a mesh for which $\Delta x = \Delta y$, Eq. (1.77) reduces to:

$$T_{m+1,n} + T_{m-1,n} + T_{m,n+1} + T_{m,n-1} - 4T_{m,n} = 0 \tag{1.78}$$

Thus, this algebraic equation is the approximate equivalent form of the differential equation given by Eq. (1.54) for node (m, n) and it can be applied to any interior node. It simply says that the temperature at node (m, n) is the arithmetic mean of the temperature of its four neighbouring points as:

$$T_{m,n} = \frac{T_{m+1,n} + T_{m-1,n} + T_{m,n+1} + T_{m,n-1}}{4} \tag{1.79}$$

On the other hand, if the node of interest is located on the boundary of the object and is subject to the convection boundary conditions, Fig. 1.29, the values of $T_{m,n}$, $T_{m,n-1}$, etc. must be calculated differently. If the boundary surface AB is exposed to atmosphere (at temperature T_∞) and is losing heat by convection (heat transfer coefficient, h), one can then write energy balance for node (m, n) as follows:

$$-k \cdot \Delta y \cdot \frac{T_{m,n} - T_{m-1,n}}{\Delta x} - k \cdot \frac{\Delta x}{2} \frac{T_{m,n} - T_{m,n-1}}{\Delta y} - k \cdot \frac{\Delta x}{2} \frac{T_{m,n} - T_{m,n+1}}{\Delta y} + h\Delta y(T_\infty - T_{m,n}) = 0$$

$$\tag{1.80}$$

Once again for a square grid, $\Delta x = \Delta y$:

$$-T_{m,n}\left(2 + \frac{h\Delta x}{k}\right) + \frac{h\Delta x}{k}T_\infty + \frac{1}{2}(2T_{m-1,n} + T_{m,n-1} + T_{m,n+1}) = 0 \tag{1.81}$$

Similar equations can be set up for the other exterior nodes like $(m, n+1)$, $(m, n-1)$ as well as along the x-axis, as the case may be. One can set up similar equations for an insulated boundary, or for a corner exposed to a convection condition. A summary of some commonly encountered situations is summarised in Table 1.3.

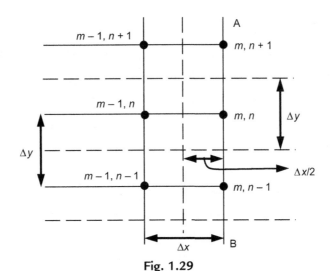

Fig. 1.29
Exterior nodes exposed to convection boundary condition.

Table 1.3 Summary of nodal formulas for finite-difference calculations (dashed lines indicate element volume)[a]

Physical Situation	Nodal Equation for Equal Increments in x- and y-Direction (Second Equation in Situation is in Form for Gauss-Seidel Iteration)
(a) Convection boundary node 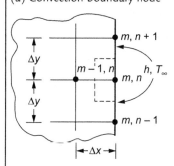	Eq. (1.81)
(b) Exterior corner with convection boundary 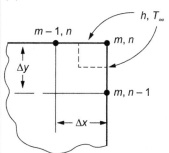	$2\dfrac{h\Delta x}{k}T_\infty + (T_{m-1,n} + T_{m,n-1}) - 2\left(\dfrac{h\Delta x}{k} + 1\right)T_{m,n} = 0$
(c) Interior corner with convection boundary 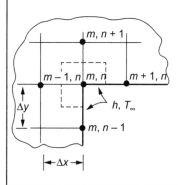	$2\dfrac{h\Delta x}{k}T_\infty + 2T_{m-1,n} + T_{m,n+1} + T_{m+1,n} + T_{m,n-1} - 2\left(\dfrac{h\Delta x}{k} + 3\right)T_{m,n} = 0$

Table 1.3 **Summary of nodal formulas for finite-difference calculations (dashed lines indicate element volume)—cont'd**

Physical Situation	Nodal Equation for Equal Increments in x- and y-Direction (Second Equation in Situation is in Form for Gauss-Seidel Iteration)
(d) Insulated boundary	$$T_{m,n+1} + T_{m,n-1} + 2T_{m-1,n} - 4T_{m,n} = 0$$
(e) Interior node near curved boundary[b]	$$\frac{2}{b(b+1)}T_2 + \frac{2}{a+1}T_{m+1,n} + \frac{2}{b+1}T_{m,n-1} + \frac{2}{a(a+1)}T_1 -$$ $$2\left(\frac{1}{a} + \frac{1}{b}\right)T_{m,n} = 0$$
(f) Boundary node with convection along curved boundary—node 2 for (e) above[c]	$$\frac{b}{\sqrt{a^2+b^2}}T_1 + \frac{b}{c^2+1}T_3 + \frac{a+1}{b}T_{m,n} + \frac{h\Delta x}{k}\left(\sqrt{c^2+1} + \sqrt{a^2+b^2}\right)T_\infty$$ $$-\left[\frac{b}{\sqrt{a^2+b^2}} + \frac{b}{c^2+1} + \frac{a+1}{b} + \frac{h\Delta x}{k}\left(\sqrt{c^2+1} + \sqrt{a^2+b^2}\right)\right]T_2 = 0$$

[a]Convection boundary may be converted to insulated surface by setting $h = 0$.
[b]This equation is obtained by multiplying the resistance by $4/(a+1)(b+1)$.
[c]This relation is obtained by dividing the resistance formulation by 2.

Example 1.16

For the square cross-section (30 mm × 30 mm) whose sides are at different temperatures as shown below, calculate the intermediate temperatures using $\Delta x = \Delta y = 10$ mm and $\Delta x = \Delta y = 5$ mm.

Solution

For $\Delta x = \Delta y = 10$ mm, the grid is shown below:

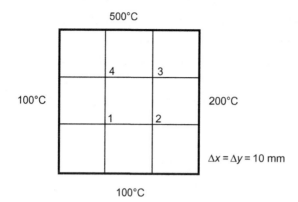

Since all the four nodes 1, 2, 3, and 4 are interior nodes, one can apply Eq. (1.78) to each one of them as:

$$\text{Node 1}: \quad T_2 + T_4 + 100 + 100 - 4T_1 = 0$$
$$\text{Node 2}: \quad T_3 + T_1 + 100 + 200 - 4T_2 = 0$$
$$\text{Node 3}: \quad T_4 + T_2 + 500 + 200 - 4T_3 = 0$$
$$\text{Node 4}: \quad T_3 + T_1 + 100 + 500 - 4T_4 = 0$$

The solution of these four equations gives:

$$T_1 = 162.5°C, \ T_2 = 187.5°C, \ T_3 = 287.5°C, \ T_4 = 262.5°C.$$

For the case of $\Delta x = \Delta y = 5\,mm$, the grid is as shown below:

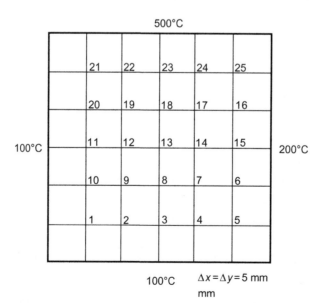

One can now apply Eq. (1.78) to nodes 1–25 resulting in the following 25 algebraic equations:

$$T_2 + T_{10} + 100 + 100 - 4T_1 = 0$$
$$T_1 + T_9 + T_3 + 100 - 4T_2 = 0$$
$$T_2 + T_8 + T_4 + 100 - 4T_3 = 0$$
$$T_3 + T_7 + T_5 + 100 - 4T_4 = 0$$
$$T_4 + T_6 + 100 + 200 - 4T_5 = 0$$
$$T_5 + T_7 + T_{15} + 200 - 4T_6 = 0$$
$$T_6 + T_4 + T_8 + T_{14} - 4T_7 = 0$$
$$T_7 + T_3 + T_{13} + T_9 - 4T_8 = 0$$
$$T_8 + T_2 + T_{10} + T_{12} - 4T_9 = 0$$
$$T_{11} + T_9 + T_1 + 100 - 4T_{10} = 0$$
$$T_{10} + T_{20} + T_{12} + 100 - 4T_{11} = 0$$
$$T_{11} + T_{19} + T_9 + T_{13} - 4T_{12} = 0$$
$$T_{12} + T_{18} + T_8 + T_{14} - 4T_{13} = 0$$
$$T_{13} + T_{17} + T_7 + T_{15} - 4T_{14} = 0$$
$$T_{14} + T_{16} + T_6 + 200 - 4T_{15} = 0$$
$$T_{15} + T_{25} + T_{17} + 200 - 4T_{16} = 0$$
$$T_{16} + T_{24} + T_{14} + T_{18} - 4T_{17} = 0$$
$$T_{17} + T_{23} + T_{13} + T_{19} - 4T_{18} = 0$$
$$T_{20} + T_{22} + T_{12} + T_{18} - 4T_{19} = 0$$
$$T_{19} + T_{21} + T_{11} + 100 - 4T_{20} = 0$$
$$T_{20} + 500 + 100 + T_{22} - 4T_{21} = 0$$
$$T_{21} + 500 + T_{19} + T_{23} - 4T_{22} = 0$$
$$T_{22} + 500 + T_{24} + T_{18} - 4T_{23} = 0$$
$$T_{23} + 500 + T_{17} + T_{25} - 4T_{24} = 0$$
$$T_{24} + T_{16} + 500 + 200 - 4T_{25} = 0$$

Needless to say that the solution of these 25 algebraic equations by hand calculations is not at all convenient and hence numerical methods are frequently used. The results are:

$T_1 = 115.66°C$, $T_2 = 128.57°C$, $T_3 = 138°C$, $T_4 = 145.94°C$, $T_5 = 159.39°C$, $T_6 = 191.64°C$, $T_7 = 186.36°C$, $T_8 = 177.49°C$, $T_9 = 160.61°C$, $T_{10} = 134.06°C$, $T_{11} = 159.98°C$, $T_{12} = 202.31°C$, $T_{13} = 225°C$, $T_{14} = 230.39°C$, $T_{15} = 220.79°C$, $T_{16} = 261.13°C$, $T_{17} = 289.39°C$, $T_{18} = 289.82°C$, $T_{19} = 263.64°C$, $T_{20} = 203.56°C$, $T_{21} = 290.61°C$, $T_{22} = 358.87°C$, $T_{23} = 381.23°C$, $T_{24} = 376.24°C$, $T_{25} = 334.34°C$

Example 1.17

An experimental furnace is supported on a brick column ($k = 0.056$ W/m K) of square cross-section (100 mm $\times$ 100 mm). The installation is such that the three sides of the column are exposed to a constant temperature of 300°C while the fourth side is exposed to atmosphere (at 25°C) and loses heat by convection ($h = 7$ W/m^2 K). Using a square grid of $\Delta x = \Delta y = 20$ mm, calculate the nodal temperatures.

Solution

The grid is shown below:

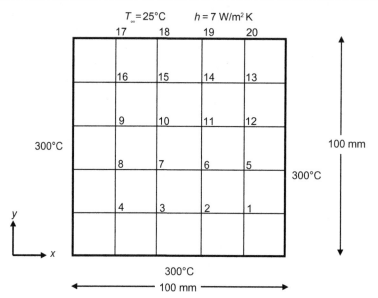

In this case, nodes 1–16 are interior nodes and one can apply Eq. (1.78). Nodes 17–20 are exposed to convective boundary condition and one can use Eq. (1.81) for these nodes. Symmetry considerations further suggest that $T_1 = T_4$, $T_2 = T_3$, $T_8 = T_5$, $T_6 = T_7$, $T_9 = T_{12}$, $T_{10} = T_{11}$, $T_{14} = T_{15}$, $T_{13} = T_{16}$, $T_{18} = T_{19}$, and $T_{17} = T_{20}$ thereby reducing the number of unknowns significantly. Introducing these ideas, the grid now looks as follows:

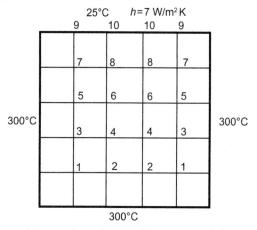

One can now set up the nodal equations for the 10 nodes as follows:

$$300 + 300 + T_3 + T_2 - 4T_1 = 0$$
$$T_1 + T_2 + T_4 + 300 - 4T_2 = 0$$
$$T_1 + T_5 + 300 + T_4 - 4T_3 = 0$$
$$T_3 + T_6 + T_2 + T_4 - 4T_4 = 0$$
$$T_7 + T_3 + T_6 + 300 - 4T_5 = 0$$
$$T_5 + T_8 + T_4 + T_6 - 4T_6 = 0$$
$$300 + T_5 + T_8 + T_9 - 4T_7 = 0$$
$$T_7 + T_{10} + T_6 + T_8 - 4T_8 = 0$$

Using $h\Delta x/k = 7 \times 0.02/0.056 = 2.5$, we can use Eq. (1.81) for the exterior nodes 9 and 10:

$$-T_9(2+2.5) + 2.5T_\infty + (1/2)(2T_7 + T_{10} + 300) = 0$$
$$-T_{10}(2+2.5) + 2.5T_\infty + (1/2)(2T_8 + T_9 + T_{10}) = 0$$

Solving these equations, we obtain the following results:

$T_1 = 283.54°C$, $T_2 = 281.32°C$, $T_3 = 273.75°C$, $T_4 = 260.41°C$, $T_5 = 251.14°C$, $T_6 = 226.15°C$,
$T_7 = 204.62°C$, $T_8 = 166.89°C$, $T_9 = 100.46°C$, $T_{10} = 69.91°C$

1.4 Heat Transfer by Convection

1.4.1 Natural and Forced Convection

Heat transfer by convection occurs as a result of the movement of fluid on a macroscopic scale in the form of eddies or circulating currents. If the currents arise from the heat transfer process itself, *natural convection* occurs, such as in the heating of a vessel containing liquid by means of a heat source situated beneath it. The liquid at the bottom of the vessel becomes heated, expands, and rises because its density has become less than that of the remaining liquid. Cold liquid of higher density takes its place and a circulating current is thus set up.

In *forced convection*, circulating currents are produced by an external agency such as an agitator in a reaction vessel or as a result of turbulent flow in a pipe. In general, the magnitude of the circulation in forced convection is greater, and higher rates of heat transfer are obtained than in natural convection.

In most cases where convective heat transfer takes place from a surface to a fluid, the circulating currents die out in the immediate vicinity of the surface and a film of fluid, free of turbulence, covers the surface. In this film, heat transfer is by thermal conduction and, as the thermal conductivity of most fluids is low, the main resistance to transfer lies there. Thus an increase in the velocity of the fluid over the surface gives rise to improved heat transfer mainly because the thickness of the film is reduced. As a guide, the film coefficient increases as (fluid velocity)^n, where $0.5 < n < 0.8$, depending upon the geometry and the nature of flow (laminar or turbulent, for instance).

If the resistance to transfer is regarded as lying within the film covering the surface, the rate of heat transfer Q is given by Eq. (1.12) as:

$$Q = kA\frac{(T_1 - T_2)}{x}$$

The effective thickness x is not generally known and therefore the equation is usually rewritten in the form:

$$Q = hA(T_1 - T_2) \tag{1.82}$$

where h is the heat transfer coefficient for the film and $(1/h)$ is the corresponding thermal resistance.

1.4.2 Application of Dimensional Analysis to Convection

So many factors influence the value of h that it is almost impossible to determine their individual effects by direct experimental methods. By arranging the variables in a series of dimensionless groups, however, the problem is made more manageable in that the number of groups is significantly less than the number of parameters. It is found that the heat transfer rate per unit area q is dependent on those physical properties which affect flow pattern (viscosity μ and density ρ), the thermal properties of the fluid (the specific heat capacity C_p and the thermal conductivity k), a linear dimension of the surface l, the velocity of flow u of the fluid over the surface, the temperature difference ΔT, and a factor determining the natural circulation effect caused by the expansion of the fluid on heating (the product of the coefficient of volumetric expansion β and the acceleration due to gravity g). Writing this as a functional relationship:

$$q = \phi\left[u, l, \rho, \mu, C_p, \Delta T, \beta g, k\right] \tag{1.83}$$

Noting the dimensions of the variables in terms of length L, mass M, time T, temperature θ, heat H:

q	Heat transferred/unit area and unit time	$H L^{-2} T^{-1}$
u	Velocity	$L T^{-1}$
l	Linear dimension	L
μ	Viscosity	$M L^{-1} T^{-1}$
ρ	Density	$M L^{-3}$
k	Thermal conductivity	$H T^{-1} L^{-1} \theta^{-1}$
C_p	Specific heat capacity at constant pressure	$H M^{-1} \theta^{-1}$
ΔT	Temperature difference	θ
(βg)	The product of the coefficient of thermal expansion and the acceleration due to gravity	$L T^{-2} \theta^{-1}$

It may be noted that both temperature and heat are taken as fundamental units as heat is not expressed here in terms of M, L, and T.

With nine parameters and five primary dimensions, Eq. (1.83) may be rearranged in terms of four dimensionless groups for a given geometrical configuration.

Using the Π-theorem for solution of the equation, and taking as the recurring set: $l, \rho, \mu, \Delta T$, and k

The nonrecurring variables are:

$$q, u, (\beta g), C_p$$

Then:

$$l \equiv L \qquad\qquad L = l$$
$$\rho \equiv M\,L^{-3} \qquad\qquad M = \rho\,L^3 = \rho l^3$$
$$\mu \equiv M\,L^{-1}\,T^{-1} \qquad\qquad T = M\,L^{-1}\mu^{-1} = \rho l^3 l^{-1}\mu^{-1} = \rho l^2 \mu^{-1}$$
$$\Delta_T \equiv \theta \qquad\qquad \theta = \Delta T$$
$$k \equiv H\,L^{-1}\,T^{-1}\,\theta^{-1} \qquad\qquad H = kLT\theta = kl\rho l^2 \mu^{-1}\Delta T = kl^3 \rho \mu^{-1}\Delta T$$

The Π groups are then:

$$\Pi_1 = qH^{-1}L^2 T = qk^{-1}l^{-3}\rho^{-1}\mu\Delta T^{-1}l^2\rho l^2\mu^{-1} = qk^{-1}l\Delta T^{-1}$$
$$\Pi_2 = uL^{-1}T = ul^{-1}\rho l^2\mu^{-1} = u\rho l\mu^{-1}$$
$$\Pi_3 = C_p H^{-1}M\theta = C_p k^{-1}l^{-3}\rho^{-1}\mu\Delta T^{-1}\rho l^3 \Delta T = C_p k^{-1}\mu$$
$$\Pi_4 = \beta g L^{-1}T^2\theta = \beta g l^{-1}\rho^2 l^4\mu^{-2}\Delta T = \beta g\Delta T\rho^2\mu^{-2}l^3$$

The relation in Eq. (1.83) becomes:

$$\frac{ql}{k\Delta T} = \frac{hl}{k} = \phi\left[\left(\frac{lu\rho}{\mu}\right)\left(\frac{C_p\mu}{k}\right)\left(\frac{\beta g\Delta Tl^3\rho^2}{\mu^2}\right)\right] \qquad (1.84)$$

or:

$$Nu = \phi[Re, Pr, Gr]$$

This general equation involves the use of four dimensionless groups, although it may frequently be simplified for design purposes. In Eq. (1.84):

hl/k is known as the *Nusselt* group Nu (already referred to in Eq. 1.46),

$lu\rho/\mu$ the *Reynolds* group Re,

$C_p\mu/k$ the *Prandtl* group Pr, and

$\beta g\Delta Tl^3\rho^2/\mu^2$ the *Grashof* group Gr

It is convenient to define other dimensionless groups which are also used in the analysis of heat transfer. These are:

$lu\rho C_p/k$ the *Peclet* group, $Pe = RePr$,

GC_p/kl the *Graetz* group Gz, and

$h/C_p\rho u$ the *Stanton* group, $St = Nu/(RePr)$

It may be noted that many of these dimensionless groups are ratios. For example, the Nusselt group $h/(k/l)$ is the ratio of the actual heat transfer to that by conduction over a thickness l, whilst the Prandtl group, $(\mu/\rho)/(k/C_p\rho)$ is the ratio of the kinematic viscosity (or momentum diffusivity) to the thermal diffusivity.

For conditions in which only natural convection occurs, the velocity is dependent on the buoyancy effects alone, represented by the Grashof number, and the Reynolds group may be omitted. Again, when forced convection occurs the effects of natural convection are usually negligible and the Grashof number may be omitted. Thus:

For natural convection:

$$Nu = f(Gr, Pr) \tag{1.85}$$

And for forced convection:

$$Nu = f(Re, Pr) \tag{1.86}$$

For most gases over a wide range of temperature and pressure, $C_p\mu/k$ is constant and the Prandtl group may often be omitted, simplifying the design equations for the calculation of film coefficients with gases. The relative importance of the natural—and forced convection effects in a given situation is ascertained by introducing another dimensionless group, Richardson number, $Ri = Gr/Re^2$. Thus, the two limiting conditions of $Ri \rightarrow oo$ and $Ri \rightarrow o$ correspond to the pure natural and forced convection regimes respectively. The value of $Ri \sim 1$ indicates that the buoyancy—induced and convection velocities are of comparable magnitudes. The mixed-convection heat transfer is further classified as aiding—buoyancy (when both velocities are in the same direction), opposing—buoyancy (when the two velocities are in the opposite directions), and cross-buoyancy (when the two velocities are perpendicular to each other).

1.4.3 Forced Convection in Tubes

Turbulent flow

The results of a number of researchers who have used a variety of gases such as air, carbon dioxide, and steam and of others who have used liquids such as water, acetone, kerosene, and benzene have been correlated by Dittus and Boelter[17] who used mixed units for their variables. On converting their relations using consistent (SI, for example) units, they become:

for heating of fluids:

$$Nu = 0.0241 Re^{0.8} Pr^{0.4} \tag{1.87}$$

and for cooling of fluids:

$$Nu = 0.0264 Re^{0.8} Pr^{0.3} \tag{1.88}$$

In these equations all of the physical properties are taken at the mean bulk temperature of the fluid $(T_i + T_o)/2$, where T_i and T_o are the inlet and outlet temperatures. The difference in the value of the index for heating and cooling occurs because in the former case the film temperature will be greater than the bulk temperature and in the latter case less. Conditions in the film, particularly the viscosity of the fluid, exert an important effect on the heat transfer process.

Subsequently McAdams[18] has reexamined the available experimental data and has concluded that an exponent of 0.4 for the Prandtl number is the most appropriate one for both heating and cooling. He also slightly modified the coefficient to 0.023 (corresponding to Colburn's value, given below in Eq. 1.92) and gave the following equation, which applies for $Re > 2100$ and for fluids of viscosities not exceeding 2 mN s/m^2:

$$Nu = 0.023 Re^{0.8} Pr^{0.4} \qquad (1.89)$$

Winterton[19] has looked into the origins of the 'Dittus and Boelter' equation and has found that there is considerable confusion in the literature concerning the origin of Eq. (1.89) which is generally referred to as the Dittus–Boelter equation in the literature on heat transfer.

An alternative equation which is in many ways more convenient has been proposed by Colburn[20] and includes the Stanton number $\left(St = h/C_p\rho u\right)$ instead of the Nusselt number. This equation takes the form:

$$j_H = St Pr^{0.67} = 0.023 Re^{-0.2} \qquad (1.90)$$

where j_H is known as the *j-factor for heat transfer*.

It may be noted that:

$$\frac{h}{C_p \rho u} = \left(\frac{hd}{k}\right) \left(\frac{\mu}{ud\rho}\right) \left(\frac{k}{C_p \mu}\right)$$

or:

$$St = Nu Re^{-1} Pr^{-1} \qquad (1.91)$$

Thus, multiplying Eq. (1.90) by $Re Pr^{0.33}$:

$$Nu = 0.023 Re^{0.8} Pr^{0.33} \qquad (1.92)$$

which is a form of Eqs. (1.87), (1.88).

Again, the physical properties are evaluated at the bulk temperature, except for the viscosity in the Reynolds group which is evaluated at the mean film temperature taken as $(T_{\text{surface}} + T_{\text{bulk fluid}})/2$

Writing a heat balance for the flow through a tube of diameter d and length l with a rise in temperature for the fluid from T_i to T_o:

$$h\pi dl \Delta T = \frac{\pi d^2}{4} C_p u \rho (T_o - T_i)$$

or:

$$St = \frac{h}{C_p\rho u} = \frac{d(T_o - T_i)}{4l\Delta T} \qquad (1.93)$$

where ΔT is the mean temperature difference between the bulk fluid and the walls.

With *very viscous liquids*, there is a marked difference at any position between the viscosity of the fluid adjacent to the surface and the value at the axis or at the bulk temperature of the fluid. Sieder and Tate[21] examined the experimental data available and suggested that a term $\left(\dfrac{\mu}{\mu_s}\right)^{0.14}$ be included to account for the viscosity variation and the fact that this will have opposite effects in heating and cooling, (μ is the viscosity at the bulk temperature and μ_s the viscosity at the wall or surface). They give a logarithmic plot, but do not propose a correlating equation. However, McAdams[18] gives the following equation, based on Sieder and Tate's work:

$$Nu = 0.027Re^{0.8}Pr^{0.33}\left(\frac{\mu}{\mu_s}\right)^{0.14} \qquad (1.94)$$

This equation may also be written in the form of the Colburn equation (1.90).

When these equations are applied to *heating or cooling of gases* for which the Prandtl group usually has a value of about 0.74, substitution of $Pr = 0.74$ in Eq. (1.692) gives:

$$Nu = 0.020Re^{0.8} \qquad (1.95)$$

Water is very frequently used as the cooling medium and the effect of the variation of physical properties with temperature may be included in Eq. (1.92) to give a simplified equation, which is useful for design purposes (Section 1.9.4).

This body of knowledge concerning the prediction of the Nusselt number for the transitional and turbulent flow in circular and noncircular ducts has been evaluated by Gnielinski[22] and others.[23,24] Firstly, Gnielinski[22] observed that the difference in the values of the mean Nusselt number for the conditions of isothermal and constant heat flux prescribed on the duct walls progressively diminishes with the increasing Reynolds number. Based on the re-analysis of experimental data, Gnielinski[22] put forward the following correlations for the mean Nusselt number:

$$Nu = \frac{\phi(Re - 1000)Pr}{1 + 12.7\phi^{1/2}(Pr^{2/3} - 1)}\left[1 + \left(\frac{d}{l}\right)^{2/3}\right] \qquad (1.96)$$

where $\phi = \left(\dfrac{R}{\rho u^2}\right)$ is the friction factor introduced in Chapter 3 of Vol. 1A. Naturally, the term (d/l) becomes less significant with the increasing pipe length and it is customary to drop the term $\left\{1 + (d/l)^{2/3}\right\}$ in most practical situations thereby reducing Eq. (1.96):

$$Nu = \frac{\phi(Re - 1000)Pr}{1 + 12.7\phi^{1/2}(Pr^{2/3} - 1)} \qquad (1.97)$$

This equation is applicable over the range $2300 \leq Re \leq 5 \times 10^6$ and $0.5 \leq Pr \leq 2000$. Gnielinski[22] also presented the following simplified forms of Eq. (1.97):

For $0.5 \leq Pr \leq 1.5$ & $10^4 \leq Re \leq 5 \times 10^6$:

$$Nu = 0.0214\left(Re^{0.8} - 100\right)Pr^{0.4} \tag{1.98}$$

For $1.5 \leq Pr \leq 500$ & $3000 \leq Re \leq 10^6$:

$$Nu = 0.012\left(Re^{0.87} - 280\right)Pr^{0.4} \tag{1.99}$$

The correction for the variation of physical properties with temperature is introduced by multiplying the right hand sides of Eqs. (1.97)–(1.99) by the factor $(T/T_s)^{0.45}$ for gases and by $(Pr/Pr_s)^{0.11}$ for liquids where absolute temperatures are used in evaluating the factor $(T/T_s)^{0.45}$. None of these equations work satisfactorily for molten metals characterised by very low Prandtl numbers ($\sim < 0.1$ or so) because under these conditions the thermal boundary-layer thickness is greater than the momentum boundary layer. Sleicher et al.[25] put forward the following equations for long tubes (fully developed flow and temperature fields) for the two types of thermal boundary conditions prescribed on the wall of the pipe or channel.

Constant wall temperature:

$$Nu = 4.8 + 0.0156Re^{0.85}Pr^{0.93} \tag{1.100}$$

Constant wall heat flux:

$$Nu = 6.3 + 0.0167Re^{0.85}Pr^{0.93} \tag{1.101}$$

Both these equations are applicable over the range $2.6 \times 10^4 \leq Re \leq 3.02 \times 10^5$ and $0.004 \leq Pr \leq 0.1$. In order to develop a feel for the role of Prandtl number on the mean Nusselt number, let us consider the values of convective heat transfer coefficient for air ($Pr = 0.7$) and water ($Pr = 7$). Indeed, there is a very big difference in the values of h for water and air for the same linear velocity. This is shown in Figs 1.30–1.32 and Table 1.4, all of which are based on the work of Fishenden and Saunders.[26]

The effect of length to diameter ratio (l/d) on the value of the heat transfer coefficient may be seen in Fig. 1.33. It is important at low Reynolds numbers but ceases to be significant at a Reynolds number of about 10^4.

It is also important to note that the film coefficient varies with the distance from the entrance to the tube. This is especially important at low (l/d) ratios and an average value is given approximately by:

$$\frac{h_{\text{average}}}{h_\infty} = 1 + \left(\frac{d}{l}\right)^{0.7} \tag{1.102}$$

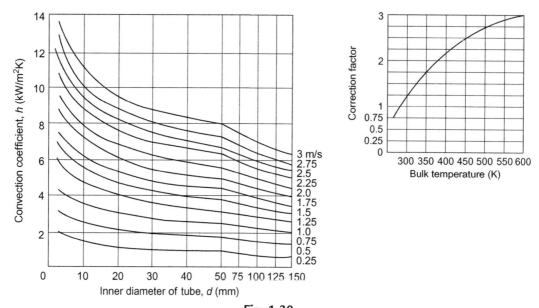

Fig. 1.30

Film coefficients of convection for flow of water through a tube at 289 K.

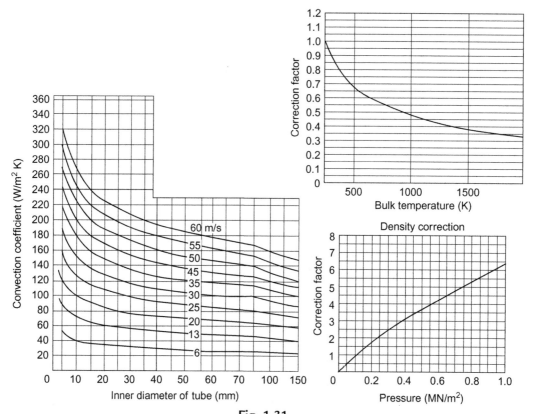

Fig. 1.31

Film coefficients of convection for flow of air through a tube at various velocities
(289 K, 101.3 kN/m^2).

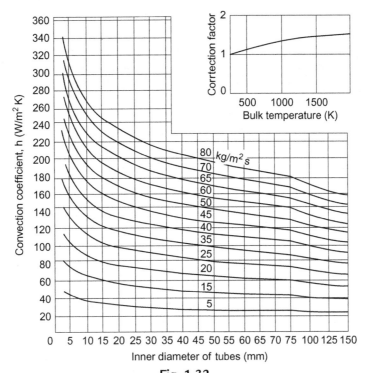

Fig. 1.32

Film coefficients of convection for flow of air through a tube for various mass velocities
(289 K, 101.3 kN/m^2).

where h_∞ is the limiting value for a very long tube.

The roughness of the surface of the inside of the pipe can have an important bearing on rates of heat transfer to the fluid, although Cope,[27] using degrees of artificial roughness ranging from 0.022 to 0.14 of the pipe diameter, found that, though the friction loss was some six times greater than for smooth tubes, the heat transfer was only 100%–120% higher. It was concluded that, for the same pressure drop, greater heat transfer was obtained from a smooth rather from a rough tube. The effect of a given scale deposit is usually less serious for gases than water because of the higher thermal resistance of the gas film, although layers of dust or of materials which sublime may seriously reduce heat transfer between gas and solid by as much as 40%.

Streamline flow

Although heat transfer to a fluid in streamline flow takes place solely by conduction, it is convenient to consider it here so that the results may be compared with those for turbulent flow.

In Chapter 3 of Vol. 1A it has been seen that, for streamline flow through a tube, the velocity distribution across a diameter is parabolic, as shown in Fig. 1.34 If a liquid enters a section

Table 1.4 Film coefficients for air and water (289 K and 101.3 kN/m^2)[26]

Inside Diameter of Tube		Velocity		Mass Velocity		Film Coefficient of Heat Transfer h	
(mm)	(in)	(m/s)	(ft/s)	(kg/m^2 s)	(lb/ft^2 h)	(W/m^2 K)[26]	(Btu/h ft^2 °F)[26]
Air							
25	1.0	5	16.4	6.11	4530	31.2	5.5
		10	32.8	12.2	9050	50.0	8.8
		20	65.6	24.5	18,100	84.0	14.8
		40	131	48.9	36,200	146	25.7
		60	197	73.4	54,300	211	37.2
50	2.0	5	16.4	6.11	4530	23.8	4.2
		10	32.8	12.2	9050	44.9	7.9
		20	65.6	24.5	18,100	77.8	13.7
		40	131	48.9	36,200	127	22.4
		60	197	73.4	54,300	181	31.9
75	3.0	5	16.4	6.11	4530	21.6	3.8
		10	32.8	12.2	9050	39.7	7.0
		20	65.6	24.5	18,100	71.0	12.5
		40	131	48.9	36,200	119	21.0
		60	197	73.4	54,300	169	29.8
Water							
25	1.0	0.5	1.64	488	361,000	2160	380
		1.0	3.28	975	722,000	3750	660
		1.5	4.92	1460	1,080,000	5250	925
		2.0	6.55	1950	1,440,000	6520	1150
		2.5	8.18	2440	1,810,000	7780	1370
50	2.0	0.5	1.64	488	361,000	1870	330
		1.0	3.28	975	722,000	3270	575
		1.5	4.92	1460	1,080,000	4540	800
		2.0	6.55	1950	1,440,000	5590	985
		2.5	8.18	2440	1,810,000	6700	1180
75	3.0	0.5	1.64	488	361,000	1760	310
		1.0	3.28	975	722,000	3070	540
		1.5	4.92	1460	1,080,000	4200	740
		2.0	6.55	1950	1,440,000	5220	920
		2.5	8.18	2440	1,810,000	6220	1100

heated on the outside, the fluid near the wall will be at a higher temperature than that in the centre and its viscosity will be lower. The velocity of the fluid near the wall will therefore be greater in the heated section, and correspondingly less at the centre. The velocity distribution will therefore be altered, as shown. If the fluid enters a section where it is cooled, the same reasoning will show that the distribution in velocity will be altered to that shown. With a gas the conditions are reversed, because of the increase in viscosity with temperature. The heat transfer problem is therefore complex and it is not amenable to a rigorous analytical treatment even in streamline flow conditions.

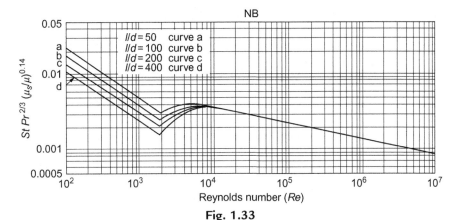

Fig. 1.33

Effect of length-to-diameter ratio of tubes on heat transfer coefficient.

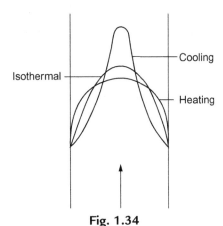

Fig. 1.34

Effect of heat transfer on the velocity distribution for a liquid in streamline flow in a pipe.

For the common problem of heat transfer between a fluid and a tube wall, the boundary layers are limited in thickness to the radius of the pipe and, furthermore, the effective area for heat flow decreases with distance from the surface. The problem can conveniently be divided into two parts. Firstly, heat transfer in the entry length in which the boundary layers are developing, and secondly, heat transfer under conditions of fully developed flow, both hydrodynamically and thermally. Boundary layer flow is discussed in Chapter 3.

For the region of fully developed flow in a pipe of length L, diameter d, and radius r, the rate of flow of heat Q through a cylindrical surface in the fluid at a distance y from the wall is given by:

$$Q = -k2\pi L(r-y)\frac{d\theta}{dy} \tag{1.103}$$

Close to the wall, the fluid velocity is low and a negligible amount of heat is carried along the pipe by the fluid flowing in this region and Q is independent of y.

Thus:

$$\frac{d\theta}{dy} = -\frac{Q}{k2\pi L}(r-y)^{-1} \text{ and } \left(\frac{d\theta}{dy}\right)_{y=0} = -\frac{Q}{2\pi kLr}$$

$$\frac{d^2\theta}{dy^2} = -\frac{Q}{k2\pi L}(r-y)^{-2} \text{ and } \left(\frac{d^2\theta}{dy^2}\right)_{y=0} = -\frac{Q}{2\pi kLr^2}$$

Thus:

$$\left(\frac{d^2\theta}{dy^2}\right)_{y=0} = r^{-1}\left(\frac{d\theta}{dy}\right)_{y=0} \qquad (1.104)$$

Assuming that the temperature of the walls remains constant at the datum temperature and that the temperature at any distance y from the walls is given by a polynomial, then:

$$\theta = a_0 y + b_0 y^2 + c_0 y^3 \qquad (1.105)$$

Thus:

$$\frac{d\theta}{dy} = a_0 + 2b_0 y + 3c_0 y^2 \text{ and } \left(\frac{d\theta}{dy}\right)_{y=0} = a_0$$

$$\frac{d^2\theta}{dy^2} = 2b_0 + 6c_0 y \text{ and } \left(\frac{d^2\theta}{dy^2}\right)_{y=0} = 2b_0$$

Thus:

$$2b_0 = \frac{a_0}{r} \quad \text{(from Eq. 1.93)}$$

and:

$$b_0 = \frac{a_0}{2r}$$

If the temperature of the fluid at the axis of the pipe is θ_s and the temperature gradient at the axis from symmetry is zero, then:

$$0 = a_0 + 2r\left(\frac{a_0}{2r}\right) + 3c_0 r^2$$

giving:

$$c_0 = -\frac{2a_0}{3r^2}$$

and:

$$\theta_s = a_0 r + r^2 \left(\frac{a_0}{2r}\right) + r^3 \left(\frac{-2a_0}{3r^2}\right)$$

$$= \frac{5}{6} a_0 r$$

$$a_0 = \frac{6\theta_s}{5 r}$$

$$b_0 = \frac{3\theta_s}{5 r^2}$$

and:

$$c_0 = -\frac{4\theta_s}{5 r^3}$$

Thus:

$$\frac{\theta}{\theta_s} = \frac{6y}{5r} + \frac{3}{5}\left(\frac{y}{r}\right)^2 - \frac{4}{5}\left(\frac{y}{r}\right)^3 \qquad (1.106)$$

Thus the rate of heat transfer per unit area at the wall:

$$q = -k\left(\frac{d\theta}{dy}\right)_{y=0}$$

$$= -\frac{6k\theta_s}{5 \ r} \qquad (1.107)$$

In general, the temperature θ_s at the axis is not known, and the heat transfer coefficient is related to the temperature difference between the walls and the bulk fluid. The bulk temperature of the fluid is defined as the ratio of the heat content to the heat capacity of the fluid flowing at any section. Thus the bulk temperature θ_B is given by:

$$\theta_B = \frac{\displaystyle\int_0^r C_p\rho\theta u_x 2\pi(r-y)\,dy}{\displaystyle\int_0^r C_p\rho u_x 2\pi(r-y)\,dy}$$

$$= \frac{\displaystyle\int_0^r \theta u_x(r-y)\,dy}{\displaystyle\int_0^r u_x(r-y)\,dy} \qquad (1.108)$$

From Poiseuille's law (Eq. 3.30 of Vol. 1A):

$$u_x = \frac{-\Delta P}{4\mu L}\left[r^2 - (r-y)^2\right] = \frac{-\Delta P}{4\mu L}(2ry - y^2)$$

Hence:

$$u_s = \frac{-\Delta P}{4\mu L}r^2 \tag{1.109}$$

where u_s is the velocity at the pipe axis, and:

$$\frac{u_x}{u_s} = \frac{2y}{r} - \left(\frac{y}{r}\right)^2 \tag{1.110}$$

Thus:

$$\int_0^r u_x(r-y)\,dy = r^2 u_s \int_0^1 \left[2\frac{y}{r} - \left(\frac{y}{r}\right)^2\right]\left(1 - \frac{y}{r}\right)d\left(\frac{y}{r}\right)$$

$$= r^2 u_s \int_0^1 \left[2\left(\frac{y}{r}\right) - 3\left(\frac{y}{r}\right)^2 + \left(\frac{y}{r}\right)^3\right]d\left(\frac{y}{r}\right) \tag{1.111}$$

$$= \frac{1}{4}r^2 u_s$$

Since:

$$\frac{\theta}{\theta_s} = \frac{6y}{5r} + \frac{3}{5}\left(\frac{y}{r}\right)^2 - \frac{4}{5}\left(\frac{y}{r}\right)^3 \quad (\text{Eq. 1.106})$$

$$\int_0^r \theta u_x(r-y)\,dy = r^2 u_s \theta_s \int_0^1 \left[\frac{6y}{5r} + \frac{3}{5}\left(\frac{y}{r}\right)^2 - \frac{4}{5}\left(\frac{y}{r}\right)^3\right]\left[2\left(\frac{y}{r}\right) - 3\left(\frac{y}{r}\right)^2 + \left(\frac{y}{r}\right)^3\right]d\left(\frac{y}{r}\right)$$

$$= r^2 u_s \theta_s \int_0^1 \left[\frac{12}{5}\left(\frac{y}{r}\right)^2 - \frac{12}{5}\left(\frac{y}{r}\right)^3 - \frac{11}{5}\left(\frac{y}{r}\right)^4 + 3\left(\frac{y}{r}\right)^5 - \frac{4}{5}\left(\frac{y}{r}\right)^6\right]d\left(\frac{y}{r}\right)$$

$$= r^2 u_s \theta_s \left(\frac{4}{5} - \frac{3}{5} - \frac{11}{25} + \frac{1}{2} - \frac{4}{35}\right) \tag{1.112}$$

$$= \frac{51}{350}r^2 u_s \theta_s$$

Substituting from Eqs. (1.111), (1.112) in Eq. (1.108):

$$\theta_B = \frac{\dfrac{51}{350}r^2 u_s \theta_s}{\dfrac{1}{4}r^2 u_s} \tag{1.113}$$

$$= \frac{102}{175}\theta_s = 0.583\theta_s$$

The heat transfer coefficient h is then given by:

$$h = -\frac{q}{\theta_B}$$

where q is the rate of heat transfer per unit area of tube.

Thus, from Eqs. (1.107), (1.113):

$$h = \frac{6k\theta_s/5r}{0.583\theta_s} = \frac{2.06k}{r} = 4.1\frac{k}{d}$$

and:

$$Nu = \frac{hd}{k} = 4.1 \tag{1.114}$$

This expression is applicable only to the region of fully developed flow. The approximate value of 4.1 compares favourably with the expected value of 3.66 for the Nusselt number in the fully developed regime.[28] The heat transfer coefficient for the inlet length can be calculated approximately, using the expressions given in Chapter 3 for the development of the boundary layers for the flow over a plane surface. It should be borne in mind that it has been assumed throughout that the physical properties of the fluid are not appreciably dependent on temperature and therefore the expressions will not be expected to hold accurately if the temperature differences are large and if the properties vary widely with temperature.

For values of ($RePr\ d/l$) greater than 12, the following empirical equation is applicable:

$$Nu = 1.62\left(RePr\frac{d}{l}\right)^{1/3} = 1.75\left(\frac{GC_p}{kl}\right)^{1/3} \tag{1.115}$$

where $G = (\pi d^2/4)\rho u$, i.e. the mass rate of flow.

The product $RePr$ is termed the Peclet number Pe.

Thus:

$$P_e = \frac{ud\rho}{\mu}\frac{C_p\mu}{k} = \frac{C_p\rho ud}{k} \tag{1.116}$$

Eq. (1.115) may then be written as:

$$Nu = 1.62\left(Pe\frac{d}{l}\right)^{1/3} \tag{1.117}$$

In this equation the temperature difference is taken as the arithmetic mean of the terminal values, that is:

$$\frac{(T_w - T_1) + (T_w - T_2)}{2}$$

where T_w is the temperature of the tube wall which is taken as constant.

If the liquid is heated almost to the wall temperature T_w (that is when GC_p/kl is very small) then, on equating the heat gained by the liquid to that transferred from the pipe:

$$GC_p(T_2 - T_1) = \pi dlh \frac{T_2 - T_1}{2}$$

or:

$$h = \frac{2}{\pi} \frac{GC_p}{dl} \tag{1.118}$$

For values of $(RePr\, d/l)$ less than about 17, the Nusselt group becomes approximately constant at 4.1; the value given in Eq. (1.114).

Experimental values of h for *viscous oils* are greater than those given by Eq. (1.115) for heating and less for cooling. This is due to the large variation of viscosity with temperature and the correction introduced for turbulent flow may also be used here, giving:

$$Nu\left(\frac{\mu_s}{\mu}\right)^{0.14} = 1.86\left(RePr\frac{d}{l}\right)^{1/3} = 2.01\left(\frac{GC_p}{kl}\right)^{1/3} \tag{1.119}$$

or:

$$Nu\left(\frac{\mu_s}{\mu}\right)^{0.14} = 1.86\left(Pe\frac{d}{l}\right)^{1/3} \tag{1.120}$$

When $(GC_p/kl) < 10$, the outlet temperature closely approaches that of the wall and Eq. (1.118) applies. These equations have been obtained with tubes about 10–40 mm in diameter, and the length of unheated tube preceding the heated section is important. The equations are not entirely consistent since for very small values of ΔT the constants in Eqs. (1.115), (1.119) would be expected to be the same. It is important to note, when using these equations for design purposes, the error may be as much as $\pm 25\%$ for turbulent flow and even greater for streamline conditions.

With laminar flow there is a marked influence of tube length and the curves shown in Fig. 1.33 show the parameter l/d from 50 to 400. Similarly, one can obtain the analogous results when the tube wall is subject to a prescribed heat flux as opposed to the isothermal condition used above. For long tubes and the fully developed conditions, the Nusselt number reaches a limiting value of $Nu = 48/11$ which is slightly higher than that given by Eq. (1.120). Also, an approximate result can be obtained near the tube entrance by assuming that the heat penetrates into the fluid only up to small distances from the tube wall. Thus, one can neglect the curvature effects.[28]

Whenever possible, streamline conditions of flow are avoided in heat exchangers because of the very low heat transfer coefficients which are obtained. With very viscous liquids, however, turbulent conditions can be produced only if a very high pressure drop across the plant is permissible. In the processing industries, streamline flow in heat exchangers is most commonly experienced with heavy oils, thermal treatment of foodstuffs, and brines at low temperatures. Since the viscosity of these materials is critically dependent on temperature, the equations would not be expected to apply with a high degree of accuracy. More detailed treatments of laminar heat transfer in Newtonian and non-Newtonian fluids flowing in circular and noncircular ducts are available in the literature.[29–31] This body of information is particularly relevant to microscale heat transfer in novel heat exchangers and/or in microfluidic devices.

1.4.4 Forced Convection Outside Tubes

Flow across single cylinders

If a fluid passes at right angles across a single tube, the distribution of velocity around the tube will not be uniform. In the same way the rate of heat flow around a hot pipe across which air is passed is not uniform but is a maximum at the front and rear, and a minimum at the sides, where the rate is only some 40% of the maximum. The general picture is shown in Fig. 1.35 but for design purposes reference is made to the average value.

A number of researchers, including, Reiher,[32] Hilpert,[33] Griffiths, and Awbery,[34] have studied the flow of a hot gas past a single cylinder, varying from a thin wire to a tube of 150 mm diameter. Temperatures up to 1073 K and air velocities up to 30 m/s have been used with Reynolds numbers $(d_o u \rho / \mu)$ from 1000 to 100,000 (where d_o is the cylinder diameter, or the outside tube diameter). The data obtained may be expressed by:

$$Nu = 0.26 Re^{0.6} Pr^{0.3} \tag{1.121}$$

Taking Pr as 0.74 for gases, this reduces to

$$Nu = 0.24 Re^{0.6} \tag{1.122}$$

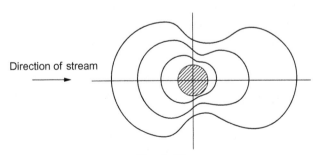

Direction of stream

Fig. 1.35

Distribution of the film heat transfer coefficient around a cylinder with flow normal to the axis for three different values of *Re*.

Davis[35] has also worked with water, paraffin, and light oils and obtained similar results. For very low values of Re (from 0.2 to 200) with liquids the data are better represented by the equation:

$$Nu = 0.86 Re^{0.43} Pr^{0.3} \qquad (1.123)$$

In each case the physical properties of the fluid are measured at the mean film temperature T_f, taken as the average of the surface temperature T_w and the mean fluid temperature T_m; where $T_m = (T_1 + T_2)/2$.

Much of the literature in this field has been reviewed by Morgan,[36] Zukauskas,[37] and Zdravkovich.[38] Zukauskas[37] recommends the following correlation for estimating the average Nusselt number for a single (long) cylinder:

$$Nu = C_0 Re^{C_1} Pr^{C_2} (Pr/Pr_w)^{C_3} \qquad (1.124)$$

where the values of C_0, C_1, and C_2 depend on the range of Reynolds number and are summarised in Table 1.5.

The ratio (Pr/Pr_w) in Eq. (1.124) accounts for the variation of physical properties with temperature and $C_3 = 0.25$ is used for the heating of fluid and $C_3 = 0.20$ is used for the cooling of fluid medium.

Another equally good, but somewhat restricted in terms of the range of Reynolds number correlation but which captures the effect of Prandtl number on heat transfer somewhat better than Eq. (1.124) is due to Churchill and Bernstein[39] as:

$Re < 10^4$:

$$Nu = 0.3 + \frac{0.62 Re^{1/2} Pr^{1/3}}{\left[1 + (0.4/Pr)^{2/3} \right]^{1/4}} \qquad (1.125)$$

$10^4 \leq Re \leq 4 \times 10^5$:

$$Nu = 0.3 + \frac{0.62 Re^{1/2} Pr^{1/3}}{\left[1 + (0.4/Pr)^{2/3} \right]^{1/4}} \left[1 + \left(\frac{Re}{2.82 \times 10^5} \right)^{1/2} \right] \qquad (1.126)$$

Table 1.5 Values of constants in Eq. (1.124)

Range of Re	C_0	C_1	C_2
1–40	0.76	0.4	0.37
40–10^3	0.52	0.5	0.37
10^3–2×10^5	0.26	0.6	0.37
2×10^5–10^7	0.023	0.8	0.4

$Re \geq 4 \times 10^5$:

$$Nu = 0.3 + \frac{0.62 Re^{1/2} Pr^{1/3}}{\left[1 + (0.4/Pr)^{2/3}\right]^{1/4}} \left[1 + \left(\frac{Re}{2.82 \times 10^5}\right)^{5/8}\right]^{4/5} \tag{1.127}$$

In Eqs. (1.124)–(1.127), the properties are evaluated at the mean film temperature if these do not show a strong temperature dependence. Also, Eq. (1.127) is applicable for $Pe > 0.2$. The two predictions are within 10% of each other up to $Re = 10^6$, at least for air and water.

For molten metals with very small Prandtl numbers, Ishiguro et al.[40] recommends the following expression (valid for $1 \leq Pe \leq 100$):

$$Nu = 1.125 (Re \cdot Pr)^{0.413} \tag{1.128}$$

Flow at right angles to tube bundles

One of the great difficulties with this geometry is that the area for flow is continually changing. Moreover the degree of turbulence is considerably less for banks of tubes in line, as at (*a*), than for staggered tubes, as at (*b*) in Fig. 1.36. With the small bundles which are common in the processing industries, the selection of the true mean area for flow is further complicated by the change in number of tubes in the rows.

The results of a number of researchers for heat transfer to and from gases flowing across tube banks may be expressed by the equation:

$$Nu = 0.33 C_h Re_{max}^{0.6} Pr^{0.3} \tag{1.129}$$

where C_h depends on the geometrical arrangement of the tubes, as shown in Table 1.6. Grimison[41] proposed this form of expression to correlate the data of Huge[42] and Pierson[43] who worked with small electrically heated tubes in rows of ten deep. Other researchers have used similar equations. Some correction factors have been given by Pierson[43] for bundles with less than ten rows although there are insufficient reported data from commercial exchangers to fix these values with accuracy. Thus, for five rows a factor of 0.92 and for eight rows a factor of 0.97 is suggested.

These equations are based on the maximum velocity through the bundle. Thus for an in-line arrangement as shown in Fig. 1.36A, $G'_{max} = G'Y/(Y - d_o)$, where Y is the pitch of the pipes at right-angle to the direction of flow; it is more convenient here to use the mass flowrate per unit area G' in place of velocity. For staggered arrangements the maximum velocity may be based on the distance between the tubes in a horizontal line or on the diagonal of the tube bundle, whichever is the less. It has been suggested that, for in-line arrangements, the constant in Eq. (1.129) should be reduced to 0.26, but there is insufficient evidence from commercial exchangers to confirm this.

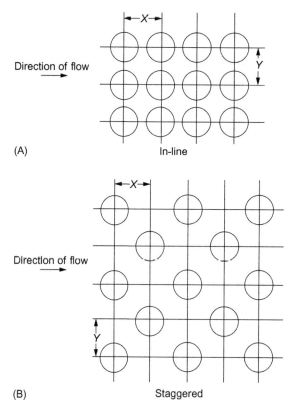

Direction of flow

(A)   In-line

Direction of flow

(B)   Staggered

Fig. 1.36
Arrangements of tubes in heat exchangers.

Table 1.6 Values of C_h and C_f[26]

| Re_{max} | $X = 1.25\,d_o$ | | | | $X = 1.5\,d_o$ | | | |
| | In-line | | Staggered | | In-line | | Staggered | |
	C_h	C_f	C_h	C_f	C_h	C_f	C_h	C_f
$Y = 1.25\,d_o$								
2000	1.06	1.68	1.21	2.52	1.06	1.74	1.16	2.58
20,000	1.00	1.44	1.06	1.56	1.00	1.56	1.05	1.74
40,000	1.00	1.20	1.03	1.26	1.00	1.32	1.02	1.50
$Y = 1.5\,d_o$								
2000	0.95	0.79	1.17	1.80	0.95	0.97	1.15	1.80
20,000	0.96	0.84	1.04	1.10	0.96	0.96	1.02	1.16
40,000	0.96	0.74	0.99	0.88	0.96	0.85	0.98	0.96

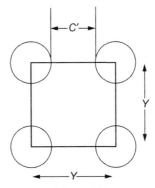

Fig. 1.37
Clearance and pitch for tube layouts.

An alternative approach has been suggested by Kern[44] who worked in terms of the hydraulic mean diameter d_e for flow *parallel* to the tubes:

$$\text{i.e. } d_e = 4 \times \frac{\text{Free area for flow}}{\text{Wetted perimeter}}$$

$$= 4\left(\frac{Y^2 - (\pi d_0^2/4)}{\pi d_0}\right)$$

for a square pitch as shown in Fig. 1.37. The maximum cross-flow area A_s is then given by:

$$A_s = \frac{d_s l_B C'}{Y}$$

where C' is the clearance, l_B the baffle spacing, and d_S the internal diameter of the shell.

The mass rate of flow per unit area G_s' is then given as rate of flow divided by A_s, and the film coefficient is obtained from a Nusselt type expression of the form:

$$\frac{h_0 d_e}{k} = 0.36 \left(\frac{d_e G_s'}{\mu}\right)^{0.55} \left(\frac{C_p \mu}{k}\right)^{1/3} \left(\frac{\mu}{\mu_s}\right)^{0.14} \tag{1.130}$$

There are insufficient published data to assess the relative merits of Eqs. (1.129), (1.130).

Thus, for instance, for 19 mm tubes on 25 mm square pitch:

$$d_e = 4\frac{\left[25^2 - (\pi/4)19^2\right]}{\pi \times 19}$$

$$= 22.8 \text{mm or } 0.023 \text{m}$$

Zukauskas[37] has reviewed bulk of the literature on forced convection heat transfer for cross flow over in-line and staggered arrays. Based on the literature data, he put forward the predictive expressions of the following forms:

$$Nu = C_0 \left(\frac{Y}{X}\right)^{C_1} Re^{C_2} Pr^{C_3} \left(\frac{Pr}{Pr_w}\right)^{0.25} \tag{1.131}$$

The values of $C_0, C_1, C_2,$ and C_3 for in-line (Fig. 1.36A) and staggered tube bundles (Fig. 1.36B) are presented in Tables 1.7 and 1.8 respectively along with the relevant range of Reynolds number.

The fluid velocity ($\langle u \rangle$) appearing in the Reynolds number used in Eq. (1.131) corresponds to its maximum value and is related to the free stream velocity (u_∞) as follows:

For an in-line arrangement:

$$\langle u \rangle = \frac{u_\infty Y}{Y - d} \tag{1.132}$$

For a staggered arrangement:

$$S_d > \frac{Y + d}{2}, \quad \langle u \rangle = \frac{u_\infty Y}{Y - d} \quad \text{(Eq. 1.132)}$$

$$S_d < \frac{Y + d}{2}, \quad \langle u \rangle = \frac{u_\infty Y}{2(S_d - d)} \tag{1.133}$$

where

$$S_d = \left[X^2 + \left(\frac{Y}{2}\right)^2 \right]^{1/2}$$

Furthermore, Zukauskas[37] asserted that equations (1.131) to (1.133) work well for tube bundles with 16 or more rows. For bundles with number of rows fewer than 16, a correction must be

Table 1.7 Values of $C_0, C_1, C_2,$ and C_3 for in-line arrays[37]

Range of Re	C_0	C_1	C_2	C_3
1–100	0.9	0	0.40	0.36
100–10^3	0.52	0	0.50	0.36
10^3–2×10^5	0.27	0	0.63	0.36
2×10^5–2×10^6	0.033	0	0.80	0.40

Table 1.8 Values of $C_0, C_1, C_2,$ and C_3 in Eq. (1.131) for staggered arrays

Range of Re	C_0	C_1	C_2	C_3
1–500	1.04	0	0.4	0.36
500–10^3	0.71	0	0.5	0.36
10^3–2×10^5	0.35	0.2	0.6	0.36
2×10^5–2×10^6	0.031	0.2	0.8	0.36

applied to the value of the heat transfer coefficient estimated using Eq. (1.131). Naturally, fewer the rows in the bundle, more severe is the correction.[37] Thus, for instance, for a bundle consisting of 5 rows, the correction factor is 0.93 which rises to the value of 0.99 for a bundle with 13 rows.

Example 1.18

14.4 tonne/h (4.0 kg/s) of nitrobenzene is to be cooled from 400 to 315 K by heating a stream of benzene from 305 to 345 K.

Two tubular heat exchangers are available each with a 0.44 m i.d. shell fitted with 166 tubes, 19.0 mm o.d., and 15.0 mm i.d., each 5.0 m long. The tubes are arranged in two passes on 25 mm square pitch with a baffle spacing of 150 mm. There are two passes on the shell side and operation is to be countercurrent. With benzene passing through the tubes, the anticipated film coefficient on the tube side is 1000 W/m² K.

Assuming that true cross-flow prevails in the shell, what value of scale resistance could be allowed if these units were used?

For nitrobenzene: $C_p = 2380\,\text{J/kg\,K}$, $k = 0.15\,\text{W/m\,K}$, $\mu = 0.70\,\text{mN\,s/m}^2$, $\rho = 1200\,\text{kg/m}^3$

Solution
(i) Tube side coefficient.

$$h_i = 1000\,\text{W/m}^2\,\text{K based on inside area}$$

or:

$$\frac{1000 \times 15.0}{19.0} = 790\,\text{W/m}^2\,\text{K based on outside area}$$

(ii) Shell side coefficient.

$$\text{Area for flow} = \text{shell diameter} \times \text{baffle spacing} \times \text{clearance/pitch}$$

$$= \frac{0.44 \times 0.150 \times 0.006}{0.025} = 0.0158\,\text{m}^2$$

Hence:

$$G'_s = \frac{4.0}{0.0158} = 253.2\,\text{kg/m}^2\text{s}$$

Taking $\mu/\mu_s = 1$ in Eq. (1.130):

$$h_0 = 0.36 \frac{k}{d_e} \left(\frac{d_e G'_s}{\mu} \right)^{0.55} \left(\frac{C_p \mu}{k} \right)^{0.33}$$

The hydraulic mean diameter,

$$d_e = 4 \left[\left(25^2 - \frac{\pi \times 19.0^2}{4} \right) / (\pi \times 19.0) \right] = 22.8\,\text{mm or } 0.023\,\text{m}$$

and here:

$$h_0 = \left(\frac{0.15}{0.023}\right) 0.36 \left(\frac{0.023 \times 253.2}{0.70 \times 10^{-3}}\right)^{0.55} \left(\frac{2380 \times 0.70 \times 10^{-3}}{0.15}\right)^{0.33}$$

$$= 2.35 \times 143 \times 2.23 = 750 \, \text{W/m}^2 \, \text{K}$$

(iii) Overall coefficient.

The logarithmic mean temperature difference is given by:

$$\Delta T_m = \frac{(400 - 345) - (315 - 305)}{\ln(400 - 345)/(315 - 305)}$$

$$= 26.4 \, \text{deg K}$$

The corrected mean temperature difference is then $\Delta T_m \times F = 26.4 \times 0.8 = 21.1 \, \text{deg K}$

(Details of the correction factor for ΔT_m are given in Section 1.9.3)

Heat load:

$$Q = 4.0 \times 2380(400 - 315) = 8.09 \times 10^5 \, \text{W}$$

The surface area of each tube $= \pi d = 3.14 \times 0.019 = 0.0598 \, \text{m}^2/\text{m}$

Thus:

$$U_0 = \frac{Q}{A_0 \Delta T_m F} = \frac{8.09 \times 10^5}{2 \times 166 \times 5.0 \times 0.0598 \times 21.1}$$

$$= 386.2 \, \text{W/m}^2 \, \text{K}$$

(iv) Scale resistance.

If scale resistance is R_d, then:

$$R_d = \frac{1}{386.2} - \frac{1}{750} - \frac{1}{1000} = 0.00026 \, \text{m}^2 \, \text{K/W}$$

This is a rather low value, though the heat exchangers would probably be used for this duty.

As an illustration, the value of h_0 is recalculated here using Eq. (1.131) presented in the preceding section. In this case, one can approximate the free stream velocity by the superficial velocity in the shell, i.e.

$$u_\infty = \frac{4}{1200 \times (\pi/4) \times 0.44^2} = 0.022 \, \text{m/s}$$

In this case, the value of S_d is calculated as:

$$S_d = \left\{0.025^2 + \left(\frac{0.025}{2}\right)^2\right\}^{1/2} = 0.028 \, \text{m}$$

Assuming in-line array arrangement of tubes:

$$\langle u \rangle = \frac{0.0220 \times 0.025}{0.025 - 0.019} = 0.091 \, \text{m/s}$$

Since in this case $S_d > (0.025 + 0.019)/2 \, \text{m}$, the effective velocity $\langle u \rangle$ will be the same as above even for the staggered square arrangement:

$$Re = \frac{1200 \times 0.091 \times 0.019}{0.7 \times 10^{-3}} = 2965$$

For in-line array (From Table 1.7):

$$Nu = 0.27(Re)^{0.63}(Pr)^{0.36}$$

Substituting values:

$$h_0 = \left(\frac{0.15}{0.019}\right) 0.27(2965)^{0.63} \left(\frac{2380 \times 0.7 \times 10^{-3}}{0.15}\right)^{0.36}$$
$$= 782 \, \text{W/m}^2 \text{K}$$

Similarly, if it were a staggered array (From Table 1.8):

$$Nu = 0.35(Re)^{0.6}(Pr)^{0.36}$$

Neglecting the effect due to the variation in properties.

This yields $h_0 = 798 \, \text{W/m}^2 \text{K}$

Both these values are comparable to the value of 750 W/m^2 K calculated using Eq. (1.130).

As discussed in Section 1.9 it is common practice to fit *baffles* across the tube bundle in order to increase the velocity over the tubes. The commonest form of baffle is shown in Fig. 1.38 where it is seen that the cut-away section is about 25% of the total area. With such an arrangement, the

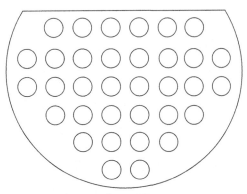

Fig. 1.38
Baffle for heat exchanger.

flow pattern becomes more complex and the extent of leakage between the tubes and the baffle, and between the baffle and the inside of the shell of the exchanger, complicates the problem, as discussed further in Section 1.9.6. Reference may also be made to Volume 6 and to the work of Short,[45] Donohue,[46] Tinker,[47] and Bell.[48] The various methods are all concerned with developing a method of calculating the true area of flow and of assessing the probable influence of leaks. When using baffles, the value of h_0, as found from Eq. (1.123), is commonly multiplied by 0.6 to allow for leakage although more accurate approaches have been developed as discussed in Section 1.9.6.

The *drop in pressure* for the flow of a fluid across a tube bundle may be important because of the small pressure head available and because by good design it is possible to get a better heat transfer for the same drop in pressure. The drop in pressure $-\Delta P_f$ depends on the velocity u_t through the minimum area of flow and in Chapter 3 of Vol. 1A an equation proposed by Grimison[39] is given as:

$$-\Delta P_f = \frac{C_f j \rho u_t^2}{6} \quad \text{(Eq. 3.83)}$$

where C_f depends on the geometry of the tube layout and j is the number of rows of tubes. It is found that the ratio of C_h, the heat transfer factor in Eq. (1.129), to C_f is greater for the in-line arrangement but that the actual heat transfer is greater for the staggered arrangement, as shown in Table 1.9.

The drop in pressure $-\Delta P_f$ over the tube bundles of a heat exchanger is also given by:

$$-\Delta P_f = \frac{f' G_s'^2 (n+1) d_v}{2 \rho d_e} \tag{1.134}$$

where f' is the friction factor given in Fig. 1.39, G'_s the mass velocity through bundle, n the number of baffles in the unit, d_v the inside shell diameter, ρ the density of fluid, d_e the equivalent diameter, and $-\Delta P_f$ the drop in pressure. Zukauskas[37] has further refined the available techniques for estimating the value of $(-\Delta P_f)$ in tube bundles.

Table 1.9 Ratio of heat transfer to friction for tube bundles ($Re_{max} = 20{,}000$)[26]

	$X = 1.25\, d_o$			$X = 1.5\, d_o$		
	C_h	C_f	C_h/C_f	C_h	C_f	C_h/C_f
In-line						
$Y = 1.25 d_o$	1	1.44	0.69	1	1.56	0.64
$Y = 1.5 d_o$	0.96	0.84	1.14	0.96	0.96	1.0
Staggered						
$Y = 1.25 d_o$	1.06	1.56	0.68	1.05	1.74	0.60
$Y = 1.5 d_o$	1.04	1.10	0.95	1.02	1.16	0.88

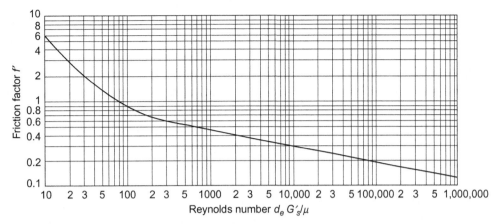

Fig. 1.39

Friction factor for flow over tube bundles.

Example 1.19

54 tonne/h (15 kg/s) of benzene is cooled by passing the stream through the shell side of a tubular heat exchanger, 1 m i.d., fitted with 5 m tubes, 19 mm o.d. arranged on a 25 mm square pitch with 6 mm clearance. If the baffle spacing is 0.25 m (19 baffles), what will be the pressure drop over the tube bundle? ($\mu = 0.5$ mN s/m^2).

Solution

$$\text{Cross} - \text{flow area}: \quad A_s = \frac{1.0 \times 0.25 \times 0.006}{0.025} = 0.06\,\text{m}^2$$

$$\text{Mass flow}: \quad G'_s = \frac{15}{0.06} = 250\,\text{kg/m}^2\text{s}$$

$$\text{Equivalent diameter}: \quad d_e = \frac{4[0.025^2 - (\pi/4)0.019^2]}{\pi \times 0.019} = 0.0229\,\text{m}$$

$$\text{Reynolds number through the tube bundle} = \frac{250 \times 0.0229}{0.5 \times 10^{-3}} = 11,450$$

From Fig. 1.39:

$$f' = 0.280$$

$$\text{Density of benzene} = 881\,\text{kg/m}^3$$

From Eq. (1.134):

$$-\Delta P_f = \frac{0.280 \times 250^2 \times 20 \times 1.0}{2 \times 881 \times 0.0229} = \underline{\underline{8674\,\text{N/m}^2}}$$

or:

$$\frac{8674}{881 \times 9.81} = \underline{\underline{1.00\,\text{m of benzene}}}$$

1.4.5 Flow in Noncircular Sections

Rectangular ducts

For the heat transfer for fluids flowing in noncircular ducts, such as rectangular ventilating ducts, the equations developed for turbulent flow inside a circular pipe may be used if an equivalent diameter, such as the hydraulic mean diameter d_e discussed previously, is used in place of d.

The data for heating and cooling water in turbulent flow in rectangular ducts are reasonably well correlated by the use of Eq. (1.87) in the form:

$$\frac{hd_e}{k} = 0.023 \left(\frac{d_e G'}{\mu} \right)^{0.8} \left(\frac{C_p \mu}{k} \right)^{0.4} \tag{1.135}$$

Whilst the experimental data of Cope and Bailey[49] are somewhat low, the data of Washington and Marks[50] for heating air in ducts are well represented by this equation.

Annular sections between concentric tubes

Concentric tube heat exchangers are widely used because of their simplicity of construction and the ease with which additions may be made to increase the area. They also give turbulent conditions at low volumetric flowrates.

In presenting equations for the film coefficient in the annulus, one of the difficulties is in selecting the best equivalent diameter to use. When considering the film on the outside of the inner tube, Davis[51] has proposed the equation:

$$\frac{hd_1}{k} = 0.031 \left(\frac{d_1 G'}{\mu} \right)^{0.8} \left(\frac{C_p \mu}{k} \right)^{0.33} \left(\frac{\mu}{\mu_s} \right)^{0.14} \left(\frac{d_2}{d_1} \right)^{0.15} \tag{1.136}$$

where d_1 and d_2 are the outer diameter of the inner tube, and the inner diameter of the outer tube, respectively.

Carpenter et al.[52] suggest using the hydraulic mean diameter $d_e = (d_2 - d_1)$ in the Sieder and Tate equation (1.94) and recommend the equation:

$$\frac{hd_e}{k} \left(\frac{\mu_s}{\mu} \right)^{0.14} = 0.027 \left(\frac{d_e G'}{\mu} \right)^{0.8} \left(\frac{C_p \mu}{k} \right)^{0.33} \tag{1.137}$$

Their data, which were obtained using a small annulus, are somewhat below those given by Eq. (1.137) for values of $d_e G'/\mu$ less than 10,000, although this may be because the flow was not fully turbulent: with an index on the Reynolds group of 0.9, the equation fitted the data much better. There is little to choose between these two equations, but they both give rather high values for h.

For the viscous region, Carpenter's results are reasonably well correlated by the equation:

$$\frac{hd_e}{k}\left(\frac{\mu_s}{\mu}\right)^{0.14} = 2.01\left(\frac{GC_p}{kl}\right)^{0.33} \tag{1.138}$$

$$= 1.86\left[\left(\frac{d_e G'}{\mu}\right)\left(\frac{C_p\mu}{k}\right)\left(\frac{d_1+d_2}{l}\right)\right]^{1/3} \tag{1.139}$$

Eqs. (1.138), (1.139) are the same as Eqs. (1.119), (1.120), with d_e replacing d.

These results have all been obtained with small units and mainly with water as the fluid in the annulus.

Flow over flat plates

For the turbulent flow of a fluid over a flat plate, the Colburn type of equation may be used with a different constant:

$$j_h = 0.037Re_x^{-0.2} \tag{1.140}$$

where the physical properties are taken as for Eq. (1.92) and the characteristic dimension in the Reynolds group is the actual distance x along the plate. This equation therefore gives a point value for j_h.

More detailed discussions of the laminar, transitional, and turbulent flow pressure drop and heat transfer predictions in ducts of noncircular cross-sections, are available in the literature.[29,53]

1.4.6 Convection to Spherical Particles

In Section 1.3.4, consideration is given to the problem of heat transfer by conduction through a surrounding fluid to spherical particles or droplets. Relative motion between the fluid and particle or droplet causes an increase in heat transfer, much of which may be due to convection. Many investigators have correlated their data in the form:

$$Nu' = 2 + \beta'' Re'^n Pr^m \tag{1.141}$$

where values of β'', a numerical constant, and exponents n and m are found by experiment. In this equation, $Nu' = hd/k$ and $Re' = du\rho/\mu$, the Reynolds number for the particle, u is the relative velocity between particle and fluid, and d is the particle diameter. As the relative velocity approaches zero, Re' tends to zero and the equation reduces to $Nu' = 2$ for pure conduction.

Rowe et al.[54] having analysed a large number of previous studies in this area and provided further experimental data, have concluded that for particle Reynolds numbers in the range 20–2000, Eq. (1.141) may be written as:

$$Nu' = 2.0 + \beta'' Re'^{0.5} Pr^{0.33} \tag{1.142}$$

where β'' lies between 0.4 and 0.8 and has a value of 0.69 for air and 0.79 for water. In some practical situations the relative velocity between particle and fluid may change due to particle acceleration or deceleration, and the value of Nu' can then be time-dependent.

The other widely used heat transfer correlations for an isothermal sphere spanning much wider ranges of conditions than those of Eqs. (1.141), (1.142) are due to Whitaker,[55] and Achenbach[56] for ordinary fluids, and due to Witte[57] for low Prandtl number fluids like molten metal. These are presented here along with their range of validity. Whitaker[55] reanalyzed much of the literature data extending over $3.5 \leq Re = Re' \leq 7.6 \times 10^4$, $0.71 \leq Pr \leq 380$ and $1 \leq (\mu/\mu_w) \leq 3.2$ and put forward the following correlation:

$$Nu' = 2 + \left(0.4Re^{1/2} + 0.06Re^{2/3}\right)Pr^{2/5}(\mu/\mu_w)^{1/4} \tag{1.143}$$

In Eq. (1.143), the physical properties are evaluated at free stream fluid temperature except μ_w which is evaluated at the surface temperature of the sphere. In contrast, the correlation of Achenbach[56] is limited to air $(Pr = 0.71)$ but spans Reynolds numbers in the range $100 \leq Re = Re' \leq 5 \times 10^6$ as follows:

$$100 \leq Re \leq 2 \times 10^5 : \ Nu' = 2 + \left(0.25Re + 3 \times 10^{-4}Re^{1.6}\right)^{1/2} \tag{1.144a}$$

$$4 \times 10^5 \leq Re \leq 5 \times 10^6 : \ Nu' = 430 + 5 \times 10^{-3}Re + 2.5 \times 10^{-10}Re^2 - 3.1 \times 10^{-17}Re^3 \tag{1.144b}$$

In the overlapping range of Reynolds numbers, the predicted values of Nu' seldom differ from each other by more than 2% for these two correlations, i.e. Eqs. (1.143), (1.144a), (1.144b).

Based on the experimental data for molten sodium, Witte[57] proposed the following correlation for a sphere:

$$Nu' = 2 + 0.386(Re \cdot Pr)^{1/2} \tag{1.145}$$

Eq. (1.145) embraces the range $3.6 \times 10^4 \leq Re \leq 1.5 \times 10^5$.

For mass transfer, which is considered in more detail in Chapter 2, an analogous relation (Eq. 2.233) applies, with the Sherwood number replacing the Nusselt number and the Schmidt number replacing the Prandtl number.

1.4.7 Natural Convection

If a beaker containing water rests on a hot plate, the water at the bottom of the beaker becomes hotter than that at the top. Since the density of the hot water is lower than that of the cold, the water in the bottom rises and heat is transferred by natural convection. In the same way air in contact with a hot plate will be heated by natural convection currents, the air near the surface being hotter and of lower density than that at some distance away. In both of these cases there is no external agency providing forced convection currents, and the transfer of heat occurs at a correspondingly lower rate since the natural convection currents move rather slowly.

For these processes, which depend on buoyancy effects, the rate of heat transfer might be expected to follow a relation of the form:

$$Nu = f(Gr, Pr) \quad \text{(Eq. 1.85)}$$

Measurements by Schmidt[58] of the upward air velocity near a 300 mm vertical plate show that the velocity rises rapidly to a maximum at a distance of about 2 mm from the plate and then falls rapidly. However, the temperature evens out at about 10 mm from the plate. Temperature measurements around horizontal cylinders have been made by Ray.[59] An excellent album of photographs illustrating the nature of natural convection flow patterns for a range of configurations is available in the literature.[60]

Natural convection from horizontal surfaces, cylinders, and spheres to air, nitrogen, hydrogen, and carbon dioxide, and to liquids (including water, aniline, carbon tetrachloride, glycerol) has been studied by several researchers, including Davis,[61] Ackermann,[62] Fishenden and Saunders,[26] Saunders,[63] Fand and colleagues[64,65] and Amato and Tien.[66,67] Most of the results are for thin wires and tubes up to about 50 mm diameter; the temperature differences used are up to about 1100 deg K with gases and about 85 deg K with liquids. The general form of the results is shown in Fig. 1.40, where log Nu is plotted against log $(Pr\ Gr)$ for streamline conditions. The curve can be represented by a relation of the form:

$$Nu = C'(GrPr)^n \tag{1.146}$$

Numerical values of C' and n, determined experimentally for various geometries, are given in Table 1.10[68] and are also summarised in Ref. 60. Values of coefficients may then be predicted using the equation:

$$\frac{hl}{k} = C'\left(\frac{\beta g \Delta T l^3 \rho^2}{\mu^2}\frac{C_p \mu}{k}\right)^n \quad \text{or} \quad h = C'\left(\frac{\Delta T}{l}\right)^n k\left(\frac{\beta g \rho^2 C_p}{\mu k}\right)^n \tag{1.147}$$

where the physical properties are at the mean of the surface and bulk temperatures and for ideal gases, the coefficient of cubical expansion β is taken as $1/T$, where T is the absolute temperature.

For vertical plates and cylinders, Kato et al.[69] have proposed the following equations for situations where $1 < Pr < 40$:

$$\text{For } Gr > 10^9: \quad Nu = 0.138Gr^{0.36}\left(Pr^{0.175} - 0.55\right) \tag{1.148}$$

and

$$\text{for } Gr < 10^9: \quad Nu = 0.683Gr^{0.25}Pr^{0.25}\left(\frac{Pr}{0.861 + Pr}\right)^{0.25} \tag{1.149}$$

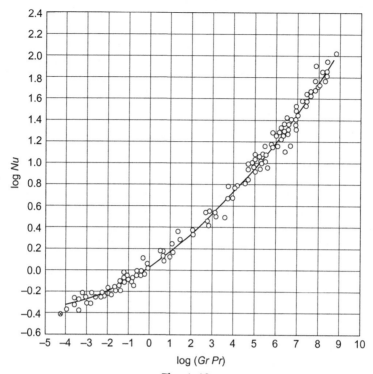

Fig. 1.40

Natural convection from horizontal tubes.

Table 1.10 Values of C', C'', and n for use in Eqs. (1.147), (1.150)[68]

Geometry	GrPr	C'	n	C'' (SI Units) (For Air at 294 K)
Vertical surfaces *l* = vertical dimension < 1 m	$<10^4$	1.36	0.20	
	10^4-10^9	0.59	0.25	1.37
	$>10^9$	0.13	0.33	1.24
Horizontal cylinders *l* = diameter < 0.2 m	$10^{-5}-10^{-3}$	0.71	0.04	
	$10^{-3}-1.0$	1.09	0.10	
	$1.0-10^4$	1.09	0.20	
	10^4-10^9	0.53	0.25	1.32
	$>10^9$	0.13	0.33	1.24
Horizontal flat surfaces Facing upwards	10^5 to 2×10^7	0.54	0.25	1.86
Facing upwards	2×10^7 to 3×10^{10}	0.14	0.33	
Facing downwards	3×10^5 to 3×10^{10}	0.27	0.25	0.88

Natural convection to air

Simplified dimensional equations have been derived for air, water, and organic liquids by grouping the fluid properties into a single factor in a rearrangement of Eq. (1.147) to give:

$$h = C''(\Delta T)^n l^{3n-1} \quad (\text{W/m}^2\,\text{K}) \tag{1.150}$$

Values of C'' (in SI units) are also given in Table 1.10 for air at 294 K. Typical values for water and organic liquids are 127 and 59 respectively.

Example 1.20

Estimate the heat transfer coefficient for natural convection from a horizontal pipe 0.15 m diameter, with a surface temperature of 400 K to air at 294 K.

Solution
Over a wide range of temperature, $k^4(\beta g \rho^2 Cp/\mu k) = 36.0$

For air at a mean temperature of $0.5(400 + 294) = 347$K, $k = 0.0310$W/mK (Table 6, Appendix A1)

Thus:

$$\frac{\beta g \rho^2 C_p}{\mu k} = \frac{36.0}{0.0310^4} = 3.9 \times 10^7$$

From Eq. (1.147):

$$GrPr = 3.9 \times 10^7 (400 - 294) \times 0.15^3$$
$$= 1.39 \times 10^7$$

From Table 1.10:

$$n = 0.25 \quad \text{and} \quad C'' = 1.32$$

Thus, in Eq. (1.149):

$$h = 1.32(400 - 294)^{0.25}(0.15)^{(3 \times 0.25)-1}$$
$$= 1.32 \times 106^{0.25} \times 0.15^{-0.25}$$
$$= 6.81\,\text{W/m}^2\,K$$

Much of the literature on free convection has been reviewed by Martynenko and Kharamstov.[60] Churchill and Usagi[70] reexamined the literature data for a vertical isothermal plate and proposed the following expression for the average Nusselt number (based on the length of the plate):

$$Nu = \frac{hL}{k} = \frac{0.67 Ra_L^{1/4}}{\left\{1 + (0.492/Pr)^{9/16}\right\}^{8/27}} \tag{1.151}$$

Eq. (1.151) is restricted to the laminar flow conditions $\left(10^5 \leq Ra_L \leq 10^9\right)$ but is valid for all Prandtl numbers. At a critical value of the Rayleigh number, the laminar flow conditions cease to exist. The critical value of the Rayleigh number depends upon the Prandtl number. For air $(Pr = 0.71)$, it is about 4×10^8 whereas for high Prandtl numbers, $Pr = 10^3 - 10^4$, it occurs at about $Ra_L \sim 10^{13}$.

Similarly, for an isothermal horizontal cylinder, the following expression[71] has gained wide acceptance in the literature:

$$Nu = \left[0.60 + 0.387Ra^{1/6} \left\{ 1 + \left(\frac{0.559}{Pr} \right)^{9/16} \right\}^{-8/27} \right]^2 \tag{1.152}$$

Eq. (1.152) can be used for both laminar and turbulent flow regimes for a long cylinder.

For an isothermal sphere, the average Nusselt number is given by the following expression due to Jafarpur and Yovanovich[72]:

$$Nu = \frac{hd}{k} = \frac{0.589Ra^{1/4}}{\left[1 + \left({}^{0.5}/_{Pr} \right)^{9/16} \right]^{4/9}} \tag{1.153}$$

Example 1.21

Rework the Example 1.20 using Eq. (1.151).

Solution

Using the values of $Ra = GrPr = 1.39 \times 10^7$ and $Pr = 0.71$ in Eq. (1.151):

$$Nu = \frac{hd}{k} = 31.2$$

$$\therefore \ h = \frac{31.2 \times 0.031}{0.15} = 6.45 \text{W/m}^2 \text{K}$$

This value is quite close to that calculated above in Example 1.20.

Example 1.22

A steel pipe of internal and outer diameters of 150 and 170 mm is used to carry superheated steam (at 4000 kPa and 350°C) at a velocity of 2.5 m/s. In order to reduce the heat loss to the atmosphere (ambient temperature of 25°C), the pipe is insulated with a 35 mm thick layer of a material (thermal conductivity of 0.075 W/m K). The pipe is situated in an underground passage so that the effect of wind velocity can be neglected, but the pipe is losing heat by free convection. Estimate the temperature of the exterior of the insulating layer and the rate of heat loss from the pipe.

Solution

In this case, there are four thermal resistances in series: forced convection inside the pipe (R_1), conduction through pipe wall (R_2), conduction through the insulation layer (R_3), and free convection from the outer surface to surroundings (R_4). The electrical circuit is:

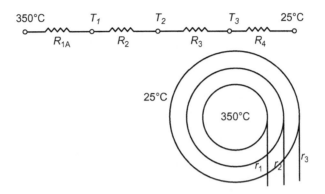

The corresponding radii are shown here:

$$r_1 = 150/2 = 75\,\text{mm}$$
$$r_2 = 170/2 = 85\,\text{mm}$$
$$r_3 = 85 + 35 = 120\,\text{mm}$$

Using the outermost area (A_0) as the reference area to calculate the overall heat transfer coefficient, U_o, the individual resistances R_1 to R_4 are expressed as follows:

$$R_1 = \frac{A_o}{A_i h_i} = \frac{r_3}{r_1 h_i}; \quad R_2 = \frac{A_o}{2\pi k_{pipe}}\ln\frac{r_2}{r_1}; \quad R_3 = \frac{A_o}{2\pi k_{in}}\ln\frac{r_3}{r_2}; \quad R_4 = \frac{A_o}{A_o h_o}$$

$$A_o = 2\pi r_3; \quad A_i = 2\pi r_1$$

In order to estimate h_i, we must calculate Re and Pr for steam at 4000 kPa and 350°C. The values of the physical properties are:

$$\text{heat capacity} = 2.50\,\text{kJ/kgK}; \quad \text{density} = 15.05\,\text{kg/m}^3$$
$$\text{viscosity} = 2.22 \times 10^{-5}\,\text{Pa s}; \quad \text{thermal conductivity} = 5.36 \times 10^{-2}\,\text{W/mK}$$
$$Pr = 1.04$$
$$Re = (15.05 \times 2.5 \times 0.15)/(2.22 \times 10^{-5}) = 2.55 \times 10^5$$

Thus, the flow is turbulent and we can use Eq. (1.92):

$$Nu = 0.023 Re^{0.8} Pr^{0.33}$$

Substituting values and solving for Nu:

$$Nu = 0.023\left(2.55 \times 10^5\right)^{0.8}(1.04)^{0.33}$$
$$= 492$$

$$\therefore \; h_i = \frac{Nu\,k}{d} = \frac{492 \times 5.36 \times 10^{-2}}{0.15} = 175.6\,\text{W/m}^2\text{K}$$

Similarly, we must evaluate h_o by using either Eq. (1.146) or Eq. (1.152); However, since temperature T_3 is unknown, the value of Gr cannot be calculated. Also, both Eqs. (1.146), (1.152) require the physical properties of air at the mean temperature $(T_3 + 25)/2$. Thus, an iterative procedure is used here. Let us assume $h_o = 6\ \text{W/m}^2\ \text{K}$ (based on our experience in Examples 1.20 and 1.21). We can now evaluate various thermal resistances as:

$$R_1 = \frac{r_3}{r_1 h_i} = \frac{0.12}{0.075 \times 175.6} = 9.11 \times 10^{-3}\frac{\text{m}^2\,\text{K}}{\text{W}}$$

$$R_2 = \frac{r_3}{k_{pipe}}\ln\frac{r_2}{r_1} = \frac{0.12}{53}\ln\frac{0.085}{0.075} = 2.83 \times 10^{-4}\frac{\text{m}^2\,\text{K}}{\text{W}}$$

$$\left(k_{pipe} = 53\,\text{W/m}^2\,\text{K}\right)$$

$$R_3 = \frac{r_3}{k_{in}}\ln\frac{r_3}{r_2} = \frac{0.12}{0.075}\ln\frac{0.120}{0.085} = 0.552\frac{\text{m}^2\,\text{K}}{\text{W}}$$

$$R_4 = \frac{1}{h_o} = \frac{1}{6} = 0.167\frac{\text{m}^2\,\text{K}}{\text{W}}$$

Note that R_3 and R_4 are the dominant resistances here. The overall heat transfer coefficient (based on the outermost area) is obtained as:

$$U_o = \frac{1}{R_1 + R_2 + R_3 + R_4} = \frac{1}{9.11 \times 10^{-3} + 2.83 \times 10^{-4} + 0.552 + 0.167}$$

or

$$U_o = 1.38\,\text{W/m}^2\,\text{K}$$

One can now write the rate of heat transfer per unit pipe length using U_o and h_o which must be equal:

$$6 \times \pi \times 0.24 \times (T_3 - 25) = 1.38 \times \pi \times 0.24(350 - 25)$$

$$\text{i.e. } T_3 = 99.75°C$$

Now we can proceed to use Eq. (1.152) to calculate the value of h_o:

At the mean temperature $(99.75 + 25)/2 = 62.4°C$, the properties of air are:

$$\rho = 1.052\,\text{kg/m}^3;\ \mu = 2.02 \times 10^{-5}\,\text{Pa s};\ Cp = 1008\,\text{J/kgK}$$

$$k = 0.029\,\text{W/mK};\ Pr = 0.709$$

$$\therefore\ Ra = Gr \cdot Pr = \left((g\beta\Delta Td^3)/(\mu/\rho)^2\right)(0.709)$$

$$\beta = 1/(273 + 25) = (1/298)\,\text{K}^{-1}$$

$$Ra = 0.709\left(\frac{9.81 \times \dfrac{1}{298} \times (99.75 - 25) \times 0.24^3}{\left(2.02 \times 10^{-5}/1.052\right)^2}\right)$$

$$Ra = 6.54 \times 10^7$$

Substituting these values in Eq. (1.152) and solving for Nu gives:

$$Nu = 49.7$$

Or,

$$h_o = (49.7 \times 0.029)/0.24 = 6.0 W/m^2 K$$

This value matches with the assumed value of $h_o = 6$ W/m^2 K and therefore, further iteration is not needed in this case.

Thus, the temperature of the outer pipe surface is 99.75 i.e. $\sim$100°C and the rate of heat loss per m:

$$Q = 1.38 \times \pi \times 0.24 \times (350 - 25) = \underline{\underline{338 W/m}}$$

Fluids between two surfaces

For the transfer of heat from a hot surface across a thin layer of fluid to a parallel cold surface:

$$\frac{Q}{Q_k} = \frac{h\Delta T}{(k/x)\Delta T} = \frac{hx}{k} = Nu \tag{1.154}$$

where Q_k is the rate at which heat would be transferred by pure thermal conduction between the layers, a distance x apart, and Q is the actual rate.

For $(Gr\,Pr) = 10^3$, the heat transferred is approximately equal to that due to conduction alone, though for $10^4 < Gr\,Pr < 10^6$, the heat transferred is given by:

$$\frac{Q}{Q_k} = 0.15(Gr\,Pr)^{0.25} \tag{1.155}$$

which is noted in Fig. 1.41. In this equation the characteristic dimension to be used for the Grashof and Nusselt numbers is x, the distance between the planes, and the heat transfer is independent of surface area, provided that the linear dimensions of the surfaces are large compared with x. For higher values of $(Gr\,Pr)$, Q/Q_k is proportional to $(Gr\,Pr)^{1/3}$, showing that the heat transferred is not entirely by convection and is not influenced by the distance x between the surfaces.

A similar form of analysis has been given by Kraussold[73] for air between two concentric cylinders. It is important to note from this general analysis that a single layer of air will not be a good insulator because convection currents set in before it becomes 25 mm thick. The good insulating properties of porous materials are attributable to the fact that they offer a series of very thin layers of air in which convection currents are not present.

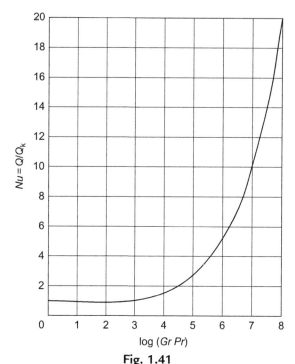

Fig. 1.41
Natural convection between surfaces.

1.5 Heat Transfer by Radiation

1.5.1 Introduction

It has been seen that heat transfer by conduction takes place through either a solid or a stationary fluid and heat transfer by convection takes place as a result of either forced or natural movement of a hot fluid. The third mechanism of heat transfer, radiation, can take place without either a solid or a fluid being present, that is through a vacuum, although many fluids are transparent to radiation, and it is generally assumed that the emission of thermal radiation is by 'waves' of wavelengths in the range 0.1–100 μm which travel in straight lines. This means that direct radiation transfer, which is the result of an interchange between various radiating bodies or surfaces, will take place only if a straight line can be drawn between the two surfaces; a situation which is often expressed in terms of one surface 'seeing' another. Having said this, it should be noted that opaque surfaces sometimes cast shadows which inhibit radiation exchange and that indirect transfer by radiation can take place as a result of partial reflection from other surfaces. Although all bodies at temperatures in excess of absolute zero radiate energy in all directions, radiation is of special importance from bodies at high temperatures such as those encountered in furnaces, boilers, and high temperature reactors, where in addition to radiation from hot surfaces, radiation from reacting flame gases may also be a consideration.

1.5.2 Radiation From a Black Body

In thermal radiation, a so-called *black body* absorbs all the radiation falling upon it, regardless of wavelength and direction, and for a given temperature and wavelength, no surface can emit more energy than a black body. The radiation emitted by a black body, whilst a function of wavelength and temperature, is regarded as *diffuse*, that is, it is independent of direction. In general, most rough surfaces and indeed most engineering materials may be regarded as being diffuse. A black body, because it is a perfect emitter or absorber, provides a standard against which the radiation properties of real surfaces may be compared.

If the *emissive power E* of a radiation source—that is the energy emitted per unit area per unit time—is expressed in terms of the radiation of a single wavelength λ, then this is known as the *monochromatic* or *spectral emissive power* E_λ, defined as that rate at which radiation of a particular wavelength λ is emitted per unit surface area, per unit wavelength in all directions. For a black body at temperature T, the spectral emissive power of a wavelength λ is given by *Planck's Distribution Law*:

$$E_{\lambda,b} = C_1 / \left[\lambda^5 \left(\exp\left(C_2 / \lambda T \right) - 1 \right) \right] \tag{1.156}$$

where, in SI units, $E_{\lambda,b}$ is in W/m^3 and $C_1 = 3.742 \times 10^{-16}$ W/m^2 and $C_2 = 1.439 \times 10^{-2}$ m K are the respective radiation constants. Eq. (1.156) permits the evaluation of the emissive power from a black body for a given wavelength and absolute temperature and values obtained from the equation are plotted in Fig. 1.42 which is based on the work of Incropera and de Witt[74] and of others.[75] It may be noted that, at a given wavelength, the radiation from a black body increases with temperature and that in general, short wavelengths are associated with high temperature sources.

Example 1.23

What is the temperature of a surface coated with carbon black if the emissive power at a wavelength of 1.0×10^{-6} m is 1.0×10^9 W/m^3? How would this be affected by a +2% error in the emissive power measurement?

Solution
From Eq. (1.156)

$$\exp\left(C_2 / \lambda T \right) = \left[\left(C_1 / E_{\lambda,b} \lambda^5 \right) + 1 \right]$$

or

$$\exp\left(1.439 \times 10^{-2} / \left(1.0 \times 10^{-6} T \right) \right) = \left[3.742 \times 10^{-16} / \left(1 \times 10^9 \times \left(1.0 \times 10^{-6} \right)^5 \right) \right] + 1$$
$$= 3.742 \times 10^5$$

Thus:

$$\left(1.439 \times 10^4 \right) T = \ln\left(3.742 \times 10^5 \right) = 12.83$$

and:

$$T = \left(1.439 \times 10^4\right)/12.83 = \underline{1121\,\text{K}}$$

With an error of +2%, the correct value is given by:

$$E_{\lambda,b} = (100 - 2)\left(1 \times 10^9\right)/100 = 9.8 \times 10^8\,\text{W/m}^3$$

In Eq. (1.156):

$$9.8 \times 10^8 = \left(3.742 \times 10^{-16}\right)/\left[\left(1 \times 10^{-6}\right)^5\left(\exp\left(1.439 \times 10^{-2}/\left(1.0 \times 10^{-6}\,T\right)\right) - 1\right)\right]$$

and:

$$\underline{T = 1120\,\text{K}}$$

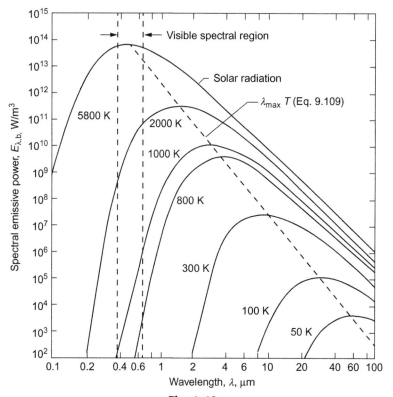

Fig. 1.42

Spectral black-body emissive power.[15,74]

Thus, the error in the calculated temperature of the surface is only 1 K.

The wavelength at which maximum emission takes place is related to the absolute temperature by *Wein's Displacement Law*, which states that the wavelength for maximum emission varies inversely with the absolute temperature of the source, or:

$$\lambda_{max}T = \text{constant}, \quad C_3 \left(= 2.898 \times 10^{-3}\,\text{mK in SI units}\right) \tag{1.157}$$

Thus, combining Eqs. (1.156), (1.157):

$$E_{\lambda\,max,b} = C_1 / \left[(C_3/T)^5 [\exp(C_2/C_3) - 1] \right]$$

or:

$$E_{\lambda\,max,b} = C_4 T^5 \tag{1.158}$$

where, in SI units, the fourth radiation constant, $C_4 = 12.86 \times 10^{-6}\,\text{W/m}^3\,\text{K}^5$. Values of the maximum emissive power are shown by the broken line in Fig. 1.42.

An interesting feature of Fig. 1.42 is that it illustrates the well-known *greenhouse effect* which depends on the ability of glass to transmit radiation from a hot source over only a limited range of wavelengths. This warms the air in the greenhouse, though to a much lower temperature than that of the external source, the sun; a temperature at which the wavelength will be much longer, as seen from Fig. 1.42, and one at which the glass will not transmit radiation. In this way, radiation outwards from within a greenhouse is considerably reduced and the air within the enclosure retains its heat. Much the same phenomenon occurs in the gases above the earth's surface which will transmit incoming radiation from the sun at a given wavelength though not radiation from the earth which, because it is at a lower temperature, emits at a longer wavelength. The passage of radiation through gases and indeed through glass is one example of a situation where the transmissivity **t**, discussed in Section 1.5.3, is not zero.

The *total emissive power E* is defined as the rate at which radiation energy is emitted per unit time per unit area of surface over all wavelengths and in all directions. This may be determined by making a summation of all the radiation at all wavelengths, by determining the area corresponding to a particular temperature under the Planck distribution curve, Fig. 1.42. In this way, from Eq. (1.156), the total emissive power is given by:

$$E_b = \int_0^\infty C_1 \mathrm{d}\lambda / \left[\lambda^5 (\exp(C_2/\lambda T) - 1) \right] \tag{1.159a}$$

which is known as the *Stefan–Boltzmann Law*. This may be integrated for any constant value of T to give:

$$E_b = \sigma T^4 \tag{1.159b}$$

where in SI units, the Stefan–Boltzmann constant $\sigma = 5.67 \times 10^{-8}\,\text{W/m}^2\,\text{K}^4$.

Example 1.24

Electrically-heated carbide elements, 10 mm in diameter and 0.5 m long, radiating essentially as black bodies, are to be used in the construction of a heater in which thermal radiation from the surroundings is negligible. If the surface temperature of the carbide is limited to 1750 K, how many elements are required to provide a radiated thermal output of 500 kW?

Solution
From Eq. (1.159b), the total emissive power is given by:

$$E_b = \sigma T^4 = (5.67 \times 10^{-8} \times 1750^4) = 5.32 \times 10^5 \, \text{W/m}^2$$

$$\text{The area of one element} = \pi(10/1000)0.5 = 1.571 \times 10^{-2} \, \text{m}^2$$

and:

$$\text{Power dissipated by one element} = (5.32 \times 10^5 \times 1.571 \times 10^{-2}) = 8.367 \times 10^3 \, \text{W}$$

Thus:

$$\text{Number of elements required} = (500 \times 1000)/(8.357 \times 10^3) = 59.8 \, \text{say} \, \underline{\underline{60}}$$

1.5.3 Radiation From Real Surfaces

The *emissivity* of a material is defined as the ratio of the radiation per unit area emitted from a 'real' or from a grey surface (one for which the emissivity is independent of wavelength) to that emitted by a black body at the same temperature. Emissivities of 'real' materials are always less than unity and they depend on the type, condition, and roughness of the material, and possibly on the wavelength and direction of the emitted radiation as well. For perfectly diffuse surfaces emissivities are independent of direction, but for real surfaces there are variations with direction and the average value is known as the *hemispherical emissivity*. For a particular wavelength λ, this is given by:

$$\mathbf{e}_\lambda = E_\lambda / E_b \tag{1.160}$$

and, similarly, the total hemispherical emissivity, an average over all wavelengths, is given by:

$$\mathbf{e} = E / E_b \tag{1.161}$$

Eq. (1.161) leads to *Kirchoff's Law* which states that the absorptivity, or fraction of incident radiation absorbed, and the emissivity of a surface are equal. If two bodies **A** and **B** of areas A_1 and A_2 are in a large enclosure from which no energy is lost, then the energy absorbed by **A** from the enclosure is $A_1 \mathbf{a}_1 I$ where I is the rate at which energy is falling on unit area of **A** and $\mathbf{a}_1$ is the absorptivity. The energy given out by **A** is $E_1 A_1$ and, at equilibrium, these two quantities will be equal or:

$$I A_1 \mathbf{a}_1 = A_1 E_1$$

and, for **B**:

$$IA_1\mathbf{a}_1 = A_2E_2$$

Thus:

$E_1/\mathbf{a}_1 = E_2/\mathbf{a}_2 = E/\mathbf{a}$ for any other body.

Since $E/\mathbf{a} = E_b/\mathbf{a}_b$, then, from Eq. (1.161):

$$\mathbf{e} = E/E_b = \mathbf{a}/\mathbf{a}_b$$

and, as by definition, $\mathbf{a}_b = 1$, the emissivity of any body is equal to its absorptivity, or:

$$\mathbf{e} = \mathbf{a} \qquad (1.162)$$

For most industrial, nonmetallic surfaces, and nonpolished metals, **e** is usually about 0.9, although values as low as 0.03 are more usual for highly polished metals such as copper or aluminium. As explained later, a small cavity in a body acts essentially as a black body with an effective emissivity of unity. The variation of emissivity with wavelength is illustrated in Fig. 1.43[74] and typical values are given in Table 1.11 which is based on the work of Hottel

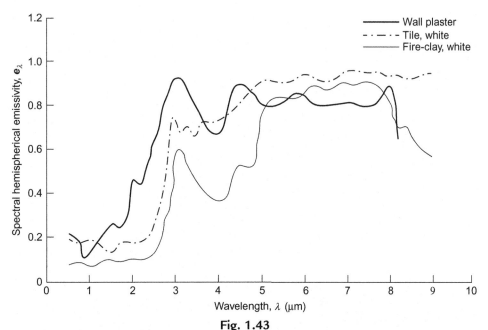

Fig. 1.43

Spectral emissivity of nonconductors as a function of wavelength.[15]

Table 1.11 Typical emissivity values[75]

Surface		T (K)	Emissivity e
(A) Metals and metallic oxides			
Aluminium	Polished plate	296	0.040
	Rough plate	299	0.055
Brass	Polished	311–589	0.096
Copper	Polished	390	0.023
	Plate, oxidised	498	0.78
Gold	Highly polished	500–900	0.018–0.35
Iron and steel	Polished iron	700–1300	0.144–0.377
	Cast iron, newly turned	295	0.435
	Smooth sheet iron	1172–1311	0.55–0.60
	Sheet steel, oxidised	295	0.657
	Iron	373	0.736
	Steel plate, rough	311–644	0.94–0.97
Lead	Pure, unoxidised	400–500	0.057–0.075
	Grey, oxidised	297	0.281
Mercury		273–373	0.09–0.12
Molybdenum	Filament	1000–2866	0.096–0.292
Monel	Metal oxidised	472–872	0.41–0.46
Nickel	Polished	500–600	0.07–0.087
	Wire	460–1280	0.096–0.186
	Plate, oxidised	472–872	0.37–0.48
Nickel alloys	Chromonickel	325–1308	0.64–0.76
	Nickelin, grey oxidised	294	0.262
Platinum	Pure, polished plate	500–900	0.054–0.104
	Strip	1200–1900	0.12–0.17
	Filament	300–1600	0.036–0.192
	Wire	500–1600	0.073–0.182
Silver	Polished	310–644	0.0221–0.0312
Tantalum	Filament	1600–3272	0.194–0.31
Tin	Bright tinned iron sheet	298	0.043–0.064
Tungsten	Filament	3588	0.39
Zinc	Pure, polished	500–600	0.045–0.053
	Galvanised sheet	297	0.276
(B) Refractories, building materials, coatings, paints, etc.			
Asbestos	Board	297	0.96
Brick	Red, rough	294	0.93
	Silica, unglazed	1275	0.80
Carbon	Filament	1311–1677	0.526
	Candle soot	372–544	0.952
	Lampblack	311–644	0.945
Enamel	White fused on iron	292	0.897
Glass	Smooth	295	0.937
Paints, lacquers	Snow-white enamel	296	0.906
	Black, shiny lacquer	298	0.875
	Black matt shellac	350–420	0.91
Plaster	Lime, rough	283–361	0.91
Porcelain	Glazed	295	0.924

Table 1.11 Typical emissivity values—cont'd

Surface		*T* (K)	Emissivity e
Refractory materials	Poor radiators	872–1272	0.65–0.75
	Good radiators	872–1272	0.80–0.90
Rubber	Hard, glossy plate	296	0.945
	Soft, grey, rough	298	0.859
Water		273–373	0.95–0.963

and Sarofim.[75] More complete data are available in Appendix Al, Table 10. If Eq. (1.160) is written as:

$$E_\lambda = \mathbf{e}_\lambda E_{\lambda,b} \tag{1.163}$$

then the spectral emissive power of a grey surface may be obtained from the spectral emissivity, $\mathbf{e}_\lambda$ and the spectral emissive power of a black body $E_{\lambda,b}$ is given by Eq. (1.156). As shown in Fig. 1.44, for a temperature of 2000 K for example, the emission curve for a real material may have

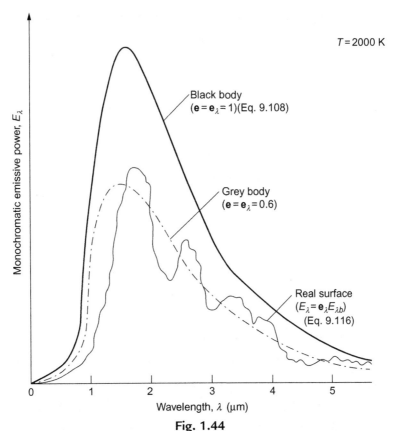

Fig. 1.44

Comparison of black body, grey body, and real surface radiation at 2000 K.[15]

a complex shape because of the variation of emissivity with wavelength, If, however, the ordinate of the black body curve $E_{\lambda,b}$ at a particular wavelength is multiplied by the spectral emissivity of the source at that wavelength, the ordinates on the curve for the real surface are obtained, and the total emissive power of the real surface is obtained by integrating E_λ over all possible wavelengths to give:

$$E = \int_0^\infty E_\lambda d\lambda = \int_0^\infty e_\lambda E_{\lambda,b} d\lambda \qquad (1.164)$$

This integration may be carried out numerically or graphically, though this approach, which has been considered in some detail by Incropera and de Witt,[12] can be difficult, especially where the spectral distribution of radiation arrives at a surface of complex structure. The amount of calculation involved cannot often be justified in practical situations and it is more usual to use a mean spectral emissivity for the surface which is assumed to be constant over a range of wavelengths. Where the spectral emissivity does not vary with wavelength then the surface is known as a *grey body* and, for a diffuse grey body, from Eqs. (1.159b), (1.161):

$$E = e E_b = e\sigma T^4 \qquad (1.165)$$

In this way, the emissive power of a grey body is a constant proportion of the power emitted by the black body, resulting in the curve shown in Fig. 1.44 where, for example, $e = 0.6$. The assumption that the surface behaves as a grey body is valid for most engineering calculations if the value of emissivity is taken as that for the dominant temperature of the radiation.

From Eq. (1.164), it is seen that the rate of heat transfer by radiation from a hot body at temperature T_1 to a cooler one at temperature T_2 is then given by:

$$q = Q/A = e\sigma \left(T_1^4 - T_2^4\right) = e\sigma (T_1 - T_2)\left(T_1^3 + T_1^2 T_2 + T_1 T_2^2 + T_2^3\right)$$

The quantity $q/(T_1 - T_2)$ is a heat transfer coefficient as used in convective heat transfer, and here it may be designated h_r, the heat transfer coefficient for radiation heat transfer where:

$$h_r = q/(T_1 - T_2) = \frac{e\sigma\left(T_1^4 - T_2^4\right)}{T_1 - T_2} = e\sigma\left(T_1^3 + T_1^2 T_2 + T_1 T_2^2 + T_3^3\right) \qquad (1.166)$$

It may be noted that if $(T_1 - T_2)$ is very small, that is T_1 and T_2 are virtually equal, then:

$$h_r = 4e\sigma T^3$$

Example 1.25

What is the emissivity of a grey surface, 10 m^2 in area, which radiates 1000 kW at 1500 K? What would be the effect of increasing the temperature to 1600 K?

Solution

The emissive power

$$E = (1000 \times 1000)/10 = 100{,}000 \, \text{W/m}^2$$

From Eq. (1.165):

$$e = E/\sigma T^4$$
$$= 100{,}000/(5.67 \times 10^{-8} \times 1500^4) = \underline{0.348}$$

At 1600 K:

$$E = \mathbf{e}\sigma T^4$$
$$= (0.348 \times 5.67 \times 10^{-8} \times 1600^4) = \underline{1295 \, \text{kW}}$$

an increase of 29.5% for a 100 deg K increase in temperature.

In a real situation, radiation incident upon a surface may be absorbed, reflected, and transmitted and the properties of *absorptivity*, *reflectivity*, and *transmissivity* may be used to describe this behaviour. In theory, these three properties will vary with the direction and wavelength of the incident radiation, although, with diffuse surfaces, directional variations may be ignored and mean, *hemispherical* properties used.

If the *absorptivity* **a**, the fraction of the incident radiation absorbed by the body is defined on a spectral basis, then:

$$\mathbf{a}_\lambda = I_{\lambda, \text{abs}}/I_\lambda \tag{1.167}$$

and the total absorptivity, the mean over all wavelengths, is defined as:

$$\mathbf{a} = I_{\text{abs}}/I \tag{1.168}$$

Since a black body absorbs all incident radiation then for a black body:

$$\mathbf{a}_\lambda = \mathbf{a} = 1$$

The absorptivity of a grey body is therefore less than unity.

In a similar way, the *reflectivity*, **r**, the fraction of incident radiation which is reflected from the surface, is defined as:

$$\mathbf{r} = I_{\text{ref}}/I \tag{1.169}$$

and the *transmissivity*, **t**, the fraction of incident radiation which is transmitted through the body, as:

$$\mathbf{t} = I_{\text{trans}}/I \tag{1.170}$$

Since, as shown in Fig. 1.45, all the incident radiation is absorbed, reflected, or transmitted, then:

$$I_{\text{abs}} + I_{\text{ref}} + I_{\text{trans}} = I$$

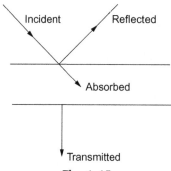

Fig. 1.45

Reflection, absorption, and transmission of radiation.

or:

$$\mathbf{a} + \mathbf{r} + \mathbf{t} = 1$$

Since most solids are opaque to thermal radiation, $\mathbf{t} = 0$ and therefore:

$$\mathbf{a} + \mathbf{r} = 1 \qquad (1.171)$$

Kirchoff's Law, discussed previously, states that, at any wavelength, the emissivity and the absorptivity are equal. If this is extended to total properties, then, at a given temperature:

$$\mathbf{e} = \mathbf{a} \quad (\text{Eq.}\,9.162)$$

For a grey body, the emissivity and the absorptivity are, by definition, independent of temperature and hence Eq. (1.162) may be applied more generally showing that, where one radiation property (**a**, **r** or **e**) is specified for an opaque body, the other two may be obtained from Eqs. (1.162), (1.171). Kirchoff's Law explains why a cavity with a small aperture approximates to a black body in that radiation entering is subjected to repeated internal absorption and reflection so that only a negligible amount of the incident radiation escapes through the aperture. In this way, $\mathbf{a} = \mathbf{e} = 1$ and at T K, the emissive power of the aperture is σT^4.

1.5.4 Radiation Transfer Between Black Surfaces

Since radiation arriving at a black surface is completely absorbed, no problems arise from multiple reflections. Radiation is emitted from a diffuse surface in all directions and therefore only a proportion of the radiation leaving a surface arrives at any other given surface. This proportion depends on the relative geometry of the surfaces and this may be taken into account by the *view factor*, *shape factor*, or *configuration F*, which is normally written as F_{ij} for radiation arriving at surface j from surface i. In this way, F_{ij}, which is, of course, completely independent of the surface temperature, is the fraction of radiation leaving surface i which is directly intercepted by surface j.

If radiant heat transfer is taking place between two black surfaces, 1 and 2, then:

$$\text{radiation emitted by surface } 1 = A_1 E_{b1}$$

where A_1 and E_{b1} are the area and black body emissive power of surface 1, respectively. The fraction of this radiation which arrives at and is totally absorbed by surface 2 is F_{12} and the heat transferred is then:

$$Q_{1\to2} = A_1 F_{12} E_{b1}$$

Similarly, the radiation leaving surface 2 which arrives at 1 is given by:

$$Q_{2\to1} = A_2 F_{21} E_{b2}$$

and the net radiation transfer between the two surfaces is $Q_{12} = (Q_{1\to2} - Q_{2\to1})$ or:

$$Q_{12} = A_1 F_{12} E_{b1} - A_2 F_{21} E_{b2}$$
$$= \sigma A_1 F_{12} T_1^4 - \sigma A_2 F_{21} T_2^4 \tag{1.172}$$

When the two surfaces are at the same temperature, $T_1 = T_2, Q_{12} = 0$ and thus:

$$Q_{12} = 0 = \sigma T_1^4 (A_1 F_{12} - A_2 F_{21})$$

Since the temperature T_1 can have any value so that, in general $T_1 \neq 0$, then:

$$A_1 F_{12} = A_2 F_{21} \tag{1.173}$$

Eq. (1.173), known as the *reciprocity relationship* or *reciprocal rule*, then leads to the equation:

$$Q_{12} = \sigma A_1 F_{12} \left(T_1^4 - T_2^4 \right) = \sigma A_2 F_{21} \left(T_1^4 - T_2^4 \right) \tag{1.174}$$

The product of an area and an appropriate view factor is known as the *exchange area*, which in SI units, is expressed in m^2. In this way, $A_1 F_{12}$ is known as exchange area 1–2.

Example 1.26

Calculate the view factor, F_{21} and the net radiation transfer between two black surfaces, a rectangle 2 m by 1 m (area A_1) at 1500 K and a disc 1 m in diameter (area A_2) at 750 K, if the view factor, $F_{12} = 0.25$.

Solution

$$A_1 = (2 \times 1) = 2\,m^2$$
$$A_2 = \left(\pi \times 1^2 \right)/4 = 0.785\,m^2$$

From Eq. (1.173):

$$A_1 F_{12} = A_2 F_{21}$$

or:

$$(2 \times 0.25) = 0.785 F_{21}$$

and:

$$F_{21} = \underline{\underline{0.637}}$$

Using Eq. (1.174):

$$Q_{12} = \sigma A_1 F_{12}\left(T_1^4 - T_2^4\right)$$
$$Q_{12} = \left(5.67 \times 10^{-8} \times 2 \times 0.25\right)\left(1500^4 - 750^4\right)$$
$$= 5.38 \times 10^5\,\text{W or } \underline{538\ \text{kW}}$$

View factors, the values of which determine heat transfer rates, are dependent on the geometrical configuration of each particular system. As a simple example, radiation may be considered between elemental areas dA_1 and dA_2 of two irregular-shaped flat bodies, well separated by a distance L between their mid-points as shown in Fig. 1.46. If α_1 and α_2 are the angles between the imaginary line joining the mid-points and the normals, the rate of heat transfer is then given by:

$$Q_{12} = \sigma\left(T_1^4 - T_2^4\right)\int_0^{A_1}\int_0^{A_2}\left(\cos\alpha_1\cos\alpha_2\,dA_1\,dA_2\right)/\pi L^2 \qquad (1.175)$$

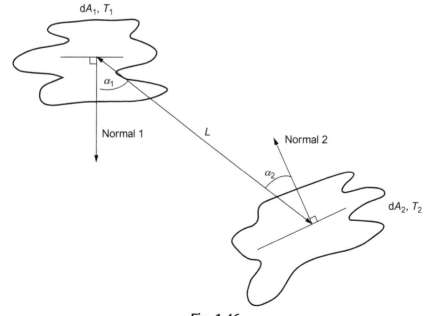

Fig. 1.46
Determination of view factor.

Eq. (1.175) may be extended to much larger surfaces by subdividing these into a series of smaller elements, each of exchange area $A_i F_{ij}$, and summing the exchange areas between each pair of elements to give:

$$A_i F_{ij} = A_j F_{ji} = \int_{Ai} \int_{Aj} \left(\cos \alpha_i \cos \alpha_j dA_i dA_j \right) / \pi L^2 \tag{1.176}$$

In this procedure, the value of the integrand can be determined numerically for every pair of elements and the double integral, approximately the sum of these values, then becomes:

$$\int_{Aj} \int_{Ai} \left(\cos \alpha_i, \cos \alpha_j dA_i dA_j \right) / \pi L^2 = \sum_{Ai} \sum_{Aj} \left(\cos \alpha_i \cos \alpha_j dA_i dA_j \right) / \pi L^2 \tag{1.177}$$

The amount of calculation involved here can be very considerable and use of a numerical method is usually required. A simpler approach is to make use of the many expressions, graphs, and tables available in the heat transfer literature. Typical data, presented by Incropera and de Witt[74] and by Hottel and Sarofim[75] are shown in Figs 1.47–1.50, where it will be seen that in many cases, the values of the view factors approach unity. This means that nearly all the radiation leaving one surface arrives at the second surface as, for example, when a sphere is contained within a second larger sphere. Note that the converse is not so, i.e. only a part of the radiation leaving the larger (outer) sphere is intercepted by the (inner) smaller sphere, and the balance falls on the larger sphere itself. Wherever a view factor approaches zero, only a negligible part of one surface can be seen by the other surface.

It is important to note here that if an element does not radiate directly to any part of its own surface, the shape factor with respect to itself, F_{11}, F_{22} and so on, is zero. This applies to any flat or convex surface for which, therefore, $F_{11} = 0$.

Example 1.27

What are the view factors, F_{12} and F_{21}, for (a) a vertical plate, 3 m high by 4 m long, positioned at right angles to one edge of a second, horizontal plate, 4 m wide and 6 m long, and (b) a 1 m diameter sphere positioned within a 2 m diameter sphere?

Solution
(a) Using the nomenclature in Fig. 1.49iii:

$$Y/X = (6/4) = 1.5 \quad \text{and} \quad Z/X = (3/4) = 0.75$$

From the figure:

$$F_{12} = \underline{\underline{0.12}}$$

From Eq. (1.173):

$$A_1 F_{12} = A_2 F_{21}$$
$$(3 \times 4)0.12 = (4 \times 6)F_{21}$$

(i) Parallel plates with mid-lines connected by perpendicular

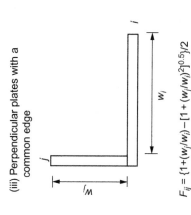

$F_{ij} = \{[(W_i + W_j)^2 + 4]^{0.5} - [(W_i - W_j)^2 + 4]^{0.5}\}/2W_i$
where: $W_i = W_i/L$ and $W_j = W_j/L$

(ii) Inclined parallel plates of equal width and a common edge

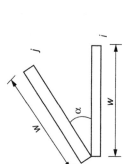

$F_{ij} = 1 - \sin(\alpha/2)$

(iii) Perpendicular plates with a common edge

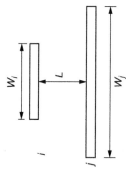

$F_{ij} = \{1 + (w_j/w_i) - [1 + (w_j/w_i)^2]^{0.5}\}/2$

(iv) Three-sided enclosure

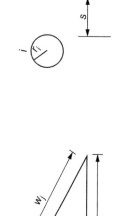

$F_{ij} = (w_i + w_j - w_k)/2w_i$

(v) Parallel cylinders of different radius

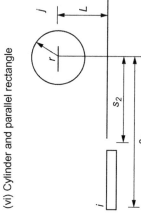

$F_{ij} = (1/2\pi)\{[-\pi + [C^2 - (R+1)^2]^{0.5} - [C^2 - (R-1)^2]^{0.5} + (R-1)\cos^{-1}[(R/C) - (1/C)] - (R+1)\cos^{-1}[(R/C) + (1/C)]\}$

where: $R = r_j/r_i$, $S = s/r_i$ and $C = 1 + R + S$

(vi) Cylinder and parallel rectangle

$F_{ij} = [r/(s_1 - s_2)][\tan^{-1}(s_1/L) - \tan^{-1}(s_2/L)]$

Fig. 1.47

View factors for two-dimensional geometries. [15]

(i) Aligned parallel rectangles

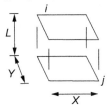

$$F_{ij} = [2/(\pi \bar{X} \bar{Y})]\{\ln[(1+\bar{X}^2)(1+\bar{Y}^2)/(1+\bar{X}^2+\bar{Y}^2)]^{0.5}$$
$$+ \bar{X}(1+\bar{Y}^2)^{0.5} \tan^{-1}[\bar{X}/(1+\bar{Y}^2)^{0.5}]$$
$$+ \bar{Y}(1+\bar{X}^2)^{0.5} \tan^{-1}[(\bar{Y}/(1+\bar{X}^2)^{0.5})] - \bar{X} \tan^{-1}\bar{X} - \bar{Y} \tan^{-1}\bar{Y}\}$$

where: $\bar{X} = X/L$ and $\bar{Y} = Y/L$

(ii) Coaxial parallel discs

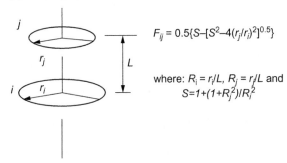

$$F_{ij} = 0.5\{S - [S^2 - 4(r_j/r_i)^2]^{0.5}\}$$

where: $R_i = r_i/L$, $R_j = r_j/L$ and
$S = 1 + (1+R_j^2)/R_i^2$

(iii) Perpendicular rectangles with a common edge

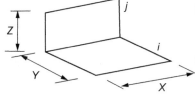

$$F_{ij} = (1/\pi W)\{W \tan^{-1}(1/W) + H \tan^{-1}(1/H) - (H^2+W^2)^{0.5}\tan^{-1}(H^2+W^2)^{-0.5}$$
$$+ 0.25\ln[(1+W^2)(1+H^2)/(1+W^2+H^2)][W^2(1+W^2+H^2)/(1+W^2)(W^2+H^2)]W^2$$
$$\times [H^2(1+H^2+W^2)/(1+H^2)(H^2+W^2)]H^2]\}$$

where: $H = Z/X$ and $W = Y/X$

Fig. 1.48
View factors for three-dimensional geometries.[15]

and:

$$F_{21} = \underline{\underline{0.06}}$$

(b) For the two spheres:

$$F_{12} = 1 \quad \text{and} \quad F_{21} = (r_1/r_2)^2 = (1/2) = \underline{\underline{0.25}}$$
$$F_{22} = 1 - (r_1/r_2)^2 = 1 - 0.25 = \underline{\underline{0.75}}$$

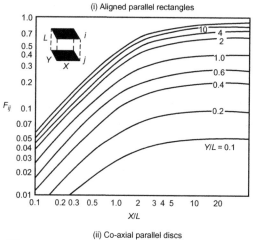

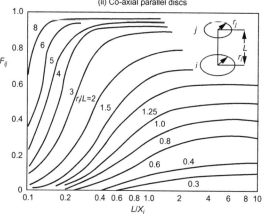

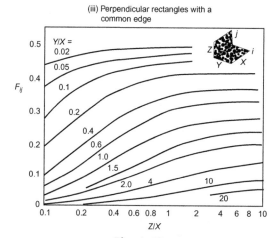

Fig. 1.49

View factors for three-dimensional geometries.[15]

(i) Two perpendicular rectangles
   — between surfaces 1 and 6

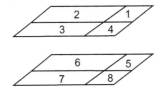

$$F_{16} = (A_6/A_1)[(1/2A_6)(A_{(1+2+3+4)}F_{(1+2+3+4)(5+6)}$$
$$+ A_6F_{6(2+4)} - A_5F_{5(1+3)} - (1/2A_6)(A_{(3+4)}F_{(3+4\times5+6)}$$
$$-A_6F_{6A} -A_5F_{53})]$$

(ii) Two parallel rectangles
   — between surfaces 1 and 7

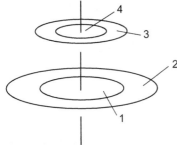

$$F_{17} = (1/4A_1)[A_{(1+2+3+4)}F_{(1+2+3+4)(5+6+7+8)} +A_1F_{15} +A_2F_{26}$$
$$+ A_3F_{37} + A_4F_{48}] - (1/4A_1)[A_{(1+2)}F_{(1+2)(5+6)} +A_{(1+4)}F_{(1+4)(5+8)}$$
$$+ A_{(3+4)}F_{(3+4)(7+8)} +A_{(2+3)}F_{(2+3)(6+7)}]$$

(iii) Two parallel circular rings
   — between surfaces 2 and 3

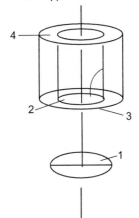

$$F_{23} = (A_{(1+2)}/A_2)[F_{(1+2)(3+4)} -F_{(1+2)4}]$$
$$+ (A_1/A_2)[F_{1(3+4)} -F_{14}]$$

(iv) A circular tube and a disc between surface 3,
    the inner wall of the tube of radius x_3 and surface 1,
    the upper surface of the disc of radius x_1.

$$F_{13} = F_{12} - F_{14}$$
$$F_{31} = (x_3^2/x_1^2)(F_{12} + F_{14})$$

Fig. 1.50
View factors obtained by using the summation rule.[75]

For a given geometry, view factors are related to each other, one example being the reciprocity relationship given in Eq. (1.173). Another important relationship is the *summation rule* which may be applied to the surfaces of a complete enclosure. In this case, all the radiation leaving one surface, say i, must arrive at all other surfaces (including itself) in the enclosure so that, for n surfaces:

$$F_{i1} + F_{i2} + F_{i3} + \cdots + F_{in} = 1$$

or:

$$F_{ij} = 1 \tag{1.178}$$

from which:

$$A_i \sum_j F_{ij} = A_i \tag{1.179}$$

This means that the sum of the exchange areas associated with a surface in an enclosure must be same as the area of that surface. The principle of the summation rule may be extended to other geometries such as, for example, radiation from a vertical rectangle (area 1) to an adjacent horizontal rectangle (area 2), as shown in Fig. 1.49iii, where they are joined to a second horizontal rectangle of the same width (area 3). In effect area 3 is an extension of area 2 but has a different view factor.

In this case:

$$A_1 F_{1(2+3)} = A_1 F_{12} + A_1 F_{13}$$

or:

$$A_1 F_{13} = A_1 F_{1(2+3)} - A_1 F_{12} \tag{1.180}$$

Eq. (1.180) allows F_{13} to be determined from the view factors F_{12} and $F_{1(2+3)}$ which can be obtained directly from Fig. 1.49iii. Typical data obtained by using this technique are shown in Fig. 1.50 which is based on the work of Howell.[76]

Example 1.28

What is the view factor F_{23} for the two parallel rings shown in Fig. 1.50iii if the inner and outer radii of the two rings are: upper$=0.2$ m and 0.3 m; lower$=0.3$ m and 0.4 m, and the rings are 0.2 m apart?

Solution
From Fig. 1.50iii:

$$F_{23} = \left(A_{(1+2)}/A_2\right)\left(F_{(1+2)(3+4)} - F_{(1+2)4}\right) - \left(A_1/A_2\right)\left(F_{1(3+4)} - F_{14}\right)$$

Laying out the data in tabular form and obtaining F from Fig. 1.49ii, then;

For.	r_i (m)	r_j (m)	L (m)	(r_j/L)	(L/r_i)	F
$F_{(1+2)(3+4)}$	0.4	0.3	0.2	1.5	0.5	0.40
$F_{(1+2)4}$	0.4	0.2	0.2	1.0	0.5	0.22
$F_{1(3+4)}$	0.3	0.3	0.2	1.5	0.67	0.55
F_{14}	0.3	0.2	0.2	1.0	0.67	0.30

$$A_{(1+2)}/A_2 = 0.4^2/(0.4^2 - 0.3^2) = 2.29$$

$$A_1/A_2 = 0.3^2/(0.4^2 - 0.3^2) = 1.29$$

and hence:

$$F_{23} = 2.29(0.40 - 0.22) + 1.29(0.55 - 0.30) = \underline{\underline{0.74}}$$

Eq. (1.174) may be extended in order to determine the net rate of radiation heat transfer from a surface in an enclosure. If the enclosure contains n black surfaces, then the net heat transfer by radiation to surface i is given by:

$$Q_i = Q_{1i} + Q_{2i} + Q_{3i} + \cdots + Q_{ni}$$

or:

$$Q_i = \sum_{j=1}^{j=n} \sigma A_j F_{ji} \left(T_j^4 - T_i^4 \right)$$

or, applying the reciprocity relationship:

$$Q_i = \sum_{j=1}^{j=n} \sigma A_i F_{ij} \left(T_j^4 - T_i^4 \right) \tag{1.181}$$

Example 1.29

A plate, 1 m in diameter at 750 K, is to be heated by placing it beneath a hemispherical dome of the same diameter at 1200 K; the distance between the plate and the bottom of the dome being 0.5 m, as shown in Fig. 1.51. If the surroundings are maintained at 290 K, the surfaces may be regarded as black bodies and heat transfer from the underside of the plate is negligible, what is the net rate of heat transfer by radiation to the plate?

Solution

Taking area 1 as that of the plate, area 2 as the underside of the hemisphere, area 3 as an imaginary cylindrical surface linking the plate and the underside of the dome which represents the black surroundings, and area 4 as an imaginary disc sealing the hemisphere and parallel to the plate then, from Eq. (1.181), the net radiation to the surface of the plate 1 is given by:

$$Q_1 = \sigma A_2 F_{21} \left(T_2^4 - T_1^4 \right) + \sigma A_3 F_{31} \left(T_3^4 - T_1^4 \right)$$

or, using the reciprocity rule:

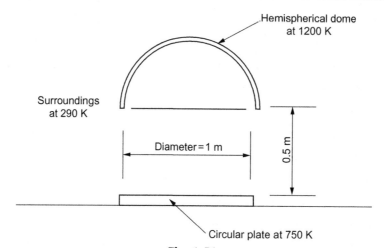

Fig. 1.51

Schematics for Example 1.29.

$$Q_1 = \sigma A_1 F_{12}\left(T_2^4 - T_1^4\right) + \sigma A_1 F_{13}\left(T_3^4 - T_1^4\right)$$

All radiation from the disc 1 to the dome 2 is intercepted by the imaginary disc 4 and hence $F_{12} = F_{14}$, which may be obtained from Fig. 1.48ii, with i and j representing areas 1 and 4 respectively. Thus:

$$R_1 = r_1/L = (0.5/0.5) = 1; R_4 = r_4/L = (0.5/0.5) = 1$$

and:

$$S = 1 + \left(1 + R_4^2\right)/\left(R_1^2\right) = 1 + (1 + 1.0)/(1.0) = 3.0$$

Thus:

$$F_{14} = 0.5\left[S - \left[S^2 - 4(r_4/r_1)^2\right]^{0.5}\right] = 0.5\left[3 - \left[(3^2 - 4(0.5/0.5)^2\right]^{0.5}\right] = 0.38$$

and:

$$F_{12} = F_{14} = 0.38$$

The summation rule states that:

$$F_{11} + F_{12} + F_{13} = 1$$

and since, for a plane surface, $F_{11} = 0$, then:

$$F_{13} = (1 - 0.38) = 0.62$$
$$A_1 = \left(\pi 1.0^2\right)/4 = 0.785\,\mathrm{m^2}$$

and hence:

$$Q_1 = \left(5.67 \times 10^{-8} \times 0.785 \times 0.38\right)\left(1200^4 - 750^4\right) + \left(5.67 \times 10^{-8} \times 0.785 \times 0.62\right)\left(290^4 - 750^4\right)$$
$$= \left(1.691 \times 10^{-8} \times 1.757 \times 10^{12}\right) - \left(2.760 \times 10^{-8} \times 3.093 \times 10^{11}\right)$$
$$= 2.12 \times 10^4\,\mathrm{W} = \underline{21.2\,\mathrm{kW}}$$

Radiation between two black surfaces may be increased considerably by introducing a third surface, which acts in effect as a reradiator. For example, if a surface 1 of area A_1 at temperature T_1 is radiating to a second surface 2 of area A_2 at temperature T_2 joined to it as shown in Fig. 1.48iii, then adding a further surface R consisting of insulating material so as to form a triangular enclosure will reduce the heat transfer to the surroundings considerably. Even though some heat will be conducted through the insulation, this will usually be small and most of the energy absorbed by the insulated surface will be reradiated back into the enclosure.

The net rate of heat transfer to surface 2 is given by:

$$Q_2 = \sigma A_1 F_{12}\left(T_1^4 - T_2^4\right) + \sigma A_R F_{R2}\left(T_R^4 - T_2^4\right) \tag{1.182}$$

where T_R is the mean temperature of the insulation, though in practice, there will be a temperature distribution across this surface. At steady-state, the net rate of radiation to surface R is equal to the heat loss from it to the surroundings, Q_{surr}, or:

$$Q_{surr} = \sigma A_1 F_{1R}\left(T_1^4 - T_R^4\right) + \sigma A_2 F_{2R}\left(T_2^4 - T_R^4\right) \tag{1.183}$$

If Q_{surr} is negligible, that is, the surface may be treated as adiabatic, then from Eq. (1.183):

$$\sigma A_1 F_{1R}\left(T_1^4 - T_R^4\right) + \sigma A_2 F_{2R}\left(T_2^4 - T_R^4\right) = 0$$

Rearranging:

$$T_R^4 = \left(A_1 F_{1R} T_1^4 + A_2 F_{2R} T_2^4\right)/\left(A_1 F_{1R} + A_2 F_{2R}\right)$$

Substituting for T_R from this equation in Eq. (1.182) and noting, from the reciprocity relationship, that $A_2 F_{2R} = A_R F_{R2}$, then:

$$Q_2 = \sigma\left(T_1^4 - T_2^4\right)\left\{A_1 F_{12} + \left[\left(1/(A_1 F_{1R}) + (1/(A_R F_{R2})\right]^{-1}\right\} \tag{1.184}$$

Example 1.30

A flat-bottomed cylindrical vessel, 2 m in diameter, containing boiling water at 373 K, is mounted on a cylindrical section of insulating material, 1 m deep and 2 m ID at the base of which is a radiant heater, also 2 m in diameter, with a surface temperature of 1500 K. If the vessel base and the heater surfaces may be regarded as black bodies and conduction though the insulation is negligible, what is the rate of radiant heat transfer to the vessel? How would this be affected if the insulation were removed so that the system was open to the surroundings at 290 K?

Solution

If area 1 is the radiant heater surface and area 2 the under-surface of the vessel, with R the insulated cylinder, then:

$$A_1 = A_2 = \left(\pi \times 2^2/4\right) = 3.14\,\mathrm{m}^2$$

and:

$$A_R = (\pi \times 2.0 \times 1.0) = 6.28\,m^2$$

From Fig. 1.49ii, with $i = 1, j = 2, r_i = 1.0\,m, r_j = 1.0\,m$ and $L = 1.0\,m$,

$$(L/r_i) = (1.0/1.0) = 1.0; \text{ and} (r_j/L) = (1.0/1.0) = 1.0$$

and:

$$F_{12} = 0.40$$

The view factor may also be obtained from Fig. 1.48ii as follows:

Using the nomenclature of Fig. 1.48:

$$R_1 = (r_1/L) = (1.0/1.0) = 1.0$$
$$R_2 = (r_2/L) = (1.0/1.0) = 1.0$$
$$S - 1 + \left[(1 + R_2^2)/R_1^2\right] = 1 + \left[(1 + 1^2)/1^2\right] = 3.0$$

and:

$$F_{12} = 0.5\left[S - \left[S^2 - 4(r_2/r_1)^2\right]^{0.5}\right] = 0.5\left[3 - \left[3^2 - \left(4 \times 1^2\right)\right]^{0.5}\right] = 0.382]$$

The summation rule states that:

$$F_{11} + F_{12} + F_{1R} = 1$$

and since, for a plane surface, $F_{11} = 0$, then: $F_{1R} = (1 - 0.382) = 0.618$

Since $A_1 = A_2$:

$$F_{21} = F_{12} \text{ and } F_{2R} = F_{1R} = 0.618$$

Also $A_R F_{R2} = A_2 F_{2R}$ and hence, from Eq. (1.184):

$$
\begin{aligned}
Q_2 &= \left[A_1 F_{12} + ((1/A_1 F_{1R}) + (1/A_2 F_{2R}))^{-1}\right]\sigma\left(T_1^4 - T_2^4\right) \\
&= \left[(3.14 \times 0.382) + [(1/(3.14 \times 0.618)] + [1/(3.14 \times 0.618)]^{-1}\right]\left(5.67 \times 10^{-8}\right)\left(1500^4 - 373^4\right) \\
&= 6.205 \times 10^5\,W \text{ or } \underline{620\,kW}
\end{aligned}
$$

If the surroundings without insulation are surface 3 at $T_3 = 290\,K$, then $F_{23} = F_{2R} = 0.618$ and, from Eq. (1.182):

$$
\begin{aligned}
Q_2 &= \sigma A_1 F_{12}\left(T_1^4 - T_2^4\right) + \sigma A_2 F_{23}\left(T_3^4 - T_2^4\right) \\
&= \left(5.67 \times 10^{-8} \times 3.14 \times 0.382\right)\left(1500^4 - 373^4\right) + \left(5.67 \times 10^{-8} \times 3.14 \times 0.618\right)\left(290^4 - 373^4\right) \\
&= 3.42 \times 10^5\,W \text{ or } \underline{342\,kW}; \text{ a reduction of } 45\%.
\end{aligned}
$$

1.5.5 Radiation Transfer Between Grey Surfaces

Since the absorptivity of a grey surface is less than unity, not all the incident radiation is absorbed and some is reflected diffusely causing multiple reflections to occur. This makes radiation between grey surfaces somewhat complex compared with black surfaces since, with grey surfaces, reflectivity as well as the geometrical configuration must be taken into account. With grey bodies, it is convenient to consider the total radiation leaving a surface Q_O, that is the radiation emitted on its own accord plus the reflected components. The equivalent flux, $Q_O/A = q_O$ is termed *radiosity* and the total radiosity Q_{Oi}, which in the SI system has the units W/m^2, is the rate at which radiation leaves per unit area of surface i over the whole span of wavelengths. If the incident radiation arriving at a grey surface i in an enclosure is Q_{Ii}, corresponding to a flux $q_{Ii} = Q_{Ii}/A_i$, then the reflected flux, that is, energy per unit area, is $r_i q_{Ii}$. The emitted flux is $e_i E_{bi} = e_i \sigma T_i^4$ and the radiosity is then given by:

$$q_{oi} = e_i E_{bi} + r_i q_{Ii} \tag{1.185}$$

The net radiation from the surface is given by:

$Q_i = $ (rate at which energy leaves the surface) $-$ (rate at which energy arrives at the surface)

or:

$$Q_i = Q_{oi} - Q_{Ii} = A_i(q_{oi} - q_{Ii}) \tag{1.186}$$

Substituting from Eq. (1.185) in Eq. (1.186) and noting that $(e_i + r_i) = 1$, then:

$$\begin{aligned} Q_i &= A_i e_i E_{bi}/r_i + (A_i/r_i)[q_{oi}(1 - e_i) - q_{oi}] \\ &= (A_i e_i/r_i)(E_{bi} - q_{oi}) \end{aligned} \tag{1.187}$$

If the temperature of a grey surface is known, then the net heat transfer to or from the surface may be determined from the value of the radiosity q_O. With regard to signs, the usual convention is that a positive value of Q_i indicates heat transfer from grey surfaces.

Example 1.31

Radiation arrives at a grey surface of emissivity 0.75 at a constant temperature of 400 K, at the rate of 3 kW/m². What is the radiosity and the net rate of radiation transfer to the surface? What coefficient of heat transfer is required to maintain the surface temperature at 300 K if the rear of the surface is perfectly insulated and the front surface is cooled by convective heat transfer to air at 295 K?

Solution
Since **e** + **r** = 1:

$$r = 0.25$$

From Eq. (1.165):

$$E_b = (5.67 \times 10^{-8} \times 400^4) = 1452\,W/m^2$$

From Eq. (1.185):

$$q_o = \mathbf{e}E_b + \mathbf{r}q_I$$
$$= (0.75 \times 1452) + (0.25 \times 3000) = \underline{\underline{1839\,W/m^2}}$$

From Eq. (1.187):

$$Q/A = q = (1.0 \times 0.75/0.25)(1452 - 1839) = \underline{\underline{-1161\,W/m^2}}$$

where the negative value indicates heat transfer to the surface.

For convective heat transfer from the surface:

$$q_c = h(T_s - T_{ambient})$$

and:

$$h_c = q_c/(T_s - T_{ambient}) = 1161/(400 - 295) = \underline{\underline{11.1 \, W/m^2 \, K}}$$

For the simplest case of a *two-surface enclosure* in which surfaces 1 and 2 exchange radiation with each other only, then, assuming $T_1 > T_2$, Q_{12} is the net rate of transfer from 1, Q_1 or the rate of transfer to 2, Q_2.

Thus:

$$Q_{12} = Q_1 = -Q_2 \tag{1.188}$$

Substituting from Eq. (1.186):

$$Q_1 = A_1(q_{O1} - q_{I1}) \tag{1.189}$$

and:

$$q_{I1}A_1 = q_{O1}A_1F_{11} + q_{O2}A_2F_{21} \tag{1.190}$$

that is:

(rate of energy incident upon surface 1) = (rate of energy arriving at surface 1 from itself)
+(rate of energy arriving at surface 1 from surface 2)

From Eqs. (1.189), (1.190) and using $A_1F_{12} = A_2F_{21}$, then:

$$Q_1 = q_{O1}(A_1 - A_1F_{11}) - q_{O2}A_1F_{12} \tag{1.191}$$

Since, by the summation rule, $(A_1 - A_1F_{11}) = A_1F_{12}$, then:

$$Q_1 = (A_1F_{12})(q_{O1} - q_{O2}) \tag{1.192}$$

From Eq. (1.187):

$$Q_1 = (A_1\mathbf{e}_1/\mathbf{r}_1)(E_{b1} - q_{O1}) \quad \text{and} \quad -Q_2 = (A_2\mathbf{e}_2/\mathbf{r}_2)(E_{b2} - q_{O2}) \tag{1.193}$$

Substituting from Eq. (1.193) into Eq. (1.192) and using the relationships in Eq. (1.188) gives:

$$Q_{12}[(1/A_1F_{12}) + (\mathbf{r}_1/A_1\mathbf{e}_1) + (\mathbf{r}_2/A_2\mathbf{e}_2)] = (E_{b1} - E_{b2})$$

and hence

$$Q_{12} = (E_{b1} - E_{b2})/[(1/A_1F_{12}) + (\mathbf{r}_1/A_1\mathbf{e}_1) + (\mathbf{r}_2/A_2\mathbf{e}_2)] \tag{1.194}$$

Since $\mathbf{r} = 1 - \mathbf{e}$, then writing $E_{b1} = \sigma T_1^4$ and $E_{b2} = \sigma T_2^4$:

$$Q_{12} = \left[\sigma\left(T_1^4 - T_2^4\right)\right] / \left[(1/A_1 F_{12}) + (1 - \mathbf{e}_1)/(A_1 \mathbf{e}_1) + (1 - \mathbf{e}_2)/(A_2 \mathbf{e}_2)\right] \qquad (1.195)$$

Eq. (1.195) is the same as Eq. (1.174) for black body exchange with two additional terms $(1 - \mathbf{e}_1)/(A_1 \mathbf{e}_1)$ and $(1 - \mathbf{e}_2)/(A_2 \mathbf{e}_2)$ introduced in the denominator for surfaces 1 and 2. One can also view the three terms in the denominator of Eq. (1.195) as the summation of three resistances in series: $1/A_1 F_{12}$ is the so-called geometric resistance whereas the other two terms $(1 - \mathbf{e}_1)/A_1 \mathbf{e}_1$ and $(1 - \mathbf{e}_2)/A_2 \mathbf{e}_2$ are known as surface substances (due to the deviation from black body radiation).

Radiation between parallel plates

For two large parallel plates of equal areas, and separated by a small distance, it may be assumed that all of the radiation leaving plate 1 falls on plate 2, and similarly all of the radiation leaving plate 2 falls on plate 1.

Thus:

$$F_{12} = F_{21} = 1$$

and:

$$A_1 = A_2$$

Substituting in Eq. (1.195):

$$Q_{12} = \frac{A_1 \sigma\left(T_1^4 - T_2^4\right)}{1 + (1 - \mathbf{e}_1)/\mathbf{e}_1 + (1 - \mathbf{e}_2)/\mathbf{e}_2}$$

and:

$$\frac{Q_{12}}{A_1} = q_{12} = \frac{\sigma\left(T_1^4 - T_2^4\right)}{\dfrac{1}{\mathbf{e}_1} + \dfrac{1}{\mathbf{e}_2} - 1} \qquad (1.196)$$

Other cases of interest include radiation between:

(i) two concentric spheres
(ii) two concentric cylinders where the length: diameter ratio is large.

In both of these cases, the inner surface 1 is convex, and all the radiation emitted by it falls on the outer surface 2.

Thus:

$$F_{12} = 1$$

and from the reciprocal rule:

$$F_{21} = F_{12}\frac{A_1}{A_2} = \frac{A_1}{A_2} \quad \text{(Eq. 9.173)}$$

Substituting in Eq. (1.195):

$$Q_{12} = \frac{A_1\sigma(T_1^4 - T_2^4)}{1 + [(1-e_1)/e_1] + [(1-e_2)/e_2]\dfrac{A_1}{A_2}}$$

$$= \frac{A_1\sigma(T_1^4 - T_2^4)}{\dfrac{1}{e_1} + \dfrac{1}{e_2}(1-e_2)\dfrac{A_1}{A_2}}$$

and:

$$\frac{Q_{12}}{A_1} = q_{12} = \frac{\sigma(T_1^4 - T_2^4)}{\dfrac{1}{e_1} + \left[\dfrac{(1-e_2)A_1}{e_2 \quad A_2}\right]} \tag{1.197}$$

For radiation from surface 1 to extensive surroundings $(A_1/A_2 \to 0)$, then:

$$q_{12} = e_1\sigma(T_1^4 - T_2^4) \tag{1.198}$$

Radiation shield

The rate of heat transfer by radiation between two surfaces may be reduced by inserting a shield, so that radiation from surface 1 does not fall directly on surface 2, but instead is intercepted by the shield at a temperature T_{sh} (where $T_1 > T_{sh} > T_2$) which then reradiates to surface 2. An important application of this principle is in a furnace where it is necessary to protect the walls from high-temperature radiation.

The principle of the radiation shield may be illustrated by considering the simple geometric configuration in which surfaces 1 and 2 and the shield may be represented by large planes separated by a small distance as shown in Fig. 1.52.

Neglecting any temperature drop across the shield (which has a surface emissivity e_{sh}), then in the steady state, the transfer rate of radiant heat to the shield from the surface 1 must equal the rate at which heat is radiated from the shield to surface 2.

Application of Eq. (1.196) then gives:

$$q_{sh} = q_{sh1}(= q_{sh2}) = \frac{\sigma(T_1^4 - T_{sh}^4)}{\left(\dfrac{1}{e_1}\right) + \left(\dfrac{1}{e_{sh}}\right) - 1} = \frac{\sigma(T_{sh}^4 - T_2^4)}{\left(\dfrac{1}{e_{sh}}\right) + \left(\dfrac{1}{e_2}\right) - 1} \tag{1.199}$$

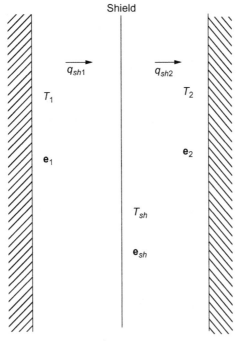

Fig. 1.52
Radiation shield.

Eliminating T_{sh}^4 in terms of T_1^4 and T_2^4 from Eq. (1.199):

$$T_1^4 - T_{sh}^4 = \frac{q_{sh}}{\sigma}\left(\frac{1}{e_1} + \frac{1}{e_{sh}} - 1\right)$$

$$T_{sh}^4 - T_2^4 = \frac{q_{sh}}{\sigma}\left(\frac{1}{e_{sh}} + \frac{1}{e_2} - 1\right)$$

Adding:

$$T_1^4 - T_2^4 = \frac{q_{sh}}{\sigma}\left(\frac{1}{e_1} + \frac{2}{e_{sh}} + \frac{1}{e_2} - 2\right)$$

and:

$$q_{sh} = \frac{\sigma\left(T_1^4 - T_2^4\right)}{\left(\dfrac{1}{e_1}\right) + \left(\dfrac{2}{e_{sh}}\right) + \left(\dfrac{1}{e_2}\right) - 2} \qquad (1.200)$$

Then from Eqs. (1.196), (1.200):

$$\frac{q_{sh}}{q_{12}} = \frac{\left(\dfrac{1}{\mathbf{e}_1}\right) + \left(\dfrac{1}{\mathbf{e}_2}\right) - 1}{\dfrac{1}{\mathbf{e}_1} + \dfrac{2}{\mathbf{e}_{sh}} + \dfrac{1}{\mathbf{e}_2} - 2} \tag{1.201}$$

For the special case where all the emissivities are equal ($\mathbf{e}_1 = \mathbf{e}_{sh} = \mathbf{e}_2$):

$$\frac{q_{sh}}{q_{12}} = 1/2$$

Similarly, it can be shown that if n shields are arranged in series, then:

$$\frac{q_{sh}}{q_{12}} = \frac{1}{n+1}$$

In practice, as a result of introducing the radiation shield, the temperature T_2 will fall because a heat balance must hold for surface 2, and the heat transfer rate from it to the surroundings will have been reduced to q_{sh}. The extent to which T_2 is reduced depends on the heat transfer coefficient between surface 2 and the surroundings.

Multisided enclosures

For the more complex case of a *multisided enclosure* formed from n surfaces, the radiosities may be obtained from an energy balance for each surface in turn in the enclosure. Thus the energy falling on a typical surface i in an enclosure formed from n surfaces is:

$$A_i q_{Ii} = q_{O1} A_1 F_{Ii} + q_{O2} A_2 F_{2i} + q_{O3} A_3 F_{3i} + \cdots + q_{On} A_n F_{ni} \tag{1.202}$$

where $A_i q_{Ii} = Q_i$ is the energy incident upon surface i, $q_{O1} A_1 F_{1i}$ is the energy leaving surface 1 which is intercepted by surface i and $q_{On} A_n F_{ni}$ is the energy leaving surface n which is intercepted by surface i. $Q_{O1} A_1 = q_{O1}$ is the energy leaving surface 1 and F_{1i} is the fraction of this which is intercepted by surface i.

From Eq. (1.185):

$$q_{Ii} = (q_{Oi} - \mathbf{e}_i E_{bi})/\mathbf{r}_i$$

Substituting for q_{Ii} into Eq. (1.202) gives:

$$\begin{aligned} A_i(q_{Oi} - \mathbf{e}_i E_{bi})/\mathbf{r}_i &= A_1 F_{1i} q_{O1} + A_2 F_{2i} q_{O2} + A_3 F_{3i} q_{O3} \\ &+ \cdots + A_j F_{ji} q_{Oi} + \cdots + A_n F_{ni} q_{On} \end{aligned} \tag{1.203}$$

Noting, for example, that for surface 2, $i = 2$, then:

$$\begin{aligned} A_2(q_{O2} - \mathbf{e}_2 E_{b2})/\mathbf{r}_2 &= A_1 F_{12} q_{O1} + A_2 F_{22} q_{O2} + A_3 F_{32} q_{O3} \\ &+ \cdots + A_j F_{j2} q_{Oj} + \cdots + A_n F_{n2} q_{On} \end{aligned} \tag{1.204}$$

Rearranging:

$$A_1 F_{12} q_{O1} + [A_2 F_{22} - (A_2/\mathbf{r}_2)] q_{O2} + A_3 F_{32} q_{O3} + \cdots + A_j F_{j2} q_{Oi}$$
$$+ \cdots + A_n F_{n2} q_{On} = (A_2 \mathbf{e}_2/\mathbf{r}_2) E_{b2} \tag{1.205}$$

Equations similar to Eq. (1.205) may be obtained for each of the surfaces in an enclosure, $i = 1, i = 2, i = 3, i = n$ and the resulting set of simultaneous equations may then be solved for the unknown radiosities, $q_{O1}, q_{O2}, \ldots, q_{On}$. The radiation heat transfer is then obtained from Eq. (1.187). This approach requires data on the areas and view factors for all pairs of surfaces in the enclosure and the emissivity, reflectivity, and the black body emissive power for each surface. Should any surface be well insulated then, in this case, $Q_i = 0$ and:

$$A_i(\mathbf{e}_i/\mathbf{r}_i)(E_{bi} - q_{Oi}) = 0$$

Since, in general, $A_i(\mathbf{e}_i/\mathbf{r}_i) \neq 0$, then $E_{bi} = q_{Oi}$.

If a surface has a specified net thermal input flux, say q_{Ii}, then, from Eq. (1.187):

$$E_{bi} = (\mathbf{r}_i/(A_i \mathbf{e}_i)) q_{Ii} + q_{Oi}.$$

It may be noted that this approach assumes that the surfaces are grey and diffuse, that emissivity and reflectivity do not vary across a surface and that the temperature, irradiation, and radiosity are constant over a surface. Since the technique uses average values over a surface, the subdivision of the enclosure into surfaces must be undertaken with care, noting that a number of surfaces may be regarded as a single surface, that it may be necessary to split one surface up into a number of smaller surfaces and also possibly to introduce an imaginary surface into the system, to represent the surroundings, for example. In a real situation, there may be both grey and black surfaces present and, for the latter, $\mathbf{r}_i$ tends to zero and $(A_i/\mathbf{r}_i)$ and $(A_i \mathbf{e}_i/\mathbf{r}_i)$ become very large.

Example 1.32

A horizontal circular plate, 1.0 m in diameter, is to be maintained at 500 K by placing it 0.20 m directly beneath a horizontal electrically heated plate, also 1.0 m in diameter, maintained at 1000 K. The assembly is exposed to black surroundings at 300 K and convection heat transfer is negligible. Estimate the electrical input to the heater and the net rate of heat transfer to the plate if the emissivity of the heater is 0.75 and the emissivity of the plate is 0.5.

Solution

Taking surface 1 as the heater, surface 2 as the heated plate, and surface 3 as an imaginary enclosure consisting of a vertical cylindrical surface representing the surroundings, then, for each surface:

Surface	$A(m^2)$	e	r	$(A/r)(m^2)$	$(Ae/r)(m^2)$
1	1.07	0.75	0.25	4.28	3.21
2	1.07	0.50	0.50	2.14	1.07
3	0.628	1.0	0		

For surface 1:

For a plane surface: $F_{11} = 0$ and $A_1 F_{11} = 0$

Using the nomenclature of Fig. 1.48:

For coaxial parallel discs with $r_1 = r_2 = 0.5$ m and $L = 0.2$ m:

$$R_1 = r_1/L = (0.5/0.20) = 2.5$$
$$R_2 = r_2/L = (0.5/0.20) = 2.5$$

and:

$$S = 1 + \left(1 + R_2{}^2\right)/R_1{}^2 = 1 + \left(1 + 2.5^2\right)/2.5^2 = 2.16$$

From Fig. 1.48ii:

$$F_{12} = 0.5\left\{S - \left[S^2 - 4(r_2/r_1)^2\right]^{0.5}\right\}$$
$$= 0.5 \times \left\{2.16 - \left[(2.16^2 - 4(0.5/0.5)^2\right]^{0.5}\right\} = 0.672$$
$$A_1 F_{12} = (1.07 \times 0.672) = 0.719\,\text{m}^2$$

and, from the summation rule:

$$A_1 F_{13} = A_1 - (A_1 F_{11} + A_1 F_{12}) = 1.07 - (0 + 0.719) = 0.350\,\text{m}^2$$

For surface 2:

For a plane surface:

$$A_2 F_{22} = 0$$

and by the reciprocity rule:

$$A_2 F_{21} = A_1 F_{12} = 0.719\,\text{m}^2$$

By symmetry:

$$A_2 F_{23} = A_1 F_{13} = 0.350\,\text{m}^2$$

For surface 3:

By the reciprocity rule:

$$A_3 F_{31} = A_1 F_{13} = 0.350\,\text{m}^2$$

and:

$$A_3 F_{32} = A_2 F_{23} = 0.350\,\text{m}^2$$

From the summation rule:

$$A_3 F_{33} = A_3 - (A_3 F_{31} + A_3 F_{32})$$
$$= 0.785 - (0.350 + 0.350) = 0.085\,\text{m}^2$$

From Eq. (1.159b):

$$E_{b1} = \sigma T_1^4 = \left(5.67 \times 10^{-8} \times 1000^4\right)$$
$$= 5.67 \times 10^4 \, W/m^2 \text{ or } 56.7 \, kW/m^2$$
$$E_{b2} = \sigma T_2^4 = \left(5.67 \times 10^{-8} \times 500^4\right)$$
$$= 3.54 \times 10^3 \, W/m^2 \text{ or } 3.54 \, kW/m^2$$
$$E_{b3} = \sigma T_3^4 = \left(5.67 \times 10^{-8} \times 300^4\right)$$
$$= 0.459 \times 10^3 \, W/m^2 \text{ or } 0.459 \, kW/m^2$$

Since surface 3 is a black body,

$$q_{O3} = E_{b3} = 0.459 \, kw/m^2$$

From Eqs. (1.204), (1.205):

$$(A_1 F_{11} - A_1/r_1)q_{O1} + A_2 F_{21}q_{O2} + A_3 F_{31}q_{O3} = -E_{b1}A_1 e_1/r_1$$
$$\times (0 - 4.28)q_{O1} + 0.719 q_{O2} + (0.350 \times 0.459)$$
$$= -(56.7 \times 1.07 \times 0.75)/0.25$$

or:

$$0.719 q_{O2} - 4.28 q_{O1} = -182 \tag{1}$$

and:

$$(A_1 F_{12}q_{O1}) +)(A_2 F_{22} - A_2/r_2)q_{O2} = -E_{b2}A_2 e_2/r_2$$
$$0.719 q_{O1} + (0 - 1.07/0.5)q_{O2} = -(3.54 \times 1.07 \times 0.5)/0.5$$

or:

$$0.719 q_{O1} - 2.14 q_{O2} = -3.79 \tag{2}$$

Solving Eqs. (1), (2) simultaneously gives:

$$q_{O_1} = 45.42 \, kW/m^2 \text{ and } q_{O_2} = 17.16 \, kW/m^2$$

power input to the heater = rate of heat transfer from the heater

From Eq. (1.187):

$$Q_1 = (A_1 e_1/r_1)(E_{b1} - q_{O1}) = (1.07 \times 0.75/0.25)(56.7 - 45.42) = \underline{36.2 \, kW}$$

Again, from Eq. (1.187), the rate of heat transfer to the plate is:

$$Q_2 = (A_2 e_2/r_2)(E_{b2} - q_{O2}) = (1.07 \times 0.5/0.25)(3.54 - 17.16) = \underline{-14.57 \, kW}$$

where the negative sign indicates heat transfer to the plate.

This section is concluded by outlining a method based on an analogy with electrical circuits when the radiation heat transfer involves two, three, or more bodies or surfaces (grey and diffuse). Based on Eq. (1.195), one can draw an equivalent electrical circuit for exchange between two radiating surfaces maintained at temperatures T_1 and T_2 respectively as (Fig. 1.53):

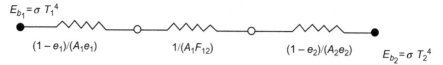

$E_{b_1} = \sigma T_1^4$

$(1-e_1)/(A_1 e_1)$ $1/(A_1 F_{12})$ $(1-e_2)/(A_2 e_2)$ $E_{b_2} = \sigma T_2^4$

Fig. 1.53
Equivalent electrical circuit for two radiating surfaces.

It is immediately seen that the heat current is given by Eq. (1.195) where $(E_{b1} - E_{b2})$ is the potential difference and $[((1-e_1)/A_1 e_1) + (1/A_1 F_{12}) + ((1-e_2)/A_2 e_2)]$ is the overall resistance. This idea can now be generalised to n—surfaces by using the ideas of radiosity (total radiation leaving a surface, q_o) and of irradiation G. The irradiation G of a surface is defined as the rate of energy received by the surface (from all directions and of all wavelengths) per unit surface area. Evidently, the value of G does not depend on the temperature of the surface, except when it receives some of its own radiation. Thus, the radiosity of the ith surface, q_{Oi}, is given as:

$$q_{Oi} = e_i \sigma T_i^4 + r_i G_i \qquad (1.206)$$

Noting that for a grey and opaque (to radiation) surface, $r_i = 1 - e_i$, Eq. (1.206) becomes:

$$q_{Oi} = e_i \sigma T_i^4 + (1-e_i) G_i \qquad (1.207)$$

If Q_i is the (nonradiative) heat supplied to maintain the ith surface at its steady temperature T_i, the overall energy balance yields:

$$Q_i = A_i(q_{Oi} - G_i) \qquad (1.208)$$

The irradiation G_i can now be obtained by considering contributions from all other surfaces of the enclosure. The energy leaving the j surface is $A_j q_{Oj}$; of this the fraction F_{ji} is intercepted by the ith surface. Thus, the total radiation incident on the ith surface (including from itself):

$$G_i A_i = \sum_{j=1}^{n} A_j q_{Oj} F_{ji} \qquad (1.209)$$

Recognising that $A_i F_{ij} = A_j F_{ji}$ and $\sum_{j=1}^{n} F_{ij} = 1$, Eq. (1.208) can now be rearranged as:

$$Q_i = A_i q_{Oi} \sum_{j=1}^{n} F_{ij} - \sum_{j=1}^{n} A_i F_{ij} q_{Oj}$$

Since $\sum_{j-1}^{n} F_{ij} = 1$, its introduction above is only for convenience.

$$\therefore \quad Q_i = \sum_{j=1}^{n} (q_{Oi} - q_{Oj}) A_i F_{ij} \qquad (1.210)$$

Elimination of G_i between Eqs. (1.207), (1.208) yields:

$$Q_i = \left(\sigma T_i^4 - q_{0i}\right)\left(\frac{e_i A_i}{1 - e_i}\right) \tag{1.211}$$

and finally, using Eqs. (1.210), (1.211):

$$\frac{\sigma T_i^4 - q_{0i}}{((1 - e_i)/A_i e_i)} = \sum_{j=1}^{n} \frac{(q_{0i} - q_{0j})}{(1/A_i F_{ij})} \tag{1.212}$$

This equation is the equivalent of the so-called Kirchoff's law for the conservation of current in electrical networks. The term on the left hand side gives the current flow from an equivalent black body (surface i) to a grey surface of radiosity q_{0i} while the right hand side is the summation of currents leaving the node q_{0i} to all other nodes. Now one can proceed to draw the corresponding electrical circuit for an enclosure consisting of three radiating surfaces 'visible' to each other as shown below in Fig. 1.54:

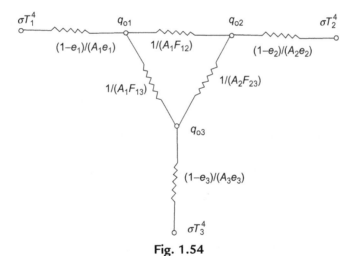

Fig. 1.54
Equivalent electrical circuit for three radiating surfaces.

Example 1.33

A circular plate (of radius 1 m) heated to a temperature of 500°C has an emissivity of $e_1 = 0.63$. It is directly placed above another plate of the same size maintained at 1000°C with an emissivity $e_2 = 0.87$. The two plates are situated 2 m away from each other. Calculate the net heat flow at each plate (only for the two facing sides) when (i) these are placed in a radiation free environment (ii) these are connected by a single adiabatic surface.

Solution

This is an enclosure consisting of 3 surfaces as shown below:

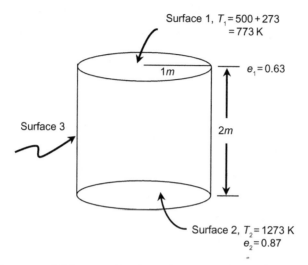

Since all three surfaces are visible to each other, the equivalent electrical circuit is:

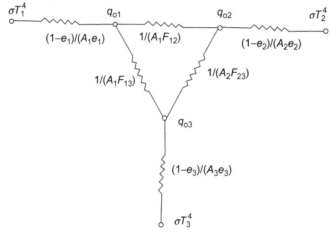

First, we need to evaluate the three view factors F_{12}, F_{13}, and F_{23} appearing in the circuit. Due to symmetry, $F_{12} = F_{21}$.

Using Fig. 1.49ii:

$$F_{12} = F_{21} = \sim 0.79$$

Now $F_{11} + F_{12} + F_{13} = 1$

$F_{11} = 0$ and hence $F_{13} = 1 - F_{12} = 1 - 0.79 = 0.21$

Similarly, one can write:

$$F_{21} + F_{22} + F_{23} = 1$$

Note $F_{22} = 0$ and $F_{21} = F_{12} = 0.79$ and

$\therefore F_{23} = 0.21$ (this result is also obvious from symmetry arguments because $F_{23} = F_{13}$).

(i) Surface 3 is nonradiative, i.e. $q_{03} = 0$, $T_3 = 0$
Applying Eq. (1.212) for three nodes:

$$\frac{\sigma T_1^4 - q_{01}}{\dfrac{1 - e_1}{A_1 e_1}} = \frac{q_{01} - q_{02}}{\dfrac{1}{A_1 F_{12}}} + \frac{q_{01} - q_{03}}{\dfrac{1}{A_1 F_{13}}}$$

$$\frac{\sigma T_2^4 - q_{02}}{\dfrac{1 - e_2}{A_2 e_2}} = \frac{q_{02} - q_{01}}{\dfrac{1}{A_2 F_{21}}} + \frac{q_{02} - q_{03}}{\dfrac{1}{A_2 F_{23}}}$$

$$q_{03} = 0$$

Now calculating various resistances:

$$A_1 = A_2 = \pi (1)^2 = 3.14 \, \text{m}^2$$

$$\therefore \quad \frac{1 - e_1}{A_1 e_1} = \frac{1 - 0.63}{3.14 \times 0.63} = 0.1870$$

$$\frac{1 - e_2}{A_2 e_2} = \frac{1 - 0.87}{3.14 \times 0.87} = 0.0475$$

$$\frac{1}{A_2 F_{23}} = \frac{1}{A_1 F_{13}} = \frac{1}{3.14 \times 0.21} = 1.516$$

$$\frac{1}{A_1 F_{12}} = \frac{1}{A_2 F_{21}} = \frac{1}{3.14 \times 0.79} = 0.4031$$

Using $\sigma = 5.669 \times 10^{-8} \, \text{W/m}^2 \, \text{K}^4$

Solving for q_{01} and q_{02}:

$$q_{01} = 52,179 \, \text{W/m}^2$$
$$q_{02} = 134,900 \, \text{W/m}^2$$

Heat flow at each surface is now given by Eq. (1.211):

$$Q_1 = \frac{5.669 \times 10^{-8} \times 773^4 - 52179}{(1 - 0.63)/(3.14 \times 0.63)} = -1.71 \times 10^5 \, \text{W}$$

$$Q_2 = \frac{5.669 \times 10^{-8} \times 1273^4 - 134,900}{(1 - 0.87)/(3.14 \times 0.87)} = 2.94 \times 10^5 \, \text{W}$$

These two values are not equal because some heat is lost to the environment (environmental surfaces).

(ii) Surface 3 is adiabatic
In this case, the net current at node 3 is 0, but q_{03} itself is unknown. The last equation in part (i) is replaced by:

$$\frac{q_{01} - q_{03}}{\dfrac{1}{A_1 F_{13}}} + \frac{q_{02} - q_{03}}{\dfrac{1}{A_2 F_{23}}} = 0$$

Solving for q_{01}, q_{02}, and q_{03}:

$$q_{01} = 6.1 \times 10^4 \, W/m^2; \; q_{02} = 1.39 \times 10^5 \, W/m^2 \text{ and } q_{03} = 9.98 \times 10^4 \, W/m^2$$

$$Q_1 = \frac{5.669 \times 10^{-8} \times 773^4 - 61,000}{(1 - 0.63)/(3.14 \times 0.63)} = -2.18 \times 10^5 \, W$$

$$Q_2 = \frac{5.669 \times 10^{-8} \times 1273^4 - 139,000}{(1 - 0.87)/(3.14 \times 0.87)} = 2.18 \times 10^5 \, W$$

As expected, the net heat flow from node 3 is zero.

1.5.6 Radiation From Gases

In the previous discussion, surfaces have been considered which are *isothermal, opaque*, and *grey* which *emit* and *reflect diffusely* and are characterised by *uniform surface radiosity* and the medium separating the surfaces has been assumed to be *nonparticipating*, in that it neither absorbs nor scatters the surface radiation nor does it emit radiation itself. Whilst, in most cases, such assumptions are valid and permit reasonably accurate results to be calculated, there are occasions where such assumptions do not hold and more refined techniques are required such as those described in the specialist literature.[74,75,77–82] For *nonpolar gases* such as N_2 and O_2, the foregoing assumptions are largely valid, since the gases do not emit radiation and they are essentially transparent to incident radiation. This is not the case with *polar molecules* such as CO_2 and H_2O vapour, NH_3, and hydrocarbon gases however, since these not only emit and absorb over a wide temperature range, but also radiate in specific wavelength intervals called *bands*. Furthermore, gaseous radiation is a volumetric rather than a surface phenomenon.

Whilst the calculation of the radiant heat flux from a gas to an adjoining surface embraces inherent spectral and directional effects, a simplified approach has been developed by Hottel and Manglesdorf,[83] which involves the determination of radiation emission from a hemispherical mass of gas of radius L, at temperature T_g to a surface element, dA_1, near the centre of the base of the hemisphere. Emission from the gas per unit area of the surface is then:

$$E_g = \mathbf{e}_g \sigma T_g^4 \tag{1.213}$$

where the gas emissivity $\mathbf{e}_g$ is a function of T_g, the total pressure of the gas P, the partial pressure of the radiating gas P_g and the radius of the hemisphere L.

Data on the emissivity of water vapour at a total pressure of 101.3 kN/m^2 are plotted in Fig. 1.55 for different values of the product of the vapour partial pressure P_w and the hemisphere radius L. For other values of the total pressure, the correction factor C_w also given in the figure must be used. Similar data for carbon dioxide are given in Fig. 1.56. Although these data refer to water vapour or carbon dioxide alone in a mixture of nonradiating gases, they may be extended to situations where both are present in such a mixture by expressing the total emissivity as:

$$\mathbf{e}_g = \mathbf{e}_w + \mathbf{e}_c - \Delta \mathbf{e} \tag{1.214}$$

where $\Delta \mathbf{e}$ is a correction factor, shown in Fig. 1.57, which allows for the reduction in emission associated with mutual absorption of radiation between the two species.

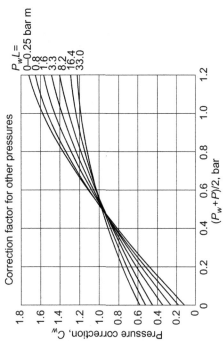

Fig. 1.55

Emissivity of water vapour in a mixture of nonradiating gases at 101.3 kN/m^2.

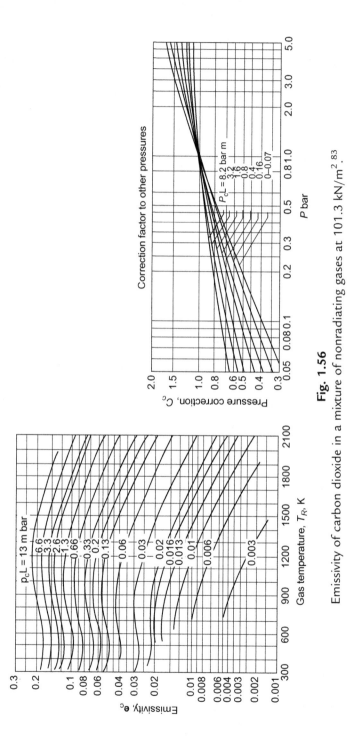

Fig. 1.56

Emissivity of carbon dioxide in a mixture of nonradiating gases at 101.3 kN/m^{2}.

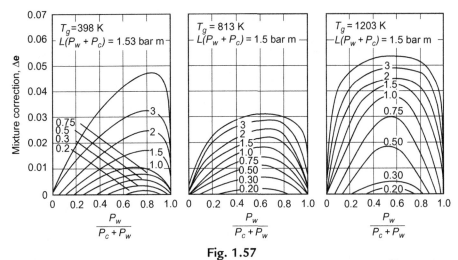

Fig. 1.57

Correction factor for water vapour-carbon dioxide mixtures.[83]

Although these data provide the emissivity of a hemispherical gas mass of radius L radiating to an element at the centre of the base, they may be extended to other geometries by using the concept of *mean beam length L_e* which correlates the dependence of gas emissivity with both the size and shape of the gas geometry in terms of a single parameter. Essentially the mean beam length is the radius of the hemisphere of gas whose *emissivity* is equivalent to that in the particular geometry considered, and typical values of L_e, which are then used to replace L in Figs 1.47–1.49 are shown in Table 1.12. Using these data and Figs 1.47–1.49, the rate of transfer of radiant heat to a surface of area A_s due to emission from an adjoining gas is given by:

$$Q = e_g A_s \sigma T_g^4 \tag{1.215}$$

Table 1.12 Mean beam lengths for various geometries[83]

Geometry	Characteristic Length	Mean Beam Length, L_e
Sphere—radiation to surface	Diameter, D	$0.65D$
Infinite circular cylinder—radiation to curved surface	Diameter, D	$0.95D$
Semiinfinite cylinder—radiation to base	Diameter, D	$0.65D$
Cylinder of equal height and diameter— radiation to entire surface	Diameter, D	$0.60D$
Infinite parallel planes—radiation to planes	Spacing between planes, L	$1.80L$
Cube—radiation to any surface	Side, L	$0.66L$
Shape of volume, V—radiation to surface of area, A	Ratio:	
	volume/area, (V/A)	$3.6(V/A)$

A black surface will not only absorb all of this radiation but will also emit radiation, and the net rate at which radiation is exchanged between the gas and the surface at temperature T_s is given by:

$$Q_{net} = A_s \sigma \left(\mathbf{e}_g T_g^4 - \mathbf{a}_g T_s^4 \right) \tag{1.216}$$

In this equation, the absorptivity $\mathbf{a}_g$ may be obtained from the emissivity using expressions of the form[83]:

Water:

$$\mathbf{a}_w = C_w \mathbf{e}_w \left(T_g / T_s \right)^{0.45} \tag{1.217}$$

Carbon dioxide:

$$\mathbf{a}_c = C_c \mathbf{e}_c \left(T_g / T_s \right)^{0.65} \tag{1.218}$$

where $\mathbf{e}_w$ and C_w and $\mathbf{e}_c$ and C_c are obtained from Figs 1.47 and 1.48 respectively, noting that T_g is replaced by T_s and $(P_w L_e)$ or $(P_c L_e)$ by $[P_w L_e (T_s/T_g)]$ or $[P_c L_e (T_s/T_g)]$ respectively. It may be noted that, in the presence of both water vapour and carbon dioxide, the total absorptivity is given by:

$$\mathbf{a}_g = \mathbf{a}_w + \mathbf{a}_c - \Delta \mathbf{a} \tag{1.219}$$

where $\Delta \mathbf{a} = \Delta \mathbf{e}$ is obtained from Fig. 1.56. If the surrounding surface is grey, some of the radiation may be reflected and Eq. (1.216) may be modified by a factor, $e_s/[1 - (1 - \mathbf{a}_g)(1 - \mathbf{e}_g)]$ to take this into account. This leads to the following equation for the heat transferred per unit time from the gas to the surface:

$$Q = \sigma e_s A_s \left(\mathbf{e}_g T_g^4 - \mathbf{a}_g T_s^4 \right) / \left[1 - \left(1 - \mathbf{a}_g \right) \left(1 - \mathbf{e}_g \right) \right] \tag{1.220}$$

Example 1.34

The walls of a combustion chamber, 0.5 m in diameter and 2 m long, have an emissivity of 0.5 and are maintained at 750 K. If the combustion products containing 10% carbon dioxide and 10% water vapour are at 150 kN/m^2 and 1250 K, what is the net rate of radiation to the walls?

Solution

The partial pressures of carbon dioxide (P_c) and of water (P_w) are:

$$P_c = P_w = (10/100)150 = 15.0 \text{kN/m}^2 \quad \text{or} \quad (15.0/100) = 0.15 \text{bar}$$

From Table 1.12:

$$L_e = 3.6 V/A = 3.6 \left(\pi/4 \times 0.5^2 \times 2 \right) / \left[2\pi/4 \times 0.5^2 \right) + (0.5\pi \times 2.0) \right] = 0.4 \text{m}$$

For water vapour:

$$P_w L_e = (0.15 \times 0.4) = 0.06 \text{bar m}$$

and from Fig. 1.54, $e_w = 0.075$

$$P = (150/100) = 1.5 \text{ bar}, \ P_w = 0.15 \text{ bar} \text{ and } 0.5(P_w + P) = 0.825 \text{ bar}$$

Since $P_w L_e = 0.06$ bar m, then from Fig. 1.54:

$$C_w = 1.4 \text{ and } e_w = (1.4 \times 0.075) = 0.105$$

For carbon dioxide:

$$P_c L_e = (0.15 \times 0.4) = 0.06 \text{ bar m}$$

and from Fig. 1.55, $e_c = 0.037$

Since $P = 1.5$ bar, $P_c = 0.15$ bar, and $P_c L_e = 0.06$ bar m, then, from Fig. 1.47:

$$C_c = 1.2 \text{ and } e_c = (1.2 \times 0.037) = 0.044$$
$$(P_w + P_c)L_e = (0.15 + 0.15)0.4 = 0.12 \text{ bar m}$$

and:

$$P_c/(P_c + P_w) = 0.15/(0.15 + 0.15) = 0.5$$

Thus, from Fig. 1.55 for $T_g > 1203$ K, $\Delta e = 0.001$ and from Eq. (1.214):

$$e_g = e_w - \Delta e = (0.105 + 0.044 - 0.001) = 0.148$$

For water vapour:

$$P_w L_e (T_s/T_g) = 0.06(750/1250) = 0.036 \text{ bar m}$$

and from Fig. 1.54 at 750 K, $e_w = 0.12$

Since $0.5(P_w + P) = 0.825$ bar and $P_w L_e (T_s/T_g) = P_c L_e (T_s/T_g) = 0.036$ bar m,

Then from Fig. 1.54:

$$C_w = 1.40 \text{ and } e_w = (0.12 \times 1.40) = 0.168$$

and the absorptivity, from Eq. (1.217) is:

$$a_w = e_w (T_g/T_s)^{0.45} = 0.168(1250/750)^{0.45} = 0.212$$

For carbon dioxide:

From Fig. 1.55 at 750 K, $e_c = 0.08$

From Fig. 1.55 at $P = 1.5$ bar and $P_c L_e (T_s/T_g) = 0.036$ bar m and:

$$C_c = 1.02 \text{ and } e_c = (0.08 \times 1.02) = 0.082$$

and the absorptivity, from Eq. (1.218) is:

$$a_c = e_c (T_g/T_s)^{0.65} = 0.082(1250/750)^{0.65} = 0.114$$
$$P_w/(P_c + P_w) = 0.5 \text{ and } (P_c + P_w)L_e (T_s/T_g) = (0.036 + 0.036) = 0.072 \text{ bar m}$$

Thus, from Fig. 1.56, for $T_g = 813$ K, $\Delta e = \Delta a < 0.01$ and this may be neglected.

Thus:

$$\mathbf{a}_g = \mathbf{a}_w + \mathbf{a}_c - \Delta\mathbf{a} = (0.212 + 0.114 - 0) = 0.326$$

If the surrounding surface is black, then:

$$Q = \sigma A_s \left(\mathbf{e}_g T_g^4 - \alpha_g T_s^4 \right) \quad \text{(Eq. 9.216)}$$

$$= \left(5.67 \times 10^{-8} \left[\left(2(\pi/4)0.5^2 \right) + (0.5\pi \times 2.0) \right] \right) \left[\left(0.148 \times 1250^4 \right) - \left(0.326 \times 750^4 \right) \right]$$

$$= 5.03 \times 10^4 \, \text{W} = \underline{50.3 \, \text{kW}}$$

For grey walls, the correction factor allowing for multiple reflection of incident radiation is:

$$C_g = \mathbf{e}_s / \left[1 - \left(1 - \mathbf{a}_g \right) \left(1 - \mathbf{e}_g \right) \right] = 0.5 / \left[1 - (1 - 0.326)(1 - 0.5) \right] = 0.754$$

and hence: net radiation to the walls, $Q_w = (50.3 \times 0.754) = \underline{37.9 \, \text{kW}}$

Radiation from gases containing suspended particles

The estimation of the radiation from pulverised-fuel flames, from dust particles in flames, and from flames made luminous as a result of the thermal decomposition of hydrocarbons to soot, involves an evaluation of radiation from clouds of particles. In pulverised-fuel flames, the mean particle size is typically 25 μm and the composition varies from a very high carbon content to virtually pure ash. In contrast, the suspended matter in luminous flames, resulting from soot formation due to incomplete mixing of hydrocarbons with air before being heated, consists of carbon together with very heavy hydrocarbons with an initial particle size of some 0.3 μm. In general, pulverised-fuel particles are sufficiently large to be substantially opaque to incident radiation, whilst the particles in a luminous flame are so small that they act as semitransparent bodies with respect to thermal or long wavelength radiation.

According to Schack,[84] a single particle of soot transmits approximately 95% of the incident radiation and a cloud must contain a very large number of particles before an appreciable emission can occur. If the concentration of particles is K', then the product of K' and the thickness of the layer L is equivalent to the product $P_g L_e$ in the radiation of gases. For a known or measured emissivity of the flame e_f, the heat transfer rate per unit time to a wall is given by:

$$Q = \mathbf{e}_f \mathbf{e}_s \sigma \left(T_f^4 - T_w^4 \right) \tag{1.221}$$

where $\mathbf{e}_s$ is the effective emissivity of the wall, and T_f and T_w are the temperatures of the flame and wall respectively, e_f varies, not only from point to point in a flame, but also depends on the type of fuel, the shape of the burner and combustion chamber, and on the air supply and the degree of preheating of the air and fuel.

1.6 Heat Transfer in the Condensation of Vapours

1.6.1 Film Coefficients for Vertical and Inclined Surfaces

When a saturated vapour is brought into contact with a cool surface, heat is transferred from the vapour to the surface and a film of condensate is produced.

In considering the heat that is transferred, the method first put forward by Nusselt[85] and later modified by others is followed. If the vapour condenses on a vertical surface, the condensate film flows downwards under the influence of gravity, although it is retarded by the viscosity of the liquid. The flow will normally be streamline and the heat flows through the film by conduction. In Nusselt's work it is assumed that the temperature of the film at the cool surface is equal to that of the surface, and the other side was at the temperature of the vapour. In practice, there must be some small difference in temperature between the vapour and the film, although this may generally be neglected except where noncondensable gas is present in the vapour.

It is shown in Chapter 3 of Vol. 1A that the mean velocity of a fluid flowing down a surface inclined at an angle ϕ to the horizontal is given by:

$$u = \frac{\rho g \sin \phi s^2}{3\mu} \quad \text{(Eq. 3.87)}$$

For a vertical surface:

$$\sin \phi = 1 \quad \text{and} \quad u = \frac{\rho g s^2}{3\mu}$$

The maximum velocity u_s which occurs at the free surface is:

$$u_s = \frac{\rho g \sin \phi s^2}{2\mu} \quad \text{(Eq. 3.88)}$$

and this is 1.5 times the mean velocity of the liquid.

Since the liquid is produced by condensation, the thickness of the film will be zero at the top and will gradually increase towards the bottom. Under steady conditions the difference in the mass rates of flow at distances x and $x+dx$ from the top of the surface will result from condensation over the small element of the surface of length dx and width w, as shown in Fig. 1.58.

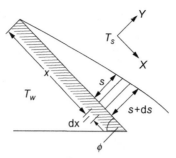

Fig. 1.58
Condensation on an inclined surface.

If the thickness of the liquid film increases from s to $s+ds$ in that distance, the increase in the mass rate of flow of liquid dG is given by:

$$\frac{d}{ds}\left(\frac{\rho^2 g \sin\phi s^3 w}{3\mu}\right)ds = \frac{\rho^2 g \sin\phi}{\mu} ws^2 ds$$

If the vapour temperature is T_s and the wall temperature is T_w, the heat transferred by thermal conduction to an element of surface of length dx is:

$$\frac{k(T_s - T_w)}{S} w dx$$

where k is the thermal conductivity of the condensate.

Thus the mass rate of condensation on this small area of surface is:

$$\frac{k(T_s - T_w)}{S\lambda} w dx$$

where λ is the latent heat of vapourisation of the liquid.

Thus:

$$\frac{k(T_s - T_w)}{s\lambda} w dx = \frac{\rho^2 g \sin\phi}{\mu} ws^2 ds$$

On integration:

$$\mu k(T_s - T_w)x = \frac{1}{4}\rho^2 g \sin\phi S^4 \lambda$$

since $s=0$ when $x=0$.

Thus:

$$s = \left(\frac{4\mu kx(T_s - T_w)}{\lambda \rho^2 g \sin\phi}\right)^{1/4} \tag{1.222}$$

Now the heat transfer coefficient h at $x=x$, $=k/s$, and hence:

$$h = \left(\frac{\rho^2 g \sin \phi \lambda k^3}{4 \mu x (T_s - T_w)} \right)^{1/4}$$

(1.223)

and:

$$Nu = \frac{hx}{k} = \left(\frac{\rho^2 g \sin \phi \lambda x^3}{4 \mu k (T_s - T_w)} \right)^{1/4}$$

(1.224)

These expressions give point values of h and Nu_x at $x=x$. It is seen that the coefficient decreases from a theoretical value of infinity at the top as the condensate film thickens. The mean value of the heat transfer coefficient over the whole surface, between $x=0$ and $x=x$ is given by:

$$h_m = \frac{1}{x} \int_0^x h \, dx = \frac{1}{x} \int_0^x K x^{-1/4} \, dx \quad \text{(where } K \text{ is independent of } x)$$

$$= \frac{1}{x} K \frac{x^{3/4}}{\frac{3}{4}} = \frac{4}{3} K x^{-1/4} = \frac{4}{3} h$$

(1.225)

$$= 0.943 \left(\frac{\rho^2 g \sin \phi \lambda k^3}{\mu x \Delta T_f} \right)^{1/4}$$

where ΔT_f is the temperature difference across the condensate film.

For a vertical surface, $\sin \phi = 1$ and:

$$h_m = 0.943 \left(\frac{\rho^2 g \lambda k^3}{\mu x \Delta T_f} \right)^{1/4}$$

(1.226)

1.6.2 Condensation on Vertical and Horizontal Tubes

The Nusselt equation

If vapour condenses on the outside of a vertical tube of diameter d_o, then the hydraulic mean diameter for the film is:

$$\frac{4 \times \text{flow area}}{\text{wetted perimeter}} = \frac{4S}{b} \text{(say)}$$

If G is the mass rate of flow of condensate, the mass rate of flow per unit area G' is G/S and the Reynolds number for the condensate film is then given by:

$$Re = \frac{(4S/b)(G/S)}{\mu} = \frac{4G}{\mu b} = \frac{4M}{\mu}$$

(1.227)

where M is the mass rate of flow of condensate per unit length of perimeter, or:

$$M = \frac{G}{\pi d_o}$$

For streamline conditions in the film, $4M/\mu \leq 2100$ and:

$$h_m = \frac{Q}{A\Delta T_f} = \frac{G\lambda}{bl\Delta T_f} = \frac{\lambda M}{l\Delta T_f}$$

From Eq. (1.226):

$$h_m = 0.943 \left(\frac{k^3 \rho^2 g}{\mu} \frac{\lambda}{l\Delta T_f} \right)^{1/4} = 0.943 \left(\frac{k^3 \rho^2 g}{\mu} \frac{h_m}{M} \right)^{1/4}$$

and hence:

$$h_m \left(\frac{\mu^2}{k^3 \rho^2 g} \right)^{1/3} = 1.47 \left(\frac{4M}{\mu} \right)^{-1/3} \tag{1.228}$$

For horizontal tubes, Nusselt proposed the equation:

$$h_m = 0.72 \left(\frac{k^3 \rho^2 g \lambda}{d_o \mu \Delta T_f} \right)^{1/4} \tag{1.229}$$

This may be rearranged to give:

$$h_m \left(\frac{\mu^2}{k^3 \rho^2 g} \right)^{1/3} = 1.51 \left(\frac{4M}{\mu} \right)^{-1/3} \tag{1.230}$$

where M is the mass rate of flow per unit length of tube.

This is approximately the same as Eq. (1.227) for vertical tubes and is a universal equation for condensation, noting that for vertical tubes $M = G/\pi d_o$ and for horizontal tubes $M = G/l$, where l is the length of the tube. Comparison of the two equations shows that, provided the length is more than three times the diameter, the horizontal tube will give a higher transfer coefficient for the same temperature conditions.

For j vertical rows of horizontal tubes, Eq. (1.229) may be modified to give:

$$h_m = 0.72 \left(\frac{k^3 \rho^2 g \lambda}{j d_o \mu \Delta T_f} \right)^{1/4} \tag{1.231}$$

Kern[44] suggests that, based on the performance of commercial exchangers, this equation is too conservative and that the exponent of j should be nearer to $-\frac{1}{6}$ than to $-\frac{1}{4}$. This topic is discussed in Chapter 4.

Table 1.13 Average values of film coefficients h_m for condensation of pure saturated vapours on horizontal tubes

Vapour	Value of h_m (W/m^2 K)	Value of h_m (Btu/h ft^2 °F)	Range of ΔT_f (deg K)
Steam	10,000–28,000	1700–5000	1–11
Steam	18,000–37,000	3200–6500	4–37
Benzene	1400–2200	240–380	23–37
Diphenyl	1300–2300	220–400	4–15
Toluene	1100–1400	190–240	31–40
Methanol	2800–3400	500–600	8–16
Ethanol	1800–2600	320–450	6–22
Propanol	1400–1700	250–300	13–20
Oxygen	3300–8000	570–1400	0.08–2.5
Nitrogen	2300–5700	400–1000	0.15–3.5
Ammonia	6000	1000	–
Freon-12	1100–2200	200–400	–

Experimental results

In testing Nusselt's equation it is important to ensure that the conditions comply with the requirements of the theory. In particular, it is necessary for the condensate to form a uniform film on the tubes, for the drainage of this film to be by gravity, and the flow streamline. Although some of these requirements have probably not been entirely fulfilled, results for pure vapours such as steam, benzene, toluene, diphenyl, ethanol, and so on, are sufficiently close to give support to the theory. Some data obtained by Haselden and Prosad[86] for condensing oxygen and nitrogen vapours on a vertical surface, where precautions were taken to see that the conditions were met, are in very good agreement with Nusselt's theory. The results for most of the researchers are within 15% for horizontal tubes, although they tend to be substantially higher than the theoretical results for vertical tubes. Typical values are given in Table 1.13 taken from McAdams[87] and elsewhere.

When considering commercial equipment, there are several factors which prevent the true conditions of Nusselt's theory from being met. The temperature of the tube wall will not be constant, and for a vertical condenser with a ratio of ΔT at the bottom to ΔT at the top of five, the film coefficient should be increased by about 15%.

Influence of vapour velocity

A high vapour velocity upwards tends to increase the thickness of the film and thus reduce h though the film may sometimes be disrupted mechanically as a result of the formation of small waves. For the downward flow of vapour, Ten Bosch[88] has shown that h increases considerably at high vapour velocities and may increase to two or three times the value given by the Nusselt equation. It must be remembered that when a large fraction of the vapour is condensed, there may be a considerable change in velocity over the surface.

Under conditions of high vapour velocity Carpenter and Colburn[89] have shown that turbulence may occur with low values of the Reynolds number, in the range 200–400. When the vapour velocity is high, there will be an appreciable drag on the condensate film and the expression obtained for the heat transfer coefficient is difficult to manage.

Carpenter and Colburn[89] have put forward a simple correlation of their results for condensation at varying vapour velocities on the inner surface of a vertical tube which takes the form:

$$h_m = 0.065 G'_m \sqrt{\frac{C_p \rho k (R'/\rho_v u^2)}{\mu \rho_v}} \tag{1.232}$$

where:

$$G'_m = \sqrt{\frac{(G'^2_1 + G'_1 G'_2 + G'^2_2)}{3}}$$

and u is the velocity calculated from G'_m. In this equation C_p, k, ρ, and μ are properties of the condensate and ρ_v refers to the vapour. G'_1 is the mass rate of flow per unit area at the top of the tube and G'_2 the corresponding value at the bottom. R' is the shear stress at the free surface of the condensate film.

As pointed out by Colburn,[90] the group $C_p \rho k/\mu \rho_v$ does not vary very much for a number of organic vapours so that a plot of h_m and G'_m will provide a simple approximate correlation with separate lines for steam and for organic vapours as shown in Fig. 1.59.[90,91] Whilst this must be regarded as an empirical approximation it is very useful for obtaining a good indication of the effect of vapour velocity.

Turbulence in the film

If Re is greater than 2100 during condensation on a vertical tube, the mean coefficient h_m will increase as a result of turbulence. The data of Kirkbride[92] and Badger[93,94] for the condensation of diphenyl vapour and Dowtherm on nickel tubes are expressed in the form:

$$h_m \left(\frac{\mu^2}{k^3 \rho^2 g} \right)^{1/3} = 0.0077 \left(\frac{4M}{\mu} \right)^{0.4} \tag{1.233}$$

Comparing Eq. (1.230) for streamline flow of condensate and Eq. (1.233) for turbulent flow, it is seen that, with increasing Reynolds number, h_m decreases with streamline flow but increases with turbulent flow. These results are shown in Fig. 1.60.

Design equations are given in Volume 6, Chapter 12, for condensation both inside and outside horizontal and vertical tubes, and the importance of avoiding flooding in vertical tubes is stressed.

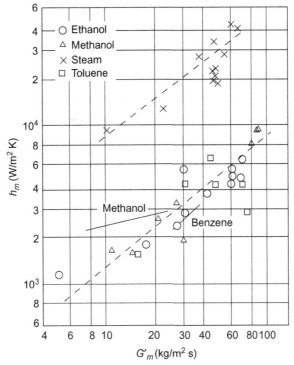

Fig. 1.59

Average heat transfer data of Carpenter and Colburn[89] (shown as points) compared with those of Tepe and Mueller[91] (shown as *solid lines*). *Dashed lines* represent Eq. (1.232).

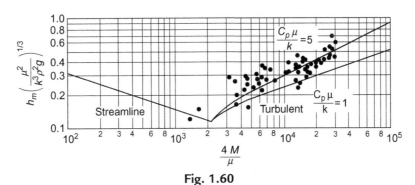

Fig. 1.60

Effects of turbulence in condensate film.

1.6.3 Dropwise Condensation

In the discussion so far, it is assumed that the condensing vapour, on coming into contact with the cold surface, wets the tube so that a continuous film of condensate is formed. If the droplets initially formed do not wet the surface, after growing slightly they will fall from the tube exposing fresh condensing surface. This is known as dropwise condensation and, since the heat does not have to flow through a film by conduction, much higher transfer coefficients are obtained. Steam is the only pure vapour for which definite dropwise condensation has been obtained, and values of h from 40 to 114 kW/m^2 K have been obtained, with much higher values on occasions. This question has been discussed by Drew, Nagle, and Smith[95] who have shown that there are many materials which make the surface nonwettable although of these, only those which are firmly held to the surface are of any practical use. Mercaptans and oleic acid have been used to promote dropwise condensation, but at present there is little practical application of this technique. Exceptionally high values of h will not give a corresponding increase in the overall coefficient, since for a condenser with steam, a value of about 11 kW/m^2 K can be obtained with film condensation. On the other hand, it may be helpful in experimental work to reduce the thermal resistance on one side of a surface to a negligible value.

1.6.4 Condensation of Mixed Vapours

In the previous discussion it has been assumed that the vapour is a pure material, such as steam or organic vapour. If it contains a proportion of noncondensable gas and is cooled below its dew point, a layer of condensate is formed on the surface with a mixture of noncondensable gas and vapour above it. The heat flow from the vapour to the surface then takes place in two ways. Firstly, sensible heat is passed to the surface because of the temperature difference. Secondly, since the concentration of vapour in the main stream is greater than that in the gas film at the condensate surface, vapour molecules diffuse to the surface and condense there, giving up their latent heat. The actual rate of condensation is then determined by the combination of these two effects, and its calculation requires a knowledge of mass transfer by diffusion, as discussed in Chapter 2.

In the design of a cooler-condenser for a mixture of vapour and a permanent gas, the method of Colburn and Hougen[96] is considered. This requires a point-to-point calculation of the condensate-vapour interface conditions T_c and P_s. A trial and error solution is required of the equation:

$$q_v + q_\lambda = q_c = U\Delta T \tag{1.234}$$

$$h_g(T_s - T_c) + k_G\lambda(P_g - P_s) = h_o(T_c - T_{cm}) = U\Delta T \tag{1.235}$$

where the first term q_v represents the sensible heat transferred to the condensing surface, the second term q_λ the latent heat transferred by the diffusing vapour molecules, and the third term q_c the heat transferred from the condensing surface through the pipe wall, dirt and scales, and water film to the cooling medium. h_g is the heat transfer coefficient over the gas film; h_o the conductance of the combined condensate film, tube wall, dirt and scale films, and the cooling medium film; and U the overall heat transfer coefficient. T_s is the vapour temperature, T_c the temperature of the condensate, T_{cm} the cooling medium temperature, ΔT the overall temperature difference $=(T_s - T_{cm})$, P_g is the partial pressure of diffusing vapour, P_s

the vapour pressure at T_c, λ the latent heat of vapourisation per unit mass, and k_G the mass transfer coefficient in mass per unit time, unit area, and unit partial pressure difference.

To evaluate the required condenser area, point values of the group $U\Delta T$ as a function of q_c must be determined by a trial and error solution of Eq. (1.235). Integration of a plot of q_c against $1/U\Delta T$ will then give the required condenser area. This method takes into account point variations in temperature difference, overall coefficient, and mass velocities and consequently produces a reasonably accurate value for the surface area required.

The individual terms in Eq. (1.235) are now examined to enable a trial solution to proceed. Values for h_g and k_G are most conveniently obtained from the Chilton and Colburn[97] analogy discussed in Chapter 2.

Thus:

$$h_g = \frac{j_h G' C_p}{\left(C_p \mu / k\right)^{0.67}} \tag{1.236}$$

$$k_G = \frac{j_d G'}{P_{Bm}\left(\mu / \rho D\right)^{0.67}} \tag{1.237}$$

Values of j_h and j_d are obtained from a knowledge of the Reynolds number at a given point in the condenser. The combined conductance h_0 is evaluated by determining the condensate film coefficient h_c from the Nusselt equation and combining this with the dirt and tube wall conductances and a cooling medium film conductance predicted from the Sieder–Tate relationships. Generally, h_0 may be considered to be a constant throughout the exchanger.

From a knowledge of h_g, k_G, and h_0 and for a given T_s and T_{cm} values of the condensate surface temperature T_c are estimated until Eq. (1.235) is satisfied. The calculations are repeated, and in this manner several point values of the group $U\Delta T$ throughout the condenser may be obtained.

The design of a cooler condenser for the case of condensation of two vapours is more complicated than the preceding single vapour-permanent gas case,[98] and an example has been given by Jeffreys.[99]

For the condensation of a vapour in the presence of a noncondensable gas, the following example is considered which is based on the work of Kern.[44]

Example 1.35

A mixture of 0.57 kg/s of steam and 0.20 kg/s of carbon dioxide at 308 kN/m^2 and its dew point enters a heat exchanger consisting of 246 tubes, 19 mm o.d., wall thickness 1.65 mm, 3.65 m long, arranged in four passes on 25 mm square pitch in a 0.54 m diameter shell and leaves at 322 K. Condensation is effected by cooling water entering and leaving the unit at 300 and 319 K respectively. If the diffusivity of steam-carbon dioxide mixtures is

1.1×10^{-5} m^2/s and the group $(\mu/\rho D)^{0.67}$ may be taken to be constant at 0.62, estimate the overall coefficient of heat transfer and the dirt factor for the condenser.

Solution

In the steam entering the condenser, there is

$$\left. \begin{array}{l} \dfrac{0.57}{18}=0.032\,\text{kmol water} \\[2mm] \text{and}\,\dfrac{0.20}{44}=0.0045\,\text{kmol}\,CO_2 \end{array} \right\} \text{total}=0.0365\,\text{kmol.}$$

Hence the partial pressure of water $=(308 \times 0.032/0.0365)=270$ kN/m^2 and from Table 11A in the Appendix, the dew point $=404$ K.

Mean molecular weight of the mixture $=(0.57+0.20)/0.0365=21.1$ kg/kmol.

At the inlet:
$$\left. \begin{array}{l} \text{vapour pressure of water}=270\,\text{kN/m}^2 \\ \text{inert pressure}=(308-270)=38\,\text{kN/m}^2 \end{array} \right\} \text{total}=308\,\text{kN/m}^2$$

At the outlet:
$$\left. \begin{array}{l} \text{partial pressure of water at }322\,\text{K}=11.7\,\text{kN/m}^2 \\ \text{inert pressure}=(308-11.7)=296.3\,\text{kN/m}^2 \end{array} \right\} \text{total}=308\,\text{kN/m}^2$$

$$\therefore \text{ steam at the outlet}=\frac{0.0045 \times 11.7}{296.3}=0.000178\,\text{kmol}$$

and:

$$\text{steam condensed}=(0.032-0.000178)=0.03182\,\text{kmol.}$$

The heat load is now estimated at each interval between the temperatures 404, 401, 397, 380, 339, and 322 K.

For the interval 404–401 K

From Table 11A in the Appendix, the partial pressure of steam at 401 K $=252.2$ kN/m^2 and hence the partial pressure of $CO_2=(308-252.2)=55.8$ kN/m^2. Steam remaining $=(0.0045 \times 252.2/55.8)=0.0203$ kmol.

∴ Steam condensed $=(0.032-0.0203)=0.0117$ kmol
Heat of condensation $=(0.0117 \times 18)(2180+1.93(404-401))=466$ kW
Heat from uncondensed steam $=(0.0203 \times 18 \times 1.93(404-401))=1.9$ kW
Heat from carbon dioxide $=(0.020 \times 0.92(404-401))=0.5$ kW
and the total for the interval $=468.4$ kW

Repeating the calculation for the other intervals of temperature gives the following results:

Interval (K)	Heat Load (kW)
404–401	468.4
401–397	323.5
397–380	343.5
380–339	220.1
339–322	57.9
Total	1407.3

and the flow of water $=1407.3/(4.187(319-300))=17.7$ kg/s.

With this flow of water and a flow area per pass of 0.0120 m², the mass velocity of water is 1425 kg/m² s, equivalent to a velocity of 1.44 m/s at which $h_i = 6.36$ kW/m² K. Basing this on the outside area, $h_{io} = 5.25$ kW/m² K.

Shell-side coefficient for entering gas mixture:

The mean specific heat, $C_p = \dfrac{(0.20 \times 0.92) + (0.57 \times 1.93)}{0.77} = 1.704$ kJ/kgK

Similarly, the mean thermal conductivity $k = 0.025$ kW/m K and the mean viscosity $\mu = 0.015$ mN s/m²

The area for flow through the shell $= 0.0411$ m² and the mass velocity on the shell side

$$= \frac{0.20 + 0.57}{0.0411} = 18.7 \text{ kg/m}^2\text{s}$$

Taking the equivalent diameter as 0.024 m, $Re = 29{,}800$ and:

$$h_g = 0.107 \text{ kW/m}^2\text{K} \quad \text{or} \quad 107 \text{W/m}^2\text{K}.$$

Now:

$$\left(\frac{\mu}{\rho D}\right)^{0.67} = 0.62, \quad \left(\frac{C_p \mu}{k}\right)^{0.67} = 1.01$$

and:

$$k_G = \frac{h_g \left(C_p \mu / k\right)^{0.67}}{C_p P_{sF} \left(\mu / \rho D\right)^{0.67}} = \frac{107 \times 1.01}{1704 P_{sF} \times 0.62}$$

$$= \frac{0.102}{P_{sF}}$$

At point 1

Temperature of the gas $T = 404$ K, partial pressure of steam $P_g = 270$ kN/m², partial pressure of the inert $P_s = 38$ kN/m², water temperature $T_w = 319$ K, and $\Delta T = (404 - 319) = 85$ K. An estimate is now made for the temperature of the condensate film of $T_c = 391$ K. In this case $P_s = 185.4$ kN/m² and $P'_s = (308 - 185.4) = 122.6$ kN/m².

Thus:

$$P_{sF} = \frac{122.6 - 38}{\ln(122.6/38)} = 72.2 \text{ kN/m}^2$$

In Eq. (1.235):

$$h_g(T_s - T_c) + k_G \lambda (p_g - P_s) = h_{i0}(T_c - T_{cm})$$

$$0.107(404 - 391) + \left(\frac{0.102}{724}\right) 2172(270 - 185.4) = 5.25(391 - 319)$$

$$259 = 378 \quad \text{i.e. there is no balance.}$$

Try $T_c = 378$ K, $P_s = 118.5$ kN/m², $P_g = (308 - 118.5) = 189.5$ kN/m² and:

$$P_{sF} = \frac{189.5 - 38}{\ln(189.5/38)} = 94.2\,\text{kN/m}^2$$

Substituting in Eq. (1.235)

$$\therefore\ 0.107(404 - 378) + \left(\frac{0.102}{94.2}\right)2172(270 - 118.5) = 5.25(378 - 319)$$

$$310 = 308\ \text{which agrees well}$$

$$\therefore\ U\Delta T = 309\,\text{kW/m}^2$$

and

$$U = \frac{309}{(404 - 319)}$$
$$= 3.64\,\text{kW/m}^2\,\text{K}$$

Repeating this procedure at the various temperature points selected, the heat-exchanger area may then be obtained as the area under a plot of Σq versus $1/U\Delta T$, or as $A = \Sigma q/U\Delta T$ according to the following tabulation:

Point	T_s (K)	T_c (K)	$U\Delta T$ (kW/ m^2)	$U\Delta T_{ow}$ (kW/ m^2)	Q (kW)	$A = Q/(U\Delta T)_{ow}$ (m^2)	ΔT (K)	ΔT_{ow} (K)	$Q/\Delta T_{ow}$ (kW/K)
1	404	378	309	–	–	–	84.4	–	–
2	401	356	228	268.5	468.4	1.75	88.1	86.3	5.42
3	397	336	145	186.5	323.5	1.74	88.6	88.4	3.66
4	380	312	40.6	88.1[a]	343.5	3.89	76.7	82.7	4.15
5	339	302	5.4	17.5[a]	220.1	12.58	38.1	55.2[a]	4.00
6	322	300	2.1	3.5[a]	51.9	14.83	22.2	29.6[a]	1.75
				Total:	1407.3	34.8			18.98

[a]Based on LMTD.

If no condensation takes place, the logarithmic mean temperature difference is 46.6 K. In practice the value is $(1407.3/18.98) = 74.2$ K.

Assuming no scale resistance, the overall coefficient is $\dfrac{1407.3}{34.8 \times 74.2} = 0.545\,\text{kW/m}^2\,\text{k}$

The available surface area on the outside of the tubes $= 0.060$ m^2/m or $(246 \times 3.65 \times 0.060) = 53.9$ m^2

The actual coefficient is therefore $\dfrac{1407.3}{53.9 \times 74.2} = 0.352\,\text{kW/m}^2\,\text{K}$

And the dirt factor is $\dfrac{(0.545 - 0.352)}{(0.545 \times 0.352)} = 1.01\,\text{m}^2\,\text{K/kW}$.

As shown in Fig. 1.61, the clean coefficient varies from $3.64\,\text{kW/m}^2\,\text{K}$ at the inlet to $0.092\,\text{kW/m}^2\,\text{K}$ at the outlet.

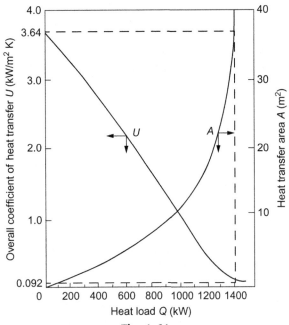

Fig. 1.61

Results for Example 1.35.

Condensation of mixed vapours is considered further in Chapter 4, where it is suggested that the local heat transfer coefficient may be expressed in terms of the local gas-film and condensate-film coefficients. For partial condensation where:

(i) *noncondensables < 0.5%*; their effect can be ignored,
(ii) *noncondensables > 70%*; the heat transfer can be taken as being by forced convection alone, and
(iii) *noncondensables 0.5%–70%*; both mechanisms are effective.

1.7 Boiling Liquids

1.7.1 Conditions for Boiling

In processing units, liquids are boiled either on submerged surfaces or on the inside of vertical tubes. Mechanical agitation may be applied in the first case and, in the second, the liquid may be driven through the tubes by means of an external pump. The boiling of liquids under either of these conditions normally leads to the formation of vapour first in the form of bubbles and later as a distinct vapour phase above a liquid interface. The conditions for boiling on the submerged surface are discussed here and the problems arising with boiling inside tubes are considered in Volume 2B. Much of the fundamental work on the ideas of boiling has been presented by Westwater[100] and Jakob,[101] Rohsenow and Clark,[102] Rohsenow,[103] and Forster[104] and by others.[105–107] The boiling of solutions in which a solid phase is separated after evaporation has proceeded to a sufficient extent is considered in Volume 2B.

For a bubble to be formed in a liquid, such as steam in water, for example, it is necessary for a surface of separation to be produced. Kelvin has shown that, as a result of the surface tension between the liquid and vapour, the vapour pressure on the inside of a concave surface will be less than that at a plane surface. As a result, the vapour pressure P_r inside the bubble is less than the saturation vapour pressure P_s at a plane surface. The relation between P_r and P_s is:

$$P_r = P_s - \left(\frac{2\sigma}{r}\right) \qquad (1.238)$$

where r is the radius of curvature of the bubble, and σ is the surface tension.

Hence the liquid must be superheated near the surface of the bubble, the extent of the superheat increasing with decrease in the radius of the bubble. On this basis it follows that very small bubbles are difficult to form without excessive superheat. The formation of bubbles is made much easier by the fact that they will form on curved surfaces or on irregularities on the heating surface, so that only a small degree of superheat is normally required.

Nucleation at much lower values of superheat is believed to arise from the presence of existing nuclei such as noncondensing gas bubbles, or from the effect of the shape of the cavities in the surface. Of these, the current discussion on the influence of cavities is the most promising. In many cavities the angle θ will be greater than 90° and the effective contact angle, which includes the contact angle of the cavity β, will be considerably greater $[=\theta + (180 - \beta)/2]$, so that a much-reduced superheat is required to give nucleation. Thus the size of the mouth of the cavity and the shape of the cavity plays a significant part in nucleation.[108]

It follows that for boiling to occur a small difference in temperature must exist between the liquid and the vapour. Jakob and Fritz[109] have measured the temperature distribution for water boiling above an electrically heated hot plate. The temperature dropped very steeply from about 383 K on the actual surface of the plate to 374 K about 0.1 mm from it. Beyond this point the temperature was reasonably constant until the water surface was reached. The mean superheat of the water above the temperature in the vapour space was about 0.5 deg K and this changed very little with the rate of evaporation. At higher pressures this superheating became smaller becoming 0.2 deg K at 5 MN/m² and 0.05 deg K at 101 MN/m². The temperature drop from the heating surface depends, however, very much on the rate of heat transfer and on the nature of the surface. Thus in order to maintain a heat flux of about 25.2 kW/m², a temperature difference of only 6 deg K was required with a rough surface as against 10.6 deg K with a smooth surface. The heat transfer coefficient on the boiling side is therefore dependent on the nature of the surface and on the difference in temperature available. For water boiling on copper plates, Jakob and Fritz[109] give the following coefficients for a constant temperature difference of 5.6 deg K, with different surfaces:

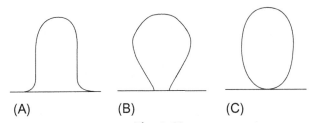

Fig. 1.62

Shapes of bubbles (A) screen surface—thin oil layer (B) chromium plated and polished surface, and (C) screen surface—clean.

(1) Surface after 8 h (28.8 ks) use and 48 h (172.8 ks) $h = 8000$ W/m^2 K
     immersion in water
(2) Freshly sandblasted $h = 3900$ W/m^2 K
(3) Sandblasted surface after long use $h = 2600$ W/m^2 K
(4) Chromium plated $h = 2000$ W/m^2 K

The initial surface, with freshly cut grooves, gave much higher figures than case (1).

The nature of the surface will have a marked effect on the physical form of the bubble and the area actually in contact with the surface, as shown in Fig. 1.62.

The three cases are:

(a) *Nonwettable surface*, where the vapour bubbles spread out thus reducing the area available for heat transfer from the hot surface to the liquid.
(b) *Partially wettable surface*, which is the commonest form, where the bubbles rise from a larger number of sites and the rate of transfer is increased.
(c) *Entirely wetted surface*, such as that formed by a screen. This gives the minimum area of contact between vapour and surface and the bubbles leave the surface when still very small. It therefore follows that if the liquid has detergent properties this may give rise to much higher rates of heat transfer.

1.7.2 Types of Boiling

Interface evaporation

In boiling liquids on a submerged surface, it is found that the heat transfer coefficient depends very much on the temperature difference between the hot surface and the boiling liquid. The general relation between the temperature difference and heat transfer coefficient was first presented by Nukiyama[110] who boiled water on an electrically heated wire. The results obtained have been confirmed and extended by others, and Fig. 1.63 shows the data of Farber and Scorah.[111] The relationship here is complex and is best considered in stages.

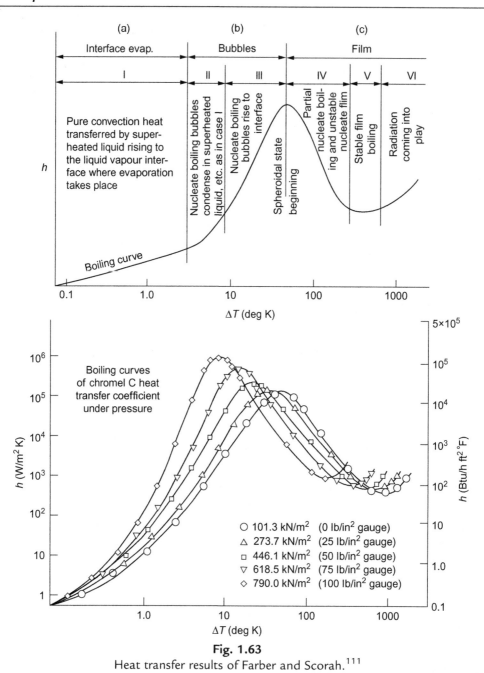

Fig. 1.63
Heat transfer results of Farber and Scorah.[111]

In interface evaporation, the bubbles of vapour formed on the heated surface move to the vapour–liquid interface by natural convection and exert very little agitation on the liquid. The results are given by:

$$Nu = 0.61(GrPr)^{1/4} \qquad (1.239)$$

which may be compared with the expression for natural convection:

$$Nu = C'(Gr\,Pr)^n \quad \text{(Eq. 9.146)}$$

where $n = 0.25$ for streamline conditions and $n = 0.33$ for turbulent conditions.

Nucleate boiling

At higher values of ΔT, the bubbles form more rapidly and form more centres of nucleation. Under these conditions the bubbles exert an appreciable agitation on the liquid and the heat transfer coefficient rises rapidly. This is the most important region for boiling in industrial equipment.

Film boiling

With a sufficiently high value of ΔT, the bubbles are formed so rapidly that they cannot get away from the hot surface, and they therefore form a blanket over the surface. This means that the liquid is prevented from flowing on to the surface by the bubbles of vapour and the heat transfer coefficient falls. The maximum coefficient occurs during nucleate boiling although this is an unstable region for operation. In passing from the *nucleate boiling* region to the *film boiling* region, two critical changes occur in the process. The first manifests itself in a decrease in the heat flux, the second is the prelude to stable film boiling. The intermediate region is generally known as the *transition* region. It may be noted that the first change in the process is an important hydrodynamic phenomenon, which is common to other two-phase systems, such as flooding in countercurrent gas–liquid or vapour–liquid systems, for example.

With very high values of ΔT, the heat transfer coefficient rises again because of heat transfer by radiation. These very high values are rarely achieved in practice and usually the aim is to operate the plant at a temperature difference a little below the value giving the maximum heat transfer coefficient.

1.7.3 Heat Transfer Coefficients and Heat Flux

The values of the heat transfer coefficients for low values of temperature difference are given by Eq. (1.239). Fig. 1.64 shows the values of h and q for water boiling on a submerged surface. Whilst the actual values vary somewhat between investigations, they all give a maximum for a temperature difference of about 22 deg K. The maximum value of h is about 50 kW/m^2 K and the maximum flux is about 1100 kW/m^2.

Similar results have been obtained by Bonilla and Perry,[112] Insinger and Bliss,[113] and others, for a number of organic liquids such as benzene, alcohols, acetone, and carbon tetrachloride. The data in Table 1.14 for liquids boiling at atmospheric pressure show that the maximum heat flux is much smaller with organic liquids than with water and the temperature difference at this condition is rather higher. In practice the critical value of ΔT may be exceeded. Sauer

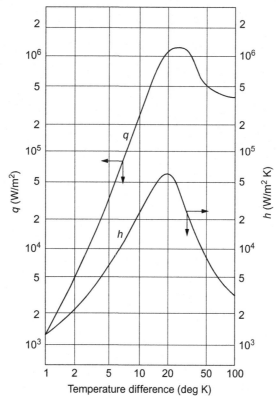

Fig. 1.64

Effect of temperature difference on heat flux and heat transfer coefficient to water boiling at 373 K on a submerged surface.

Table 1.14 Maximum heat flux for various liquids boiling at atmospheric pressure

Liquid	Surface	Critical ΔT (deg K)	Maximum flux (kW/m^2)
Water	Chromium	25	910
50 mol% ethanol-water	Chromium	29	595
Ethanol	Chromium	33	455
n-Butanol	Chromium	44	455
iso-Butanol	Nickel	44	370
Acetone	Chromium	25	455
iso-Propanol	Chromium	33	340
Carbon tetrachloride	Copper	–	180
Benzene	Copper	–	170–230

et al.[114] found that the overall transfer coefficient U for boiling ethyl acetate with steam at 377 kN/m^2 was only 14% of that when the steam pressure was reduced to 115 kN/m^2.

In considering the problem of nucleate boiling, the nature of the surface, the pressure, and the temperature difference must be taken into account as well as the actual physical properties of the liquid.

Apart from the question of scale, the nature of the clean surface has a pronounced influence on the rate of boiling. Thus Bonilla and Perry[112] boiled ethanol at atmospheric pressure and a temperature difference of 23 deg K, and found that the heat flux at atmospheric pressure was 850 kW/m^2 for polished copper, 450 for gold plate, and 370 for fresh chromium plate, and only 140 for old chromium plate. This wide fluctuation means that care must be taken in anticipating the heat flux, since the high values that may be obtained initially may not persist in practice because of tarnishing of the surface.

Effect of temperature difference

Cryder and Finalborgo[115] boiled a number of liquids on a horizontal brass surface, both at atmospheric and at reduced pressure. Some of their results are shown in Fig. 1.65, where the coefficient for the boiling liquid h is plotted against the temperature difference between the hot surface and the liquid. The points for the various liquids in Fig. 1.65 lie on nearly parallel straight lines, which may be represented by:

$$h = \text{constant} \times \Delta T^{2.5} \tag{1.240}$$

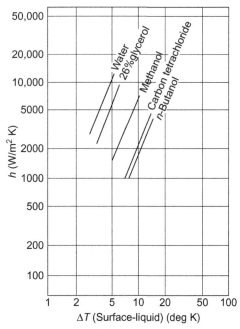

Fig. 1.65

Effect of temperature difference on the heat transfer coefficient for boiling liquids (Cryder and Finalborgo[115]).

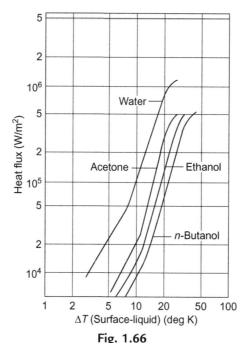

Fig. 1.66
Effect of temperature difference on heat flux to boiling liquids (Bonilla and Perry[112]).

This value for the index of ΔT has been found by other researchers, although Jakob and Linke[116] found values as high as 4 for some of their work. It is important to note that this value of 2.5 is true only for temperature differences up to 19 deg K.

In some ways it is more convenient to show the results in the form of heat flux versus temperature difference, as shown in Fig. 1.66, where some results from a number of researchers are given.

Effect of pressure

Cryder and Finalborgo[115] found that h decreased uniformly as the pressure and hence the boiling point was reduced, according to the relation $h = \text{constant} \times B^{T''}$, where T'' is numerically equal to the temperature in K and B is a constant. Combining this with Eq. (1.240), their results for h were expressed in the empirical form:

$$h = \text{constant} \times \Delta T^{2.5} B^{T''}$$

or, using SI units:

$$\log\left(\frac{h}{5.67}\right) = a' + 2.5 \log \Delta T + b'(T'' - 273) \tag{1.241}$$

where $(T'' - 273)$ is in °C.

If a' and b' are given the following values, h is expressed in W/m^2 K:

	a'	b'		a'	b'
Water	-0.96	0.025	Kerosene	-4.13	0.022
Methanol	-1.11	0.027	10% Na$_2$SO$_4$	-1.47	0.029
CCl$_4$	-1.55	0.022	24% NaCl	-2.43	0.031

The values of a' will apply only to a particular apparatus although a value of b' of 0.025 is of more general application. If h_n is the coefficient at some standard boiling point T_n, and h at some other temperature T, Eq. (1.241) may be rearranged to give:

$$\log \frac{h}{h_n} = 0.025 \left(T'' - T_n'' \right) \tag{1.242}$$

for a given material and temperature difference.

As the pressure is raised above atmospheric pressure, the film coefficient increases for a constant temperature difference. Cichelli and Bonilla[117] have examined this problem for pressures up to the critical value for the vapour, and have shown that ΔT for maximum rate of boiling decreases with the pressure. They obtained a single curve, shown in Fig. 1.67, by plotting q_{max}/P_c against P_R, where P_c is the critical pressure and P_R the reduced pressure $= P/P_c$. This curve represents the data for water, ethanol, benzene, propane, n-heptane, and several mixtures with water. For water, the results cover only a small range of P/P_c because of the high value of P_c. For the organic liquids investigated, it was shown that the maximum value of heat flux q occurs at a pressure P of about one-third of the critical pressure P_c. As shown in Table 1.15, the range of physical properties of the organic liquids is not wide and further data are required to substantiate the previous relation.

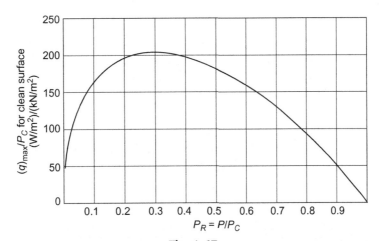

Fig. 1.67
Effect of pressure on the maximum heat flux in nucleate boiling.

Table 1.15 Typical heat transfer coefficients for boiling liquids

Liquid	Boiling Point (deg K)	$\triangle T$ (deg K)	h (W/m² K)	h (Btu/h ft² °F)
Water	372	4.7	9000	1600
	372	2.9	2700	500
	326	8.8	4700	850
	326	6.1	1300	250
Methanol	337	8.9	4800	850
	337	5.6	1500	250
	306	14.4	3000	500
	306	9.3	900	150
Carbon tetrachloride	349	12.6	3500	600
	349	7.2	1100	200
	315	20.1	2000	400
	315	11.8	700	100

1.7.4 Analysis Based on Bubble Characteristics

It is a matter of speculation as to why such high values of heat flux are obtained with the boiling process. It was once thought that the bubbles themselves were carriers of latent heat which was added to the liquid by their movement. It has now been shown, by determining the numbers of bubbles that this mechanism would result in the transfer of only a moderate part of the heat that is actually transferred. The current views are that the high flux arises from the agitation produced by the bubbles, and two rather different explanations have been put forward. Rohsenow and Clark[102] and Rohsenow[103] base their argument on the condition of the bubble on leaving the hot surface. By calculating the velocity and size of the bubble an expression may be derived for the heat transfer coefficient in the form of a Nusselt type equation, relating the Nusselt group to the Reynolds and Prandtl groups. Forster and Zuber,[118,119] however, argue that the important velocity is that of the growing bubble, and this is the term used to express the velocity. In either case the bubble movement is vital in obtaining a high flux. The liquid adjacent to the surface is agitated and exerts a mixing action by pushing hot liquid from the surface to the bulk of the stream.

Considering in more detail the argument proposed by Rohsenhow and Clark[102] and Rohsenhow,[103] the size of a bubble at the instant of breakaway from the surface has been determined by Fritz[120] who has shown that d_b is given by:

$$d_b = C_1 \phi \left(\frac{2\sigma}{g(\rho_l - \rho_v)} \right)^{1/2} \tag{1.243}$$

where σ is the surface tension, ρ_l and ρ_v the density of the liquid and vapour, ϕ is the contact angle, and C_1 is a constant depending on conditions.

The flowrate of vapour per unit area as bubbles u_b is given by:

$$u_b = \frac{fn\pi d_b^3}{6} \tag{1.244}$$

where f is the frequency of bubble formation at each bubble site and n is the number of sites of nucleation per unit area.

The heat transferred by the bubbles q_b is to a good approximation given by:

$$q_b = \frac{1}{6}\pi d_b^3 fn\rho_v \lambda \tag{1.245}$$

where λ is the latent heat of vapourisation.

It has been shown that for heat flux rates up to 3.2 kW/m^2, the product $f\,d_b$ is constant and that the total heat flow per unit area q is proportional to n. From Eq. (1.245), it is seen that q_b is proportional to n at a given pressure, so that $q \propto q_b$.

Hence:

$$q = C_2\frac{\pi}{6}d_b^3 fn\rho_v \lambda \tag{1.246}$$

Substituting from Eqs. (1.245), (1.246), the mass flow per unit area:

$$\rho_v u_b = fn\frac{\pi}{6}d_b^3\rho_v = \frac{q}{C_2\lambda} \tag{1.247}$$

A Reynolds number for the bubble flow, which represents the term for agitation may be defined as:

$$Re_b = \frac{d_b\rho_v u_b}{\mu_l}$$

$$= C_1\phi\left(\frac{2\sigma}{g(\rho_l - \rho_v)}\right)^{1/2}\left(\frac{q}{C_2\lambda}\right)\frac{1}{\mu_l} \tag{1.248}$$

$$= C_3\phi\frac{q}{\lambda\mu_l}\left(\frac{\sigma}{g(\rho_l - \rho_v)}\right)^{1/2}$$

The Nusselt group for bubble flow,

$$Nu_b = h_b C_1 \frac{\phi_1}{k_l}\left(\frac{2\sigma}{g(\rho_l - \rho_v)}\right)^{1/2}$$

$$= C_4 h_b \frac{\phi}{k_l}\left(\frac{\sigma}{g(\rho_l - \rho_v)}\right)^{1/2} \tag{1.249}$$

and hence a final correlation is obtained of the form:

$$Nu_b = \text{constant} Re_b^n Pr^m \tag{1.250}$$

or:

$$Nu_b = \text{constant} \left[\frac{C_3\phi q}{\mu_l \lambda} \left(\frac{\sigma}{g(\rho_l - \rho_v)}\right)^{1/2}\right]^n \left(\frac{C_l\mu_l}{k_l}\right)^m \tag{1.251}$$

where n and m have been found experimentally to be 0.67 and -0.7 respectively and the constant, which depends on the metal surface, ranges from 67 to 100 for polished chromium, 77 for platinum wire, and 166 for brass.[103]

A comprehensive study of nucleate boiling of a wide range of liquids on thick plates of copper, aluminium, brass, and stainless steel has been carried out by Pioro[121] who has evaluated the constants in Eq. (1.251) for different combinations of liquid and surface.

Forster and Zuber[118,119] who employed a similar basic approach, although the radial rate of growth dr/dt was used for the bubble velocity in the Reynolds group, showed that:

$$\frac{dr}{dt} = \frac{\Delta T C_l \rho_l}{2\lambda \rho_v} \left(\frac{\pi D_{Hl}}{t}\right)^{1/2} \tag{1.252}$$

where D_{Hl} is the thermal diffusivity ($k_l/C_l\rho_l$) of the liquid. Using this method, a final correlation in the form of Eq. (1.250) has been presented.

Although these two forms of analysis give rise to somewhat similar expressions, the basic terms are evaluated in quite different ways and the final expressions show many differences. Some data fit the Rohsenow equation reasonably well,[121] and other data fit Forster's equation somewhat better.

These expressions all indicate the importance of the bubbles on the rate of transfer, although as yet they have not been used for design purposes. Insinger and Bliss[113] made the first approach by dimensional analysis and Mcnelly[122] has subsequently obtained a more satisfactory result. The influence of ΔT is taken into account by using the flux q, and the last term allows for the change in volume when the liquid vapourises. The following expression was obtained in which the numerical values of the indices were deduced from existing data:

$$\frac{hd}{k_l} = 0.225 \left(\frac{C_l\mu_l}{k_l}\right)^{0.69} \left(\frac{qd}{\lambda\mu}\right)^{0.69} \left(\frac{Pd}{\sigma}\right)^{0.31} \left(\frac{\rho_l}{\rho_v} - 1\right)^{0.33} \tag{1.253}$$

1.7.5 Subcooled Boiling

If bubbles are formed in a liquid, which is much below its boiling point, then the bubbles will collapse in the bulk of the liquid. Thus if a liquid flows over a very hot surface then the bubbles formed are carried away from the surface by the liquid and subcooled boiling occurs. Under these conditions a very large number of small bubbles are formed and a very high heat flux is obtained. Some results for these conditions are given in Fig. 1.68.

If a liquid flows through a tube heated on the outside, then the heat flux q will increase with ΔT as shown in Fig. 1.68. Beyond a certain value of ΔT, the increase in q is very rapid. If the velocity through the tube is increased, then a similar plot is obtained with a higher value of q at low values of ΔT and then the points follow the first line. Over the first section, forced convection boiling exists where an increase in Reynolds number does not bring about a very great increase in q because the bubbles are themselves producing agitation in the boundary layer near the wall. Over the steep section, subcooled boiling exists where the velocity is not

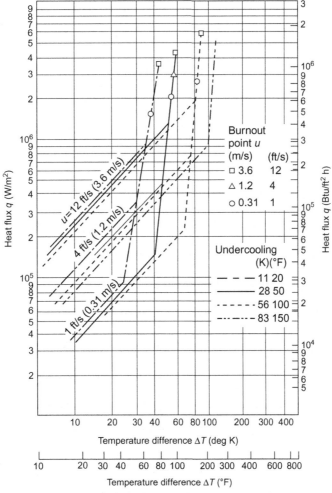

Fig. 1.68

Heat flux in subcooled boiling.

important provided it is sufficient to remove the bubbles rapidly from the surface. In the same way, mechanical agitation of a liquid boiling on a submerged surface will not markedly increase the heat flux.

1.7.6 Design Considerations

In the design of vapourisers and reboilers, two types of boiling are important—nucleate boiling in a pool of liquid as in a kettle-type reboiler or a jacketed vessel, and convective boiling which occurs where the vapourising liquid flows over a heated surface and heat transfer is by both forced convection and nucleate boiling as, for example, in forced circulation or thermosyphon reboilers. The discussion here is a summary of that given in Volume 6 where a worked example is given.

In the absence of experimental data, the correlation given by Forster and Zuber[119] may be used to estimate *pool boiling* coefficients, although the following reduced pressure correlation given by Mostinski[123] is much simpler to use and gives reliable results for h (in W/m^2 K):

$$h = 0.104 P_c^{0.69} q^{0.7} \left[1.8 \left(\frac{P}{P_c} \right)^{0.17} + 4 \left(\frac{P}{P_c} \right)^{1.2} + 10 \left(\frac{P}{P_c} \right)^{10} \right] \tag{1.254}$$

In this equation, P_c and P are the critical and operating pressures (bar), respectively, and q is the heat flux (W/m^2). Both equations are for single component fluids, although they may also be used for close-boiling mixtures and for wider boiling ranges with a factor of safety. In reboiler and vapouriser design, it is important that the heat flux is well below the critical value. A correlation is given for the heat transfer coefficient for the case where film-boiling takes place on tubes submerged in the liquid.

Convective boiling, which occurs when the boiling liquid flows through a tube or over a tube bundle (such as in evaporators), depends on the state of the fluid at any point. The effective heat transfer coefficient can be considered to be made up of the convective and nucleate boiling components. The convective boiling coefficient is estimated using an equation for single-phase forced-convection heat transfer (Eq. 1.92, for example) modified by a factor to allow for the effects of two-phase flow. Similarly, the nucleate boiling coefficient is obtained from the Forster and Zuber or Mostinski correlation, modified by a factor dependent on the liquid Reynolds number and on the effects of two-phase flow. The estimation of convective boiling coefficients is illustrated by means of an example in Volume 6 and in design handbooks.[107]

One of the most important areas of application of heat transfer to boiling liquids is in the use of evaporators to affect an increase in the concentration of a solution. This topic is considered in Volume 2.

For vapourising the liquid at the bottom of a distillation column, a reboiler is used, as shown in Fig. 1.69. The liquid from the still enters the boiler at the base and, after flowing over the

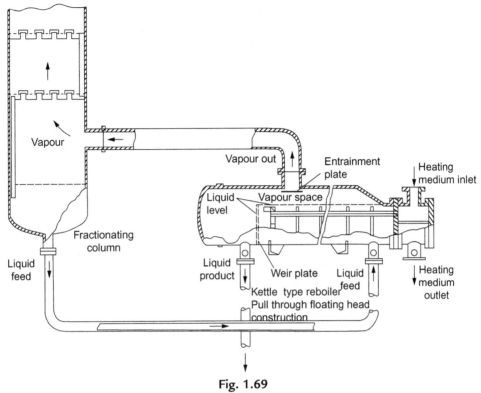

Fig. 1.69
Reboiler installed on a distillation column.

tubes, passes out over a weir. The vapour formed, together with any entrained liquid, passes from the top of the unit to the column. The liquid flow may be either by gravity or by forced circulation. In such equipment, provision is made for expansion of the tubes either by having a floating head as shown, or by arranging the tubes in the form of a hairpin bend (Fig. 1.70). A vertical reboiler may also be used with steam condensing on the outside of the tube bundle. With all systems, it is undesirable to vapourise more than a small percentage of the feed since a good liquid flow over the tubes is necessary to avoid scale formation.

In the design of *forced convection* reboilers, the normal practice is to calculate the heat transfer coefficient on the assumption that heat is transferred by forced convection only, and this gives safe values. Kern[44] recommends that the heat flux should not exceed 60 kW/m^2 for organics and 90 kW/m^2 for dilute aqueous solutions. In *thermosyphon reboilers*, the fluid circulates at a rate at which the pressure losses in the system are just balanced by the hydrostatic head and the design involves an iterative procedure based on an assumed circulation rate through the exchanger. *Kettle reboilers*, such as that shown in Fig. 1.70, are essentially pool boiling devices and their design, based on nucleate boiling data, uses the Zuber equation for single tubes, modified by a tube-density factor. This general approach is developed further in Volume 6.

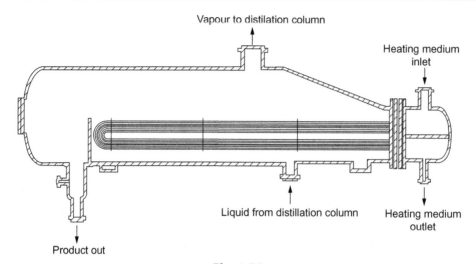

Vapour to distilation column

Heating medium inlet

Liquid from distillation column

Heating medium outlet

Product out

Fig. 1.70
Kettle reboiler with hairpin tubes.

1.8 Heat Transfer in Reaction Vessels

1.8.1 Helical Cooling Coils

A simple jacketed pan or kettle is very commonly used in the processing industries as a reaction vessel. In many cases, such as in nitration or sulphonation reactions, heat has to be removed or added to the mixture in order to either control the rate of reaction or to bring it to completion. The addition or removal of heat is conveniently arranged by passing steam or water through a jacket fitted to the outside of the vessel or through a helical coil fitted inside the vessel. In either case, some form of agitator is used to obtain even distribution in the vessel. This may be of the anchor type for very thick pastes or a propeller or turbine if the contents are not too viscous, as discussed in Chapter 7 of Vol. 1A.

In such a vessel, the thermal resistances to heat transfer arise from the water film on the inside of the coil, the wall of the tube, the film on the outside of the coil, and any scale that may be present on either surface. The overall transfer coefficient may be expressed by:

$$\frac{1}{UA} = \frac{1}{h_i A_i} + \frac{x_w}{k_w A_w} + \frac{1}{h_o A_o} + \frac{R_o}{A_o} + \frac{R_i}{A_i} \tag{1.255}$$

where R_o and R_i are the scale resistances and the other terms have the usual definitions.

Inside film coefficient

The value of h_i may be obtained from a form of Eq. (1.92):

$$\frac{h_i d}{k} = 0.023 \left(\frac{du\rho}{\mu}\right)^{0.8} \left(\frac{C_p\mu}{k}\right)^{0.33} \tag{1.256}$$

if water is used in the coil, and the Sieder and Tate equation (Eq. 1.94) if a viscous brine is used for cooling.

These equations have been obtained for straight tubes; with a coil somewhat greater transfer is obtained for the same physical conditions. Jeschke[124] cooled air in a 31 mm steel tube wound in the form of a helix and expressed his results in the form:

$$h_i(\text{coil}) = h_i(\text{straight pipe}) \left(1 + 3.5\frac{d}{d_c}\right) \tag{1.257}$$

where d is the inside diameter of the tube and d_c the diameter of the helix. Pratt[125] has examined this problem in greater detail for liquids and has given almost the same result. Combining Eqs. (1.256), (1.257), the inside film coefficient h_i for the coil may be calculated.

Outside film coefficient

The value of h_o is determined by the physical properties of the liquor and by the degree of agitation achieved. This latter quantity is difficult to express in a quantitative manner and the group L^2N ρ/μ has been used both for this problem and for the allied one of power used in agitation, as discussed in Chapter 7 of Vol. 1A. In this group L is the length of the paddle and N the revolutions per unit time. Chilton, Drew and Jebens,[126] working with a small tank only 0.3 m in diameter d_v, expressed their results by:

$$\frac{h_o d_v}{k} \left(\frac{\mu_s}{\mu}\right)^{0.14} = 0.87 \left(\frac{C_p\mu}{k}\right)^{1/3} \left(\frac{L^2N\rho}{\mu}\right)^{0.62} \tag{1.258}$$

where the factor $(\mu_s/\mu)^{0.14}$ allows for the difference between the viscosity adjacent to the coil (μ_s) and that in the bulk of the liquor. A wide range of physical properties was achieved by using water, two oils, and glycerol.

Pratt[125] used both circular and square tanks of up to 0.6 m in size and a series of different arrangements of a simple paddle as shown in Fig. 1.71. The effect of altering the arrangement of the coil was investigated and the tube diameter d_o, the gap between the turns d_g, the diameter of the helix d_c, the height of the coil d_p, and the width of the stirrer W were all varied. The final equations for tanks were:

For cylindrical tanks

$$\frac{h_o d_v}{k} = 34 \left(\frac{L^2N\rho}{\mu}\right)^{0.5} \left(\frac{C_p\mu}{k}\right)^{0.3} \left(\frac{d_g}{d_p}\right)^{0.8} \left(\frac{W}{d_c}\right)^{0.25} \left(\frac{L^2d_v}{d_o^3}\right)^{0.1} \tag{1.259}$$

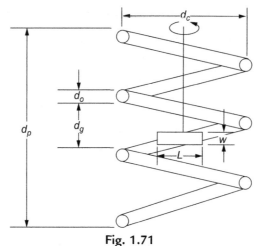

Fig. 1.71
Arrangement of coil in Pratt's work.[125]

For square tanks:

$$\frac{h_o l_v}{k} = 39 \left(\frac{L^2 N \rho}{\mu}\right)^{0.5} \left(\frac{C_p \mu}{k}\right)^{0.3} \left(\frac{d_g}{d_p}\right)^{0.8} \left(\frac{W}{d_c}\right)^{0.25} \left(\frac{L^2 l_v}{d_o^3}\right)^{0.1} \tag{1.260}$$

where l_v is the length of the side of the vessel.

These give almost the same results as the earlier equations over a wide range of conditions. Cummings and West[127] have tested these results with a much larger tank of 0.45 m^3 capacity and have given an expression similar to Eq. (1.258) but with a constant of 1.01 instead of 0.87. A retreating blade turbine impeller was used, and in many cases a second impeller was mounted above the first, giving an agitation which is probably more intense than that attained by the other researchers. A constant of 0.9 seems a reasonable average from existing work.

Example 1.36

Toluene is continuously nitrated to mononitrotoluene in a cast-iron vessel, 1 m diameter, fitted with a propeller agitator 0.3 m diameter rotating at 2.5 Hz. The temperature is maintained at 310 K by circulating 0.5 kg/s cooling water through a stainless steel coil 25 mm o.d. and 22 mm i.d. wound in the form of a helix, 0.80 m in diameter. The conditions are such that the reacting material may be considered to have the same physical properties as 75% sulphuric acid. If the mean water temperature is 290 K, what is the overall coefficient of heat transfer?

Solution

The overall coefficient U_o based on the outside area of the coil is given by Eq. (1.255):

$$\frac{1}{U_o} = \frac{1}{h_o} + \frac{x_w d_o}{k_w d_w} + \frac{d_o}{h_i d} + R_o + \frac{R_i d_o}{d}$$

where d_w is the mean diameter of the pipe.

From Eqs. (1.256), (1.257), the inside film coefficient for the water is given by:

$$h_i = \frac{k}{d}\left(1 + 3.5\frac{d}{d_c}\right)0.023\left(\frac{du\rho}{\mu}\right)^{0.8}\left(\frac{C_p\mu}{k}\right)^{0.4}$$

In this equation:

$$\rho u = \frac{0.5}{(\pi/4) \times 0.022^2} = 1315\,kg/m^2\,s$$

$d = 0.022$ m, $d_c = 0.80$ m, $k = 0.59$ W/m K, $\mu = 1.08$ mN s/m^2 or 1.08×10^{-3} N s/m^2, and $C_p = 4.18 \times 10^3$ J/kg K

Thus:

$$h_i = \frac{0.59}{0.022}\left(1 + 3.5 \times \frac{0.022}{0.80}\right)0.023\left(\frac{0.022 \times 1315}{1.08 \times 10^{-3}}\right)^{0.8}\left(\frac{4.18 \times 10^3 \times 1.08 \times 10^{-3}}{0.59}\right)^{0.4}$$

$$= 0.680(26,780)^{0.8}(7.65)^{0.4} = 5490\,W/m^2\,K$$

The external film coefficient is given by Eq. (1.258):

$$\frac{h_o d_v}{k}\left(\frac{\mu_s}{\mu}\right)^{0.14} = 0.87\left(\frac{C_p\mu}{k}\right)^{0.33}\left(\frac{L^2 N\rho}{\mu}\right)^{0.62}$$

For 75% sulphuric acid:

$k = 0.40$ W/m K, $\mu_s = 8.6 \times 10^{-3}$ N s/m^2 at 300 K, $\mu = 6.5 \times 10^{-3}$ N s/m^2 at 310 K, $C_p = 1.88 \times 10^3$ J/kg K, and $p = 1666$ kg/m^3

Thus:

$$\frac{h_o \times 1.0}{0.40}\left(\frac{8.6}{6.5}\right)^{0.14} = 0.87\left(\frac{1.88 \times 10^3 \times 6.5 \times 10^{-3}}{0.40}\right)\left(\frac{0.3^2 \times 2.5 \times 1666}{6.5 \times 10^{-3}}\right)^{0.62}$$

$$2.5h_o \times 1.4 = 0.87 \times 3.09 \times 900$$

and:

$$h_o = 930\,W/m^2\,K$$

Taking $k_w = 15.9$ W/m K and R_o and R_i as 0.0004 and 0.0002 m^2 K/W, respectively: and:

$$\frac{1}{U_o} = \frac{1}{930} + \frac{0.0015 \times 0.025}{15.9 \times 0.0235} + \frac{0.025}{5490 \times 0.022} + 0.0004 + \frac{0.0002 \times 0.025}{0.022}$$

$$= 0.00107 + 0.00010 + 0.00021 + 0.00040 + 0.00023 = 0.00201\,m^2\,K/W$$

and:

$$\underline{U_o = 498\,W/m^2\,K}$$

In this calculation a mean area of surface might have been used with sufficient accuracy. It is important to note the great importance of the scale terms which together form a major part of the thermal resistance.

1.8.2 Jacketed Vessels

In many cases, heating or cooling of a reaction mixture is most satisfactorily achieved by condensing steam in a jacket or passing water through it—an arrangement which is often used for organic reactions where the mixture is too viscous for the use of coils and a high-speed agitator. Chilton et al.[126] and Cummings and West[127] have measured the transfer coefficients for this case by using an arrangement as shown in Fig. 1.72, where heat is supplied to the jacket and simultaneously removed by passing water through the coil. Chilton measured the temperatures of the inside of the vessel wall, the bulk liquid, and the surface of the coil by means of thermocouples and thus obtained the film heat transfer coefficients directly. Cummings and West[127] used an indirect method to give the film coefficient from measurements of the overall coefficients.

Chilton et al.[126] expressed their results by:

$$\frac{h_b d_v}{k}\left(\frac{\mu_s}{\mu}\right)^{0.14} = 0.36\left(\frac{L^2 N \rho}{\mu}\right)^{0.67}\left(\frac{C_p \mu}{k}\right)^{0.33}\tag{1.261}$$

where h_b is the film coefficient for the liquor adjacent to the wall of the vessel. Cummings and West[127] used the same equation although the coefficient was 0.40. Considering that Chilton's vessel was only 0.3 m in diameter and fitted with a single paddle of 150 mm length, and that Cummings and West used a 0.45 m^3 vessel with two turbine impellers, agreement between their results is remarkably good. The group $(\mu_s/\mu)^{0.14}$ is again used to allow for the difference in the viscosities at the surface and in the bulk of the fluid.

Brown et al.[128] have given data on the performance of 1.5 m diameter sulphonators and nitrators of 3.4 m^3 capacity as used in the dyestuffs industry. The sulphonators were of cast iron and had a wall thickness of 25.4 mm; the annular space in the jacket being also 25.4 mm. The agitator of the sulphonator was of the anchor type with a 127 mm clearance at the walls and was

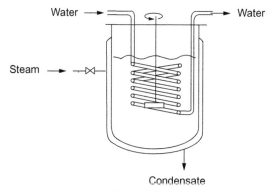

Fig. 1.72
Reaction vessel with jacket and coil.

Table 1.16 Data on common agitators for use in Eqs. (1.261), (1.262)

Type of Agitator	Constant	Index
Flat blade disc turbine		
Unbaffled, or baffled vessel, $Re < 400$	0.54	0.67
Baffled, $Re > 400$	0.74	0.67
Retreating-blade turbine with three blades, jacketed and baffled vessel, $Re = 2 \times 10^4$ to 2×10^6		
Glassed steel impeller	0.33	0.67
Alloy steel impeller	0.37	0.67
Propeller with three blades		
Baffled vessel, $Re = 5500–37,000$	0.64	0.67
Flat blade paddle		
Baffled or unbaffled vessel, $Re > 4000$	0.36	0.67
Anchor		
$Re = 30–300$	1.00	0.50
$Re = 300–5000$	0.38	0.67

driven at 0.67 Hz. The nitrators were fitted with four-blade propellers of 0.61 m diameter driven at 2 Hz. For cooling, the film coefficient h_b for the inside of the vessel was given by:

$$\frac{h_b d_v}{k} \left(\frac{\mu_s}{\mu}\right)^{0.14} = 0.55 \left(\frac{L^2 N \rho}{\mu}\right)^{0.67} \left(\frac{C_p \mu}{k}\right)^{0.25} \tag{1.262}$$

which is very similar to that given by Eq. (1.261).

The film coefficients for the water jacket were in the range 635–1170 W/m² K for water rates of 1.44–9.23 1/s, respectively. It may be noted that 7.58 1/s corresponds to a vertical velocity of only 0.061 m/s and to a Reynolds number in the annulus of 5350. The thermal resistance of the wall of the pan was important, since with the sulphonator it accounted for 13% of the total resistance at 323 K and 31% at 403 K. The change in viscosity with temperature is important when considering these processes since, for example, the viscosity of the sulphonation liquors ranged from 340 mN s/m² at 323 K to 22 mN s/m² at 403 K.

In discussing Eqs. (1.261), (1.262) Fletcher[129] has summarised correlations obtained for a wide range of impeller and agitator designs in terms of the constant before the Reynolds number and the index on the Reynolds number as shown in Table 1.16.

1.8.3 Time Required for Heating or Cooling

It is frequently necessary to heat or cool the contents of a large batch reactor or storage tank. In this case the physical constants of the liquor may alter and the overall transfer coefficient may change during the process. In practice, it is often possible to assume an average value of the transfer coefficient so as to simplify the calculation of the time required for heating or cooling a batch of liquid. The heating of the contents of a storage tank is commonly effected by condensing steam, either in a coil or in some form of hairpin tube heater.

In the case of a storage tank with liquor of mass m and specific heat C_p, heated by steam condensing in a helical coil, it may be assumed that the overall transfer coefficient U is constant. If T_s is the temperature of the condensing steam, T_1 and T_2 the initial and final temperatures of the liquor, A the area of heat transfer surface, and T is the temperature of the liquor at any time t, then the rate of transfer of heat is given by:

$$Q = mC_p\frac{dT}{dt} = UA(T_s - T)$$

$$\therefore \quad \frac{dT}{dt} = \frac{UA}{mC_p}(T_s - T)$$

$$\therefore \quad \int_{T_1}^{T_2}\frac{dT}{T_s - T} = \frac{UA}{mC_p}\int_0^t dt$$

$$\therefore \quad \ln\frac{T_s - T_1}{T_s - T_2} = \frac{UA}{mC_p}t \tag{1.263}$$

From this equation, the time t of heating from T_1 to T_2, may be calculated. The same analysis may be used if the steam condenses in a jacket of a reaction vessel.

This analysis does not allow for any heat losses during the heating, or, for that matter, cooling operation. Obviously the higher the temperature of the contents of the vessel, the greater are the heat losses and, in the limit, the heat supplied to the vessel is equal to the heat losses, at which stage no further rise in the temperature of the contents of the vessel is possible. This situation is illustrated in Example 1.37.

The heating-up time can be reduced, by improving the rate of heat transfer to the fluid, by agitation of the fluid for example, and by reducing heat losses from the vessel by insulation. In the case of a large vessel there is a limit to the degree of agitation possible, and circulation of the fluid through an external heat exchanger is an attractive alternative.

Example 1.37

A vessel contains 1 tonne (1 Mg) of a liquid of specific heat capacity 4.0 kJ/kg K. The vessel is heated by steam at 393 K which is fed to a coil immersed in the agitated liquid and heat is lost to the surroundings at 293 K from the outside of the vessel. How long does it take to heat the liquid from 293 to 353 K and what is the maximum temperature to which the liquid can be heated? When the liquid temperature has reached 353 K, the steam supply is turned off for 2 h (7.2 ks) and the vessel cools. How long will it take to reheat the material to 353 K? The surface area of the coil is 0.5 m² and the overall coefficient of heat transfer to the liquid may be taken as 600 W/m² K. The outside area of the vessel is 6 m² and the coefficient of heat transfer to the surroundings may be taken as 10 W/m² K.

Solution

If T K is the temperature of the liquid at time t s, then a heat balance on the vessel gives:

$$(1000 \times 4000)\frac{dT}{dt} = (600 \times 0.5)(393 - T) - (10 \times 6)(T - 293)$$

or:

$$4,000,000\frac{dT}{dt} = 135,480 - 360T$$

and:

$$11,111\frac{dT}{dt} = 376.3 - T$$

The equilibrium temperature occurs when $dT/dt = 0$, that is when:

$$\underline{T = 376.3\,\text{K.}}$$

In heating from 293 to 353 K, the time taken is:

$$t = 11,111\int_{293}^{353}\frac{dT}{(376.3 - T)}$$

$$= 11,111\ln\left(\frac{83.3}{23.3}\right)$$

$$= 14,155\,\text{s (or 3.93 h)}.$$

The steam is turned off for 7200 s and during this time a heat balance gives:

$$(1000 \times 4000)\frac{dT}{dt} = -(10 \times 6)(T - 293)$$

$$66,700\frac{dT}{dt} = 293 - T$$

The change in temperature is then given by:

$$\int_{353}^{T}\frac{dT}{(293 - T)} = \frac{1}{66,700}\int_{0}^{7200}dt$$

$$\ln\frac{-60}{293 - T} = \frac{7200}{66,700} = 0.108$$

and:

$$T = 346.9\,\text{K.}$$

The time taken to reheat the liquid to 353 K is then given by:

$$t = 11,111\int_{346.9}^{353}\frac{dT}{(376.3 - T)}$$

$$= 11,111\ln\left(\frac{29.4}{23.3}\right)$$

$$= \underline{\underline{2584\,\text{s (0.72 h)}}}$$

1.9 Shell and Tube Heat Exchangers

1.9.1 General Description

Since shell and tube heat exchangers can be constructed with a very large heat transfer area in a relatively small volume, fabricated from alloy steels to resist corrosion and be used for heating, cooling, and for condensing a very wide range of fluids, they are the most widely used form of heat transfer equipment. Figs 1.73–1.75 show various forms of construction. The simplest type of unit, shown in Fig. 1.73, has fixed tube plates at each end into which the tubes are expanded. The tubes are connected so that the internal fluid makes several passes up and down the exchanger thus enabling a high velocity of flow to be obtained for a given heat transfer area and throughput of fluid. The fluid flowing in the shell is made to flow first in one sense and then in the opposite sense across the tube bundle by fitting a series of baffles along the length. These baffles are frequently of the segmental form with about 25% cut away, as shown in Fig. 1.80 to provide the free space to increase the velocity of flow across the tubes, thus giving higher rates of heat transfer. One problem with this type of

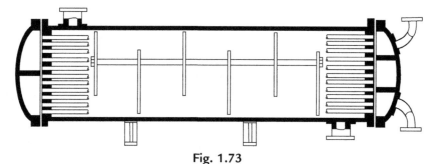

Fig. 1.73
Heat exchanger with fixed tube plates (four tube, one shell-pass).

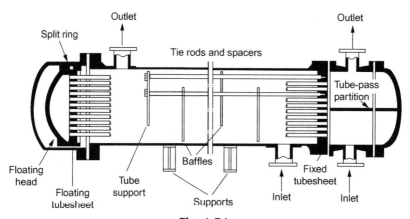

Fig. 1.74
Heat exchanger with floating head (two tube-pass, one shell-pass).

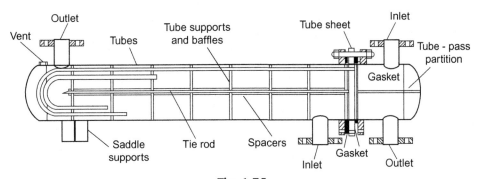

Fig. 1.75
Heat exchanger with hairpin tubes.

construction is that the tube bundle cannot be removed for cleaning and no provision is made to allow for differential (thermal) expansion between the tubes and the shell, although an expansion joint may be fitted to the shell.

In order to allow for the removal of the tube bundle and for considerable expansion of the tubes, a floating head exchanger is used, as shown in Fig. 1.74. In this arrangement one tube plate is fixed as before, but the second is bolted to a floating head cover so that the tube bundle can move relative to the shell. This floating tube sheet is clamped between the floating head and a split backing flange in such a way that it is relatively easy to break the flanges at both ends and to draw out the tube bundle. It may be noted that the shell cover at the floating head end is larger than that at the other end. This enables the tubes to be placed as near as possible to the edge of the fixed tube plate, leaving very little unused space between the outer ring of tubes and the shell.

Another arrangement which provides for expansion involves the use of hairpin tubes, as shown in Fig. 1.75. This design is very commonly used for the reboilers on large fractionating columns where steam is condensed inside the tubes.

In these designs, there is one pass for the fluid on the shell-side and a number of passes on the tube-side. It is often an advantage to have two or more shell-side passes, although this considerably increases the difficulty of construction and, very often therefore, several smaller exchangers are connected together to obtain the same effect.

The essential requirements in the design of a heat exchanger are, firstly, the provision of a unit which is reliable and has the desired capacity, and secondly, the need to provide an exchanger at minimum overall cost. In general, this involves using standard components and fittings and making the design as simple as possible. In most cases, it is necessary to balance the capital cost in terms of the depreciation against the operating cost. Thus in a condenser, for example, a high heat transfer coefficient is obtained and hence a small exchanger is required if a higher water velocity is used in the tubes. Against this, the cost of pumping increases rapidly with

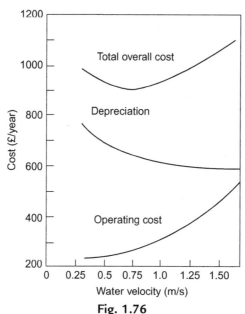

Fig. 1.76

Effect of water velocity on annual operating cost of condenser.

increase in velocity and an economic balance must be struck. A typical graph showing the operating costs, depreciation, and the total cost plotted as a function of the water velocity in the tubes is shown in Fig. 1.76.

1.9.2 Basic Components

The various components, which make up a shell and tube heat exchanger are shown in Figs 1.74 and 1.75 and these are now considered. Many different mechanical arrangements are used and it is convenient to use a basis for classification. The standard published by the Tubular Exchanger Manufacturer's Association (TEMA[130]) is outlined here. It should be noted that Saunders[131] has presented a detailed discussion of design codes and problems in fabrication.

Of the various *shell types* shown in Fig. 1.77, the simplest, with entry and exit nozzles at opposite ends of a single pass exchanger, is the TEMA E-type on which most design methods are based, although these may be adapted for other shell types by allowing for the resulting velocity changes. The TEMA F-type has a longitudinal baffle giving two shell passes and this provides an alternative arrangement to the use of two shells required in order to cope with a close temperature approach or low shell-side flowrates. The pressure drop in two shells is some eight times greater than that encountered in the E-type design, although any potential leakage between the longitudinal baffle and the shell in the F-type design may restrict the range of application. The so-called 'split-flow' type of unit with a longitudinal baffle is classified as the

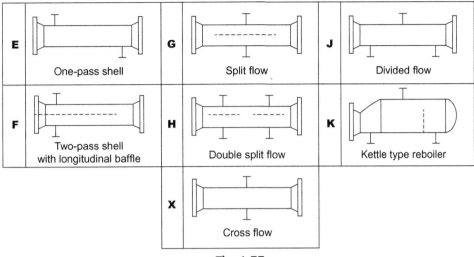

Fig. 1.77
TEMA shell types.

TEMA G-type whose performance is superior although the pressure drop is similar to the E-type. This design is used mainly for reboilers and only occasionally for systems where there is no change of phase. The so-called 'divided-flow' type, the TEMA J-type, has one inlet and two outlet nozzles and, with a pressure drop some one-eighth of the E-type, finds application in gas coolers and condensers operating at low pressures. The TEMA X-type shell has no cross baffles and hence the shell-side fluid is in pure counterflow giving extremely low pressure drops and again, this type of design is used for gas cooling and condensation at low pressures.

The *shell* of a heat exchanger is commonly made of carbon steel and standard pipes are used for the smaller sizes and rolled welded plate for the larger sizes (say 0.4–1.0 m). The thickness of the shell may be calculated from the formula for thin-walled cylinders and a minimum thickness of 9.5 mm is used for shells over 0.33 m o.d. and 11.1 mm for shells over 0.9 m o.d. Unless the shell is designed to operate at very high pressures, the calculated wall thickness is usually less than these values, although a corrosion allowance of 3.2 mm is commonly added to all carbon steel parts and thickness is determined more by rigidity requirements than simply internal pressure. The minimum shell thickness for various materials is given in BS3274.[132] A shell diameter should be such as to give as close a fit to the tube bundle as practical in order to reduce bypassing around the outside of the bundle. Typical values for the clearance between the outer tubes in the bundle and the inside diameter of the shell are given in Fig. 1.78 for various types of exchanger.

The detailed design of the *tube bundle* must take into account both shell-side and tube-side pressures since these will both affect any potential leakage between the tube bundle and the shell which cannot be tolerated where high purity or uncontaminated materials are required.

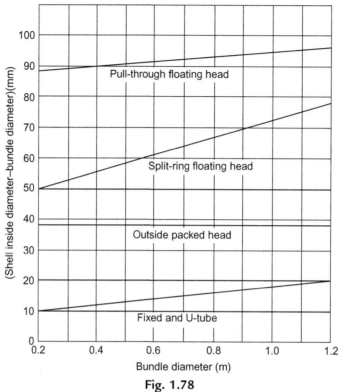

Fig. 1.78

Shell-bundle clearance.

In general, tube bundles make use of a fixed tubesheet, a floating-head or U-tubes which are shown in Figs 1.73, 1.74, and 1.75 respectively. It may be noted here that the thickness of the fixed tubesheet may be obtained from a relationship of the form:

$$d_t = d_G \sqrt{(0.25P/f)} \tag{1.264}$$

where d_G is the diameter of the gasket (m), P the design pressure (MN/m^2), f the allowable working stress (MN/m^2) and d_t the thickness of the sheet measured at the bottom of the partition plate grooves. The thickness of the floating head tubesheet is very often calculated as $\sqrt{2d_t}$.

In selecting a *tube diameter*, it may be noted that smaller tubes give a larger heat transfer area for a given shell, although 19 mm o.d. tubes are normally the minimum size used in order to permit adequate cleaning. Although smaller diameters lead to shorter tubes, more holes have to be drilled in the tubesheet, which adds to the cost of construction and increases the likelihood of tube vibration. Heat exchanger tubes are usually in the range 16 mm (5/8 in.) to 50 mm (2 in.) O.D.; the smaller diameter usually being preferred as these give more compact and therefore cheaper units. Against this, larger tubes are easier to clean especially by mechanical methods and are therefore widely used for heavily fouling fluids. The tube

Table 1.17 Standard dimensions of steel tubes

Outside Diameter d_o		Wall Thickness		Cross Sectional Area for Flow		Surface are Per Unit Length	
(mm)	(in)	(mm)	(in)	(m^2)	(ft^2)	(m^2/m)	(ft^2/ft)
16	0.630	1.2	0.047	0.000145	0.00156	0.0503	0.165
		1.6	0.063	0.000129	0.00139		
		2.0	0.079	0.000113	0.00122		
20	0.787	1.6	0.063	0.000222	0.00239	0.0628	0.206
		2.0	0.079	0.000201	0.00216		
		2.6	0.102	0.000172	0.00185		
25	0.984	1.6	0.063	0.000373	0.00402	0.0785	0.258
		2.0	0.079	0.000346	0.00373		
		2.6	0.102	0.000308	0.00331		
		3.2	0.126	0.000272	0.00293		
30	1.181	1.6	0.063	0.000564	0.00607	0.0942	0.309
		2.0	0.079	0.000531	0.00572		
		2.6	0.102	0.000483	0.00512		
		3.2	0.126	0.000437	0.00470		
38	1.496	2.0	0.079	0.000908	0.00977	0.1194	0.392
		2.6	0.102	0.000845	0.00910		
		3.2	0.126	0.000784	0.00844		
50	1.969	2.0	0.079	0.001662	0.01789	0.1571	0.515
		2.6	0.102	0.001576	0.01697		
		3.2	0.126	0.001493	0.01607		

thickness or gauge must be such as to withstand the internal pressure and also to provide an adequate corrosion allowance. Details of steel tubes used in heat exchangers are given in BS3606[133] and summarised in Table 1.17, and standards for other materials are given in BS3274.[132]

In general, the larger the *tube length*, the lower is the cost of an exchanger for a given surface area due to the smaller shell diameter, the thinner tube sheets and flanges and the smaller number of holes to be drilled, and the reduced complexity. Preferred tube lengths are 1.83 m (6 ft), 2.44 m (8 ft), 3.88 m (12 ft), and 4.88 m (16 ft); larger sizes are used where the total tube-side flow is low and fewer, longer tubes are required in order to obtain a required velocity. With the number of tubes per tube-side pass fixed in order to obtain a required velocity, the total length of tubes per tube-side pass is determined by the heat transfer surface required. It is then necessary to fit the tubes into a suitable shell to give the desired shell-side velocity. It may be noted that with long tube lengths and relatively few tubes in a shell, it may be difficult to arrange sufficient baffles for adequate support of the tubes. For good all-round performance, the ratio of tube length to shell diameter is usually in the range 5–10.

Tube layout and pitch, considered in Section 1.4.4 and shown in Fig. 1.79, make use of equilateral triangular, square, and staggered square arrays. The triangular layout gives a robust

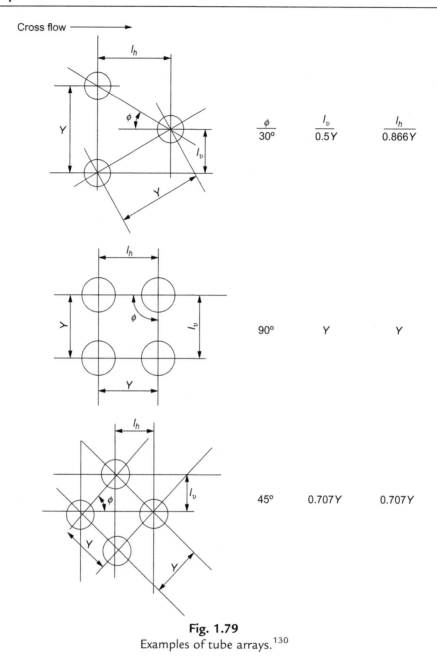

Fig. 1.79
Examples of tube arrays.[130]

tube sheet although, because the vertical and horizontal distances between adjacent tubes is generally greater in a square layout compared with the equivalent triangular pitch design, the square array simplifies maintenance and particularly cleaning on the shell-side. Good practice requires a minimum pitch of 1.25 times the tube diameter and/or a minimum web thickness

Table 1.18 Constants for use with Eq. (1.265)

Number of Passes		1	2	4	6	8
Triangular pitch[a]	a	0.319	0.249	0.175	0.0743	0.0365
	b	2.142	2.207	2.285	2.499	2.675
Square pitch[a]	a	0.215	0.156	0.158	0.0402	0.0331
	b	2.207	2.291	2.263	1.617	2.643

[a]Pitch $= 1.25d_0$.

between tubes of about 3.2 mm to ensure adequate strength for tube rolling. In general, the smallest pitch in triangular 30° layout is used for clean fluids in both laminar and turbulent flow and a 90° or 45° layout with a 6.4 mm clearance where mechanical cleaning is required. The bundle diameter, d_b, may be estimated from the following empirical equation, which is based on standard tube layouts:

$$\text{Number of tubes, } N_t = a(d_b/d_o)^b \tag{1.265}$$

where the values of the constants a and b are given in Table 1.18. Tables giving the number of tubes that can be accommodated in standard shells using various tube sizes, pitches, and number of passes for different exchanger types are given, for example, in Kern,[44] Ludwig[134] and others.[135,136]

Various *baffle designs* are shown in Fig. 1.80. The cross-baffle is designed to direct the flow of the shell-side fluid across the tube bundle and to support the tubes against sagging and possible vibration, and the most common type is the segmental baffle which provides a baffle window. The ratio, baffle spacing/baffle cut, is very important in maximising the ratio of heat transfer rate to pressure drop. Where very low pressure drops are required, double segmental or 'disc and doughnut' baffles are used to reduce the pressure drop by some 60%. Triple segmental baffles and designs in which all the tubes are supported by all the baffles provide for low-pressure drops and minimum tube vibration.

With regard to *baffle spacing*, TEMA[130] recommends that segmental baffles should not be spaced closer than 20% of the shell inside diameter or 50 mm whichever is the greater and that the maximum spacing should be such that the unsupported tube lengths, given in Table 1.19, are not exceeded. It may be noted that the majority of failures due to vibration occur when the unsupported tube length is in excess of 80% of the TEMA maximum; the best solution is to avoid having tubes in the baffle window.

1.9.3 Mean Temperature Difference in Multipass Exchangers

In an exchanger with one shell pass and several tube-side passes, the fluids in the tubes and shell will flow cocurrently in some of the passes and countercurrently in the others. For given inlet and outlet temperatures, the mean temperature difference for countercurrent flow

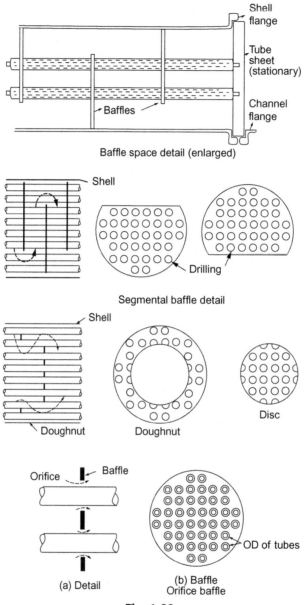

Fig. 1.80
Baffle designs.

Table 1.19 Maximum unsupported spans for tubes

Approximate Tube OD (mm)	Maximum Unsupported Span (mm)	
	Materials Group A	Materials Group B
19	1520	1321
25	1880	1626
32	2240	1930
38	2540	2210
50	3175	2794

Materials: Group A: carbon and high alloy steel, low alloy steel, nickel–copper, nickel, nickel–chromium–iron. Group B: aluminium and aluminium alloys, copper and copper alloys, titanium and zirconium.

is greater than that for cocurrent or parallel flow, and there is no easy way of finding the true temperature difference for the unit. The problem has been investigated by Underwood[137] and by Bowman et al.[138] who have presented graphical methods for calculating the true mean temperature difference in terms of the value of θ_m which would be obtained for countercurrent flow, and a correction factor F. Provided the following conditions are maintained or assumed, F can be found from the curves shown in Figs 1.81–1.84.

(a) The shell fluid temperature is uniform over the cross-section considered as constituting a pass.
(b) There is equal heat transfer surface in each pass.
(c) The overall heat transfer coefficient U is constant throughout the exchanger.
(d) The heat capacities of the two fluids are constant over the temperature range.
(e) There is no change in phase of either fluid.
(f) Heat losses from the unit are negligible.

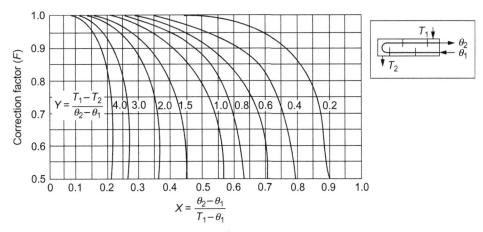

Fig. 1.81
Correction for logarithmic mean temperature difference for single shell pass exchanger.

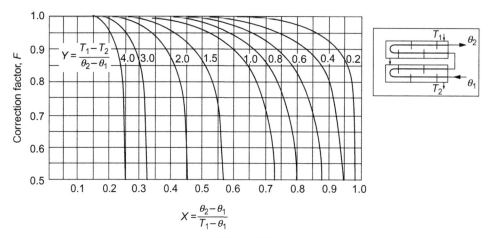

Fig. 1.82

Correction for logarithmic mean temperature difference for double shell pass exchanger.

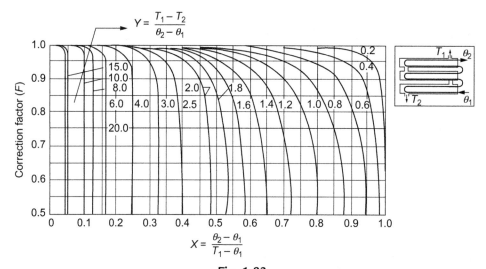

Fig. 1.83

Correction for logarithmic mean temperature difference for three shell pass exchanger.

Then:

$$Q = UAF\theta_m \tag{1.266}$$

F is expressed as a function of two parameters:

$$X = \frac{\theta_2 - \theta_1}{T_1 - \theta_1} \quad \text{and} \quad y = \frac{T_1 - T_2}{\theta_2 - \theta_1} \tag{1.267}$$

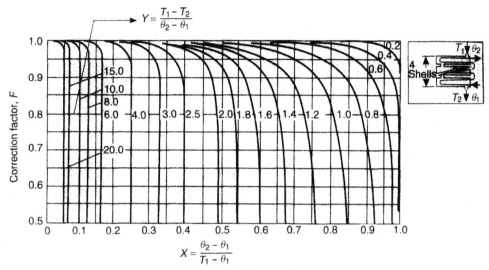

Fig. 1.84

Correction for logarithmic mean temperature difference for four shell pass exchanger.

If a one shell-side system is used Fig. 1.81 applies, for two shell-side passes Fig. 1.82, for three shell-side passes Fig. 1.83, and for four shell-passes Fig. 1.84. For the case of a single shell-side pass and two tube-side passes illustrated in Fig. 1.85A and B the temperature profile is as shown. Because one of the passes constitutes a parallel flow arrangement, the exit temperature of the cold fluid θ_2 cannot closely approach the hot fluid temperature T_1. This is true for the conditions shown in Fig. 1.85A and B and Underwood[137] has shown that F is the same in both cases.

If, for example, an exchanger is required to operate over the following temperatures:

$$T_1 = 455\,\text{K}, \quad T_2 = 372\,\text{K}$$
$$\theta_1 = 283\,\text{K}, \quad \theta_2 = 388\,\text{K}$$

Then:

$$X = \frac{\theta_2 - \theta_1}{T_1 - \theta_1} = \frac{388 - 283}{455 - 283} = 0.6$$

and:

$$Y = \frac{T_1 - T_2}{\theta_2 - \theta_1} = \frac{455 - 372}{388 - 283} = 0.8$$

For a single shell pass arrangement, from Fig. 1.81, F is 0.65 and, for a double shell pass arrangement, from Fig. 1.82, F is 0.95. On this basis, a two shell-pass design would be used.

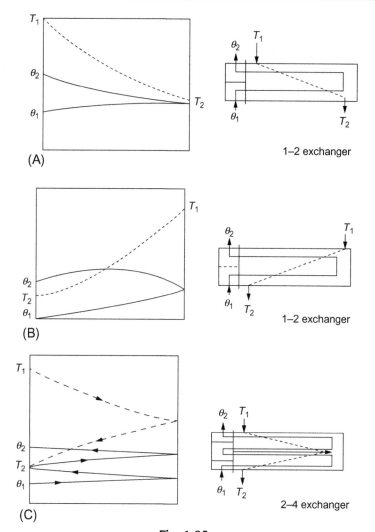

Fig. 1.85
Temperature profiles in single (A) and double shell pass (B) exchangers.

In order to obtain maximum heat recovery from the hot fluid, θ_2 should be as high as possible. The difference $(T_2 - \theta_2)$ is known as the *approach temperature* and, if $\theta_2 > T_2$, then a *temperature cross* is said to occur; a situation where the value of F decreases very rapidly when there is but a single pass on the shell-side. This implies that, in parts of the heat exchanger, heat is actually being transferred in the wrong direction. This may be illustrated by taking as an example the following data where equal ranges of temperature are considered:

Case	T_1	T_2	θ_1	θ_2	Approach $(T_2 - \theta_2)$	X	Y	F
1	613	513	363	463	50	0.4	1	0.92
2	573	473	373	473	0	0.5	1	0.80
3	543	443	363	463	Cross of 20	0.55	1	0.66

If a temperature cross occurs with a single pass on the shell-side, a unit with two shell passes should be used. It is seen from Fig. 1.85B that there may be some point where the temperature of the cold fluid is greater than θ_2 so that beyond this point the stream will be cooled rather than heated. This situation may be avoided by increasing the number of shell passes. The general form of the temperature profile for a two shell-side unit is as shown in Fig. 1.85C. Longitudinal shell-side baffles are rather difficult to fit and there is a serious chance of leakage. For this reason, the use of two exchangers arranged in series, one below the other, is to be preferred. It is even more important to employ separate exchangers when three passes on the shell-side are required. On the very largest installations, it may be necessary to link up a number of exchangers in parallel arranged as, say, sets of three in series as shown in Fig. 1.86. This arrangement is preferable for any very large unit, which would be unwieldy as a single system. When the total surface area is much greater than 250 m^2, consideration should be given to using multiple smaller units even though the initial cost may be higher and/or the so-called compact heat exchangers.

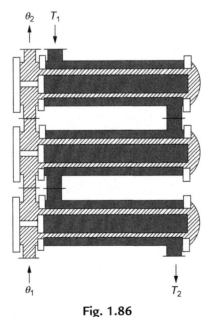

Fig. 1.86
Set of three heat exchangers in series.

In many processing operations, there may be a large number of process streams, some of which need to be heated and others cooled. An overall heat balance will indicate whether, in total, there is a net surplus or deficit of heat available. It is of great economic importance to achieve the most effective match of the hot and cold streams in the heat exchanger network so as to reduce to a minimum both the heating and cooling duties placed on the work utilities, such as supplies of steam and cooling water. This necessitates making the best use of the temperature driving forces. In considering the overall requirements, there will be some point where the temperature difference between the hot and cold streams is a minimum and this is referred to as the *pinch*. The lower the temperature difference at the pinch point, the lower will be the demand on the utilities, although it must be remembered that a greater area, (and hence cost) will be involved and an economic balance must therefore be struck. Heat exchanger networks are discussed in Volume 6 in detail, in the User Guide published by the Institution of Chemical Engineers.[139] Subsequently, Linnhoff,[140] and others[141] have given an overview of the industrial application of pinch analysis to the design of heat exchanger networks in order to reduce both capital costs and energy requirements.

Example 1.38

Using the data of Example 1.1, calculate the surface area required to effect the given duty using a multipass heat exchanger in which the cold water makes two passes through the tubes and the hot water makes a single pass through the shell.

Solution

As in Example 1.1, the heat load $= 1672$ kW

With reference to Fig. 1.81:

$$T_1 = 360K, \quad T_2 = 340K, \quad \theta_1 = 300K, \quad \theta_2 = 316K$$

and hence:

$$X = \frac{\theta_2 - \theta_1}{T_1 - \theta_1} = \frac{316 - 300}{360 - 300} = 0.267$$

and:

$$Y = \frac{T_1 - T_2}{\theta_2 - \theta_1} = \frac{360 - 340}{316 - 300} = 1.25$$

from Fig. 1.81:

$$F = 0.97$$

and hence:

$$F\theta_m = (41.9 \times 0.97) = 40.6K$$

The heat transfer area is then:

$$A = \frac{1672}{2 \times 40.6} = 20.6 m^2$$

1.9.4 Film Coefficients

Practical values

In any item of heat transfer equipment, the required area of heat transfer surface for a given load is determined by the overall temperature difference θ_m, the overall heat transfer coefficient U, and the correction factor F in Eq. (1.266). The determination of the individual film coefficients, which determine the value of U has proved difficult even for simple cases, and it is quite common for equipment to be designed on the basis of practical values of U rather than from a series of film coefficients. For the important case of the transfer of heat from one fluid to another across a metal surface, two methods have been developed for measuring film coefficients. The first requires a knowledge of the temperature difference across each film and therefore involves measuring the temperatures of both fluids and the surface of separation. With a concentric tube system, it is very difficult to insert a thermocouple into the thin tube and to prevent the thermocouple connections from interfering with the flow of the fluid. Nevertheless, this method is commonly adopted, particularly when electrical heating is used. It must be noted that when the heat flux is very high, as with boiling liquids, there will be an appreciable temperature drop across the tube wall and the position of the thermocouple is then important. For this reason, working with stainless steel, which has a relatively low thermal conductivity, is difficult.

The second method uses a technique proposed by Wilson.[142] If steam is condensing on the outside of a horizontal tube through which water is passed at various velocities, then the overall and film transfer coefficients are related by:

$$\frac{1}{U} = \frac{1}{h_o} + \frac{x_w}{k_w} + R_i + \frac{1}{h_i} \qquad \text{(from Eq. 9.255)}$$

provided that the transfer area on each side of the tube is approximately the same.

For conditions of turbulent flow, the transfer coefficient for the water side, $h_i = \varepsilon u^{0.8}$, R_i the scale resistance is constant, and h_o the coefficient for the condensate film is almost independent of the water velocity. Thus, Eq. (1.255) reduces to:

$$\frac{1}{U} = (\text{constant}) + \frac{1}{\varepsilon u^{0.8}}$$

If $1/U$ is plotted against $1/u^{0.8}$ a straight line, known as a Wilson plot, is obtained with a slope of $1/\varepsilon$ and an intercept equal to the value of the constant. For a clean tube, R_i should be zero, and hence h_0 can be found from the value of the intercept, as x_w/k_w will generally be small for a metal tube, h_i may also be obtained at a given velocity from the difference between $1/U$ at that velocity and the intercept.

This technique has been applied by Rhodes and Younger[143] to obtain the values of h_0 for condensation of a number of organic vapours, by Pratt[125] to obtain the inside coefficient for coiled tubes, and by Coulson and Mehta[144] to obtain the coefficient for an annulus. If the results are repeated over a period of time, the increase in the value of R_i can also be obtained by this method.

Typical values of thermal resistances and individual and overall heat transfer coefficients are given in Tables 1.20–1.23.

Correlated data

Heat transfer data for turbulent flow *inside* conduits of uniform cross-section are usually correlated by a form of Eq. (1.94):

$$Nu = CRe^{0.8}Pr^{0.33}(\mu/\mu_s)^{0.14} \tag{1.268}$$

Table 1.20 Thermal resistance of heat exchanger tubes

| Gauge (BWG) | Thickness (mm) | Copper | Values of x_w/k_w (m^2 K/kW) | | | |
			Steel	Stainless Steel	Admiralty Metal	Aluminium
18	1.24	0.0031	0.019	0.083	0.011	0.0054
16	1.65	0.0042	0.025	0.109	0.015	0.0074
14	2.10	0.0055	0.032	0.141	0.019	0.0093
12	2.77	0.0072	0.042	0.176	0.046	0.0123
			Values of x_w/k_w (ft^2 °h°F/Btu)			
18	0.049	0.000018	0.00011	0.00047	0.000065	0.000031
16	0.065	0.000024	0.00014	0.00062	0.000086	0.000042
14	0.083	0.000031	0.00018	0.0008	0.00011	0.000053
12	0.109	0.000041	0.00024	0.001	0.00026	0.000070

Table 1.21 Thermal resistances of scale deposits from various fluids

	m^2°K/kW	ft^2°h°F/Btu		m^2°K/kW	ft^2°h°F/Btu
Water[a]			Steam		
Distilled	0.09	0.0005	Good quality, oil-		
sea	0.09	0.0005	free	0.052	0.0003
Clear river	0.21	0.0012	Poor quality, oil-		
Untreated			Free	0.09	0.0005
cooling tower	0.58	0.0033	Exhaust from		
Treated cooling			Reciprocating		
tower	0.26	0.0015	Engines	0.18	0.001
Treated boiler	0.26	0.0015			
feed					
Hard well	0.58	0.0033	*Liquids*		
			Treated brine	0.27	0.0015
Gases			Organics	0.18	0.001
Air	0.25–0.50	0.0015–0.003	Fuel oils	1.0	0.006
Solvent vapours	0.14	0.0008	Tars	2.0	0.01

[a]For a velocity of 1 m/s ($\approx$3 ft/s) and temperatures of less than 320 K (122°F).

Table 1.22 Approximate overall heat transfer coefficients *U* for shell and tube equipment

Hot Side	Cold Side	Overall *U*	
		W/m² K	Btu/h ft² °F
Condensers			
Steam (pressure)	Water	2000–4000	350–750
Steam (vacuum)	Water	1700–3400	300–600
Saturated organic solvents (atmospheric)	Water	600–1200	100–200
Saturated organic solvents (vacuum some noncondensable)	Water-brine	300–700	50–120
Organic solvents (atmospheric and high noncondensable)	Water-brine	100–500	20–80
Organic solvents (vacuum and high noncondensable)	Water-brine	60–300	10–50
Low boiling hydrocarbons (atmospheric)	Water	400–1200	80–200
High boiling hydrocarbons (vacuum)	Water	60–200	10–30
Heaters			
Steam	Water	1500–4000	250–750
Steam	Light oils	300–900	50–150
Steam	Heavy oils	60–400	10–80
Steam	Organic solvents	600–1200	100–200
Steam	Gases	30–300	5–50
Dowtherm	Gases	20–200	4–40
Dowtherm	Heavy oils	50–400	8–60
Evaporators			
Steam	Water	2000–4000	350–750
Steam	Organic solvents	600–1200	100–200
Steam	Light oils	400–1000	80–180
Steam	Heavy oils (vacuum)	150–400	25–75
Water	Refrigerants	400–900	75–150
Organic solvents	Refrigerants	200–600	30–100
Heat exchangers (no change of state)			
Water	Water	900–1700	150–300
Organic solvents	Water	300–900	50–150
Gases	Water	20–300	3–50
Light oils	Water	400–900	60–160
Heavy oils	Water	60–300	10–50
Organic solvents	Light oil	100–400	20–70
Water	Brine	600–1200	100–200
Organic solvents	Brine	200–500	30–90
Gases	Brine	20–300	3–50
Organic solvents	Organic solvents	100–400	20–60
Heavy oils	Heavy oils	50–300	8–50

Table 1.23 Approximate film coefficients for heat transfer

	h_i or h_o	
	W/m² K	**Btu/ft² h °F**
No change of state		
Water	1700–11,000	300–2000
Gases	20–300	3–50
Organic solvents	350–3000	60–500
Oils	60–700	10–120
Condensation		
Steam	6000–17,000	1000–3000
Organic solvents	900–2800	150–500
Light oils	1200–2300	200–400
Heavy oils (vacuum)	120–300	20–50
Ammonia	3000–6000	500–1000
Evaporation		
Water	2000–12,000	30–200
Organic solvents	600–2000	100–300
Ammonia	1100–2300	200–400
Light oils	800–1700	150–300
Heavy oils	60–300	10–50

where, based on the work of Sieder and Tate,[21] the index for the viscosity correction term is usually 0.14 although higher values have been reported. Using values of C of 0.021 for gases, 0.023 for nonviscous liquids, and 0.027 for viscous liquids, Eq. (1.268) is sufficiently accurate for design purposes, and any errors are far outweighed by uncertainties in predicting shell-side coefficients. Rather more accurate tube-side data may be obtained by using correlations given by the Engineering Sciences Data Unit and, based on this work, Butterworth[145] offers the equation:

$$St = ERe^{-0.205} Pr^{-0.505} \qquad (1.269)$$

where:

The Stanton Number $St = NuRe^{-1}Pr^{-1}$

and

$$E = 0.22\exp\left[-0.0225(\ln Pr)^2\right]$$

Eq. (1.269) is valid for Reynolds Numbers in excess of 10,000. Where the Reynolds Number is less than 2000, the flow will be laminar and, provided natural convection effects are negligible, film coefficients may be estimated from a form of Eq. (1.119) modified to take account of the variation of viscosity over the cross-section:

$$Nu = 1.86(RePr)^{0.33}(d/l)^{0.33}(\mu/\mu_s)^{0.14} \qquad (1.270)$$

The minimum value of the Nusselt Number for which Eq. (1.270) applies is 3.5. Reynolds Numbers in the range 2000–10,000 should be avoided in designing heat exchangers as the flow is then unstable and coefficients cannot be predicted with any degree of accuracy. If this cannot be avoided, the lesser of the values predicted by Eqs. (1.268), (1.270) should be used.

As discussed in Section 1.4.3, heat transfer data are conveniently correlated in terms of a heat transfer factor j_h, again modified by the viscosity correction factor:

$$j_h = StPr^{0.67}(\mu/\mu_s)^{-0.14} \tag{1.271}$$

which enables data for laminar and turbulent flow to be included on the same plot, as shown in Fig. 1.87. Data from Fig. 1.87 may be used together with Eq. (1.271) to estimate coefficients with heat exchanger tubes and commercial pipes although, due to a higher roughness, the values for commercial pipes will be conservative. Eq. (1.271) is rather more conveniently expressed as:

$$Nu = (hd/k) = j_h RePr^{0.33}(\mu/\mu_s)^{-0.14} \tag{1.272}$$

It may be noted that whilst Fig. 1.87 is similar to Fig. 1.33, the values of j_h differ due to the fact that Kern[44] and other researchers define the heat transfer factor as:

$$j_H = NuPr^{-0.33}(\mu/\mu_s)^{-0.14} \tag{1.273}$$

Thus the relationship between j_h and j_H is:

$$j_h = j_H/Re \tag{1.274}$$

As discussed in Section 1.4.3, by incorporating physical properties into Eqs. (1.268), (1.270), correlations have been developed specifically for water and Eq. (1.275), based on data from Eagle and Ferguson,[146] may be used:

$$h = 4280(0.00488T - 1)u^{0.8}/d^{0.2} \tag{1.275}$$

which is in SI units, with h (film coefficient) in W/m^2 K, T in K, u in m/s, and d in m.

Example 1.39

Estimate the heat transfer area required for the system considered in Examples 1.1 and 1.38, assuming that no data on the overall coefficient of heat transfer are available.

Solution

As in the previous examples,

$$\text{heat load} = 1672\,kW$$

and:

$$\text{corrected mean temperature difference, } F\theta_m = 40.6\,deg\,K$$

In the tubes;

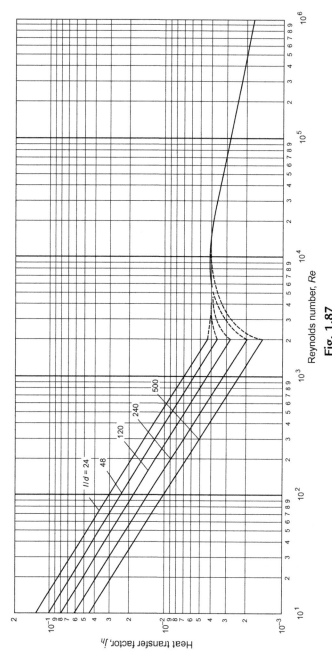

Fig. 1.87

Heat transfer factor for flow inside tubes.

mean water temperature, $T = 0.5(360 + 340) = 350 \, \text{K}$

Assuming a tube diameter, $d = 19$ mm or 0.0019 m and a water velocity, $u = 1$ m/s, then, in Eq. (1.275):

$$h_i = 4280((0.00488 \times 350) - 1)1.0^{0.8}/0.0019^{0.2} = 10610 \, \text{W/m}^2 \, \text{K} \, \text{or} \, 10.6 \, \text{kW/m}^2 \, \text{K}$$

From Table 1.23, an estimate of the shell-side film coefficient is:

$$h_0 = 0.5(1700 + 11000) = 6353 \, \text{W/m}^2 \, \text{K} \, \text{or} \, 6.35 \, \text{kW/m}^2 \, \text{K}$$

For steel tubes having a wall thickness of 1.6 mm, the thermal resistance of the wall, from Table 1.20 is:

$$x_w/k_w = 0.025 \, \text{m}^2 \, \text{K/kW}$$

and the thermal resistance for treated water, from Table 1.21, is 0.26 m^2 K/kW for both layers of scale. Thus, in Eq. (1.255):

$$(1/U) = (1/h_o) + (x_w/k_w) + R_i + R_o + (1/h_i)$$
$$= (1/6.35) + 0.025 + 0.52 + (1/10.6) = 0.797 \, \text{m}^2 \, \text{K/kW}$$

and:

$$U = 1.25 \, \text{kW/m}^2 \, \text{K}$$

The heat transfer area required is then:

$$A = Q/F\theta_m U = 1672/(40.6 \times 1.25) = \underline{32.9 \, \text{m}^2}$$

As discussed in Section 1.4.4, the complex flow pattern on the *shell-side* and the great number of variables involved make the prediction of coefficients and pressure drop very difficult, especially if leakage and bypass streams are taken into account. Until about 1960, empirical methods were used to account for the difference in the performance of real exchangers as compared with that for cross-flow over ideal tube banks. The methods of Kern[44] and Donohue[147] are typical of these 'bulk flow' methods and their approach, together with more recent methods involving an analysis of the contribution to heat transfer by individual streams in the shell, are discussed in Section 1.9.6.

Special correlations have also been developed for liquid metals, used in recent years in the nuclear industry with the aim of reducing the volume of fluid in the heat transfer circuits. Such fluids have high thermal conductivities, though in terms of heat capacity per unit volume, liquid sodium, for example, which finds relatively widespread application, has a value of $C_p\rho$ of only 1275 kJ/m^3 K.

Although water has a much greater value, it is unsuitable because of its high vapour pressure at the desired temperatures and the corresponding need to use high-pressure piping. Because

Table 1.24 Prandtl numbers of liquid metals

Metal	Temperature (K)	Prandtl Number Pr
Potassium	975	0.003
Sodium	873–1073	0.004
Na/K alloy (56:44)	975	0.06
Lead	673	0.02
Mercury	575	0.008
Lithium	475	0.065
	673	0.036
	973	0.025

of their high thermal conductivities, liquid metals have particularly low values of the Prandtl number (about 0.01) and they behave rather differently from normal fluids under conditions of forced convection. Some values for typical liquid metals are given in Table 1.24.

The results of work on sodium, lithium, and mercury for forced convection in a pipe have been correlated by the expression:

$$Nu = 0.625(RePr)^{0.4} \tag{1.276}$$

although the accuracy of the correlation is not very good. With values of Reynolds number of about 18,000, it is quite possible to obtain a value of h of about 11 kW/m^2 K for flow in a pipe.

1.9.5 Pressure Drop in Heat Exchangers

Tube-side

Pressure drop on the tube-side of a shell and tube exchanger is made up of the friction loss in the tubes and losses due to sudden contractions and expansions and flow reversals experienced by the tube-side fluid. The friction loss may be estimated by the methods outlined in Section 3.4.3 of Vol. 1A from which the basic equation for isothermal flow is given by Eq. (3.18) of Vol. 1A which can be written as:

$$-\Delta P_t = 4j_f(l/d_i)(\rho u^2) \tag{1.277}$$

where j_f is the dimensionless friction factor. Clearly the flow is not isothermal and it is usual to incorporate an empirical correction factor to allow for the change in physical properties, particularly viscosity, with temperature to give:

$$-\Delta P_t = 4j_f(l/d_i)(\rho u^2)(\mu/\mu_s)^m \tag{1.278}$$

where $m = -0.25$ for laminar flow ($Re < 2100$) and -0.14 for turbulent flow ($Re > 2100$). Values of j_f for heat exchanger tubes are given in Fig. 1.88, which is based on Fig. 3.7 of Vol. 1A.

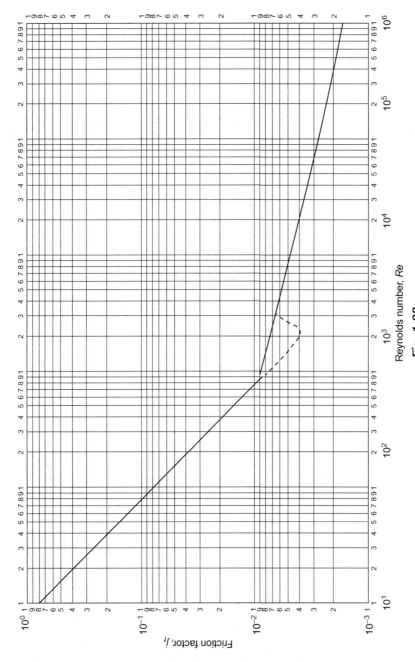

Reynolds number, *Re*

Fig. 1.88

Tube-side friction factors.[44] *Note*: The friction factor j_f is the same as the friction factor for nines $\varphi(=(R/\rho u^2))$ defined in Chapter 3 of Vol. 1A.

There is no entirely satisfactory method for estimating losses due to contraction at the tube inlets, expansion at the exits and flow reversals, although Kern[44] suggests adding four velocity heads per pass, Frank[148] recommends 2.5 velocity heads and Butterworth[149] recommends a figure of 1.8. Lord et al.[150] suggests that the loss per pass is equivalent to a tube length of 300 diameters for straight tubes and 200 for U-tubes, whilst Evans[151] recommends the addition of 67 tube diameters per pass. Another approach is to estimate the number of velocity heads by using factors for pipe-fittings as discussed in Section 3.4.4 of Vol. 1A and given in Table 3.2 of Vol. 1A. With four tube passes, for example, there will be four contractions equivalent to a loss of $(4 \times 0.5) = 2$ velocity heads, four expansions equivalent to a loss of $(4 \times 1.0) = 4$ velocity heads and three $180°$ bends equivalent to a loss of $(3 \times 1.5) = 4.5$ velocity heads. In this way, the total loss is 10.5 velocity heads, or 2.6 per pass, giving support to Frank's proposal of 2.5. Using this approach, Eq. (1.278) becomes:

$$-\Delta P_{\text{total}} = N_P \left[4j_f (l/d_i)(\mu/\mu_s)^m + 1.25 \right] \left(\rho u^2 \right) \qquad (1.279)$$

where N_p is the number of tube-side passes. Additionally, there will be expansion and contraction losses at the inlet and outlet nozzles respectively, and these losses may be estimated by adding one velocity head for the inlet, and 0.5 of a velocity head for the outlet, based on the nozzle velocities. Losses in the nozzles are only significant for gases at pressures below atmospheric pressure.

Shell-side

As discussed in Section 1.4.4, the prediction of pressure drop, and indeed heat transfer coefficients in the shell is very difficult due to the complex nature of the flow pattern in the segmentally baffled unit. Whilst the baffles are intended to direct fluid across the tubes, the actual flow is a combination of cross-flow between the baffles and axial or parallel flow in the baffle windows as shown in Fig. 1.89, although even this does not represent the actual flow pattern because of leakage through the clearances necessary for the fabrication and assembly of the unit. This more realistic flow pattern is shown in Fig. 1.90 which is based on the work of Tinker[152] who identifies the various streams in the shell as follows:

A—fluid flowing through the clearance between the tube and the hole in the baffle.
B—the actual cross-flow stream.
C—fluid flowing through the clearance between the outer tubes and the shell.
E—fluid flowing through the clearance between the baffle and the shell.
F—fluid flowing through the gap between the tubes because of any pass-partition plates. This is especially significant with a vertical gap.

Because stream A does not bypass the tubes, it is the pressure drop rather than the heat transfer, which is affected. Streams C, E, and F bypass the tubes, thus reducing the effective heat transfer area. Stream C, the main bypass stream, is most significant in pull-through bundle units

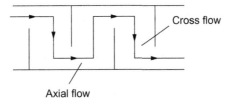

Fig. 1.89
Idealised main stream flow.

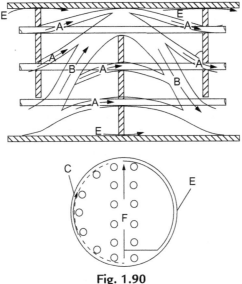

Fig. 1.90
Shell-side leakage and by-pass paths.[152]

where there is of necessity a large clearance between the bundle and the shell, although this can be reduced by using horizontal sealing strips. In a similar way, the flow of stream F may be reduced by fitting dummy tubes. As an exchanger becomes fouled, clearances tend to plug and this increases the pressure drop. The whole question of shell-side pressure drop estimation in relation to design procedures is now discussed.

1.9.6 Heat Exchanger Design

Process conditions

A first-stage consideration in the design process is the allocation of fluids to either shell or tubes and, by and large, the more corrosive fluid is passed through the tubes to reduce the costs of expensive alloys and clad components. Similarly, the fluid with the greatest fouling tendency is also usually passed through the tubes where cleaning is easier. Furthermore, velocities through the tubes are generally higher and more readily controllable and can be adjusted to reduce fouling. Where special alloys are in contact with hot fluids, the fluids should be passed through the tubes to reduce costs. In addition the shell temperature is lowered, thus reducing lagging costs. Passing hazardous materials through the tubes leads to greater safety and, because high-pressure tubes are cheaper than a high-pressure shell, streams at high pressure are best handled on the tube-side. In a similar way, where a very low pressure drop is required as in vacuum operation for example, the fluids involved are best handled on the tube-side where higher film heat transfer coefficients are obtained for a given pressure drop. Provided the flow is turbulent, a higher heat transfer coefficient is usually obtained with a more viscous liquid in the shell because of the more complex flow patterns although, because the tube-side coefficient can be predicted with greater accuracy, it is better to place the fluid in the tubes if turbulent flow in the shell is not possible. Normally, the most economical design is achieved with the fluid with the lower flowrate in the shell.

In selecting a design velocity, it should be recognised that at high velocities, high rates of heat transfer are achieved and fouling is reduced, but pressure drops are also higher. Normally, the velocity must not be so high as to cause erosion, which can be reduced at the tube inlet by fitting plastic inserts, and yet be such that any solids are kept in suspension. For process liquids, velocities are usually 0.3–1.0 m/s in the shell and 1.0–2.0 m/s in the tubes, with a maximum value of 4.0 m/s when fouling must be reduced. Typical water velocities are 1.5–2.5 m/s. For vapours, velocities lie in the range of 5–10 m/s with high-pressure fluids and 50–70 m/s with vacuum operation, the lower values being used for materials of high molecular weight.

In general, the higher of the temperature differences between the outlet temperature of one stream and the inlet temperature of the other should be 20 deg K and the lower temperature difference should be 5–7 deg K for water coolers and 3–5 deg K when using refrigerated brines, although optimum values can only be determined by an economic analysis of alternative designs.

Similar considerations apply to the selection of pressure drops where there is freedom of choice, although a full economic analysis is justified only in the case of very expensive units. For liquids, typical values in optimised units are 35 kN/m² where the viscosity is less than 1 mN s/m² and 50–70 kN/m² where the viscosity is 1–10 mN s/m²; for gases, 0.4–0.8 kN/m² for high vacuum operation, 50% of the system pressure at 100–200 kN/m², and 10% of the system pressure above 1000 kN/m². Whatever pressure drop is used, it is important that erosion and flow-induced tube vibration caused by high velocity fluids are avoided.

Design methods

It is shown in Section 1.9.5 that, with the existence of various bypass and leakage streams in practical heat exchangers, the flow patterns of the shell-side fluid, as shown in Figs. 1.89 and 1.90, are complex in the extreme and far removed from the idealised cross-flow situation discussed in Section 1.4.4. One simple way of using the equations for cross-flow presented in Section 1.4.4, however, is to multiply the shell-side coefficient obtained from these equations by the factor 0.6 in order to obtain at least an estimate of the shell-side coefficient in a practical situation. The pioneering work of Kern[44] and Donohue,[147] who used correlations based on the total stream flow and empirical methods to allow for the performance of real exchangers compared with that for cross-flow over ideal tube banks, went much further and, although their early design method does not involve the calculation of bypass and leakage streams, it is simple to use and quite adequate for preliminary design calculations.

The method, which is based on experimental work with a great number of commercial exchangers with standard tolerances, gives a reasonably accurate prediction of heat transfer coefficients for standard designs, although predicted data on pressure drop is less satisfactory as it is more affected by leakage and bypassing. Using a similar approach to that for tube-side flow, shell-side heat transfer and friction factors are correlated using a hypothetical shell diameter and shell-side velocity where, because the cross-sectional area for flow varies across the shell diameter, linear and mass velocities are based on the maximum area for cross-flow; that is at the shell equator. The shell equivalent diameter is obtained from the flow area between the tubes taken parallel to the tubes, and the wetted perimeter, as outlined in Section 1.9.4 and illustrated in Fig. 1.37. The shell-side factors, j_h and j_f, for various baffle cuts and tube arrangements based on the data given by Kern[44] and Ludwig[134] are shown in Figs 1.91 and 1.92.

The general approach is to calculate the area for cross-flow for a hypothetical row of tubes at the shell equator from the equation given in Section 1.4.4:

$$A_s = d_s l_B C'/Y \tag{1.280}$$

where d_s is the shell diameter, l_B is the baffle length and (C'/Y) is the ratio of the clearance between the tubes and the distance between tube centres. The mass flow divided by the area A_s gives the mass velocity G'_s, and the linear velocity on the shell-side u_s is obtained by dividing the mass velocity by the mean density of the fluid. Again, using the equations in Section 1.4.4, the shell-side equivalent or hydraulic diameter is given by:

For square pitch:

$$d_e = 4\left(Y^2 - \pi d_o^2/4\right)/\pi d_o = 1.27\left(Y^2 - 0.785 d_o^2\right)/d_o \tag{1.281}$$

and for triangular pitch:

$$d_e = 4[(0.87Y \times Y/2) - (0.5\pi d_o^2/4]/(\pi d_o/2)$$
$$= 1.15\left(Y^2 - 0.910 d_o^2\right)/d_o \tag{1.282}$$

Using this equivalent diameter, the shell-side Reynolds number is then:

$$Re_s = G'_s d_e/\mu = u_s d_e \rho/\mu \tag{1.283}$$

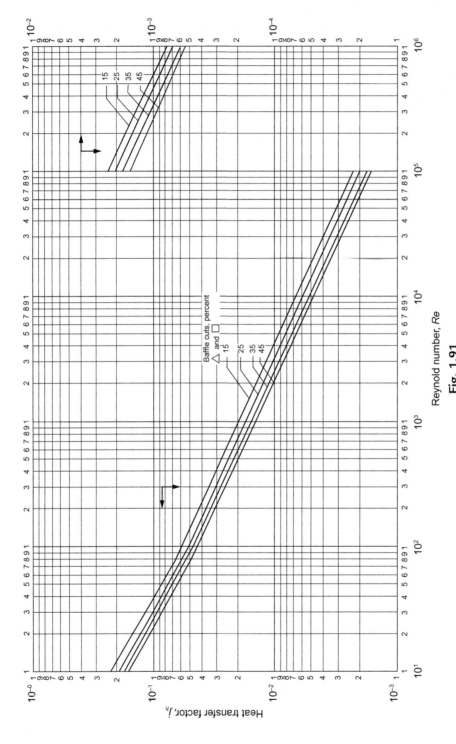

Reynold number, Re

Fig. 1.91

Shell-side heat-transfer factors with segmental baffles.[44]

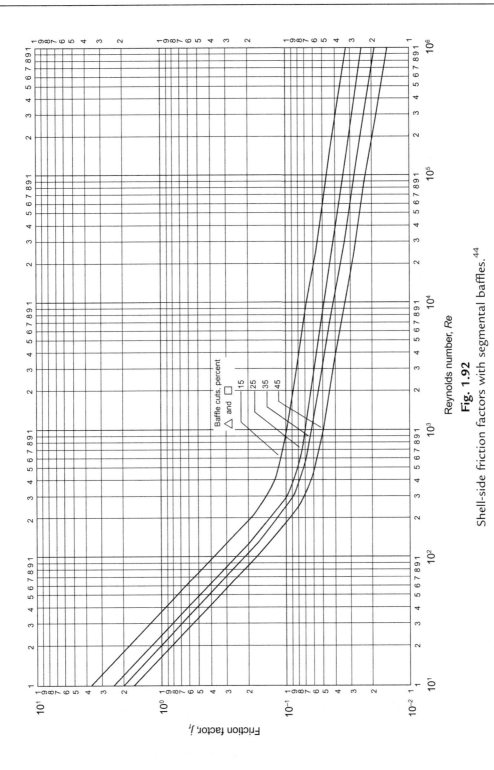

Reynolds number, *Re*

Fig. 1.92

Shell-side friction factors with segmental baffles.[44]

where G' is the mass flowrate per unit area. Hence j_h may be obtained from Fig. 1.91. The shell-side heat transfer coefficient is then obtained from a rearrangement of Eq. (1.272):

$$Nu = \left(h_s d_e / k_f\right) = j_h Re Pr^{0.33} (\mu/\mu ls)^{0.14} \tag{1.284}$$

In a similar way, the factor j_f is obtained from Fig. 1.92 and the pressure drop estimated from a modified form of Eq. (1.278):

$$-\Delta P_s = 4 j_f (d_s/d_e)(l/l_B)\left(\rho u_s^2\right)(\mu/\mu_s)^{-0.14} \tag{1.285}$$

where (l/l_B) is the number of times the flow crosses the tube bundle $= (n+1)$.

The pressure drop over the shell nozzles should be added to this value, although this contribution is usually only significant with gases. In general, the nozzle pressure loss is 1.5 velocity heads for the inlet and 0.5 velocity heads for the outlet, based on the nozzle area or the free area between the tubes in the row adjacent to the nozzle, whichever is the least. Kern's method is now illustrated in the following example.

Example 1.40

Using Kern's method, design a shell and tube heat exchanger to cool 30 kg/s of butyl alcohol from 370 to 315 K using treated water as the coolant. The water will enter the heat exchanger at 300 K and leave at 315 K.

Solution

Since it is corrosive, the water will be passed through the tubes.

At a mean temperature of $0.5(370 + 315) = 343$ K, from Table 3, Appendix A1, the thermal capacity of butyl alcohol $= 2.90$ kJ/kg K and hence:

$$\text{Heat load} = (30 \times 2.90)(370 - 315) = 4785\,\text{kW}$$

If the heat capacity of water is 4.18 kJ/kg K, then:

$$\text{Flow rate of cooling water} = 4785/(4.18(315 - 300)) = 76.3\,\text{kg/s}$$

The logarithmic mean temperature difference,

$$\theta_m = [(370 - 315) - (315 - 300)]/\ln[(370 - 315)/(315 - 300)] = 30.7\,\text{deg K}$$

With one shell-side pass and two tube-side passes, then from Eq. (1.267):

$$Y = (370 - 315)/(315 - 300) = 3.67 \quad \text{and} \quad X = (315 - 300)/(370 - 300) = 0.21$$

and from Fig. 1.85:

$$F = 0.85 \text{ and } F\theta_m = (0.85 \times 30.7) = 26.1\,\text{deg K}$$

From Table 1.22, an estimated value of the overall coefficient is $U = 500$ W/m^2 K and hence, the provisional area, from Eq. (1.266), is:

$$A = \left(4785 \times 10^3\right)/(26.1 \times 500) = 367\,\text{m}^2$$

It is convenient to use 20 mm OD, 16 mm ID tubes, 4.88 m long which, allowing for the tube-sheets, would provide an effective tube length of 4.83 m. Thus:

$$\text{Surface area of one tube} = \pi(20/1000)(4.83) = 0.303\,\text{m}^2$$

and:

$$\text{Number of tubes required} = (367/0.303) = 1210$$

With a clean shell-side fluid, 1.25 triangular pitch may be used and, from Eq. (1.265):

$$1210 = 0.249(d_b/20)^{2.207}$$

from which:

$$d_b = 937\,\text{mm}$$

Using a split-ring floating head unit then, from Fig. 1.81, the diametrical clearance between the shell and the tubes $= 68$ mm and:

$$\text{Shell diameter,} \quad d_s = (937 + 68) = 1005\,\text{mm}$$

which approximates to the nearest standard pipe size of 1016 mm.

Tube-side coefficient

The water-side coefficient may now be calculated using Eq. (1.272), although here, use will be made of the j_h factor.

$$\text{Cross} - \text{sectional area of one tube} = (\pi/4) \times 16^2 = 201\,\text{mm}^2$$

$$\text{Number of tubes/pass} = (1210/2) = 605$$

Thus:

$$\text{Tube} - \text{side flow area} = \left(605 \times 201 \times 10^{-6}\right) = 0.122\,\text{m}^2$$

$$\text{Mass velocity of the water} = (76.3/0.122) = 625\,\text{kg/m}^2\text{s}$$

Thus, for a mean water density of 995 kg/m^3:

$$\text{Water velocity,} \quad u = (625/995) = 0.63\,\text{m/s}$$

At a mean water temperature of $0.5(315 + 300) = 308$ K, viscosity, $\mu = 0.8$ mN s/m^2 and thermal conductivity, $k = 0.59$ W/m K.

Thus:

$$Re = du\rho/\mu = \left(16 \times 10^{-3} \times 0.63 \times 995\right)/\left(0.8 \times 10^{-3}\right) = 12540$$

$$Pr = C_p\mu/k = \left(4.18 \times 10^3 \times 0.8 \times 10^{-3}\right)/0.59 = 5.67$$

$$l/d_i = 4.83/\left(16 \times 10^{-3}\right) = 302$$

Thus, from Fig. 1.87, $j_h = 3.7 \times 10^{-3}$ and, in Eq. (1.272), neglecting the viscosity term:

$$\left(h_i \times 16 \times 10^{-3}\right)/0.59 = \left(3.7 \times 10^{-3} \times 12540 \times 5.67^{0.33}\right)$$

and:

$$h_i = 3030 \, \text{W/m}^2 \, \text{K}$$

Shell-side coefficient

The baffle spacing will be taken as 20% of the shell diameter or $(1005 \times 20/100) = 201$ mm

The tube pitch $= (1.25 \times 20) = 25$ mm and, from Eq. (1.280):

$$\text{Cross} - \text{flow area,} \quad A_s = \left[(25 - 20)/25\right]\left(1005 \times 201 \times 10^{-6}\right) = 0.040 \text{m}^2$$

Thus:

$$\text{Mass velocity in the shell,} \quad G_s = (30/0.040) = 750 \text{kg/m}^2 \text{s}$$

From Eq. (1.282):

$$\text{Equivalent diameter,} \quad d_e = 1.15\left[25^2 - \left(0.917 \times 20^2\right)\right]/20 = 14.2 \text{mm}$$

At a mean shell-side temperature of $0.5(370 + 315) = 343$ K, from Appendix A1: density of butyl alcohol, $\rho = 780$ kg/m^3, viscosity, $\mu = 0.75$ mN s/m^2, heat capacity, $C_p = 3.1$ kJ/kg K, and thermal conductivity, $k = 0.16$ W/m K.

Thus from Eq. (1.283):

$$Re = G_s d_e/\mu = \left(750 \times 14.2 \times 10^{-3}\right)/\left(0.75 \times 10^{-3}\right) = 14,200$$

$$Pr = C_p\mu/k = \left(3.1 \times 10^3 \times 0.75 \times 10^{-3}\right)/0.16 = 14.5$$

Thus, with a 25% segmental cut, from Fig. 1.91: $j_h = 5.0 \times 10^{-3}$

Neglecting the viscosity correction term in Eq. (1.284):

$$\left(h_s \times 14.2 \times 10^{-3}\right)/0.16 = 5.0 \times 10^{-3} \times 14200 \times 14.5^{0.33}$$

and:

$$h_s = 1933 \, \text{W/m}^2 \, \text{K}$$

The mean butanol temperature $=343$ K, the mean water temperature $=308$ K and hence the mean wall temperature may be taken as $0.5(343+308)=326$ K at which $\mu_s=1.1$ mN s/m^2

Thus:

$$(\mu/\mu_s)^{0.14} = (0.75/1.1)^{0.14} = 0.95$$

showing that the correction for a low viscosity fluid is negligible.

Overall coefficient

The thermal conductivity of cupro-nickel alloys $=50$ W/m K and, from Table 1.20, scale resistances will be taken as 0.00020 m^2 K/W for the water and 0.00018 m^2 K/W for the organic liquid.

Based on the outside area, the overall coefficient is given by:

$$1/U = 1/h_o + R_o + x_w/k_w + R_i/(d_i/d_o) + (1/h_i)(d_i/d_o)$$
$$= (1/1933) + 0.00020 + \left[0.5(20-16) \times 10^{-3}/50\right] + (0.00015 \times 20)/16 + 20/(3030 \times 16)$$
$$= 0.00052 + 0.00020 + 0.00004 + 0.000225 + 0.00041 = 0.00140 \mathrm{m}^2 \mathrm{K/W}$$

and:

$$U = 717 \mathrm{W/m}^2 \mathrm{K}$$

which is well in excess of the assumed value of 500 W/m^2 K.

Pressure drop

On the *tube-side*. $Re=12{,}450$ and from Fig. 1.88, $j_f=4.5 \times 10^{-3}$

Neglecting the viscosity correction term, Eq. (1.279) becomes:

$$\Delta P_t = 2\left(4 \times 4.5 \times 10^{-3}(4830/16) + 1.25\right)\left(995 \times 0.63^2\right) = 5279 \mathrm{N/m}^2 \text{ or } 5.28 \mathrm{kN/m}^2$$

which is low, permitting a possible increase in the number of tube passes.

On the *shell-side*, the linear velocity, $(G_s/\rho)=(750/780)=0.96$ m/s

From Fig. 1.92, when $Re=14{,}200$, $j_f=4.6 \times 10^{-2}$

Neglecting the viscosity correction term, in Eq. (1.285):

$$-\Delta P_s = \left(4 \times 4.6 \times 10^{-2}\right)(1005/14.2)(4830/201)\left(780 \times 0.96^2\right)$$
$$= 224950 \mathrm{N/m}^2 \text{ or } 225 \mathrm{kN/m}^2$$

This value is very high and thought should be given to increasing the baffle spacing. If this is doubled, this will reduce the pressure drop by approximately $(1/2)^2 = 1/4$ and:

$-\Delta P_S = (225/4) = 56.2 \text{ kN/m}^2$ which is acceptable.

Since $h_0 \propto Re^{0.8} \propto w^{0.8}$,

$$h_0 = 1933(1/2)^{0.8} = 1110 \text{ W/m}^2 \text{K}$$

which gives an overall coefficient of $\underline{561 \text{ W/m}^2 \text{K}}$ which is still in excess of the assumed value of 500 W/m^2 K.

Further detailed discussion of Kern's method together with a worked example is presented in Volume 6.

Whilst Kern's method provides a simple approach and one which is quite adequate for preliminary design calculations, much more reliable predictions may be achieved by taking into account the contribution to heat transfer and pressure drop made by the various idealised flow streams shown in Fig. 1.90. Such an approach was originally taken by Tinker[152] and many of the methods subsequently developed have been based on his model which unfortunately is difficult to follow and tedious to use. The approach has been simplified by Devore,[153] however, who, in using standard tolerances for commercial exchangers and a limited number of baffle designs, gives nomographs which enable the method to be used with simple calculators. Devore's method has been further simplified by Mueller[154] who gives an illustrative example. Palen and Taborek[155] and Grant[156] have described how both Heat Transfer Inc. and Heat Transfer and Fluid Flow Services have used Tinker's method to develop proprietary computer-based methods.

Using Tinker's approach, Bell[157,158] has described a semianalytical method, based on work at the University of Delaware, which allows for the effects of major bypass and leakage streams, and which is suitable for use with calculators. In this approach, the heat transfer coefficient and the pressure drop are obtained from correlations for flow over ideal tube banks, applying correction factors to allow for the effects of leakage, bypassing and flow in the window zone. This method gives more accurate predictions than Kern's method and can be used to determine the effects of constructional tolerances and the use of sealing tapes. This method is discussed in some detail in Volume 6, where an illustrative example is offered.

A more recent approach is that offered by Wills and Johnston[159] who have developed a simplified version of Tinker's method. This has been adopted by the Engineering Sciences Data Unit, ESDU,[160] and it gives a useful calculation technique for providing realistic checks on 'black box' computer predictions. The basis of this approach is shown in Fig. 1.93 which shows fluid flowing from A to B in two streams—over the tubes in cross-flow, and bypassing the tube bundle—which then combine to form a single stream. In addition, leakage occurs between the tubes and the baffle and between the baffle and the shell, as shown. For each of these

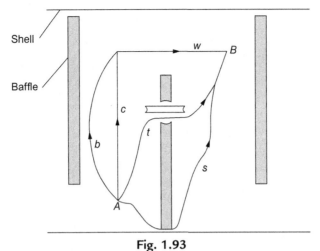

Fig. 1.93
Flow streams in the Wills and Johnston method.[159]

streams, a coefficient is defined which permits the pressure drop for each stream to be expressed in terms of the square of the mass velocity for that stream. A knowledge of the total mass velocity and the sum of the pressure drops in each zone enables the coefficients for each stream to be estimated by an iterative procedure, and the flowrate of each stream to be obtained. The estimation of the heat transfer coefficient and the pressure drop is then a relatively simple operation. This method is of especial value in investigating the effect of various shell-to-baffle and baffle-to-tube tolerances on the performance of a heat exchanger, both in terms of heat transfer rates and the pressure losses incurred.

1.9.7 Heat Exchanger Performance

One of the most useful methods of evaluating the performance of an existing heat exchanger or to assess a proposed design is to determine its *effectiveness η*, which is defined as the ratio of the actual rate of heat transfer Q to the maximum rate Q_{max} that is thermodynamically possible or:

$$\eta = \frac{Q}{Q_{max}} \tag{1.286}$$

Q_{max} is the heat transfer rate, which would be achieved if it were possible to bring the outlet temperature of the stream with the lower heat capacity, to the inlet temperature of the other stream. Using the nomenclature in Fig. 1.94, and taking stream 1 as having the lower value of GC_P, then:

$$Q_{max} = G_1 C_{p1}(T_{11} - T_{21}) \tag{1.287}$$

An overall heat balance gives:

$$Q = G_1 C_{p1}(T_{11} - T_{12}) = G_2 C_{p2}(T_{22} - T_{21})$$

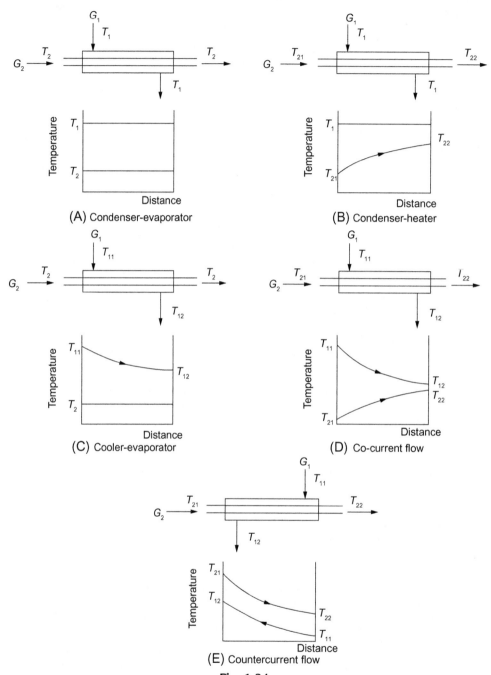

Fig. 1.94

Nomenclature for effectiveness of heat exchangers.

Thus, based on stream 1:

$$\eta = \frac{G_1 C_{p1}(T_{11} - T_{12})}{G_1 C_{p1}(T_{11} - T_{21})} = \frac{T_{11} - T_{12}}{T_{11} - T_{21}} \tag{1.288}$$

and, based on stream 2:

$$\eta = \frac{G_2 C_{p2}(T_{22} - T_{21})}{G_1 C_{p1}(T_{11} - T_{21})} \tag{1.289}$$

In calculating temperature differences, the positive value should always be taken.

Example 1.41

A flow of 1 kg/s of an organic liquid of heat capacity 2.0 kJ/kg K is cooled from 350 to 330 K by a stream of water flowing countercurrently through a double-pipe heat exchanger. Estimate the effectiveness of the unit if the water enters the exchanger at 290 K and leaves at 320 K.

Solution
Heat load,

$$Q = 1 \times 2.0(350 - 330) = 40 \, kW$$

Flow of water,

$$G_{cool} = \frac{40}{4.187(320 - 290)} = 0.318 \, kg/s$$

For organic:

$$\left(G C_p\right)_{hot} = (1 \times 2.0) = 2.0 \, kW/K \left(= G_2 C_{p2}\right)$$

For water:

$$\left(G C_p\right)_{cold} = (0.318 \times 4.187) = 1.33 \, kW/K \left(= G C_p\right)_{min} = \left(G_1 C_{p1}\right)$$

From Eq. (1.289):

$$\text{Effectiveness } \eta = \frac{2.0(350 - 330)}{1.33(350 - 290)}$$
$$= 0.50$$

1.9.8 Transfer Units

The concept of a *transfer unit* is useful in the design of heat exchangers and in assessing their performance, since its magnitude is less dependent on the flowrate of the fluids than the heat transfer coefficient which has been used so far. The number of transfer units N is defined by:

$$N = \frac{UA}{(GC_p)_{min}} \tag{1.290}$$

where $(GC_p)_{min}$ is the lower of the two values G_1C_{P1} and G_2C_{P2}. N is the ratio of the heat transferred for a unit temperature driving force to the heat absorbed by the fluid stream when its temperature is changed by 1 deg K. Thus, the number of transfer units gives a measure of the amount of heat which the heat exchanger can transfer. The relation for the effectiveness of the heat exchanger in terms of the heat capacities of the streams is now given for a number of flow conditions. The relevant nomenclature is given in Fig. 1.94.

Transfer units are also used extensively in the calculation of mass transfer rates in countercurrent columns and reference should be made to Chapter 2.

Considering *cocurrent* flow as shown in Fig. 1.94D, for an elemental area dA of a heat exchanger, the rate of transfer of heat dQ is given by:

$$dQ = UdA(T_1 - T_2) = UdA\theta \tag{1.291}$$

where T_1 and T_2 are the temperatures of the two streams and θ is the point value of the temperature difference between the streams.

In addition:

$$dQ = G_2C_{p2}dT_2 = -G_1C_{p1}dT_1$$

Thus:

$$dT_2 = \frac{dQ}{G_2C_{p2}} \quad \text{and} \quad dT_1 = \frac{-dQ}{G_1C_{p1}}$$

and:

$$dT_1 - dT_2 = d(T_1 - T_2) = d\theta = -dQ\left(\frac{1}{G_1C_{p1}} + \frac{1}{G_2C_{p2}}\right)$$

Substituting from Eq. (1.291) for dQ:

$$\frac{d\theta}{\theta} = -UdA\left[\frac{1}{G_1C_{p1}} + \frac{1}{G_2C_{p2}}\right] \tag{1.292}$$

Integrating:

$$\ln \frac{\theta_2}{\theta_1} = -UA\left[\frac{1}{G_1 C_{p1}} + \frac{1}{G_2 C_{p2}}\right]$$

or:

$$\ln \frac{T_{12} - T_{22}}{T_{11} - T_{21}} = \frac{UA}{G_1 C_{p1}}\left[1 + \frac{G_1 C_{p1}}{G_2 C_{p2}}\right] \tag{1.293}$$

If $G_1 C_{P1} < G_2 C_{P2}, G_1 C_{P1} = (GC_p)_{min}$

From Eq. (1.290):

$$N = \frac{UA}{G_1 C_{p1}}$$

Thus:

$$\frac{T_{12} - T_{22}}{T_{11} - T_{21}} = \exp\left[-N\left(1 + \frac{G_1 C_{p1}}{G_2 C_{p2}}\right)\right] \tag{1.294}$$

From Eqs. (1.288), (1.289):

$$T_{11} - T_{12} = \eta(T_{11} - T_{21})$$

$$T_{22} - T_{21} = \eta\frac{G_1 C_{p1}}{G_2 C_{p2}}(T_{11} - T_{21})$$

Adding:

$$T_{11} - T_{12} + T_{22} - T_{21} = \eta\left(1 + \frac{G_1 C_{p1}}{G_2 C_{p2}}\right)(T_{11} - T_{21})$$

$$1 - \frac{T_{12} - T_{22}}{T_{11} - T_{21}} = \eta\left(1 + \frac{G_1 C_{p1}}{G_2 C_{p2}}\right)$$

Substituting in Eq. (1.294):

$$\eta = \frac{1 - \exp\left[-N\left(1 + \dfrac{G_1 C_{p1}}{G_2 C_{p2}}\right)\right]}{1 + \dfrac{G_1 C_{p1}}{G_2 C_{p2}}} \tag{1.295}$$

For the particular case where $G_1 C_{P1} = G_2 C_{P2}$:

$$\eta = 0.5[1 - \exp(-2N)] \tag{1.296}$$

For a very large exchanger $(N \rightarrow \infty)$, $\eta \rightarrow 0.5$.

A similar procedure may be followed for *countercurrent flow* (Fig. 1.94E), although it should be noted that, in this case, $\theta_1 = T_{11} - T_{22}$ and $\theta_2 = T_{12} - T_{21}$.

The corresponding equation for the effectiveness factor η is then:

$$\eta = \frac{1 - \exp\left[-N\left(1 - \frac{G_1 C_{p1}}{G_2 C_{p2}}\right)\right]}{1 - \frac{G_1 C_{p1}}{G_2 C_{p2}} \exp\left[-N\left(1 - \frac{G_1 C_{p1}}{G_2 C_{p2}}\right)\right]} \tag{1.297}$$

For the case where $G_1 C_{P1} = G_2 C_{P2}$, it is necessary to expand the exponential terms to give:

$$\eta = \frac{N}{1 + N} \tag{1.298}$$

In this case, for a very large exchanger ($N \to \infty$), $\eta \to 1$.

If one component is merely undergoing *a phase change at constant temperature* (Fig. 1.94B and C) $G_1 C_{P1}$ is effectively zero and both Eqs. (1.295), (1.297) reduce to:

$$\eta = 1 - \exp(-N) \tag{1.299}$$

Effectiveness factors η are plotted against number of transfer units N with $(G_1 C_{P1}/G_2 C_{P2})$ as parameter for a number of different configurations by Kays and London.[161] Examples for countercurrent flow (based on Eq. 1.289) and an exchanger with one shell pass and two tube passes are plotted in Fig. 1.95A and B respectively.

Example 1.42

A process requires a flow of 4 kg/s of purified water at 340 K to be heated from 320 K by 8 kg/s of untreated water which can be available at 380, 370, 360, or 350 K. Estimate the heat transfer surfaces of one shell pass, two tube pass heat exchangers suitable for these duties. In all cases, the mean heat capacity of the water streams is 4.18 kJ/kg K and the overall coefficient of heat transfer is 1.5 kW/m² K.

Solution

For the untreated water:

$$GC_P = (8.0 \times 4.18) = 33.44 \, \text{kW/K}$$

For the purified water:

$$GC_P = (4.0 \times 4.18) = 16.72 \, \text{kW/K}$$

Thus:

$$\left(GC_p\right)_{\min} = 16.72 \, \text{kW/K} = G_1 C_{P_1}$$

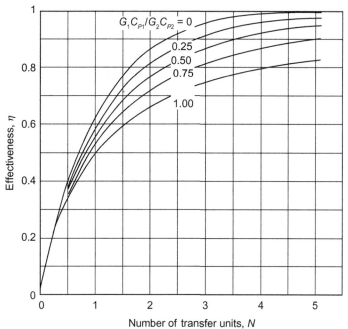

(A) True countercurrent flow

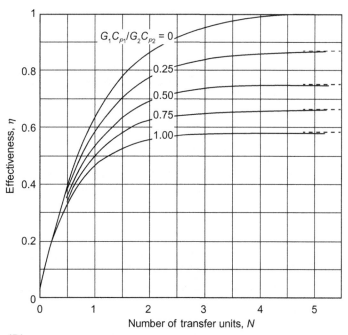

(B) One shell pass, two-tube pass exchanger

Fig. 1.95

Effectiveness of heat exchangers as a function of number of transfer units.[161]

and:

$$\frac{G_1 C_{p1}}{G_2 C_{p2}} = \frac{16.72}{33.44} = 0.5$$

From Eq. (1.289):

$$\eta = \frac{4.0 \times 4.18(340 - 320)}{4.0 \times 4.18(T_{11} - 320)}$$

$$= \frac{20}{(T_{11} - 320)}.$$

Thus η may be calculated from this equation using values of $T_{11} = 380, 370, 360,$ or 350 K and then N obtained from Fig. 1.95B. The area required is then calculated from:

$$A = \frac{N(GC_p)_{min}}{(U)} \quad (Eq.\,9.290)$$

to give the following results:

T_{11} (K)	$\eta(-)$	N (−)	A (m²)
380	0.33	0.45	5.0
370	0.4	0.6	6.6
360	0.5	0.9	10.0
350	0.67	1.7	18.9

Obviously, the use of a higher untreated water temperature is attractive in minimising the area required, although in practice any advantages would be offset by increased water costs, and an optimisation procedure would be necessary to obtain the most effective design.

Example 1.43

An existing single shell pass and two tube pass heat exchanger (area $= 10$ m²) is available for the following application:

Tube side fluid:

$$G = 2,98,00\,\text{kg/h}; \quad C_p = 3.25\,\text{kJ/kgK}$$

Shell side fluid:

$$G = 9510\,\text{kg/h}; \quad C_p = 4\,\text{kJ/kgK}$$

The inlet temperatures on the tube side and shell side respectively are 30 and 260°C. The past experience suggests the overall heat transfer coefficient to be of the order of 1600 W/m² K. Determine the two outlet temperatures and the rate of heat transfer.

Solution

Since both exit temperatures are not known, the use of Eq. (1.266) here entails a trial and error method. On the other hand, it is straightforward to use the effectiveness—NTU method as

$$\left(G_1 C_{p1}\right)_{hot} = \frac{9510}{3600} \times 4000 = 10,567\,\text{W/K}$$

$$\left(G_1 C_{p1}\right)_{cold} = \frac{2,98,00}{3600} \times 3250 = 26,903\,\text{W/K}$$

$$\therefore \ \left(G_1 C_{p1}\right)_{hot} = \left(G C_p\right)_{min}$$

$$\therefore \ \frac{\left(G_1 C_{p1}\right)_{hot}}{\left(G_1 C_{p1}\right)_{cold}} = \frac{10567}{26903} = 0.392$$

$$NTU = N = \frac{UA}{\left(G_1 C_{p1}\right)_{min}} = \frac{1600 \times 10}{10,567} = 1.514$$

Now using Fig. 1.94 for $N = 1.514$, $G_1 C_{p1}/G_2 C_{p2} = 0.392$,

$$\eta = 0.66 = \frac{260 - T_o}{260 - 30}$$

$$\therefore \ T_o = 108.2°\text{C (hot stream)}$$

From the overall energy balance:

$$\frac{9510}{3600} \times 4000(260 - 108.2) = \frac{2,98,00}{3600} \times 3250 \times (t_o - 30)$$

Solving for t_o:

$$t_o = 89.6°\text{C}$$

and the heat duty:

$$Q = \frac{9510}{3600} \times 4000 \times (260 - 108.2) = 1604\,\text{kW}$$

1.10 Other Forms of Equipment

1.10.1 Finned-Tube Units

Film coefficients

When viscous liquids are heated in a concentric tube or standard tubular exchanger by condensing steam or hot liquid of low viscosity, the film coefficient for the viscous liquid is much smaller than that on the hot side and it therefore controls the rate of heat transfer. This condition also arises with air or gas heaters where the coefficient on the gas side will be very low compared with that for the liquid or condensing vapour on the other side. It is often possible to obtain a much better performance by increasing the area of surface on the side with the limiting coefficient. This may be done conveniently by using a finned tube as in Fig. 1.96, which shows one typical form of such equipment which may have either longitudinal (Fig. 1.96A) or transverse (Fig. 1.96B) fins.

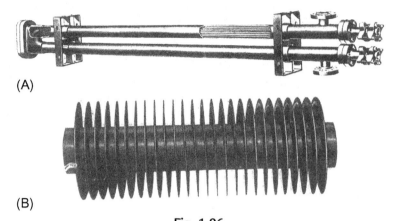

(A)

(B)

Fig. 1.96
(A) Heat exchanger showing tubes with longitudinal fins. (B) Tube with radial fins.

The calculation of the film coefficients on the fin side is complex because each unit of surface on the fin is less effective than a unit of surface on the tube wall. This arises because there will be a temperature gradient along the fin so that the temperature difference between the fin surface and the fluid will vary along the fin. To calculate the film coefficients, it is convenient to consider firstly the extended surface as shown in Fig. 1.97. A cylindrical rod of length L and cross-sectional area A and perimeter b is heated at one end by a surface at temperature T_1 and cooled throughout its length by a medium at temperature T_G so that the cold end is at a temperature T_2.

A heat balance over a length dx at distance x from the hot end gives:

$$\text{heat in} = \text{heat out along rod} + \text{heat lost to surroundings}$$

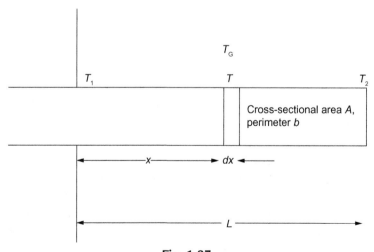

Fig. 1.97
Heat flow in rod with heat loss to surroundings.

or:

$$-kA\frac{dT}{dx} = \left[-kA\frac{dT}{dx} + \frac{d}{dx}\left(-kA\frac{dT}{dx}\right)dx\right] + hbdx(T - T_G)$$

where h is the film coefficient from fin to surroundings.

Writing the temperature difference $T - T_G$ equal to θ:

$$kA\frac{d^2T}{dx^2}dx = hbdx\theta$$

Since T_G is constant, $d^2T/dx^2 = d^2\theta/dx^2$.

Thus:

$$\frac{d^2\theta}{dx^2} = \frac{hb}{kA}\theta = m^2\theta \quad \left(\text{where } m^2 = \frac{hb}{kA}\right) \tag{1.300}$$

and:

$$\theta = C_1e^{mx} + C_2e^{-mx} \tag{1.301}$$

In solving this equation, three important cases may be considered:

(a) A long rod with temperature falling to that of surroundings, that is $\theta = 0$ when $x = \infty$ and using $\theta = T_1\text{-}T_c = \theta_1$ when $x = 0$.
   In this case:

$$\theta = \theta_1e^{-mx} \tag{1.302}$$

(b) A short rod from which heat loss from its end is neglected.
   At the hot end:

$$x = 0, \quad \theta = \theta_1 = C_1 + C_2$$

   At the cold end:

$$x = L, \quad \frac{d\theta}{dx} = 0$$

Thus:

$$0 = C_1me^{mL} - C_2me^{-mL}$$

and:

$$\theta_1 = C_1 + C_1e^{2mL}$$

Thus:

$$C_1 = \frac{\theta_1}{1 + e^{2mL}}, \quad C_2 = \frac{\theta_1}{1 + e^{-2mL}}$$

Hence:

$$\theta = \frac{\theta_1 e^{mx}}{1+e^{2mL}} + \frac{\theta_1 e^{2mL} e^{=mx}}{1+e^{2mL}}$$

$$= \frac{\theta_1}{1+e^{2mL}} \left[e^{mx} + e^{2mL} e^{-mx} \right]$$

(1.303)

or:

$$\frac{\theta}{\theta_1} = \frac{e^{-mL} e^{mx} + e^{mL} e^{-mx}}{e^{-mL} + e^{mL}}$$

This may be written:

$$\frac{\theta}{\theta_1} = \frac{\cosh m(L-x)}{\cosh mL}$$

(1.304)

(c) More accurately, allowing for heat loss from the end:
At the hot end:

$$x=0, \quad \theta=\theta_1 = C_1 + C_2$$

At the cold end:

$$x=L, \quad Q = hA\theta_{x=L} = -kA\left(\frac{d\theta}{dx}\right)_{x=L}$$

The determination of C_1 and C_2 in Eq. (1.301) then gives:

$$\theta = \frac{\theta_1}{1+Je^{-2mL}} \left(Je^{-2mL} e^{mx} + e^{-mx} \right)$$

(1.305)

where:

$$J = \frac{km-h}{km+h}$$

or again, noting that $\cosh x = \frac{1}{2}(e^x + e^{-x})$ and $\sin x = \frac{1}{2}(e^x - e^{-x})$:

$$\frac{\theta}{\theta_1} = \frac{\cosh m(L-x) + (h/mk)\sinh m(L-x)}{\cosh mL + (h/mk)\sinh mL}$$

(1.306)

The *heat loss* from a finned tube is obtained initially by determining the heat flow into the base of the fin from the tube surface. Thus the heat flow to the root of the fin is:

$$Q_f = -kA\left(\frac{dT}{dx}\right)_{x=0} = -kA\left(\frac{d\theta}{dx}\right)_{x=0}$$

For case (a):

$$Q_f = -kA(-m\theta_1) = kA\sqrt{\frac{hb}{kA}}\theta_1 = \sqrt{hbkA}\theta_1$$

(1.307)

For case (b):

$$Q_f = -kAm\theta_1 \frac{1-e^{2mL}}{1+e^{2mL}} = \sqrt{hbkA}\theta_1 \tanh mL \qquad (1.308)$$

For case (c):

$$Q_f = \sqrt{hbkA}\theta_1 \left(\frac{1-Je^{-2mL}}{1+Je^{-2mL}}\right) \qquad (1.309)$$

These expressions are valid provided that the cross-section for heat flow remains constant, though it need not be circular. When it is not constant, as with a radial or tapered fin, for example, the temperature distribution is in the form of a *Bessel function*.[162]

If the fin were such that there was no drop in temperature along its length, then the maximum rate of heat loss from the fin would be:

$$Q_f\text{max} = bLh\theta_1$$

The *fin effectiveness* is then given by Q_f/Q_f max and, for case (b), this becomes:

$$\frac{kAm\theta_1 \tanh mL}{bLh\theta_1} = \frac{\tanh mL}{mL} \qquad (1.310)$$

Example 1.44

In order to measure the temperature of a gas flowing through a copper pipe, a thermometer pocket is fitted perpendicularly through the pipe wall, the open end making very good contact with the pipe wall. The pocket is made of copper tube, 10 mm o.d. and 0.9 mm thickwall, and it projects 75 mm into the pipe. A thermocouple is welded to the bottom of the tube and this gives a reading of 475 K when the wall temperature is at 365 K. If the coefficient of heat transfer between the gas and the copper tube is 140 W/m^2 K, calculate the gas temperature. The thermal conductivity of copper may be taken as 350 W/m K. This arrangement is shown in Fig. 1.98.

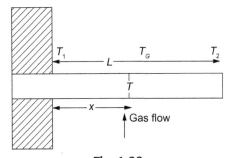

Fig. 1.98
Heat transfer to thermometer pocket.

Solution

In this case, it is reasonable to assume that the heat loss from the tip is zero, i.e. when $x=L$, $(d\theta/dx)=0$ and hence one can use Eq. (1.304).

If θ is the temperature difference $(T-T_G)$, then:

$$\theta = \theta_1 \frac{\cosh m(L-x)}{\cosh mL}$$

At $x=L$:

$$\theta = \frac{\theta_1}{\cosh mL} \quad \text{because} \ \cosh 0 = 1$$

$$m^2 = \frac{hb}{kA}$$

where the perimeter: $b=\pi \times 0.010$ m, tube i.d. $=(10-2\times0.9)=8.2$ mm or 0.0082 m.

Cross-sectional area of metal: and:

$$A = \frac{\pi}{4}\left(10.0^2 - 8.2^2\right) = 8.19\pi\,\text{mm}^2 \ \text{or} \ 8.19\pi \times 10^{-6}\,\text{m}^2$$

$$\therefore \ m^2 = \frac{hb}{kA} = \frac{(140 \times 0.010\pi)}{(350 \times 8.19\pi \times 10^{-6})} = 488\,\text{m}^{-2}$$

and:

$$m = 22.1\,\text{m}^{-1}$$

$$\theta_1 = T_G - 365, \quad \theta_2 = T_G - 475$$

$$\frac{\theta_1}{\theta_2} = \cosh mL$$

$$\therefore \ \frac{T_G - 365}{T_G - 475} = \cosh(22.1 \times 0.075) = 2.72$$

and:

$$\underline{\underline{T_G = 539\,\text{K}}}$$

Example 1.45

A steel tube fitted with transverse circular steel fins of constant cross-section has the following specifications:

tube o.d.: $d_2=54.0$ mm fin diameter $d_1=70.0$ mm
fin thickness: $w=2.0$ mm number of fins/metre run $=230$

Determine the heat loss per metre run of the tube when the surface temperature is 370 K and the temperature of the surroundings is 280 K. The heat transfer coefficient between gas and fin is 30 W/m^2 K and the thermal conductivity of steel is 43 W/m K.

Solution

Assuming that the height of the fin is small compared with its circumference and that it may be treated as a straight fin of length $(\pi/2)(d_1 + d_2)$, then:

The perimeter:

$$b = \frac{2\pi(d_1 + d_2)}{2} = \pi(d_1 + d_2)$$

The area:

$$A = \frac{\pi(d_1 + d_2)w}{2}$$

i.e. the average area at right-angles to the heat flow.

Then:

$$m = \left(\frac{hb}{kA}\right)^{0.5} = \left\{\frac{h\pi(d_1 + d_2)}{\{k\pi(d_1 + d_2)w/2\}}\right\}^{0.5}$$

$$= \left(\frac{2h}{kw}\right)^{0.5}$$

$$= \left(\frac{2 \times 30}{43 \times 0.002}\right)^{0.5}$$

$$= 26.42 \, \text{m}^{-1}$$

From Eq. (1.308), the heat flow is given for case (b) as:

$$Q_f = (hbkA)^{0.5}\theta_1 \tanh mL = mkA\theta_1 \frac{e^{2mL} - 1}{1 + e^{2mL}}$$

In this equation:

$$A = \frac{[\pi(70.0 + 54.0) \times 2.0]}{2} = 390 \, \text{mm}^2 \text{ or } 0.00039 \, \text{m}^2$$

$$L = \frac{d_1 - d_2}{2} = 8.0 \, \text{mm} \text{ or } 0.008 \, \text{m}$$

$$mL = 26.42 \times 0.008 = 0.211$$

$$\theta_1 = 370 - 280 = 90 \, \text{degK}$$

$$\therefore \quad Q_f = \frac{26.42 \times 43 \times 3.9 \times 10^{-4} \times 90(e^{0.422} - 1)}{1 + e^{0.422}}$$

$$= \frac{39.9 \times 0.525}{2.525} = 8.29 \, \text{W per fin}$$

The heat loss per metre run of tube $= 8.29 \times 230 = 1907$ W/m

or:

$$1.91 \, \text{kW/m}$$

In this case, the low value of mL indicates a fin efficiency of almost 1.0, though where mL tends to 1.0 the efficiency falls to about 0.8.

Example 1.46

A 5 mm diameter metallic rod ($k = 380$ W/m K) is 450 mm long and acts as a support joining two process vessels whose outer surface temperatures are 363 K. If the ambient temperature is 298 K and the convective heat transfer coefficient is 25 W/m^2 K. Find the temperature at the mid-point of the support and at a distance of 150 mm from one end.

Solution
This is effectively the case of a thin cylindrical rod whose two ends are maintained at 363 K:

Owing to symmetry, at the mid-point, i.e. $x = L$, $\dfrac{d\theta}{dx} = 0$ and hence it can be treated as a fin:

$L = \dfrac{450}{2} = 225$ mm; $d = 5$ mm

$h = 25$ W/m^2 K; $k = 380$ W/m K

$$\theta_1 = T|_{x=0} - T_G = 363 - 298 = 65 \, \text{K}$$

$$m^2 = \frac{hb}{kA} = \frac{25 \times \pi \times 0.005}{380 \times \dfrac{\pi}{4} \times (0.005)^2} = 52.63 \, \text{m}^{-2}$$

$$\therefore \quad m = 7.25 \, \text{m}^{-1}$$

Using Eq. (1.304):

When $x = L = 225$ mm:

$$\theta|_L = \frac{\theta_1}{\cosh mL} = \frac{65}{\cosh(7.25 \times 0.225)} = 24.5 \, \text{K}$$

$$\theta|_{x=L} = T|_{x=L} - 298 = 24.5 \, \text{K}$$

$$\therefore \quad \text{Temperature at mid-point} = 298 + 24.5 = 322.5 \, \text{K}$$

Similarly, the temperature at $x = 100$ mm:

$$\theta = \theta_1 \frac{\cosh m(L-x)}{\cosh mL} = (65) \frac{\cosh 7.25(0.225-0.1)}{\cosh (7.25 \times 0.225)}$$

$$\theta = T - 298 = 35.3\,\text{K}$$

$$\therefore \quad T = 333.3\,\text{K.}$$

Example 1.47

In an ingenious device to measure the thermal conductivity of a material, one reference rod (of known thermal conductivity = 43 W/m K) joins two large steel plates which are maintained at 373 K. The ambient temperature is 298 K and the mid-point temperature of this rod is measured to be 322 K. For an identical rod (same size) of unknown thermal conductivity, the mid-point temperature is measured to be 348 K under otherwise identical conditions, what is the thermal conductivity of this material?

Solution

This situation is similar to that in Example 1.46 in so far that at the mid-point in both rods $d\theta/dx = 0$. Thus, one can use Eq. (1.304):

$$\frac{\theta}{\theta_1} = \frac{\cosh m(L-x)}{\cosh mL}$$

At mid-point, $x = L$ and for both rods, $\theta_1 = 373 - 298 = 75$ K

For material A at $x = L$, $\theta = 322 - 298 = 24$ K

$$\therefore \quad 24 = \frac{75}{\cosh m_A L} \quad \text{or} \quad m_A L = \cosh^{-1}(75/24)$$

Similarly, for second material B:

$$(348 - 298) = \frac{75}{\cosh m_B L} \quad \text{or} \quad m_B L = \cosh^{-1} 75/50$$

Dividing one by the other:

$$\frac{m_A}{m_B} = \sqrt{\frac{k_B}{k_A}} = \frac{\cosh^{-1}(75/24)}{\cosh^{-1}(75/50)} = \frac{1.806}{0.962} = 1.876$$

Or

$$k_B/k_A = 3.52 \quad \text{and} \quad k_B = 3.52 \times 43 = 151\,\text{W/mK}$$

Practical data

A neat form of construction has been designed by the Brown Fintube Company of America. On both prongs of a hairpin tube are fitted horizontal fins, which fit inside concentric tubes, joined at the base of the hairpin. Units of this form can be conveniently arranged in

Table 1.25 Data on surface of finned tube units[a]

Pipe Size Outside Diameter		Outside Surface of Pipe		Surface of Finned Pipe				
					(m²/m) Height of Fin		(ft²/ft) Height of Fin	
mm	(in)	(m²/m)	(ft²/ft length)	Number of Fins	12.7 mm	25.4 mm	0.5 in.	1 in.
25.4	1	0.08	0.262	12	0.385	0.689	1.262	2.262
				16	0.486	0.893	1.595	2.929
				20	0.587	1.096	1.927	3.595
48.3	1.9	0.15	0.497	20	0.660	1.167	2.164	3.830
				24	0.761	1.371	2.497	4.497
				28	0.863	1.574	2.830	5.164
				36	1.066	1.980	3.498	6.497

[a]Brown Fintube Company.

banks to give large heat transfer surfaces. It is usual for the extended surface to be at least five times greater than the inside surface, so that the low coefficient on the fin side is balanced by the increase in surface. An indication of the surface obtained is given in Table 1.25.

A typical hairpin unit with an effective surface on the fin side of 9.4 m² has an overall length of 6.6 m, height of 0.34 m, and width of 0.2 m. The free area for flow on the fin side is 2.645 mm² against 1.320 mm² on the inside; the ratio of the transfer surface on the fin side to that inside the tubes is 5.93:1.

The fin side film coefficient h_f has been expressed by plotting:

$$\frac{h_f}{C_p G'}\left(\frac{C_p \mu}{k}\right)^{2/3}\left(\frac{\mu}{\mu_s}\right)^{-0.14} \quad \text{against} \quad \frac{d_e G'}{\mu}$$

where h_f is based on the total fin side surface area (fin and tube), G' is the mass rate of flow per unit area, and d_e is the equivalent diameter, or:

$$d_e = \frac{4 \times \text{cross} - \text{sectional area f or flow on fin side}}{\text{total wetted perimeter for flow (fin} + \text{outside of tube} + \text{inner surface of shell tube)}}$$

Experimental work has been carried out with exchangers in which the inside tube was 48 mm outside diameter and was fitted with 24, 28, or 36 fins (12.5 mm by 0.9 mm) in a 6.1 m length; the finned tubes were inserted inside tubes 90 mm inside diameter. With steam on the tube side, and tube oils and kerosene on the fin side, the experimental data were well correlated by plotting:

$$\frac{h_f}{C_p G'}\left(\frac{C_p \mu}{k}\right)^{2/3}\left(\frac{\mu}{\mu_s}\right)^{-0.14} \quad \text{against} \quad \frac{d_e G'}{\mu}$$

Table 1.26 Data on finned tubes

Inside diam. of tube	19 mm	25 mm	38 mm	50 mm	75 mm
Outside diam. of fin	64 mm	70 mm	100 mm	110 mm	140 mm
No. of fins/m run	Heat transferred (kW/m)				
65	0.47	0.63	0.37	1.07	1.38
80	0.49	0.64	1.02	1.12	1.44
100	0.54	0.69	1.14	1.24	1.59
Inside diam. of tube	$\frac{3}{4}$ in.	1 in.	$1\frac{1}{2}$ in.	2 in.	3 in.
Outside diam. Of fin	$2\frac{1}{2}$ in.	$2\frac{3}{4}$ in.	$3\frac{7}{8}$ in.	$4\frac{5}{16}$ in.	$5\frac{1}{2}$ in.
No. of fins/ft run	Heat transferred (Btu/h ft)				
20	485	650	1010	1115	1440
24	505	665	1060	1170	1495
30	565	720	1190	1295	1655

Data taken from catalogue of G. A. Harvey and Co. Ltd. of London.

Typical values were:

$$\frac{h_f}{C_p G'}\left(\frac{C_p \mu}{k}\right)^{2/3}\left(\frac{\mu}{\mu_s}\right)^{-0.14} = 0.25 \quad 0.055 \quad 0.012 \quad 0.004$$

$$\frac{d_e G'}{\mu} = \quad 1 \quad 10 \quad 100 \quad 1000$$

Some indication of the performance obtained with *transverse finned tubes* is given in Table 1.26. The figures show the heat transferred per unit length of pipe when heating air on the fin side with steam or hot water on the tube side, using a temperature difference of 100 deg K. The results are given for three different spacings of the fins.

1.10.2 Plate-Type Exchangers

A series of plate type heat exchangers, which present some special features was first developed by the APV Company. The general construction is shown in Fig. 1.99, which shows an Alfa-Laval exchanger and from which it is seen that the equipment consists of a series of parallel plates held firmly together between substantial head frames. The plates are one-piece pressings, frequently of stainless steel, and are spaced by rubber sealing gaskets cemented into a channel around the edge of each plate. Each plate has a number of troughs pressed out at right angles to the direction of flow and arranged so that they interlink with each other to form a channel of constantly changing direction and section. With normal construction, the gap between the plates is 1.3–1.5 mm. Each liquid flows in alternate spaces and a large surface can be obtained in a small volume.

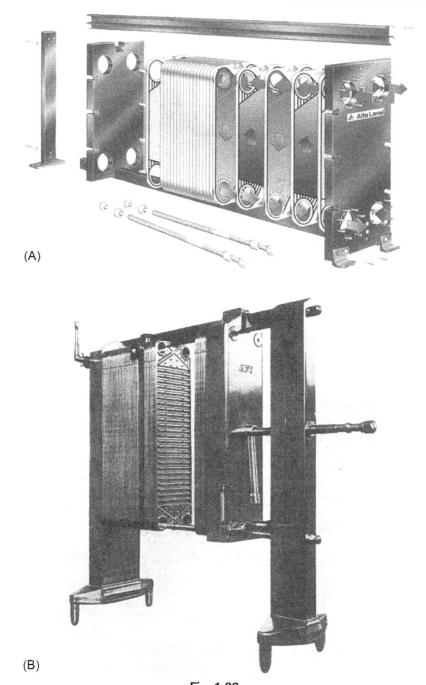

(A)

(B)

Fig. 1.99
(A) Alfa-Laval plate heat exchanger. (B) APV plate heat exchanger.

Table 1.27 Plate areas

Plate Type	Projected Area		Developed Area	
	m^2	ft^2	m^2	ft^2
HT	0.09	1.00	0.13	1.35
HX	0.13	1.45	0.17	1.81
HM	0.27	2.88	0.35	3.73
HF	0.36	3.85	0.43	4.60

Table 1.28 Performance of plate-type exchanger type HF

Heat Transferred Per Plate		Water Flow		U Based on Developed Area	
W/K	Btu/h °F	1/s	gal/h	kW/m^2 K	Btu/h ft^2 °F
1580	3000	0.700	550	3.70	650
2110	4000	1.075	850	4.94	870
2640	5000	1.580	1250	6.13	1080

Courtesy of the APV Company.

Because of the shape of the plates, the developed area of surface is appreciably greater than the projected area. This is shown in Table 1.27 for the four common sizes of plate.

A high degree of turbulence is obtained even at low flowrates and the high heat transfer coefficients obtained are illustrated by the data in Table 1.28. These refer to the heating of cold water by the equal flow of hot water in an HF type exchanger (aluminium or copper), at an average temperature of 310 K.

Using a stainless steel plate with a flow of 0.00114 m^3/s, the heat transferred is 1760 W/K for each plate.

The high transfer coefficient enables these exchangers to be operated with very small temperature differences, so that a high heat recovery is obtained. These units have been particularly successful in the dairy and brewing industries, where the low liquid capacity and the close control of temperature have been valuable features. A further advantage is that they are easily dismantled for inspection of the whole plate. The necessity for the long gasket is an inherent weakness, but the exchangers have been worked successfully up to 423 K and at pressures of 930 kN/m^2. They are now being used in the processing and gas industries with solvents, sugar, acetic acid, ammoniacal liquor, and so on.

1.10.3 Spiral Heat Exchangers

A spiral plate exchanger is illustrated in Fig. 1.100 in which two fluids flow through the channels formed between the spiral plates (Fig. 1.100A). With this form of construction, the velocity may be as high as 2.1 m/s and overall transfer coefficients of 2.8 kW/m^2 K are

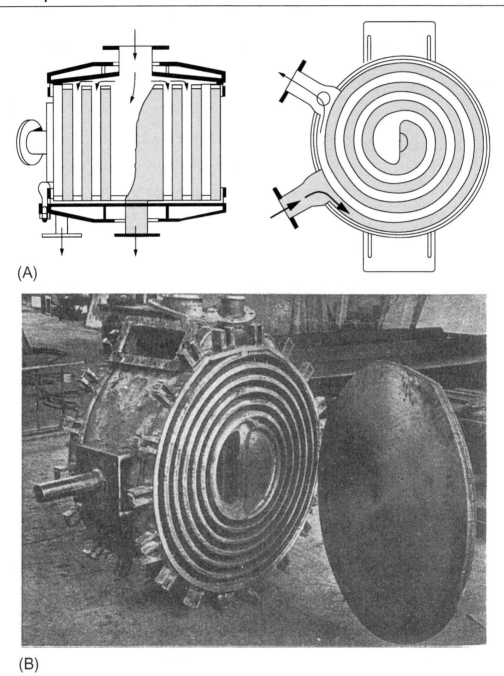

(A)

(B)

Fig. 1.100
Spiral heat exchanger (A) flow paths. Spiral plate exchanger (B) with cover removed.

frequently obtained. The size can therefore be kept relatively small and the cost becomes comparable or even less than that of shell and tube units, particularly when they are fabricated from alloy steels. Fig. 1.100B shows a spiral plate exchanger without its cover.

A further design of spiral heat exchanger, described by Neil,[163] is essentially a single pass counterflow heat exchanger with fixed tube plates distinguished by the spiral winding of the tubes, each consisting, typically of a 10 mm o.d. tube wound on to a 38 mm o.d. coil such that the inner heat transfer coefficient is 1.92 times greater than for a straight tube. The construction overcomes problems of differential expansion rates of the tubes and the shell and the characteristics of the design enable the unit to perform well with superheated steam where the combination of counterflow, high surface area per unit volume of shell and the high inside coefficient of heat transfer enables the superheat to be removed effectively in a unit of reasonable size and cost.

1.10.4 Compact Heat Exchangers

Advantages of compact units

In general, heat exchanger equipment accounts for some 10% of the cost of a plant; a level at which there is no great incentive for innovation. Trends such as the growth in energy conservation schemes, the general move from bulk chemicals to value-added products, and plant problems associated with drilling rigs have, however, all prompted the development and increased application of compact heat exchangers. Here compactness is a matter of degree, maximising the heat transfer area per unit volume of exchanger and this leads to narrow channels. Making shell and tube exchangers more compact presents construction problems and a more realistic approach is to use plate or plate and fin heat exchangers. The relation between various types of exchanger, in terms of the heat transfer area per unit volume of exchanger is shown in Fig. 1.101, taken from Redman.[164] In order to obtain a thermal effectiveness in excess of 90%, countercurrent flow is highly desirable and this is not easily achieved in shell and tube units which often have a number of tube-side passes in order to maintain reasonable velocities. Because of the baffle design on the shell-side, the flow at the best may be described as cross-flow, and the situation is only partly redeemed by having a train of exchangers, which is an expensive solution. Again in dealing with high value added products, which could well be heat sensitive, a more controllable heat exchanger than a shell and tube unit, in which not all the fluid is heated to the same extent, might be desirable. One important application of compact heat exchangers is the cooling of natural gas on offshore rigs where space (costing as much as £120,000/m^2) is of paramount importance.

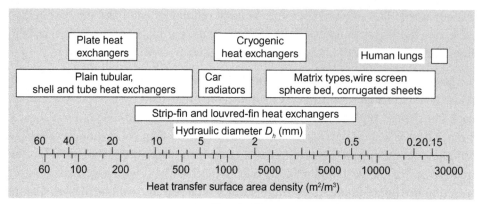

Fig. 1.101

Surface area as a function of volume of exchanger for different types.

Plate and fin exchangers

Plate and fin heat exchangers, used in the motor, aircraft, and transport industries for many years, are finding increased application in the processing industries and in particular in natural gas liquefaction, cryogenic air separation, the production of olefins, and in the separation of hydrogen and carbon monoxide. Potential applications include ammonia production, offshore processing, nuclear engineering, and syngas production. As described by Gregory,[165] the concept is that of flat plates of metal, usually aluminium, with corrugated metal, which not only holds the plates together but also acts as a secondary surface for heat transfer. Bars at the edges of the plates retain each fluid between adjacent plates, and the space between each pair of plates, apportioned to each fluid according to the heat transfer and pressure drop requirements, is known as a layer. The heights of the bars and corrugations are standardised in the UK at 3.8, 6.35, and 8.9 mm and typical designs are shown in Fig. 1.102.

There are four basic forms of corrugation: *plain* in which a sheet of metal is corrugated in the simplest way with fins at right angles to the plates; *serrated* where each cut is offset in relation to the preceding fin; *herringbone* corrugation made by displacing the fins sideways every

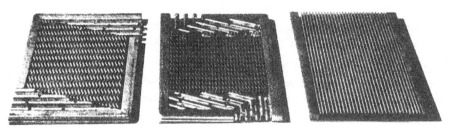

Fig. 1.102

Plate and fin exchangers.

9.5 mm to give a zig-zag path; and *perforated* corrugation, a term used for a plain corrugation made from perforated material. Each stream to be heated or cooled can have different corrugation heights, different corrugation types, different number of layers, and different entry and exit points including part length entry and exit. Because the surface in the hot fluid (or fluids) can vary widely from that in the cold fluid (or fluids), it is unrealistic to quote surface areas for either side, as in shell and tube units, though the overall surface area available to all the fluids is $1000 \, \text{m}^2/\text{m}^3$ of exchanger (Fig. 1.101).

In design, the general approach is to obtain the term (hA) for each stream, sum these for all the cold and all the hot streams and determine an overall value of (hA) given by:

$$\frac{1}{(hA)_{ov}} = \frac{1}{(hA)_w} + \frac{1}{(hA)_c} \qquad (1.311)$$

where *ov*, *w*, and *c* refer, respectively to overall, hot-side and cold-side values.

Printed-circuit exchangers

Various devices such as small diameter tubes, fins, tube inserts, and porous boiling surfaces may be used to improve the surface density and heat transfer coefficients in shell and tube heat exchangers and yet these are not universally applicable and in general, such units remain essentially bulky. Plate and fin exchangers have either limited fluid compatibility or limited fin efficiency, problems which are overcome by using printed-circuit exchangers as described by Johnston.[166] These are constructed from flat metal plates which have fluid flow passages chemically milled into them by means of much the same techniques as are used to produce electrical printed circuits. These plates are diffusion-bonded together to form blocks riddled with precisely sized, shaped, routed, and positioned passages, and these blocks are in turn welded together to form heat exchange cores of the required capacity. Fluid headers are attached to the core faces and sometimes the assembly is encapsulated. Passages are typically 0.3–1.5 mm deep giving surface areas of 1000–$5000 \, \text{m}^2/\text{m}^3$, an order of magnitude higher than surface densities in shell and tube designs; and in addition the fine passages tend to sustain relatively high heat transfer coefficients, undiminished by fin inefficiencies, and so less surface is required.

In designing a unit, each side of the exchanger is independently tailored to the duty required, and the exchanger effectiveness (discussed in Section 1.9.4) can range from 2% to 5% to values in excess of 98% without fundamental design or construction problems arising. Countercurrent, cocurrent, and cross-flow contacting can be employed individually or in combination.

A note of caution on the use of photo-etched channels has been offered by Ramshaw[167] who points out that the system is attractive in principle provided that severe practical problems such as fouling are not encountered. With laminar flow in matrices with a mean plate spacing of 0.3–1 mm, volumetric heat transfer coefficients of $7 \, \text{MW}/\text{m}^3$ K have been obtained with modest pressure drops. Such values compare with 0.2 MW/m^3 K for shell and tube exchangers and 1.2 MW/m^3 K for plate heat exchangers.

1.10.5 Scraped-Surface Heat Exchangers

In cases where a process fluid is likely to crystallise on cooling or the degree of fouling is very high or indeed the fluid is of very high viscosity, use is often made of scraped-surface heat exchangers in which a rotating element has spring-loaded scraper blades which wipe the inside surface of a tube which may typically be 0.15 m in diameter. Double-pipe construction is often employed with a jacket, say 0.20 m in diameter, and one common arrangement is to connect several sections in series or to install several pipes within a common shell. Scraped-surface units of this type are used in paraffin-wax plants and for evaporating viscous or heat-sensitive materials under high vacuum. This is an application to which the *thin-film device* is especially suited because of the very short residence times involved. In such a device, the clearance between the agitator and the wall may be either fixed or variable since both rigid and hinged blades may be used. The process liquid is continuously spread in a thin layer over the vessel wall and it moves through the device either by the action of gravity or that of the agitator or of both. A tapered or helical agitator produces longitudinal forces on the liquid.

In describing chillers for the production of wax distillates, Nelson[168] points out that the rate of cooling depends very much on the effectiveness of the scrapers, and quotes overall coefficients of heat transfer ranging from 15 W/m^2 K with a poorly fitting scraper to 90 W/m^2 K where close fitting scrapers remove the wax effectively from the chilled surface.

The *Votator* design has two or more floating scraper-agitators which are forced against the cylinder wall by the hydrodynamic action of the fluid on the agitator and by centrifugal action; the blades are loosely attached to a central shaft called the mutator. The votator is used extensively in the food processing industries and also in the manufacture of greases and detergents. As the blades are free to move, the clearance between the blades and the wall varies with operating conditions and a typical installation may be 75–100 mm in diameter and 0.6–1.2 m long. In the *spring-loaded* type of scraped-surface heat exchanger, the scrapers are held against the wall by leaf springs, and again there is a variable clearance between the agitator and the cylinder wall since the spring force is balanced by the radial hydrodynamic force of the liquid on the scraper. Typical applications of this device are the processing of heavy waxes and oils and crystallising solutions. Generally the units are 0.15–0.3 m in diameter and up to 12 m long. Some of the more specialised heat exchangers and chemical reactors employ helical ribbons, augers, or twisted tapes as agitators and, in general, these are fixed clearance devices used for high viscosity materials. There is no general rule as to maximum or minimum dimensions since each application is a special case.

One of the earliest investigations into the effectiveness of scrapers for improving heat transfer was that of Huggins[169] who found that although the improvement with water was negligible, cooling times for more viscous materials could be considerably reduced. This was confirmed by Laughlin[170] who has presented operating data on a system where the process fluid changes from a thin liquid to a paste and finally to a powder. Houlton,[171] making tests on the votator, found that back-mixing was negligible and some useful data on a number of food products have been obtained by Bolanowski and Lineberry[172] who, in addition to discussing the operation and uses of the votator, quote overall heat transfer coefficients for each food tested. Using a *liquid-full* system, Skelland et al.[173–175] have proposed the following general design correlation for the votator:

$$\frac{hd_v}{k} = c_1 \left(\frac{C_p\mu}{k}\right)^{c_2} \left(\frac{(d_v - d_r)u\rho}{\mu}\right) \left(\frac{d_v N}{u}\right)^{0.82} \left(\frac{d_r}{d_v}\right)^{0.55} (n_B)^{0.53} \tag{1.312}$$

where for cooling viscous liquids $c_1 = 0.014$ and $c_2 = 0.96$, and for thin mobile liquids $c_1 = 0.039$ and $c_2 = 0.70$. In this correlation, d_v is the diameter of the vessel, d_r is the diameter of the rotor, and u is the average axial velocity of the liquid. This correlation may only be applied to the range of experimental data upon which it is based, since h will not approach zero as d_r, $(d_v - d_r)$, N, and u approach zero. Reference to the use of the votator for crystallisation is made in Volume 2, Chapter 15.

The majority of work on heat transfer in *thin-film* systems has been directed towards obtaining data on specific systems rather than developing general design methods, although Bott et al.[176–178] have developed the following correlations for heating without change of phase:

$$\frac{hd_v}{k} = Nu = 0.018(Re'')^{0.6} Re^{0.46} Pr^{0.87} \left(\frac{d_v}{l}\right)^{0.48} (n_B)^{0.24} \tag{1.313}$$

and for evaporation:

$$\frac{hd_v}{k} = Nu = 0.65(Re'')^{0.43} Re^{0.25} Pr^{0.3} (n_B)^{0.33} \tag{1.314}$$

From both of these equations, it will be noted that the heat transfer coefficient is *not* a function of the temperature difference. Here $Re'' = (d_o^2 N\rho/\mu)$ and $Re = (ud_v\rho/\mu)$, where d_v is the tube diameter and u is the average velocity of the liquid in the film in the axial direction.

It is also of significance that the agitation suppresses nucleation in a fluid which might otherwise deposit crystals.

This section is concluded by noting that detailed descriptions about the relative merits and demerits of the currently available various designs of compact heat exchangers, their design and operation, and potential applications are available in excellent reference books.[179–181]

1.11 Thermal Insulation

1.11.1 Heat Losses Through Lagging

A hot reaction or storage vessel or a steam pipe will lose heat to the atmosphere by radiation, conduction, and convection. The loss by radiation is a function of the fourth power of the absolute temperatures of the body and surroundings, and will be small for low temperature differences but will increase rapidly as the temperature difference increases. Air is a very poor conductor, and the heat loss by conduction will therefore be small except possibly through the supporting structure. On the other hand, since convection currents form very easily, the heat loss from an unlagged surface is considerable. The conservation of heat, and hence usually of total energy, is an economic necessity, and some form of lagging should normally be applied to hot surfaces. Lagging of plant operating at high temperatures is also necessary in order to achieve acceptable working conditions in the vicinity. In furnaces, as has already been seen, the surface temperature is reduced substantially by using a series of insulating bricks which are poor conductors.

The two main requirements of a good lagging material are that it should have a low thermal conductivity and that it should suppress convection currents. The materials that are frequently used are cork, 85% magnesia, glass wool, and vermiculite. Cork is a very good insulator though it becomes charred at moderate temperatures and is used mainly in refrigerating plants. Eighty-five percent magnesia is widely used for lagging steam pipes and may be applied either as a hot plastic material or in preformed sections. The preformed sections are quickly fitted and can frequently be dismantled and reused whereas the plastic material must be applied to a hot surface and cannot be reused. Thin metal sheeting is often used to protect the lagging.

The rate of heat loss per unit area is given by:

$$\frac{\text{total temperature difference}}{\text{total thermal resistance}}$$

For the case of heat loss to the atmosphere from a lagged steam pipe, the thermal resistance is due to that of the condensate film and dirt on the inside of the pipe, that of the pipe wall, that of the lagging, and that of the air film outside the lagging. Thus for unit length of a lagged pipe:

$$\frac{Q}{l} = \sum \Delta T \left/ \left[\frac{1}{h_i \pi d} + \frac{x_w}{k_w \pi d_w} + \frac{x_1}{k_r \pi d_m} + \frac{1}{(h_r + h_c) \pi d_s} \right] \right. \tag{1.315}$$

where d is the inside diameter of pipe, d_w the mean diameter of pipe wall, d_m the logarithmic mean diameter of lagging, d_s the outside diameter of lagging, x_w, x_l are the pipe wall and lagging thickness respectively, k_w, k_r the thermal conductivity of the pipe wall and lagging, and h_i, h_r, h_c the inside film, radiation, and convection coefficients, respectively.

Example 1.48

A steam pipe, 150 mm i.d. and 168 mm o.d., is carrying steam at 444 K and is lagged with 50 mm of 85% magnesia. What is the heat loss to air at 294 K?

Solution

In this case:

$$d = 150\,\text{mm or } 0.150\,\text{m}$$

$$d_0 = 168\,\text{mm or } 0.168\,\text{m}$$

$$d_w = \frac{150 + 168}{2} = 159\,\text{mm or } 0.159\,\text{m}$$

$$d_s = 268\,\text{mm or } 0.268\,\text{m}$$

$$d_m, \text{the log mean of } d_0 \text{ and } d_s = 215\,\text{mm or } 0.215\,\text{m.}$$

The coefficient for condensing steam together with that for any scale will be taken as 8500 W/m² K, k_w as 45 W/m K, and k_l as 0.073 W/m K.

The temperature on the outside of the lagging is estimated at 314 K and $(h_r + h_c)$ will be taken as 10 W/m² K.

The thermal resistances are therefore:

$$\frac{1}{h_i \pi d} = \frac{1}{8500 \times \pi \times 0.150} = 0.00025$$

$$\frac{x_w}{k_w \pi d_w} = \frac{0.009}{45 \times \pi \times 0.159} = 0.00040$$

$$\frac{x_l}{k_l \pi d_m} = \frac{0.050}{0.073 \times \pi \times 0.215} = 1.013$$

$$\frac{1}{(h_r + h_c)\pi d_s} = \frac{1}{10 \times \pi \times 0.268} = 0.119$$

The first two terms may be neglected and hence the total thermal resistance is 1.132 m K/W.

The heat loss per metre length $= (444 - 294)/1.132 = \underline{132.5\,\text{W/m}}$ (from Eq. 1.315).

The temperature on the outside of the lagging may now be checked as follows:

$$\frac{\Delta T(\text{lagging})}{\sum \Delta T} = \frac{1.013}{1.132} = 0.895$$

$$\Delta T(\text{lagging}) = 0.895(444 - 294) = 134\,\text{deg K}$$

Thus the temperature on the outside of the lagging is $(444 - 134) = 310$ K, which approximates to the assumed value.

Taking an emissivity of 0.9, from Eq. (1.166):

$$h_r = \frac{[0.9 \times 5.67 \times 10^{-8}(310^4 - 294^4)]}{310 - 294} = 7.40\,\text{W/m}^2\,\text{K}$$

From Table 1.10 for air $(GrPr = 10^4 - 10^9)$, $n = 0.25$ and $C'' = 1.32$.

Substituting in Eq. (1.150) (putting l = diameter = 0.268 m):

$$h_c = C''(\Delta T)^n l^{3n-1} = 1.32 \left[\frac{310 - 294}{0.268} \right]^{0.25} = 3.67 \, \text{W/m}^2 \, \text{K}$$

Thus $(h_r + h_c) = 11.1 \, \text{W/m}^2$ K, which is close to the assumed value. In practice it is rare for forced convection currents to be absent, and the heat loss is probably higher than this value.

If the pipe were unlagged, $(h_r + h_c)$ for $\Delta T = 150$ K would be about 20 W/m^2 K and the heat loss would then be:

$$\frac{Q}{l} = (h_r + h_c) \pi d_o \Delta T$$

$$= (20 \times \pi \times 0.168 \times 150) = 1584 \, \text{W/m}$$

or:

$$\underline{\underline{1.58 \, \text{kW/m}}}$$

Under these conditions it is seen that the heat loss has been reduced by more than 90% by the addition of a 50 mm thickness of lagging.

1.11.2 Economic Thickness of Lagging

Increasing the thickness of the lagging will reduce the loss of heat and thus give a saving in the operating costs. The cost of the lagging will increase with thickness and there will be an optimum thickness when further increase does not save sufficient heat to justify the cost. In general the smaller the pipe, the smaller the thickness used, though it cannot be too strongly stressed that some lagging everywhere is better than excellent lagging in some places and none in others. For temperatures of 373–423 K, and for pipes up to 150 mm diameter, Lyle[182] recommends a 25 mm thickness of 85% magnesia lagging and 50 mm for pipes over 230 mm diameter. With temperatures of 470–520 K 38 mm is suggested for pipes less than 75 mm diameter and 50 mm for pipes up to 230 mm diameter.

1.11.3 Critical Thickness of Lagging

As the thickness of the lagging is increased, resistance to heat transfer by thermal conduction increases. Simultaneously the outside area from which heat is lost to the surroundings also increases, giving rise to the possibility of increased heat loss by convection. It is perhaps easiest to think of the lagging as acting as a fin of very low thermal conductivity. For a cylindrical pipe, there is the possibility of heat losses being increased by the application of lagging, only if $hr/k < 1$, where k is the thermal conductivity of the lagging, h is the outside film coefficient, and r is the outside diameter of the pipe. In practice, this situation is likely to arise only for pipes of small diameters.

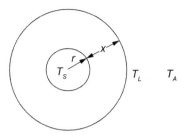

Fig. 1.103
Critical lagging thickness.

The heat loss from a pipe at a temperature T_s to surroundings at temperature T_A is considered. Heat flows through lagging of thickness x across which the temperature falls from a constant value T_S at its inner radius r, to an outside temperature T_L which is a function of x, as shown in Fig. 1.103. The rate of heat loss Q from a length l of pipe is given by Eq. (1.316), by considering the heat loss from the outside of the lagging; and by Eqs. (1.317), (1.318), which give the transfer rate by thermal conduction through the lagging of logarithmic mean radius r_m:

$$Q = 2\pi l (r+x)(T_L - T_A)h \tag{1.316}$$

$$Q = \frac{2\pi l r_m}{x} k(T_S - T_L) \tag{1.317}$$

$$= \frac{2\pi l k}{\ln \dfrac{r+x}{r}}(T_S - T_L) \tag{1.318}$$

Equating the values given in Eqs. (1.316), (1.318):

$$(r+x)h(T_L - T_A) = \frac{k}{\ln \dfrac{r+x}{r}}(T_S - T_L)$$

Then:

$$a = (r+x)\frac{h}{k} \ln \frac{r+x}{r} = \frac{T_S - T_L}{T_L - T_A}$$

$$aT_L - aT_A = T_S - T_L$$

$$T_L = \frac{T_S + aT_A}{a+1}$$

Substituting in Eq. (1.316):

$$Q = 2\pi l(r+x)\left\{\left(\frac{T_S + aT_A}{a+1} - T_A\right)\right\}h$$

$$= \frac{2\pi lh(r+x)}{a+1}(T_S - T_A)$$

$$= 2\pi lh(T_S - T_A)\left\{(r+x)\frac{1}{1+(r+x)\frac{h}{k}\ln\left(\frac{r+x}{r}\right)}\right\}$$

(1.319)

Differentiating with respect to x:

$$\frac{1}{2\pi lh(T_S - T_A)}\frac{dQ}{dx} = \frac{\left\{1+(r+x)\frac{h}{k}\ln\frac{r+x}{r} - (r-x)\left[\frac{h}{k}\ln\frac{r+x}{r} + (r+x)\frac{h}{k}\frac{r}{(r+x)}\frac{1}{r}\right]\right\}}{\left[1+(r+x)\frac{h}{k}\ln\left(\frac{r+x}{r}\right)\right]^2}$$

The maximum value of $Q(Q_{max})$ occurs when $dQ/dx = 0$.

that is, when:

$$1 - (r+x)\frac{h}{k} = 0$$

or:

$$x = \frac{k}{h} - r$$

(1.320)

When the relation between heat loss and lagging thickness exhibits a maximum for the unlagged pipe $(x=0)$, then:

$$\frac{hr}{k} = 1$$

(1.321)

When $hr/k > 1$, the addition of lagging always reduces the heat loss.

When $hr/k < 1$, thin layers of lagging increase the heat loss and it is necessary to exceed the critical thickness given by Eq. (1.320) before any benefit is obtained from the lagging.

Substituting in Eq. (1.319) gives the maximum heat loss as:

$$Q_{max} = 2\pi lh(T_S - T_A)\left\{\frac{k}{h}\frac{1}{1+\frac{kh}{hk}\ln\frac{k}{hr}}\right\}$$

$$= 2\pi l(T_S - T_A)k\frac{1}{1+\ln\frac{k}{hr}}$$

(1.322)

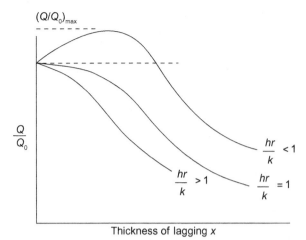

Fig. 1.104
Critical thickness of lagging.

For an unlagged pipe, $x=0$ and $T_L=T_S$. Substituting in Eq. (1.316) gives the rate of heat loss Q_o as:

$$Q_o = 2\pi l r (T_S - T_A) h \tag{1.323}$$

Thus:

$$\frac{Q_{max}}{Q_o} = \frac{k}{rh} \bigg/ \left(1 + \ln\frac{k}{rh}\right) \tag{1.324}$$

The ratio Q/Q_o is plotted as a function of thickness of lagging (x) in Fig. 1.104.

Example 1.49

A pipeline of 100 mm outside diameter, carrying steam at 420 K, is to be insulated with a lagging material which costs £10/m^3 and which has a thermal conductivity of 0.1 W/m K. The ambient temperature may be taken as 285 K, and the coefficient of heat transfer from the outside of the lagging to the surroundings as 10 W/m^2 K. If the value of heat energy is 7.5×10^{-4} £/MJ and the capital cost of the lagging is to be depreciated over 5 years with an effective simple interest rate of 10% per annum based on the initial investment, what is the economic thickness of the lagging?

Is there any possibility that the heat loss could actually be increased by the application of too thin a layer of lagging?

Solution

For a thick-walled cylinder, the rate of conduction of heat through lagging is given by Eq. (1.22):

$$Q = \frac{2\pi l k (T_i - T_o)}{\ln(d_o/d_i)} \text{ W}$$

where d_o and d_i are the external and internal diameters of the lagging and T_o and T_i, the corresponding temperatures.

Substituting $k = 0.1$ W/m K, $T_o = 420$ K (neglecting temperature drop across pipe wall), and $d_i = 0.1$ m, then:

$$\frac{Q}{l} = \frac{2\pi \times 0.1(420 - T_o)}{\ln(d_o/0.1)} \text{ W/m}$$

The term Q/l must also equal the heat loss from the outside of the lagging, or:

$$\frac{Q}{l} = h_o(T_o - 285)\pi d_o = 10(T_o - 285)\pi d_o \text{ W/m}$$

Thus:

$$T_o = \left\{ \frac{Q}{l} \frac{1}{10\pi d_o} + 285 \right\} \text{K}$$

Substituting:

$$\frac{Q}{l} = \frac{2\pi \times 0.1 \left[420 - \dfrac{Q}{l} \dfrac{1}{10\pi d_o} - 285 \right]}{\ln(d_o/0.1)}$$

or:

$$\frac{Q}{l} = \frac{2\pi \times 0.1 \times 135}{\ln(d_o/0.1) + 2\pi \times 0.1 \times \dfrac{1}{10\pi d_o}} = \frac{84.82}{\ln(d_o/0.1) + (0.02/d_o)} \text{W/m}$$

Value of heat lost $= £7.5 \times 10^{-4}$/MJ

or:

$$\frac{84.82}{\ln(d_o/0.1) + (0.02/d_o)} \times 7.5 \times 10^{-4} \times 10^{-6} = \frac{6.36 \times 10^{-8}}{\ln(d_o/2.1) + (0.02/d_o)} £/\text{m s}$$

$$\text{Volume of lagging per unit pipe length} = \frac{\pi}{4}[d_o^2 - (0.1)^2] \text{ m}^3/\text{m}$$

$$\text{Capital cost of lagging} = £10/\text{m}^3 \text{ or } \frac{\pi}{4}[d_o^2 - 0.01]10 = 7.85(d_o^2 - 0.01)£/\text{m}$$

Noting that 1 year $= 31.5$ Ms, then:

$$\text{Depreciation} = 7.85[d_o^2 - 0.01]/(5 \times 31.5 \times 10^6) = 4.98 \times 10^{-8}(d_o^2 - 0.01)£/\text{Ms}$$

$$\text{Interest charges} = (0.1 \times 7.85)(d_o^2 - 0.01)/(31.5 \times 10^6) = 2.49 \times 10^{-8}(d_o^2 - 0.01)£/\text{Ms}$$

$$\text{Total capital charges} = 7.47 \times 10^{-8}(d_o^2 - 0.01)£/\text{Ms}$$

Total cost (capital charges + value of heat lost) is given by:

$$C = \left\{ \frac{6.36}{\ln(d_o/0.1) + (0.02/d_o)} + 7.47(d_o^2 - 0.01) \right\} 10^{-8} £/\text{ms}$$

Differentiating with respect to d_o:

$$10^8 \frac{dC}{dd_o} = 6.36 \left[\frac{-1}{[\ln(d_o/0.1) + (0.02/d_o)]^2} \right] \left[\frac{1}{d_o} - \frac{0.02}{d_o^2} \right] + 7.47(2d_o)$$

In order to obtain the minimum value of C, dC/dd_o must be put equal to zero.

Then:

$$\frac{1}{[\ln(d_o/0.1) + (0.02/d_o)]^2} = \frac{(7.47 \times 2)}{6.36} \left[\frac{d_o}{(1/d_o) - (0.02/d_o^2)} \right]$$

that is:

$$\frac{1}{[\ln(d_o/0.1) + (0.02/d_o)]^2} = 2.35 \frac{d_o^3}{(d_o - 0.02)}$$

A trial and error solution gives $d_o - 0.426$ m or 426 mm

Thus, the economic thickness of lagging $= (426 - 100)/2 = 163$ mm

For this pipeline:

$$\frac{hr}{k} = \frac{10 \times (50 \times 10^{-3})}{0.1} = 5$$

From Eq. (1.321), the critical value of hr/k, below which the heat loss may be increased by a thin layer of lagging, is 1. For $hr/k > 1$, as in this problem, the situation will not arise.

1.12 Nomenclature

		Units in SI System	Dimensions in M, L, T, θ (or M, L, T, θ, H)
A	Area available for heat transfer *or* area of radiating surface	m^2	L^2
A_e	External area of body	m^2	L^2
A_s	Maximum cross-flow area over tube bundle	m^2	L^2
a	Absorptivity	—	—
$\mathbf{a}_b$	Absorptivity of a black body	—	—
$\mathbf{a}_s$	Absorptivity of a gas	—	—
b	Wetted perimeter of condensation surface or perimeter of fin	m	L
C	Constant	—	—
C_1	First radiation constant	$W\,m^2$	$M\,L^4\,T^{-3}$ (*or* $H\,L^2\,T^{-1}$)

C_2	Second radiation constant	mK	$L\theta$
C_3	Third radiation constant	mK	$L\theta$
C_4	Fourth radiation constant	W/m^3 K^5	$M\,L^{-3}\,T^{-3}\,\theta^{-5}$ (or $H\,L^{-3}\,T^{-1}\,\theta^{-5}$)
C_f	Constant for friction in flow past a tube bundle	–	–
C_h	Constant for heat transfer in flow past a tube bundle	–	–
C'	Clearance between tubes in heat exchanger	m	L
C''	Coefficient in Eq. (1.137) (SI units only)	J/m^{3n+1} s K^{n+1}	$M\,L^{1-3n}\,T^{-3}\,\theta^{-n-1}$ (or $H\,L^{-3n-1}\,T^{-1}\,\theta^{-n-1}$)
C_p	Specific heat at constant pressure	J/kg K	$L^2\,T^{-2}\,\theta^{-1}$ (or $H\,M^{-1}\,\theta^{-1}$)
D	Diffusivity of vapour	m^2/s	$L^2\,T^{-1}$
D	Diameter	m	L
D_H	Thermal diffusivity ($k/C_p\rho$)	m^2/s	$L^2\,T^{-1}$
d	Diameter (internal or of sphere)	m	L
d_1, d_2	Inner and outer diameters of annulus	m	L
d_c	Diameter of helix	m	L
d_e	Hydraulic mean diameter	m	L
d_g	Gap between turns in coil	m	L
d_m	Logarithmic mean diameter of lagging	m	L
d_o	Outside diameter of tube	m	L
d_p	Height of coil	m	L
d_r	Diameter of shaft	m	L
d_s	Outside diameter of lagging *or* inside diameter of shell	m	L
d_t	Thickness of fixed tube sheet	m	L
d_v	Internal diameter of vessel	m	L
d_w	Mean diameter of pipe wall	m	L
E	Emissive power	W/m^2	$M\,T^{-3}$ (or $H\,L^2\,T^{-1}$)
E_z'	Energy emitted per unit area and unit time per unit wavelength	W/m^3	$M\,L^{-1}\,T^{-3}$ (or $H\,L^{-3}\,T^{-1}$)
E_λ	Energy emitted per unit area per unit wavelength	W/m^3	$M\,T^{-3}$ (or $H\,L^{-3}\,T^{-1}$)
e	Emissivity	–	–
e′	$\frac{1}{2}(e_s + 1)$	–	–

$\mathbf{e}_F$	Emissivity of a flame	–	–
$\mathbf{e}_g$	Effective emissivity of a gas	–	–
$\mathbf{e}_{sh}$	Emissivity of shield	–	–
$\mathbf{e}_\lambda$	Spectral hemispherical emissivity	–	–
F	Geometric factor for radiation *or* correction factor for logarithmic mean temperature difference	–	–
f	Working stress	N/m^2	$M\,L^{-1}\,T^{-2}$
f'	Shell-side friction factor	–	–
G	Mass rate of flow	kg/s	$M\,T^{-1}$
G'	Mass rate of flow per unit area	kg/m^2 s	$M\,L^{-2}\,T^{-1}$
G_s'	Mass flow per unit area over tube bundle	kg/m^2 s	$M\,L^{-2}\,T^{-1}$
g	Acceleration due to gravity	m/s^2	$L\,T^{-2}$
H	Ratio: Z/X	–	–
H	Heat transfer coefficient	W/m^2 K	$M\,T^{-3}\,\theta^{-1}$ (*or* $H\,L^{-2}\,T^{-1}\,\theta^{-1}$)
h_b	Film coefficient for liquid adjacent to vessel	W/m^2 K	$M\,T^{-3}\,\theta^{-1}$ (*or* $H\,L^{-2}\,T^{-1}\,\theta^{-1}$)
h_c	Heat transfer coefficient for convection	W/m^2 K	$M\,T^{-3}\,\theta^{-1}$ (*or* $H\,L^{-2}\,T^{-1}\,\theta^{-1}$)
h_f	Fin-side film coefficient	W/m^2 K	$M\,T^{-3}\,\theta^{-1}$ (*or* $H\,L^{-2}\,T^{-1}\,\theta^{-1}$)
h_m	Mean value of h over whole surface	W/m^2 K	$M\,T^{-3}\,\theta^{-1}$ (*or* $H\,L^{-2}\,T^{-1}\,\theta^{-1}$)
h_n	Heat transfer coefficient for liquid boiling at T_n	W/m^2 K	$M\,T^{-3}\,\theta^{-1}$ (*or* $H\,L^{-2}\,T^{-1}\,\theta^{-1}$)
h_r	Heat transfer coefficient for radiation	W/m^2 K	$M\,T^{-3}\,\theta^{-1}$ (*or* $H\,L^{-2}\,T^{-1}\,\theta^{-1}$)
I	Intensity of radiation	W/m^2	$M\,T^{-3}$ (*or* $H\,L^{-2}\,T^{-1}$)
I'	Intensity of radiation falling on body	W/m^2	$M\,T^{-3}$ (*or* $H\,L^{-2}\,T^{-1}$)
J	For fin: $(km - h)/(km + h)$	–	–
j	Number of vertical rows of tubes	–	–
j_d	j-factor for mass transfer	–	–
j_h	j-factor for heat transfer	–	–
K	Concentration of particles in a flame	m^{-3}	L^{-3}
K'	Factor describing particle concentration	N/m^2	$M\,L^{-1}\,T^{-2}$
k	Thermal conductivity	W/m K	$M\,L\,T^{-3}\,\theta^{-1}$ (*or* $H\,L^{-1}\,T^{-1}\,\theta^{-1}$)

k_a	Arithmetic mean thermal conductivity	W/m K	$MLT^{-3}\theta^{-1}$ (*or* $HL^{-1}T^{-1}\theta^{-1}$)
k_m	Mean thermal conductivity	W/m K	$MLT^{-3}\theta^{-1}$ (*or* $HL^{-1}T^{-1}\theta^{-1}$)
k_o	Thermal conductivity at zero temperature	W/m K	$MLT^{-3}\theta^{-1}$ (*or* $HL^{-1}T^{-1}\theta^{-1}$)
k'	Constant in Eq. (1.13)	K^{-1}	θ^{-1}
k_G	Mass transfer coefficient (mass/unit area, unit time, unit partial pressure difference)	s/m	$L^{-1}T$
L	Length of paddle, length of fin, or characteristic dimension	m	L
L	Separation of surfaces, length of a side, or radius of hemispherical mass of gas	m	L
L_e	Mean beam length	m	L
l	Length of tube or plate, or distance apart of faces, *or* thickness of gas stream	m	L
l_B	Distance between baffles	m	L
lv	Length of side of vessel	m	L
M	Mass rate of flow of condensate per unit length of perimeter for vertical pipe and per unit length of pipe for horizontal pipe	kg/s m	$ML^{-1}T^{-1}$
m	Mass of liquid	kg	M
m	For fin: $(hb/kA)^{0.5}$	m^{-1}	L^{-1}
n	Number of baffles	–	–
N	Number of revolutions in unit time	Hz	T^{-1}
N	Number of general term in series, *or* number of transfer units	–	–
n	An index	–	–
n_B	Number of blades on agitator or number of baffles	–	
P	Pressure	N/m^2	$ML^{-1}T^{-2}$
P_{Bm}	Logarithmic mean partial pressure of inert gas **B**	N/m^2	$ML^{-1}T^{-2}$
P_c	Critical pressure	N/m^2	$ML^{-1}T^{-2}$
P_c	Partial pressure of carbon dioxide	N/m^2	$ML^{-1}T^{-2}$
p_g	Partial pressure of radiating gas	N/m^2	$ML^{-1}T^{-2}$
P_R	Reduced pressure (P/P_c)	–	–
P_r	Vapour pressure at surface of radius r	N/m^2	$ML^{-1}T^{-2}$
P_s	Saturation vapour pressure	N/m^2	$ML^{-1}T^{-2}$

P_w	Partial pressure of water vapour	N/m^2	$M\,L^{-1}\,T^{-2}$
p	Parameter in Laplace transform	s^{-1}	T^{-1}
Q	Heat flow *or* generation per unit time	W	$M\,L^2\,T^{-3}$ (*or* $H\,T^{-1}$)
Q_e	Radiation emitted per unit time	W	$M\,L^2\,T^{-3}$ (*or* $H\,T^{-1}$)
Q_f	Heat flow to root of fin per unit time	W	$M\,L^2\,T^{-3}$ (*or* $H\,T^{-1}$)
Q_G	Rate of heat generation per unit volume	W/m^3	$M\,L^{-1}\,T^{-3}$ (*or* $H\,L^{-3}\,T^{-1}$)
Q_I	Radiation incident on a surface	W	$M\,L^2\,T^{-3}$ (*or* $H\,T^{-1}$)
Q_i	Total incident radiation per unit time	W	$M\,L^2\,T^{-3}$ (*or* $H\,T^{-1}$)
Q_k	Heat flow per unit time by conduction in fluid	W	$M\,L^2\,T^{-3}$ (*or* $H\,T^{-1}$)
Q_o	Total radiation leaving surface per unit time	W	$M\,L^2\,T^{-3}$ (*or* $H\,T^{-1}$)
Q_r	Total heat reflected from surface per unit time	W	$M\,L^2\,T^{-3}$ (*or* $H\,T^{-1}$)
Q	Heat flow per unit time and unit area	W/m^2	$M\,T^{-3}$ (*or* $H\,L^{-2}\,T^{-1}$)
q_I	Energy arriving at unit area of a grey surface	W/m^2	$M\,T^{-3}$ (*or* $H\,L^{-2}\,T^{-1}$)
q_{sh}	Value of q with shield	W/m^2	$M\,T^{-3}$ (*or* $H\,L^{-2}\,T^{-1}$)
q_0	Radiosity–energy leaving unit area of a grey surface	W/m^2	$M\,T^{-3}$ (*or* $H\,L^{-2}\,T^{-1}$)
R	Thermal resistance	m^2 K/W	$M^{-1}\,T^3\,\theta$ (*or* $H^{-1}\,L^2\,T\theta$)
R	Ratio: r_j/r_i or r/L	–	–
R_i, R_0	Thermal resistance of scale on inside, outside of tubes	m^2 K/W	$M^{-1}\,T^3\,\theta$ (*or* $H^{-1}\,L^2\,T\theta$)
R'	Shear stress at free surface of condensate film	N/m^2	$M\,L^1\,T^{-2}$
r	Reflectivity	–	–
r	Radius	m	L
r_1, r_2	Radius (inner, outer) of annulus or tube	m	L
r_a	Arithmetic mean radius	m	L
r_m	Logarithmic mean radius	m	L
S	Flow area for condensate film	m^2	L^2
S	Ratio: s/r_i or $1 + \left(1 + R_j^2\right)/R_i^2$	–	–
s	Thickness of condensate film at a point	m	L
s	Distance between surfaces	m	L
T	Temperature	K	θ

T_c	Temperature of free surface of condensate	K	θ
T_{cm}	Temperature of cooling medium	K	θ
T_f	Mean temperature of film	K	θ
T_g	Temperature of gas	K	θ
T_G	Temperature of atmosphere surrounding fin	K	θ
T_m	Mean temperature of fluid	K	θ
T_n	Standard boiling point	K	θ
T_s	Temperature of condensing vapour	K	θ
T_{sh}	Temperature of shield	K	θ
T_w	Temperature of wall	K	θ
t	Transmissivity	–	–
t	Time	s	T
U	Overall heat transfer coefficient	W/m^2 K	$M\,T^{-3}\,\theta^{-1}$ $(or\ H\,L^{-2}\,T^{-1}\,\theta^{-1})$
u	Velocity	m/s	$L\,T^{-1}$
u_m	Maximum velocity in condensate film	m/s	$L\,T^{-1}$
u_y	Velocity at distance y from surface	m/s	$L\,T^{-1}$
V	Volume	m^3	L^3
W	Ratio: w/L or Y/X	–	–
W	Width of stirrer	m	L
w	Width of fin or surface	m	L
w_1, w_2	... Indices in equation for heat transfer by convection	–	–
X	Length of surface	m	L
$\overline{X}$	Ratio: X/L	–	–
X	Distance between centres of tubes in direction of flow	m	L
X	Ratio of temperature differences used in calculation of mean temperature difference	–	–
x	Distance in direction of transfer or along surface	m	L
Y	Distance between centres of tubes at right angles to flow direction *or* width of surface	m	L
$\overline{Y}$	Ratio: Y/L	–	–
Y	Ratio of temperature differences used in calculation of mean temperature difference	–	–
y	Distance perpendicular to surface	m	L

Z	Width of vertical surface	m	L
Z	Wavelength	m	L
Z_m	Wavelength at which maximum energy is emitted	m	L
z	Distance in third principal direction	m	L
α	Angle between two surfaces *or*	–	–
	Angle between normal and direction of radiation	–	–
β	Coefficient of cubical expansion	K^{-1}	θ^{-1}
η	Effectiveness of heat exchanger, defined by Eq. (1.200)	–	–
ϵ	Coefficient relating h to $u^{0.8}$	$J/s^{0.2} m^{2.8} K$	$M L^{-0.8} T^{-2.2} \theta^{-1}$ (*or* $H L^{-2.8} t^{-0.2} \theta^{-1}$)
λ	Wavelength *or*	m	L
	latent heat of vapourisation per unit mass	J/kg	$L^2 T^{-2}$ (*or* $H M^{-1}$)
μ	Viscosity	$N s/m^2$	$M L^{-1} T^{-1}$
μ_s	Viscosity of fluid at surface	$N s/m^2$	$M L^{-1} T^{-1}$
ρ	Density or density of liquid	kg/m^3	$M L^{-3}$
ρ_v	Density of vapour	kg/m^3	$M L^{-3}$
σ	Stefan–Boltzmann constant, *or*	$W/m^2 K^4$	$M T^{-3} \theta^{-4}$ (*or* $H L^{-2} \theta^{-4}$)
	Surface tension	N/m	$M T^{-2}$
τ	Response time for heating or cooling	s	T
θ	Temperature or temperature difference	K	θ
θ_α	Temperature difference in Schmidt method	K	θ
θ_c	Temperature of centre of body	K	θ
θ_m	Logarithmic mean temperature difference	K	θ
θ_{xt}	Temperature at $t=t$, $x=x$	K	θ
θ_o	Initial uniform temperature of body	K	θ
θ'	Temperature of source or surroundings	K	θ
Ψ	$\dfrac{G_1 C_{p1} + G_2 C_{p2}}{G_1 C_{p1} G_2 C_{p2}}$	–	–
ϕ	Angle between surface and horizontal *or* angle of contact	–	–
$\bar{\theta}$	Laplace transform of temperature	s K	$T \theta$
ω	Solid angle	–	–
F_o	Fourier number $D_H t/L^2$	–	–
Gr	Grashof number	–	–
Gz	Graetz number	–	–

Nu	Nusselt number	–	–
Nu'	Particle Nusselt number	–	–
Pr	Prandtl number	–	–
Re	Reynolds number	–	–
Re'	Particle Reynolds number	–	–
Re''	Rotational flow Reynolds number	–	–
Re_x	Reynolds number for flat plate	–	–
Δ	Finite difference in a property	–	–

Suffix, w refers to wall material

Suffixes i, o refer to inside, outside of wall or lagging; or inlet, outlet conditions

Suffixes b, l, v refer to bubble, liquid, and vapour

Suffixes g, b, and s refer to gas, black body, and surface.

References

1. Underwood AJV. Graphical computation of logarithmic mean temperature difference. *Ind Chem* 1933;**9**:167.
2. Spalding DB. *Combustion and mass transfer*. Oxford: Pergamon Press; 1979.
3. Long VD. A simple model of droplet combustion. *J Inst Fuel* 1964;**37**:522.
4. Monaghan MT, Siddal RG, Thring MW. The influence of initial diameter on the combustion of single drops of liquid fuel. *Combust Flame* 1968;**17**:45.
5. Sadhal SS, Ayyaswamy PS, Chung JS. *Transport phenomena with drops and bubbles*. New York: Springer; 1997.
6. Sirignano WA. *Fluid dynamics and transport of droplets and sprays*. 2nd ed., New York: Cambridge University Press; 2014.
7. Carslaw HS, Jaeger JC. *Conduction of heat in solids*. 2nd ed., Oxford: Oxford University Press; 1959.
8. Richardson JF. Spread of fire by thermal conduction. *Fuel* 1949;**28**:265.
9. Schmidt E. *Beitrage zur technischen Physik (Föppl Festschrift)*. Berlin: Springer-Verlag; 1924. p. 178–9.
10. Gallagher RH. *Finite element analysis. Fundamentals*. Englewood Cliffs, NJ: Prentice Hall; 1975.
11. Heisler MP. Temperature charts for induction and constant temperature heating. *Trans ASME* 1947;**69**:227.
12. Incropera FP, DeWitt DP. *Introduction to heat transfer*. 4th ed. New York: Wiley; 1996.
13. Waldie B. Review of recent work on the processing of powders in high temperature plasmas: Pt. II. Particle dynamics, heat transfer and mass transfer. *Chem Eng (Lond)* 1972;**261**:188.
14. Taylor TD. Heat transfer from single spheres in a low Reynolds number slip flow. *Phys Fluids* 1963;**6**:987.
15. Field MA, Gill DW, Morgan BB, Hawksley PGW. *Combustion of pulverised coal*. Leatherhead: BCURA; 1967.
16. Hahne E, Grigull U. Shape factor and shape resistance for steady multidimensional heat conduction. *Int J Heat Mass Transfer* 1975;**18**:751.
17. Dittus FW, Boelter LMK. Heat transfer in automobile radiators of the tubular type. *Univ Calif Berkeley Publ Eng* 1930;**2**:443. Reprinted: *Int Commun Heat Mass Transfer* 1985;**12**:3.
18. McAdams WH. *Heat transmission*. 2nd ed., New York: McGraw-Hill; 1942.
19. Winterton RHS. Where did the Dittus and Boelter equation come from? *Int J Heat Mass Transfer* 1998;**41**:809.
20. Colburn AP. A method of correlating forced convection heat transfer data and a comparison with fluid friction. *Trans Am Inst Chem Eng* 1933;**29**:174.
21. Sieder EN, Tate GE. Heat transfer and pressure drop of liquids in tubes. *Ind Eng Chem* 1936;**28**:1429.
22. Gnielinski V. New equations for heat and mass transfer in turbulent pipe and channel flow. *Int Chem Eng* 1976;**16**:359–68.

23. Sleicher CA, Rouse MW. A convective correlation for heat transfer to constant and variable property fluids in turbulent pipe flow. *Int J Heat Mass Transfer* 1975;**18**:677.

24. Shah RK, London AL. *Laminar flow forced convection in ducts. Advances in heat transfer.* New York: Academic Press; 1978.

25. Sleicher CA, Awad AS, Notter RH. Temperature and eddy diffusivity profiles in NaK. *Int J Heat Mass Transfer* 1973;**16**:1565.

26. Fishenden M, Saunders OA. *An introduction to heat transfer.* Oxford: Oxford University Press; 1950.

27. Cope WF. The friction and heat transmission coefficients of rough pipes. *Proc Inst Mech Eng* 1941;**45**:99.

28. Bird RB, Stewart WE, Lightfoot EN. *Transport phenomena.* 2nd ed., New York: Wiley; 2002.

29. Shah RK, Bhatti MS. *Laminar convective heat transfer in ducts. Handbook of single-phase convective heat transfer.* New York: Wiley; 1987. [chapter 3].

30. Kleinstreuer C. *Microfluidics and nanofluidics: theory and selected applications.* New York: Wiley; 2013.

31. Panigrahi PK. *Transport phenomena in microfluidic systems.* London: Wiley; 2015.

32. Reiher M. Wärmeübergang von strömender Luft an Röhren und Röhrenbündeln im Kreuzstrom. *Mitt Forsch* 1925;**269**:1.

33. Hilpert R. Wärmeabgabe von geheizten Drähten und Röhren. *Forsch Geb IngWes* 1933;**4**:215.

34. Griffiths E, Awbery JH. Heat transfer between metal pipes and a stream of air. *Proc Inst Mech Eng* 1933;**125**:319.

35. Davis AH. Convective cooling of wires in streams of viscous liquids. *Philos Mag* 1924;**47**:1057.

36. Morgan VT. The overall convective heat transfer from smooth circular cylinders. *Adv Heat Tran* 1975;**11**:199.

37. Zukauskas A. Convective heat transfer in cross flow. In: Zukauskas A, Kakac S, Shah RK, Aung W, editors. *Handbook of single-phase convective heat transfer.* New York: Wiley; 1987. [chapter 6].

38. Zdravkovich MM. *Flow around circular cylinders: applications.* vol. 2. New York: Oxford University Press; 2003.

39. Churchill SW, Bernstein M. A correlating equation for forced convection from gases and liquids to a circular cylinder in crossflow. *J Heat Transfer* 1977;**99**:300.

40. Ryoji I, Kenichiro S, Toshiaki K. Heat transfer around a circular cylinder in a liquid-solution crossflow. *Int J Heat Mass Transfer* 1979;**22**:1041.

41. Grimison ED. Correlation and utilization of new data on flow resistance and heat transfer for cross flow of gases over tube banks. *Trans Am Soc Mech Eng* 1937;**59**:583, and Trans Am Soc Mech Eng 1938;60:381.

42. Huge EC. Experimental investigation of effects of equipment size on convection heat transfer and flow resistance in cross flow of gases over tube banks. *Trans Am Soc Mech Eng* 1937;**59**:573.

43. Pierson OL. Experimental investigation of influence of tube arrangement on convection heat transfer and flow resistance in cross flow of gases over tube banks. *Trans Am Soc Mech Eng* 1937;**59**:563.

44. Kern DQ. *Process heat transfer.* New York: McGraw-Hill; 1950.

45. Short BE. *Heat transfer and pressure drop in heat exchangers.* Univ. of Texas Pub. No. 4324; 1943.

46. Donohue DA. Heat transfer and pressure drop in heat exchangers. *Ind Eng Chem* 1949;**41**:2499.

47. Tinker T. Analysis of the fluid flow pattern in shell and tube exchangers and the effect of flow distribution on the heat exchangers performance, In: *Proceedings of the general discussion on heat transfer, September*; 1951. p. 89. Inst. of Mech. Eng. and Am. Soc. Mech. Eng.

48. Kreith F, editor. *CRC handbook of thermal engineering.* Boca Raton: CRC Press; 1999.

49. Cope WF, Bailey A. Heat transmission through circular, square, and rectangular tubes. *Aeronaut Res Comm (Gt Brit) Tech Rept* 1933;**43**:199.

50. Washington L, Marks WM. Heat transfer and pressure drop in rectangular air passages. *Ind Eng Chem* 1937;**29**:337.

51. Davis ES. Heat transfer and pressure drop in annuli. *Trans Am Soc Mech Eng* 1943;**65**:755.

52. Carpenter FG, Colburn AP, Schoenborn EM, Wurster A. Heat transfer and friction of water in an annular space. *Trans Am Inst Chem Eng* 1946;**42**:165.

53. Bhatti MS, Shah RK. Kakac S, Shah RK, Aung W, editors. *Handbook of single-phase convective heat transfer.* New York: Wiley; 1987.

54. Rowe PN, Claxton KT, Lewis JB. Heat and mass transfer from a single sphere in an extensive flowing fluid. *Trans Inst Chem Eng* 1965;**41**:T14.

55. Whitaker S. Forced convection heat transfer correlations for flow in pipes, past flat plates, single cylinders, single spheres, and for flow in packed beds and tube bundles. *AIChE J* 1972;**18**:361.

56. Achenbach E. Heat transfer from spheres up to $Re = 6 \times 10^6$, In: *Proc. 6th Int. Heat Trans. Conf. 5*, Washington, DC: Hemisphere; 1978. p. 341.

57. Witte LC. An experimental study of forced convection heat transfer from a sphere to liquid sodium. *J Heat Transfer* 1968;**90**:9.

58. Schmidt E. Versuche über den Wärmeübergang in ruhender Luft. *Z ges Kalte-Ind* 1928;**35**:213.

59. Ray BB. Convection from heated cylinders in air. *Proc Indian Assoc Cultiv Sci* 1920;**6**:95.

60. Martynenko OG, Khramstov PP. *Free convective heat transfer*. New York: Springer; 2005.

61. Davis AH. Natural convective cooling in fluids. *Philos Mag* 1922;**44**:920.

62. Ackermann G. Die Wärmeabgabe einer horizontal geheizten Röhrean kaltes Wasser bei natürlicher Konvektion. *Forsch Geb IngWes* 1932;**3**:42.

63. Saunders OA. Effect of pressure on natural convection to air. *Proc Roy Soc A* 1936;**157**:278.

64. Fand RM, Morris EW, Lum M. Natural convection heat transfer from horizontal cylinders to air, water and silicone oils for Rayleigh numbers between 3×10^2 and 2×10^7. *Int J Heat Mass Transfer* 1977;**20**:1173.

65. Fand RM, Brucker J. A correlation for heat transfer by natural convection from horizontal cylinders that accounts for viscous dissipation. *Int J Heat Mass Transfer* 1983;**26**:709.

66. Amato WS, Tien C. Free convection heat transfer from isothermal spheres in water. *Int J Heat Mass Transfer* 1972;**15**:327.

67. Amato WS, Tien C. Free convection heat transfer from isothermal spheres in polymer solutions. *Int J Heat Mass Transfer* 1976;**19**:1257.

68. Perry RH, Green DW, editors. *Perry's chemical engineers' handbook*. 6th ed., New York: McGraw-Hill; 1984.

69. Kato H, Nishiwaki N, Hirata M. On the turbulent heat transfer by free convection from a vertical plate. *Int J Heat Mass Transfer* 1968;**11**:1117.

70. Churchill SW, Usagi R. A general expression for the correlation of rates of transfer and other phenomena. *AIChE J* 1972;**18**:1121.

71. Churchill SW, Chu HHS. Correlating equations for laminar and turbulent free convection from a vertical plate. *Int J Heat Mass Transfer* 1975;**18**:1323.

72. Jafarpur K, Yovanovich MM. Laminar free convective heat transfer from isothermal spheres: a new analytical method. *Int J Heat Mass Transfer* 1992;**35**:2195.

73. Kraussold H. Wärmeabgabe von zylindrischen Flüssigkeiten bei natürlicher Konvektion. *Forsch Geb IngWes* 1934;**5**:186.

74. Bergman TL, Lavine AS, Incropera FP, De Witt DP. *Fundamentals of heat and mass transfer*. 7th ed. New York: Wiley; 2011.

75. Hottel HC, Sarofim AF. *Radiation heat transfer*. New York: McGraw-Hill; 1967.

76. Howell JR. *A catalog of radiation configuration factors*. New York: McGraw-Hill; 1982.

77. Siegel R, Howell JR. *Thermal radiation heat transfer*. New York: McGraw-Hill; 1981.

78. Tien CL. Thermal radiation properties of gases. *Adv Heat Tran* 1968;**5**:253–324.

79. Sparrow EM. Radiant Interchange between surfaces separated by non-absorbing and non-emitting media. In: Rosenhow WM, Hartnett JP, editors. *Handbook of heat transfer*. New York: McGraw-Hill; 1973.

80. Dunkle RV. Radiation exchange in an enclosure with a participating gas. In: Rosenhow WM, Hartnett JP, editors. *Handbook of heat transfer*. New York: McGraw-Hill; 1973.

81. Sparrow EM, Cess RD. *Radiation heat transfer*. New York: Hemisphere; 1978.

82. Edwards DK. *Radiation heat transfer notes*. New York: Hemisphere; 1981.

83. Hottel HC, Mangelsdorf HG. Heat transmission by radiation from non-luminous gases. Experimental study of carbon dioxide and water vapour. *Trans Am Inst Chem Eng* 1935;**31**:517.

84. Schack A. Die Strahlung von leuchtenden Flammen. *Z Tech Phys* 1925;**6**:530.

85. Nusselt W. Die Oberflächenkondensation des Wasserdampfes. *Z Ver Deut Ing* 1916;**60**: 541 and 569.

86. Haselden GG, Prosad S. Heat transfer from condensing oxygen and nitrogen vapours. *Trans Inst Chem Eng* 1949;**27**:195.

87. McAdams WH. *Heat transmission*. 3rd ed., New York: McGraw-Hill; 1954.

88. Ten Bosch M. *Die Wärmeübertragung*. Berlin: Springer; 1936.

89. Carpenter EF, Colburn AP. The effect of vapour velocity on condensation inside tubes, In: *Proceedings of the General Discussion on Heat Transfer, September*; 1951. p. 20. Inst. of Mech. Eng. and Am. Soc. Mech. Eng.

90. Colburn AP. Problems in design and research on condensers of vapours and vapour mixtures, In: *Proceedings of the General Discussion on Heat Transfer, September*; 1951. p. 1. Inst. of Mech. Eng. and Am. Soc. Mech. Eng.

91. Tepe JB, Mueller AC. Condensation and subcooling inside an inclined tube. *Chem Eng Prog* 1947;**43**:267.

92. Kirkbride GC. Heat transfer by condensing vapours on vertical tubes. *Ind Eng Chem* 1934;**26**:425.

93. Badger WL. The evaporation of caustic soda to high concentrations by means of diphenyl vapour. *Ind Eng Chem* 1930;**22**:700.

94. Badger WL. Heat transfer coefficient for condensing Dowtherm films. *Trans Am Inst Chem Eng* 1937;**33**:441.

95. Drew TB, Nagle WM, Smith WQ. The conditions for drop-wise condensation of steam. *Trans Am Inst Chem Eng* 1935;**31**:605.

96. Colburn AP, Hougen OA. Design of cooler condenders for mixtures of vapors with non-condensing gases. *Ind Eng Chem* 1934;**26**:1178.

97. Chilton TH, Colburn AP. Mass transfer (absorption) coefficients. *Ind Eng Chem* 1934;**26**:1183.

98. Revilock JF, Hurlburt HZ, Brake DR, Lang EG, Kern DQ. Heat and mass transfer analogy: an appraisal using plant scale data. *Chem Eng Prog Symp Ser No 30* 1960;**56**:161.

99. Jeffreys GV. *The manufacture of acetic anhydride. A Problem in Chemical Engineering Design*. London: Institution of Chemical Engineers; 1961.

100. Westwater JW. Boiling of liquids. In: Drew TB, Hooper JW, editors. *Advances in chemical engineering*. New York: Academic Press; 1956.

101. Jakob M. Heat transfer in evaporation and condensation. *Mech Eng* 1936;**58**:643. 729.

102. Rohsenow WM, Clark JA. A study of the mechanism of boiling heat transfer. *Trans Am Soc Mech Eng* 1951;**73**:609.

103. Rohsenow WM. A method of correlating heat transfer data for surface boiling of liquids. *Trans Am Soc Mech Eng* 1952;**74**:969.

104. Forster HK. On the conduction of heat into a growing vapor bubble. *J Appl Phys* 1954;**25**:1067.

105. Carey VP. *Liquid-vapor phase-change phenomena*. 2nd ed., New York: Taylor & Francis; 1992.

106. Collier JG, Thome JR. *Convective boiling and condensation*. 3rd ed. Oxford: Oxford University Press; 1994.

107. Thome JR. *Encyclopedia of two-phase heat transfer and flow I. Fundamentals and methods*. Singapore: World Scientific; 2016.

108. Griffith P, Wallis JD. The role of surface conditions in nuclear boiling. *Chem Eng Prog Symp Ser No 30* 1960;**56**:49.

109. Jakob M, Fritz W. Versuche über den Verdampfungsvorgang. *Forsch Geb IngWes* 1931;**2**:435.

110. Nukiyama S. English abstract pp. S53–S54. The maximum and minimum values of the heat Q transmitted from metal to boiling water under atmospheric pressure. *J Soc Mech Eng (Jpn)* 1934;**37**:367.

111. Farber EA, Scorah RL. Heat transfer to boiling water under pressure. *Trans Am Soc Mech Eng* 1948;**70**:369.

112. Bonilla CF, Perry CH. Heat transmission to boiling binary liquid mixtures. *Trans Am Inst Chem Eng* 1941;**37**:685.

113. Insinger TH, Bliss H. Transmission of heat to boiling liquids. *Trans Am Inst Chem Eng* 1940;**36**:491.

114. Sauer ET, Cooper HBH, Akin GA, McAdams WH. Heat transfer to boiling liquids. *Mech Eng* 1938;**60**:669.

115. Cryder DS, Finalborgo AC. Heat transmission from metal surfaces to boiling liquids. *Trans Am Inst Chem Eng* 1937;**33**:346.

116. Jakob M, Linke W. Der Wärmeübergang von einer waagerechten Platte an siedendes Wasser. *Forsch Geb IngWes* 1933;**4**:75.

117. Cichelli MT, Bonilla CF. Heat transfer to liquids boiling under pressure. *Trans Am Inst Chem Eng* 1945;**41**:755.

118. Forster HK, Zuber N. Growth of a vapor bubble in a superheated liquid. *J Appl Phys* 1954;**25**:474.

119. Forster HK, Zuber N. Dynamics of vapor bubbles and boiling heat transfer. *AIChE J* 1955;**1**:531.

120. Fritz W. Berechnung des Maximalvolumens von Dampfblasen. *Physik Z* 1935;**36**:379.

121. Pioro IL. Experimental evaluation of constants for the Rohsenow pool boiling correlation. *Int J Heat Mass Transfer* 1999;**42**:2003.

122. McNelly MJ. A correlation of the rates of heat transfer to nucleate boiling liquids. *J Imp Coll Chem Eng Soc* 1953;**7**:18.

123. Mostinski IL. Calculation of boiling heat transfer coefficients, based on the law of corresponding states. *Brit Chem Eng* 1963;**8**:580.

124. Jeschke D. Wärmeübergang und Druckverlust in Röhrschlangen. *Z Ver Deut Ing* 1925;**69**:1526.

125. Pratt NH. The heat transfer in a reaction tank cooled by means of a coil. *Trans Inst Chem Eng* 1947;**25**:163.

126. Chilton TH, Drew TB, Jebens RH. Heat transfer coefficients in agitated vessels. *Ind Eng Chem* 1944;**36**:570.

127. Cummings GH, West AS. Heat transfer data for kettles with jackets and coils. *Ind Eng Chem* 1950;**42**:2303.

128. Brown RW, Scott MA, Toyne C. An investigation of heat transfer in agitated jacketed cast iron vessels. *Trans Inst Chem Eng* 1947;**25**:181.

129. Fletcher P. Heat transfer coefficients for stirred batch reactor design. *Chem Eng (Lond)* 1987;**435**:33.

130. *Standards of the tubular exchanger manufacturers association* (TEMA), 7th ed. New York, 1988.

131. Saunders EAD. *Heat exchangers selection, design and construction.* Harlow: Longman Scientific and Technical; 1988.

132. BS 3274: (British Standards Institution, London) *British Standard 3274* 1960: Tubular heat exchangers for general purposes.

133. BS 3606: (British Standards Institution, London) *British Standard 3606* 1978: Specification for steel tubes for heat exchangers.

134. Coker AK. 4th ed. *Ludwig's applied process design for chemical and petroleum plants.* vol. 3. New York: Elsevier; 2015.

135. Kakac S, Liu H, Pramuanjaroenkij A. *Heat exchangers: selection, rating and thermal design.* 4th ed., Boca Raton, FL: CRC Press; 2012.

136. Thulukkanam K. *Heat exchanger design handbook.* 2nd ed. Boca Raton: CRC Press; 2013.

137. Underwood AJV. The calculation of the mean temperature difference in multipass heat exchangers. *J Inst Pet Technol* 1934;**20**:145.

138. Bowman RA, Mueller AC, Nagle WM. Mean temperature difference in design. *Trans Am Soc Mech Eng* 1940;**62**:283.

139. Linnhoff B, Townsend DW, Boland D, Hewitt GF, Thomas BEA, Guy AR, Marsland RH. *A user guide on process integration for the efficient use of energy.* Rugby, England: IChem E; 1982.

140. Linnhoff B. Use pinch analysis to knock down capital costs and emissions. *Chem Eng Prog* 1994;**90**:32.

141. Smith R. *Chemical process design and integration.* 2nd ed. New York: Wiley; 2016.

142. Wilson EE. A basis for rational design of heat transfer apparatus. *Trans Am Soc Mech Eng* 1915;**37**:546.

143. Rhodes FH, Younger KR. Rate of heat transfer between condensing organic vapours and a metal tube. *Ind Eng Chem* 1935;**27**:957.

144. Coulson JM, Mehta RR. Heat transfer coefficients in a climbing film evaporator. *Trans Inst Chem Eng* 1953;**31**:208.

145. Butterworth D. A calculation method for shell and tube heat exchangers in which the overall coefficient varies along the length, In: *Conference on advances in thermal and mechanical design of shell and tube heat exchangers, NEL Report No. 590*, East Kilbride, Glasgow: National Engineering Laboratory; 1973.

146. Eagle A, Ferguson RM. On the coefficient of heat transfer from the internal surfaces of tube walls. *Proc Roy Soc* 1930;**127**:540.

147. Donohue DA. Heat exchanger design. *Pet. Ref.* 34 (August, 1955) 94, (Oct) 128, (Dec) 175, 35 (1956) (Jan) 155.

148. Frank O. Simplified design procedure for tubular exchangers. *Practical aspects of heat transfer*, Chem. Eng. Prog. Tech. Manual (A.I.Ch.E. 1978).

149. Butterworth D. *Introduction to heat transfer. Engineering design guide 18.* Oxford: Oxford University Press; 1978.

150. Lord RC, Minton PE, Slusser RP. Guide to trouble-free heat exchangers. *Chem Eng (Albany)* 1970;**77**:153.

151. Evans FL. *Equipment design handbook.* vol. 2. 2nd ed., Houston, TX: Gulf; 1980.

152. Tinker T. Shell-side characteristics of shell and tube heat exchangers. *Trans Am Soc Mech Eng* 1958;**80**:36.

153. Devore A. Use nomograms to speed exchanger design. *Hyd Proc Pet Ref* 1962;**41**:103.

154. Mueller AC. Heat exchangers, section 18. In: Rohsenow WM, Hartnett JP, Cho YI, editors. *Handbook of heat transfer fundamentals*. 3rd ed. New York: McGraw-Hill; 1998.

155. Palen JW, Taborak J. Solution of shell side flow pressure drop and heat transfer by stream analysis method. *Chem Eng Prog Symp Ser No 92* 1969;**65**:53.

156. Grant IDR. Flow and pressure drop with single and two phase flow on the shell-side of segmentally baffled shell and tube exchangers, In: *Conference on advances in thermal and mechanical design of shell and tube heat exchangers, NEL Report 590*, East Kilbride, Glasgow: National Engineering Laboratory; 1973.

157. Bell KJ. Exchanger design: based on the Delaware research report. *Pet Chem* 1960;**32**:C26.

158. Bell KJ. *Final report of the co-operative research program on shell and tube heat exchangers*. University of Delaware Eng Expt Sta Bull 5 (University of Delaware, 1963).

159. Wills MJN, Johnston D. A new and accurate hand calculation method for shell side pressure drop and flow distribution, In: *22nd Nat Heat Transfer Conf, HTD, 36*, New York: ASME; 1984.

160. ESDU. Baffled shell and tube heat exchangers: flow distribution, pressure drop and heat transfer on the shell side. *Engineering sciences data unit report 83038*. ESDU International, London; 1983.

161. Kays WM, London AL. *Compact heat exchangers*. 3rd ed. Malabar, FL: Krieger; 1998.

162. Mickley HS, Sherwood TK, Reed CE. *Applied mathematics in chemical engineering*. 2nd ed. New York: McGraw-Hill; 1957.

163. Neil DS. The use of superheated steam in calorifiers. *Processing* 1984;**11**:12.

164. Redman J. Compact future for heat exchangers. *Chem Eng (Lond)* 1988;**452**:12.

165. Gregory E. Plate and fin heat exchangers. *Chem Eng (Lond)* 1987;**440**:33.

166. Johnston A. Miniaturized heat exchangers for chemical processing. *Chem Eng (Lond)* 1986;**431**:36.

167. Ramshaw C. Process intensification-a game for n players. *Chem Eng (Lond)* 1985;**415**:30.

168. Nelson WL. *Petroleum refinery engineering*. 4th ed. New York: McGraw-Hill; 1958.

169. Huggins FE. Effects of scrapers on heating, cooling and mixing. *Ind Eng Chem* 1931;**23**:749.

170. Laughlin HG. Data on evaporation and drying in a jacketed kettle. *Trans Am Inst Chem Eng* 1940;**36**:345.

171. Houlton HG. Heat transfer in the votator. *Ind Eng Chem* 1944;**36**:522.

172. Bolanowski SP, Lineberry DD. Special problems of the food industry. *Ind Eng Chem* 1952;**44**:657.

173. Skelland AHP. Correlation of scraped-film heat transfer in the votator. *Chem Eng Sci* 1958;**7**:166.

174. Skelland AHP. Scale-up relationships for heat transfer in the votator. *Brit Chem Eng* 1958;**3**:325.

175. Skelland AHP, Oliver DR, Tooke S. Heat transfer in a water-cooled scraped-surface heat exchanger. *Brit Chem Eng* 1962;**7**:346.

176. Bott TR. Design of scraped-surface heat exchangers. *Brit Chem Eng* 1966;**11**:339.

177. Bott TR, Romero JJB. Heat transfer across a scraped-surface. *Can J Chem Eng* 1963;**41**:213.

178. Bott TR, Sheikh MR. Effects of blade design in scraped-surface heat transfer. *Brit Chem Eng* 1964;**9**:229.

179. Klemes JJ, Arsenyeva O, Kapustenko P, Tovazhnyanskyy L. *Compact heat exchangers for energy transfer intensification: low grade heat and fouling mitigation*. Boca Raton: CRC Press; 2015.

180. Hesselgreaves JE, Law R, Reay D. *Compact heat exchangers: selection, design and operation*. 2nd ed. Oxford: Butterworth-Heinemann; 2016.

181. Zohuri B. *Compact heat exchangers: selection, application, design and evaluation*. New York: Springer; 2017.

182. Lyle O. *Efficient use of steam*. London: HMSO; 1947.

Further Reading

1. Anderson EE. *Solar energy fundamentals for designers and engineers*. Reading, MA: Addison-Wesley; 1982.

2. Azbel D. *Heat transfer applications in process engineering*. New York: Noyes; 1984.

3. Bergman TL, Lavine AS, Incropera FP, De Witt DP. *Fundamentals of heat and mass transfer*. 7th ed. New York: Wiley; 2011.

4. Bergman TL, Lavine AS, Incropera FP, De Witt DP. *Introduction to heat transfer*. 6th ed. New York: Wiley; 2011.

5. Bird RB, Stewart WE, Lightfoot EN. *Transport phenomena*. 2nd ed. New York: Wiley; 2002.

6. Chapman AJ. *Heat transfer*. 4th ed. New York: Macmillan; 1984.

7. Cheremisinoff NP, editor. *Handbook of heat and mass transfer: vol. 1, Heat transfer operations*. Houston, TX: Gulf; 1986.

8. Collier JG, Thome JR. *Convective boiling and condensation*. 3rd ed. Oxford: Oxford University Press; 1994.

9. Edwards DK. *Radiation heat transfer notes*. New York: Hemisphere Publishing; 1981.

10. Gebhart B. *Heat transfer*. 2nd ed. New York: McGraw-Hill; 1971.

11. Grober H, Erk E, Grigull U. *Fundamentals of heat transfer*. New York: McGraw-Hill; 1961.

12. Hallstrom B, Skjoldebrand C, Tracardh C. *Heat transfer and food products*. London: Elsevier Applied Science; 1988.

13. Hewitt GF, editor. *Heat exchanger design handbook (HEDH)*. New York: Begell House; 1998.

14. Hewitt GF, Shires GL, Bott TR. *Process heat transfer*. Boca Raton: CRC Press; 1994.

15. Hottel HC, Sarofim AF. *Radiative transfer*. New York: McGraw-Hill; 1967.

16. Howell JR. *A catalog of radiation configuration factors*. New York: McGraw-Hill; 1982.

17. Jakob M. *Heat transfer*. vol. 1. New York: Wiley; 1949.

18. Jakob M. *Heat transfer*. vol. II. New York: Wiley; 1957.

19. Kakac S, Shah RK, Aung W, editors. *Handbook of single-phase convective heat transfer*. New York: Wiley; 1987.

20. Kays WM, Crawford ME. *Convective heat and mass transfer*. 3rd ed. New York: McGraw-Hill; 1994.

21. Kays WM, London AL. *Compact heat exchangers*. 2nd ed. New York: McGraw-Hill; 1964.

22. Kern DQ. *Process heat transfer*. New York: McGraw-Hill; 1950.

23. Kreith F, Manglik RK, Bohn M. *Principles of heat transfer*. 7th ed. Connecticut: Cengage; 2010.

24. McAdams WH. *Heat transmission*. 3rd ed. New York: McGraw-Hill; 1954.

25. Minkowycz WJ. *Handbook of numerical heat transfer*. New York: Wiley; 1988.

26. Minton PE. *Handbook of evaporation technology*. New York: Noyes; 1986.

27. Modest MF. *Radiative heat transfer*. 3rd ed. Boca Raton: CRC Press; 2013.

28. Ozisik MN. *Boundary value problems of heat conduction*. Stanton, PA: Int. Textbook Co.; 1965

29. Planck M. *The theory of heat radiation*. New York: Dover; 1959.

30. Rohsenhow WM, Hartnett JP, Cho YI, editors. *Handbook of heat transfer fundamentals*. 3rd ed. New York: McGraw-Hill; 1998.

31. Schack A. *Industrial heat transfer*. London: Chapman & Hall; 1965.

32. Smith RA. *Vaporisers: selection, design and operation*. London: Longman; 1987.

33. Siegel R, Howell JR. *Thermal radiation heat transfer*. 2nd ed. New York: McGraw-Hill; 1981.

34. Sparrow EM, Cess RD. *Radiation heat transfer*. New York: Hemisphere; 1978.

35. Taylor M, editor. *Plate-fin heat exchangers: guide to their specification and use*. Harwell: HTFS; 1987.

36. Touloukian YS. *Thermophysical properties of high temperature solid materials*. New York: Macmillan; 1967.

37. Wood WD, Deem HW, Lucks CF. *Thermal radiation properties*. New York: Plenum Press; 1964.

PART 2

Mass Transfer

CHAPTER 2

Mass Transfer

2.1 Introduction

The term mass transfer is used to denote the transfer of a component in a mixture from a region where its concentration is high to a region where the concentration is lower. Mass-transfer process can take place in a gas or vapour or in a liquid, and it can result from the random velocities of the molecules (*molecular diffusion*) or from the circulating or eddy currents present in a turbulent fluid (*eddy diffusion*).

In processing, it is frequently necessary to separate a mixture into its components, and in a physical process, differences in a particular property are exploited as the basis for the separation process. For example, *fractional distillation* depends on differences in volatility, *gas absorption* depends on differences in solubility of the gases in a selective absorbent, and similarly, and *liquid–liquid extraction* is based on the selectivity of an immiscible liquid solvent for one of the constituents. The rate at which the process takes place is dependent both on the *driving force* (concentration difference) and on the mass-transfer resistance. In most of these applications, mass transfer takes place across a phase boundary where the concentrations on either side of the interface are related by the phase-equilibrium relationship. Where a chemical reaction takes place during the course of the mass-transfer process, the overall transfer rate depends both on the chemical kinetics of the reaction and on the mass-transfer resistance, and it is important to understand the relative significance of these two factors in any practical application.

In this chapter, consideration will be given to the basic principles underlying mass transfer both with and without chemical reaction and to the models that have been proposed to enable the rates of transfer to be calculated. The applications of mass transfer to the design and operation of separation processes are discussed in Volume 2, and the design of reactors is dealt with in Volume 3.

A simple example of a mass-transfer process is that occurring in a box consisting of two compartments, each containing a different gas, initially separated by an impermeable partition. When the partition is removed, the gases start to mix, and the mixing process continues at a constantly decreasing rate until eventually (theoretically after the elapse of an infinite time) the whole system acquires a uniform composition. The process is one of molecular diffusion in which the mixing is attributable solely to the random motion of the molecules. The rate of

Coulson and Richardson's Chemical Engineering. https://doi.org/10.1016/B978-0-08-102550-5.00002-X

diffusion is governed by Fick's law, first proposed by Fick[1] in 1855, which expresses the mass-transfer rate as a linear function of the molar concentration gradient. In a mixture of two gases **A** and **B**, assumed ideal, Fick's law for steady-state diffusion may be written as

$$N_A = -D_{AB}\frac{dC_A}{dy} \qquad (2.1)$$

where

N_A is the molar flux of **A** (moles per unit area per unit time),
C_A is the concentration of **A** (moles of **A** per unit volume),
D_{AB} is known as the diffusivity or diffusion coefficient for **A** in **B**, and
y is distance in the direction of transfer.

An equation of exactly the same form may be written for **B**:

$$N_B = -D_{BA}\frac{dC_B}{dy} \qquad (2.2)$$

where D_{BA} is the diffusivity of **B** in **A**.

As indicated in the next section, for an ideal gas mixture, at constant pressure, $(C_A + C_B)$ is constant (Eqs 2.9a–2.9c), and hence,

$$\frac{dC_A}{dy} = -\frac{dC_B}{dy} \qquad (2.3)$$

The condition for the pressure or molar concentration to remain constant in such a system is that there should be no net transference of molecules. The process is then referred to as one of *equimolecular counterdiffusion*, and

$$N_A + N_B = 0$$

This relation is satisfied only if $D_{BA} = D_{AB}$, and therefore, the suffixes may be omitted, and Eq. (2.1) becomes

$$N_A = -D\frac{dC_A}{dy} \qquad (2.4)$$

Eq. (2.4), which describes the mass-transfer rate arising solely from the random movement of molecules, is applicable to a stationary medium or a fluid in streamline flow. If circulating currents or eddies are present, then the molecular mechanism will be reinforced, and the total mass-transfer rate may be written as

$$N_A = -(D + E_D)\frac{dC_A}{dy} \qquad (2.5)$$

whereas D is a physical property of the system and a function only of its composition, pressure, and temperature and E_D, which is known as the *eddy diffusivity*, is dependent on the flow pattern and varies with position. The estimation of E_D presents some difficulty, and this problem is considered in Chapter 4.

The molecular diffusivity D may be expressed in terms of the molecular velocity u_m and the mean free path of the molecules λ_m. In Chapter 4, it is shown that for conditions where the kinetic theory of gases is applicable, the molecular diffusivity is proportional to the product $u_m\lambda_m$. Thus, the higher the velocity of the molecules, the greater is the distance they travel before colliding with other molecules, and the higher is the diffusivity D.

Because molecular velocities increase with rise of temperature T, so also does the diffusivity, which, for a gas, is approximately proportional to T raised to the power of 1.5. As the pressure P increases, the molecules become closer together, and the mean free path is shorter, and consequently, the diffusivity is reduced, with D for a gas becoming approximately inversely proportional to the pressure.

Thus,

$$D \propto T^{1.5}/P \tag{2.6}$$

A method of calculating D in a binary mixture of gases is given later (Eq. 2.43). For *liquids*, the molecular structure is far more complex, and no such simple relationship exists, although various semiempirical predictive methods, such as Eq. (2.96), are useful.

In the discussion so far, the fluid has been considered to be a continuum, and distances on the molecular scale have, in effect, been regarded as small compared with the dimensions of the containing vessel, and thus, only a small proportion of the molecules collides directly with the walls. As the pressure of a gas is reduced, however, the mean free path may increase to such an extent that it becomes comparable with the dimensions of the vessel, and a significant proportion of the molecules may then collide directly with the walls rather than with other molecules. Similarly, if the linear dimensions of the system are reduced, as for instance when diffusion is occurring in the small pores of a catalyst particle (Section 2.7), the effects of collision with the walls of the pores may be important even at moderate pressures. Where the main resistance to diffusion arises from collisions of molecules with the walls, the process is referred to *Knudsen diffusion*, with a Knudsen diffusivity D_{Kn} that is proportional to the product $u_m l$, where l is a linear dimension of the containing vessel.

Since, from the *kinetic theory*,[2] $u_m \propto (\mathbf{R}T/M)^{0.5}$,

$$D_{Kn} \propto l(\mathbf{R}T/M)^{0.5} \tag{2.7}$$

Each resistance to mass transfer is proportional to the reciprocal of the appropriate diffusivity, and thus, when both molecular and Knudsen diffusion must be considered together, the effective diffusivity D_e is obtained by summing the resistances as

$$1/D_e = 1/D + 1/D_{Kn} \tag{2.8}$$

In *liquids*, the effective mean path of the molecules is so small that the effects of Knudsen-type diffusion need not be considered.

2.2 Diffusion in Binary Gas Mixtures

2.2.1 Properties of Binary Mixtures

If **A** and **B** are ideal gases in a mixture, the ideal gas law, Eq. (2.15) of Vol. 1A, may be applied to each gas separately and to the mixture:

$$P_A V = n_A RT \tag{2.9a}$$

$$P_B V = n_B RT \tag{2.9b}$$

$$PV = nRT \tag{2.9c}$$

where n_A and n_B are the number of moles of **A** and **B**, n is the total number of moles in a volume V; and P_A, P_B, and P are the respective partial pressures and the total pressure.

Thus,

$$P_A = \frac{n_A}{V} \mathbf{R}T = C_A \mathbf{R}T = \frac{c_A}{M_A} \mathbf{R}T \tag{2.10a}$$

$$P_B = \frac{n_B}{V} \mathbf{R}T = C_B \mathbf{R}T = \frac{c_B}{M_B} \mathbf{R}T \tag{2.10b}$$

and

$$P = \frac{n}{V} \mathbf{R}T = C_T \mathbf{R}T \tag{2.10c}$$

where c_A and c_B are mass concentrations; M_A and M_B are molecular weights; and C_A, C_B, and c_T are the molar concentrations of **A** and **B**, respectively, and the total molar concentration of the mixture.

From Dalton's law of partial pressures,

$$P = P_A + P_B = \mathbf{R}T(C_A + C_B) = \mathbf{R}T\left(\frac{c_A}{M_A} + \frac{c_B}{M_B}\right) \tag{2.11}$$

Thus,

$$C_T = C_A + C_B \tag{2.12}$$

and

$$1 = x_A + x_B \tag{2.13}$$

where x_A and x_B are the mole fractions of **A** and **B**.

Thus, for a system at constant pressure P and constant molar concentration C_T,

$$\frac{\mathrm{d}P_A}{\mathrm{d}y} = -\frac{\mathrm{d}P_B}{\mathrm{d}y} \tag{2.14}$$

$$\frac{\mathrm{d}C_A}{\mathrm{d}y} = -\frac{\mathrm{d}C_B}{\mathrm{d}y} \tag{2.15}$$

$$\frac{\mathrm{d}c_A}{\mathrm{d}y} = -\frac{\mathrm{d}c_B}{\mathrm{d}y}\frac{M_A}{M_B} \tag{2.16}$$

and

$$\frac{\mathrm{d}x_A}{\mathrm{d}y} = -\frac{\mathrm{d}x_B}{\mathrm{d}y} \tag{2.17}$$

By substituting from Eqs (2.7a), (2.7b) into Eq. (2.4), the mass-transfer rates N_A and N_B can be expressed in terms of partial pressure gradients rather than concentration gradients. Furthermore, N_A and N_B can be expressed in terms of gradients of mole fraction.

Thus,

$$N_A = -\frac{D}{RT}\frac{\mathrm{d}P_A}{\mathrm{d}y} \tag{2.18}$$

or

$$N_A = -DC_T\frac{\mathrm{d}x_A}{\mathrm{d}y} \tag{2.19}$$

Similarly,

$$N_B = -\frac{D}{RT}\frac{\mathrm{d}P_B}{\mathrm{d}y} = +\frac{D}{RT}\frac{\mathrm{d}P_A}{\mathrm{d}y} \quad \text{(from Eq.2.14)} \tag{2.20}$$

or

$$N_B = -DC_T\frac{\mathrm{d}x_B}{\mathrm{d}y} = +DC_T\frac{\mathrm{d}x_A}{\mathrm{d}y} \quad \text{(from Eq.2.17)} \tag{2.21}$$

2.2.2 Equimolecular Counterdiffusion

When the mass-transfer rates of the two components are equal and opposite, the process is said to be one of *equimolecular counterdiffusion*. Such a process occurs in the case of the box with a movable partition, referred to in Section 2.1. It occurs also in a distillation column when the molar latent heats of the two components are the same. At any point in the column, a falling stream of liquid is brought into contact with a rising stream of vapour with which it is *not* in equilibrium. The less volatile component is transferred from the vapour to the liquid, and the more volatile component is transferred in the opposite direction. If the molar latent heats of the components are equal, the condensation of a given amount of less volatile component releases exactly the amount of latent heat required to volatilise the same molar quantity of the more volatile component. Thus, at the interface and consequently throughout the liquid and vapour phases, equimolecular counterdiffusion is taking place.

Under these conditions, the differential forms of equation for N_A (Eqs 2.4, 2.18, 2.19) may be simply integrated, for constant temperature and pressure, to give respectively:

$$N_A = -D \frac{C_{A_2} - C_{A_1}}{y_2 - y_1} = \frac{D}{y_2 - y_1}(C_{A_1} - C_{A_2}) \tag{2.22}$$

$$N_A = -\frac{D}{\mathbf{R}T} \frac{P_{A_2} - P_{A_1}}{y_2 - y_1} = \frac{D}{\mathbf{R}T(y_2 - y_1)}(P_{A_1} - P_{A_2}) \tag{2.23}$$

$$N_A = -DC_T \frac{x_{A_2} - x_{A_1}}{y_2 - y_1} = \frac{DC_T}{y_2 - y_1}(x_{A_1} - x_{A_2}) \tag{2.24}$$

Similar equations apply to N_B that is equal to $-N_A$, and suffixes 1 and 2 represent the values of quantities at positions y_1 and y_2, respectively.

Eq. (2.22) may be written as

$$N_A = h_D(C_{A_1} - C_{A_2}) \tag{2.25}$$

where $h_D = D/(y_2 - y_1)$ is a *mass-transfer coefficient* with the driving force expressed as a *difference in molar concentration*; its dimensions are those of velocity (LT^{-1}).

Similarly, Eq. (2.23) may be written as

$$N_A = k'_G(P_{A_1} - P_{A_2}) \tag{2.26}$$

where $k'_G = D/[\mathbf{R}T(y_2 - y_1)]$ is a *mass-transfer coefficient* with the driving force expressed as a *difference in partial pressure*. It should be noted that its dimensions here, $\mathrm{NM}^{-1}\mathrm{L}^{-1}\mathrm{T}$, are different from those of h_D. It is always important to use the form of mass-transfer coefficient corresponding to the appropriate driving force.In a similar way, Eq. (2.24) may be written as

$$N_A = k_x(x_{A_1} - x_{A_2}) \tag{2.27}$$

where $k_x = DC_T/(y_2 - y_1)$ is a *mass-transfer coefficient* with the driving force in the form of a *difference in mole fraction*. The dimensions here are $\mathrm{NL}^{-2}\mathrm{T}^{-1}$.

2.2.3 Mass Transfer Through A Stationary Second Component

In several important processes, one component in a gaseous mixture will be transported relative to a fixed plane, such as a liquid interface, and the other will undergo no net movement. In gas absorption, a soluble gas **A** is transferred to the liquid surface where it dissolves, whereas the insoluble gas **B** undergoes no net movement with respect to the interface. Similarly, in evaporation from a free surface, the vapour moves away from the surface, but the air has no net movement. The mass-transfer process therefore differs from that described in Section 2.2.2.

The concept of a stationary component may be envisaged by considering the effect of moving the box, discussed in Section 2.1, in the opposite direction to that in which **B** is diffusing, at a velocity equal to its diffusion velocity, so that to the external observer **B** appears to be stationary. The total velocity at which **A** is transferred will then be increased to its diffusion velocity plus the velocity of the box.

For the absorption of a soluble gas **A** from a mixture with an insoluble gas **B**, the respective diffusion rates are given by

$$N_A = -D\frac{dC_A}{dy} \quad \text{(Eq.2.4)}$$

$$N_B = -D\frac{dC_B}{dy} = D\frac{dC_A}{dy} \quad \text{(from Eq.2.3)}$$

Since the total mass-transfer rate of **B** is zero, there must be a 'bulk flow' of the system towards the liquid surface exactly to counterbalance the diffusional flux away from the surface, as shown in Fig. 2.1, where

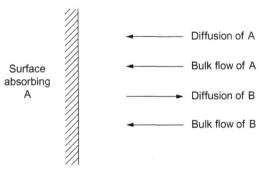

Fig. 2.1
Mass transfer through a stationary gas **B**.

Bulk flow of

$$\mathbf{B} = -D\frac{dC_A}{dy} \qquad (2.28)$$

The corresponding bulk flow of **A** must be c_A/c_B times that of **B**, since bulk flow implies that the gas moves *en masse*.

Thus, Bulk flow of

$$\mathbf{A} = -D\frac{dC_A}{dy}\frac{C_A}{C_B} \qquad (2.29)$$

Therefore the total flux of **A**, N_A', is given by

$$N_A' = -D\frac{dC_A}{dy} - D\frac{dC_A}{dy}\frac{C_A}{C_B} \qquad (2.30)$$

Eq. (2.30) is known as Stefan's law.[3] Thus, the bulk flow enhances the mass-transfer rate by a factor C_T/C_B, known as the *drift factor*. The fluxes of the components are given in Table 2.1.

Writing Eq. (2.30) as

$$N_A' = D\frac{C_T}{C_B}\frac{dC_B}{dy} \quad \text{(from Eq. 2.3)}$$

On integration,

$$N_A' = \frac{DC_T}{y_2 - y_1} \ln\frac{C_{B_2}}{C_{B_1}} \qquad (2.31)$$

By definition, C_{Bm}, the logarithmic mean of C_{B_1} and C_{B_2}, is given by

$$C_{Bm} = \frac{C_{B_2} - C_{B_1}}{\ln(C_{B_2}/C_{B_1})} \qquad (2.32)$$

Thus, substituting for $\ln(C_{B_1}/C_{B_2})$ in Eq. (2.31),

Table 2.1 Fluxes of components of a gas mixture

	Component A	Component B	A + B
Diffusion	$-D\dfrac{dC_A}{dy}$	$+D\dfrac{dC_A}{dy}$	0
Bulk flow	$-D\dfrac{dC_A}{dy}\cdot\dfrac{C_A}{C_B}$	$-D\dfrac{dC_A}{dy}$	$-D\dfrac{dC_A}{dy}\cdot\dfrac{C_A + C_B}{C_B}$
Total	$D\dfrac{dC_A}{dy}\dfrac{C_A + C_B}{C_B}$	0	$-D\dfrac{dC_A}{dy}\cdot\dfrac{C_A + C_B}{C_B}$
	$= -D\dfrac{dC_A}{dy}\cdot\dfrac{C_T}{C_B}$		$= -D\dfrac{dC_A}{dy}\cdot\dfrac{C_T}{C_B}$

$$N'_A = \left(\frac{DC_T}{y_2 - y_1}\right)\frac{C_{B_2} - C_{B_1}}{C_{Bm}}$$

$$= \left(\frac{D}{y_2 - y_1}\frac{C_T}{C_{Bm}}\right)(C_{A_1} - C_{A_2})$$

(2.33)

or in terms of partial pressures,

$$N'_A = \left(\frac{D}{RT(y_2 - y_1)}\frac{P}{P_{Bm}}\right)(P_{A_1} - P_{A_2})$$

(2.34)

Similarly, in terms of mole fractions,

$$N'_A = \left(\frac{DC_T}{y_2 - y_1}\frac{1}{x_{Bm}}\right)(x_{A_1} - x_{A_2})$$

(2.35)

Eq. (2.31) can be simplified when the concentration of the diffusing component **A** is small. Under these conditions, c_A is small compared with C_T, and Eq. (2.31) becomes

$$N'_A = \frac{DC_T}{y_2 - y_1} \ln\left[1 - \left(\frac{C_{A_2} - C_{A_1}}{C_T - C_{A_1}}\right)\right]$$

$$= \frac{DC_T}{y_2 - y_1}\left[-\left(\frac{C_{A_2} - C_{A_1}}{C_T - C_{A_1}}\right) - \frac{1}{2}\left(\frac{C_{A_2} - C_{A_1}}{C_T - C_{A_1}}\right)^2 - \cdots\right]$$

For small values of $C_A, C_T - C_{A_1} \asymp C_T$, only the first term in the series is significant.

Thus,

$$N'_A \asymp \frac{D}{y_2 - y_1}(C_{A_1} - C_{A_2})$$

(2.36)

Eq. (2.36) is identical to Eq. (2.22) for equimolecular counterdiffusion. Thus, the effects of bulk flow can be neglected at low concentrations.

Eq. (2.33) can be written in terms of a mass-transfer coefficient h_D to give

$$N'_A = h_D(C_{A_1} - C_{A_2})$$

(2.37)

where

$$h_D = \frac{D}{y_2 - y_1}\frac{C_T}{C_{Bm}}$$

(2.38)

Similarly, working in terms of partial pressure difference as the driving force, Eq. (2.34) can be written as

$$N'_A = k_G(P_{A_1} - P_{A_2})$$

(2.39)

where

$$k_G = \frac{D}{RT(y_2 - y_1)} \frac{P}{P_{Bm}}$$ (2.40)

Using mole fractions as the driving force, Eq. (2.35) becomes

$$N'_A = k_x(x_{A_1} - x_{A_2})$$ (2.41)

where

$$k_x = \frac{DC_T}{y_2 - y_1} \frac{C_T}{C_{Bm}} = \frac{DC_T}{(y_2 - y_1)x_{Bm}}$$ (2.42)

It may be noted that all the transfer coefficients here are greater than those for equimolecular counterdiffusion by the factor $(C_T/C_{Bm} = P/P_{Bm})$, which is an integrated form of the drift factor.

When the concentration c_A of the gas being transferred is low, C_T/C_{Bm} then approaches unity, and the two sets of coefficients become identical.

Example 2.1

Ammonia gas is diffusing at a constant rate through a layer of stagnant air 1 mm thick. Conditions are such that the gas contains 50% by volume ammonia at one boundary of the stagnant layer. The ammonia diffusing to the other boundary is quickly absorbed, and the concentration is negligible at that plane. The temperature is 295 K, and the atmospheric pressure and under these conditions the diffusivity of ammonia in air is 1.8×10^{-5} m^2/s. Estimate the rate of diffusion of ammonia through the layer.

Solution

If the subscripts 1 and 2 refer to the two sides of the stagnant layer and the subscripts A and B refer to ammonia and air, respectively, then the rate of diffusion through a stagnant layer is given by

$$N_A = -\frac{D}{RTx}(P/P_{Bm})(P_{A2} - P_{A1}) \quad (\text{Eq. } 2.31)$$

In this case, $x = 0.001$ m, $D = 1.8 \times 10^{-5}$ m^2/s, $R = 8314$ J/kmol K, $T = 295$ K and $P = 101.3$ kN/m^2, and hence,

$$P_{A1} = (0.50 \times 101.3) = 50.65 \text{ kN/m}^2$$
$$P_{A2} = 0$$
$$P_{B1} = (101.3 - 50.65) = 50.65 \text{ kN/m}^2 = 5.065 \times 10^4 \text{ N/m}^2$$
$$P_{B2} = (101.3 - 0) = 101.3 \text{ kN/m}^2 = 1.013 \times 10^5 \text{ N/m}^2$$

Thus,

$$P_{BM} = (101.3 \quad 50.65)/ \ln(101.3/50.65) - 73.07 \text{kN/m}^2 = 7.307 \times 10^4 \text{N/m}^2$$

and

$$P/P_{BM} = (101.3/73.07) = 1.386.$$

Thus, substituting in Eq. (2.31) gives

$$N_A = -\left[1.8 \times 10^{-5}/(8314 \times 295 \times 0.001)\right] 1.386\left(0 - 5.065 \times 10^4\right)$$

$$= 5.15 \times 10^{-4}\,\text{kmol/m}^2\text{s}$$

2.2.4 Diffusivities of Gases and Vapours

Experimental values of diffusivities are given in Table 2.2 for a number of gases and vapours in air at 298 K and atmospheric pressure. The table also includes values of the Schmidt number Sc, the ratio of the kinematic viscosity (μ/ρ) to the diffusivity (D) for very low concentrations of the diffusing gas or vapour. The importance of the Schmidt number in problems involving mass transfer is discussed in Chapter 4.

Experimental determination of diffusivities

Diffusivities of vapours are most conveniently determined by the method developed by Winkelmann[5] in which liquid is allowed to evaporate in a vertical glass tube over the top of which a stream of vapour-free gas is passed, at a rate such that the vapour pressure is maintained

Table 2.2 Diffusivities (diffusion coefficients) of gases and vapours in air at 298 K and atmospheric pressure[4]

Substance	D (m^2/s$\times 10^6$)	$\mu/\rho D$	Substance	D (m^2/s$\times 10^6$)	$\mu/\rho D$
Ammonia	28.0	0.55	Valeric acid	6.7	2.31
Carbon dioxide	16.4	0.94	i-Caproic acid	6.0	2.58
Hydrogen	71.0	0.22	Diethyl amine	10.5	1.47
Oxygen	20.6	0.75	Butyl amine	10.1	1.53
Water	25.6	0.60	Aniline	7.2	2.14
Carbon disulphide	10.7	1.45	Chlorobenzene	7.3	2.12
Ethyl ether	9.3	1.66	Chlorotoluene	6.5	2.38
Methanol	15.9	0.97	Propyl bromide	10.5	1.47
Ethanol	11.9	1.30	Propyl iodide	9.6	1.61
Propanol	10.0	1.55	Benzene	8.8	1.76
Butanol	9.0	1.72	Toluene	8.4	1.84
Pentanol	7.0	2.21	Xylene	7.1	2.18
Hexanol	5.9	2.60	Ethyl benzene	7.7	2.01
Formic acid	15.9	0.97	Propyl benzene	5.9	2.62
Acetic acid	13.3	1.16	Diphenyl	6.8	2.28
Propionic acid	9.9	1.56	n-Octane	6.0	2.58
i-Butyric acid	8.1	1.91	Mesitylene	6.7	2.31

Note: the group $(i/\rho D)$ in the above table is evaluated for mixtures composed largely of air.
In this table, the figures taken from Perry and Green[4] are based on data in *International Critical Tables 5* (1928) and Landolt-Börnstein, *Physikalische-Chemische Tabellen* (1935).

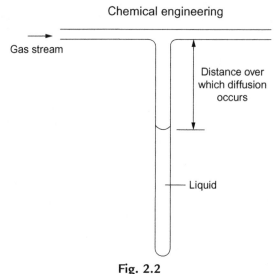

Chemical engineering

Gas stream

Distance over
which diffusion
occurs

Liquid

Fig. 2.2
Determination of diffusivities of vapours.

almost at zero (Fig. 2.2). If the apparatus is maintained at a steady temperature, there will be no eddy currents in the vertical tube, and mass transfer will take place from the surface by molecular diffusion alone. The rate of evaporation can be followed by the rate of fall of the liquid surface, and since the concentration gradient is known, the diffusivity can then be calculated.

Example 2.2

The diffusivity of the vapour of a volatile liquid in air can be conveniently determined by Winkelmann's method in which liquid is contained in a narrow-diameter vertical tube and maintained at a constant temperature, and an air stream is passed over the top of the tube sufficiently rapidly to ensure that the partial pressure of the vapour there remains approximately zero. On the assumption that the vapour is transferred from the surface of the liquid to the air stream by molecular diffusion alone, calculate the diffusivity of carbon tetrachloride vapour in air at 321 K and atmospheric pressure from the experimental data given in Table 2.3.

The vapour pressure of carbon tetrachloride at 321 K is 37.6 kN/m^2, and the density of the liquid is 1540 kg/m^3. The kilogramme molecular volume may be taken as 22.4 m^3.

Solution
From Eq. (2.33), the rate of mass transfer is given by

$$N'_A = D \frac{C_A}{L} \frac{C_T}{C_{Bm}}$$

where c_A is the saturation concentration at the interface and L is the effective distance through which mass transfer is taking place. Considering the evaporation of the liquid,

Table 2.3 Experimental data for diffusivity calculation

Time From Commencement of Experiment			Liquid Level (mm)	Time From Commencement of Experiment			Liquid Level (mm)
(h)	(min)	(ks)		(h)	(min)	(ks)	
0	0	0.0	0.0	32	38	117.5	54.7
0	26	1.6	2.5	46	50	168.6	67.0
3	5	11.1	12.9	55	25	199.7	73.8
7	36	27.4	23.2	80	22	289.3	90.3
22	16	80.2	43.9	106	25	383.1	104.8

$$N'_A = \frac{\rho_L}{M}\frac{dL}{dt}$$

where ρ_L is the density of the liquid.

Hence, $\dfrac{\rho_L}{M}\dfrac{dL}{dt} = D\dfrac{C_A}{L}\dfrac{C_T}{C_{Bm}}$

Integrating and putting $L = L_0$ at $t = 0$:

$$L^2 - L_0^2 = \frac{2MD\,C_A C_T}{\rho_L}\frac{C_T}{C_{Bm}}t$$

L_0 will not be measured accurately nor is the effective distance for diffusion, L, at time t.

Accurate values of $(L - L_0)$ are available, however, and hence,

$$(L - L_0)(L - L_0 + 2L_0) = \frac{2MD\,C_A C_T}{\rho_L}\frac{C_T}{C_{Bm}}t$$

or

$$\frac{t}{L - L_0} = \frac{\rho_L}{2MD}\frac{C_{Bm}}{C_A C_T}(L - L_0) + \frac{\rho_L C_{Bm}}{MD C_A C_T}L_0$$

If s is the slope of a plot of $t/(L - L_0)$ against $(L - L_0)$, then

$$s = \frac{\rho_L C_{Bm}}{2MDC_A C_T} \quad \text{or} \quad D = \frac{\rho_L C_{Bm}}{2MC_A C_T s}$$

From a plot of $t/(L - L_0)$ against $(L - L_0)$ as shown in Fig. 2.3:

$s = 0.0310$ ks/mm^2 or 3.1×10^7 s/m^2

and

$$C_T = \left(\frac{1}{22.4}\right)\left(\frac{273}{321}\right) = 0.0380 \text{ kmol/m}^3$$

$$M = 154 \text{ kg/kmol}$$

$$C_A = \left(\frac{37.6}{101.3}\right)0.0380 = 0.0141 \text{ kmol/m}^3$$

$$\rho_L = 1540 \text{ kg/m}^3$$

$$C_{B1} = 0.0380 \text{ kmol/m}^3, \quad C_{B2} = \left(\frac{101.3 - 37.6}{101.3}\right)0.0380 = 0.0238 \text{ kmol/m}^3$$

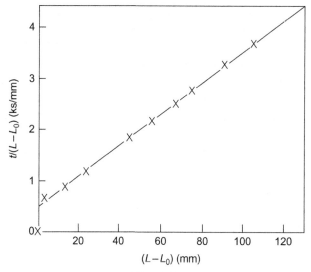

Fig. 2.3

Plot of $t/(L-L_0)$ versus $(L-L_0)$ for Example 2.1.

Thus,

$$C_{Bm} = \frac{(0.0380-0.0238)}{\ln(0.0380/0.0238)} = 0.0303 \text{ kmol/m}^3$$

and

$$D = \frac{1540 \times 0.0303}{2 \times 154 \times 0.0141 \times 0.0380 \times 3.1 \times 10^7}$$

$$= \underline{\underline{9.12 \times 10^{-6} \text{ m}^2/\text{s}}}$$

Prediction of diffusivities

Where the diffusivity D for the transfer of one gas in another is not known and experimental determination is not practicable, it is necessary to use one of the many predictive procedures. A commonly used method due to Gilliland[6] is based on the 'Stefan-Maxwell' hard-sphere model, and this takes the following form:

$$D = \frac{4.3 \times 10^{-4} T^{1.5} \sqrt{(1/M_A)+(1/M_B)}}{P\left(\mathbf{V}_A^{1/3}+\mathbf{V}_B^{1/3}\right)^2} \tag{2.43}$$

where D is the diffusivity in m²/s, T is the absolute temperature (K), M_A and M_B are the molecular masses of **A** and **B**, P is the total pressure in N/m², and v_A and V_B are the molecular volumes of **A** and **B**. The molecular volume is the volume in m^3 of 1 kmol of the material in the form of liquid at its boiling point and is a measure of the volume occupied by the molecules themselves. It may not always be known, although an approximate value can be obtained, for all but simple molecules, by application of *Kopp's law* of additive volumes. Kopp has presented a particular value for the equivalent atomic volume of each element,[7] as given in Table 2.4, such that when the atomic volumes of the elements of the molecule in question are added in the appropriate proportions, an approximate value the equivalent molecular volume is obtained. There are certain exceptions to this rule, and corrections have to be made if the elements are combined in particular ways.

Table 2.4 Atomic and structural diffusion volume increments (m^3/kmol)[7]

Antimony	0.0242	Oxygen, double-bonded	0.0074
Arsenic	0.0305	Coupled to two other elements	
Bismuth	0.0480	In aldehydes and ketones	0.0074
Bromine	0.0270	In methyl esters	0.0091
Carbon	0.0148	In ethyl esters	0.0099
Chlorine, terminal, as in R–Cl	0.0216	In higher esters and ethers	0.0110
Medial, as in R–CHC1–R$'$	0.0246	In acids	0.0120
Chromium	0.0274	In union with S, P, N	0.0083
Fluorine	0.0087	Phosphorus	0.0270
Germanium	0.0345	Silicon	0.0320
Hydrogen	0.0037	Sulphur	0.0256
Nitrogen, double-bonded	0.0156	Tin	0.0423
In primary amines	0.0105	Titanium	0.0357
In secondary amines	0.0120	Vanadium	0.0320
		Zinc	0.0204

For a three-membered ring, as in ethylene oxide	Deduct 0.0060
For a four-membered ring, as in cyclobutane	Deduct 0.0085
For a five-membered ring, as in furane	Deduct 0.0115
For a six-membered ring, as in benzene, pyridine	Deduct 0.0150
For an anthracene ring formation	Deduct 0.0475
For naphthalene	Deduct 0.0300

Diffusion volumes of simple molecules (m^3/kmol)

H_2	0.0143	CO_2	0.0340	NH_3	0.0258
O_2	0.0256	H_2O	0.0189	H_2S	0.0329
N_2	0.0312	SO_2	0.0448	Cl_2	0.0484
Air	0.0299	NO	0.0236	Br_2	0.0532
CO	0.0307	N_2O	0.0364	I_2	0.0715

It will be noted from Eq. (2.43) that the diffusivity of a vapour is inversely proportional to the pressure and varies with the absolute temperature raised to the power of 1.5, although it has been suggested that this underestimates the temperature dependence.

A method, proposed more recently by Fuller, Schettler, and Giddings,[8] is claimed to give an improved correlation. In this approach, the values of the 'diffusion volume' have been 'modified' to give a better correspondence with experimental values and have then been adjusted arbitrarily to make the coefficient in the equation equal to unity. The method does contain some anomalies, however, particularly in relation to the values of V for nitrogen, oxygen, and air. Details of this method are given in Volume 6.

Example 2.3

Ammonia is absorbed in water from a mixture with air using a column operating at 1 bar and 295 K. The resistance to transfer may be regarded as lying entirely within the gas phase. At a point in the column, the partial pressure of the ammonia is 7.0 kN/m². The back pressure at the water interface is negligible, and the resistance to transfer may be regarded as lying in a stationary gas film 1 mm thick. If the diffusivity of ammonia in air is 2.36×10^{-5} m²/s, what is the transfer rate per unit area at that point in the column? How would the rate of transfer be affected if the ammonia-air mixture were compressed to double the pressure?

Solution
Concentration of ammonia in the gas

$$= \left(\frac{1}{22.4}\right)\left(\frac{101.3}{101.3}\right)\left(\frac{273}{295}\right)\left(\frac{7.0}{101.3}\right) = 0.00285 \text{ kmol/m}^3$$

Thus,

$$\frac{C_T}{C_{Bm}} = \frac{101.3 \ln(101.3/94.3)}{101.3 - 94.3} = 1.036$$

From Eq. (2.33),

$$N'_A = \frac{D}{y_2 - y_1}\frac{C_T}{C_{Bm}}(C_{A1} - C_{A2})$$

$$= \left(\frac{2.36 \times 10^{-5}}{1 \times 10^{-3}}\right)(1.036 \times 0.00285)$$

$$= \underline{\underline{6.97 \times 10^{-5} \text{ kmol/m}^2\text{s}}}$$

If the pressure is doubled, the driving force is doubled, the C_T/C_{Bm} is essentially unaltered, and the diffusivity, being inversely proportional to the pressure (Eq. 2.43), is halved. The mass-transfer rate therefore remains the same.

2.2.5 Mass Transfer Velocities

It is convenient to express mass-transfer rates in terms of velocities for the species under consideration where

$$\text{Velocity} = \frac{\text{Flux}}{\text{Concentration}}$$

which, in the SI system, has the units $(\text{kmol/m}^2\text{s})/(\text{kmol/m}^3) = \text{m/s}$.

For diffusion according to Fick's law,

$$u_{DA} = \frac{N_A}{C_A} = -\frac{D}{C_A}\frac{dC_A}{dy} \tag{2.44a}$$

and

$$u_{DB} = \frac{N_B}{C_B} = -\frac{D}{C_B}\frac{dC_B}{dy} = \frac{D}{C_B}\frac{dC_A}{dy} \tag{2.44b}$$

Since $N_B = -N_A$, then

$$u_{DB} = -u_{DA}\frac{C_A}{C_B} = -u_{DA}\frac{x_A}{x_B} \tag{2.45}$$

As a result of the diffusional process, there is no net overall molecular flux arising from diffusion in a binary mixture, the two components being transferred at equal and opposite rates. In the process of equimolecular counterdiffusion that occurs, for example, in a distillation column when the two components have equal molar latent heats, the diffusional velocities are the same as the velocities of the molecular species relative to the walls of the equipment or the phase boundary.

If the physical constraints placed upon the system result in a bulk flow, the velocities of the molecular species relative to one another remain the same, but in order to obtain the velocity relative to a fixed point in the equipment, it is necessary to add the bulk flow velocity. An example of a system in which there is a bulk flow velocity is that in which one of the components is transferred through a second component that is undergoing no net transfer, for example, in the absorption of a soluble gas **A** from a mixture with an insoluble gas **B** (see Section 2.2.3). In this case, because there is no set flow of **B**, the sum of its diffusional velocity and the bulk flow velocity must be zero.

In this case,

Component	A	B
Diffusional velocity	$u_{DA} = -\dfrac{D}{C_A}\dfrac{dC_A}{dy}$	$u_{DA} = +\dfrac{D}{C_B}\dfrac{dC_A}{dy}$
Bulk flow velocity	$u_F = -\dfrac{D}{C_B}\dfrac{dC_A}{dy}$	$u_F = -\dfrac{D}{C_B}\dfrac{dC_A}{dy}$
Total velocity	$u_A = -D\dfrac{C_T}{C_A C_B}\dfrac{dC_A}{dy}$	$u_B = 0$
Flux	$N'_A = u_A C_A = -D\dfrac{C_T}{C_B}\dfrac{dC_A}{dy}$	$N'_B = 0$

The flux of **A** has been given as Stefan's law (Eq. 2.30).

2.2.6 General Case for Gas-Phase Mass Transfer in a Binary Mixture

Whatever the physical constraints placed on the system, the diffusional process causes the two components to be transferred at equal and opposite rates and the values of the diffusional velocities u_{DA} and u_{DB} given in Section 2.2.5 are always applicable. It is the bulk flow velocity u_F that changes with imposed conditions and gives rise to differences in overall mass-transfer rates. In equimolecular counterdiffusion, u_F is zero. In the absorption of a soluble gas **A** from a mixture, the bulk velocity must be equal and opposite to the diffusional velocity of **B** as this latter component undergoes no net transfer.

In general, for any component,

$$\text{Total transfer} = \text{transfer by diffusion} + \text{transfer by bulk flow.}$$

For component **A**:

Total transfer (moles/area time) $= N'_A$
Diffusional transfer according to Fick's law $= N_A = -D\dfrac{dC_A}{dy}$
Transfer by bulk flow $= u_F C_A$

Thus, for **A**,

$$N'_A = N_A + u_F C_A \tag{2.46a}$$

and for **B**,

$$N'_B = N_B + u_F C_B \tag{2.46b}$$

$$\text{The bulk flow velocity } u_F = \frac{\text{Total moles transferred/area time}}{\text{Total molar concentration}}$$

$$= \frac{\left(N'_A + N'_B\right)}{C_T} \tag{2.47}$$

substituting

$$N'_A = N_A + \frac{C_A}{C_T}\left(N'_A + N'_B\right)$$

$$N'_A = -D\frac{dC_A}{dy} + x_A\left(N'_A + N'_B\right)$$

$$N'_A = -DC_T\frac{dx_A}{dy} + x_A\left(N'_A + N'_B\right) \tag{2.48}$$

Similarly for **B**,

$$N'_B = DC_T\frac{dx_A}{dy} + (1 - x_A)\left(N'_A + N'_B\right) \tag{2.49}$$

For equimolecular counterdiffusion, $N'_A = -N'_B$, and Eq. (2.48) reduces to Fick's law. For a system in which **B** undergoes no net transfer, $N'_B = 0$, and Eq. (2.48) is identical to Stefan's law.

For the general case,

$$fN'_A = -N'_B \tag{2.50}$$

If in a distillation column, for example, the molar latent heat of **A** is f times that of **B**, the condensation of 1 mole of **A** (taken as the less volatile component) will result in the vaporisation of f moles of **B**, and the mass-transfer rate of **B** will be f times that of **A** in the opposite direction.

Substituting into Eq. (2.48),

$$N'_A = -DC_T \frac{dx_A}{dy} + x_A \left(N'_A - fN'_A \right) \tag{2.51}$$

Thus, $[1 - x_A(1-f)]N'_A = -DC_T \frac{dx_A}{dy}$

If x_A changes from x_{A_1} to x_{A_2} as y goes from y_1 to y_2, then

$$N'_A \int_{y_1}^{y_2} dy = -DC_T \int_{x_{A_1}}^{x_{A_2}} \frac{dx_A}{1 - x_A(1-f)}$$

Thus,

$$N'_A (y_2 - y_1) = -DC_T \frac{1}{1-f} \left[\ln \frac{1}{(1-f)^{-1} - x_A} \right]_{x_{A_1}}^{x_{A_2}}$$

or

$$N'_A = \frac{DC_T}{y_2 - y_1} \frac{1}{1-f} \ln \frac{1 - x_{A_2}(1-f)}{1 - x_{A_1}(1-f)} \tag{2.52}$$

2.2.7 Diffusion as a Mass Flux

Fick's law of diffusion is normally expressed in *molar* units or

$$N_A = -D \frac{dC_A}{dy} = -DC_T \frac{dx_A}{dy} \quad \text{(Eq. 2.4)}$$

where x_A is the mole fraction of component **A**.

The corresponding equation for component **B** indicates that there is an equal and opposite molar flux of that component. If each side of Eq. (2.4) is multiplied by the molecular weight of **A**, M_A, then

$$J_A = -D\frac{dc_A}{dy} = -DM_A\frac{dC_A}{dy} = -DC_TM_A\frac{dx_A}{dy} \tag{2.53}$$

where J_A is a flux in mass per unit area and unit time (kg/m^2 s in SI units) and c_A is a concentration in mass terms, (kg/m^3 in SI units).

Similarly, for component **B**,

$$J_B = -D\frac{dc_B}{dy} \tag{2.54}$$

Although the sum of the molar concentrations is constant in an ideal gas at constant pressure, the sum of the mass concentrations is not constant, and dc_A/dy and dc_B/dy are not equal and opposite,

Thus,

$$C_A + C_B = C_T = \frac{c_A}{M_A} + \frac{c_B}{M_B} = \text{constant} \tag{2.55}$$

or

$$\frac{1}{M_A}\frac{dc_A}{dy} + \frac{1}{M_B}\frac{dc_B}{dy} = 0$$

and

$$\frac{dc_B}{dy} = -\frac{M_B}{M_A}\frac{dc_A}{dy} \tag{2.56}$$

Thus, the diffusional process does not give rise to equal and opposite mass fluxes.

2.2.8 Thermal Diffusion

If a temperature gradient is maintained in a binary gaseous mixture, a concentration gradient is established with the light component collecting preferentially at the hot end and the heavier one at the cold end. This phenomenon, known as the *Soret effect*, may be used as the basis of a separation technique of commercial significance in the separation of isotopes.

Conversely, when mass transfer is occurring as a result of a constant concentration gradient, a temperature gradient may be generated; this is known as the *Dufour effect*.

In a binary mixture consisting of two gaseous components **A** and **B** subject to a temperature gradient, the flux due to thermal diffusion is given by Grew and Ibbs[9]:

$$(N_A)_{Th} = -D_{Th}\frac{1}{T}\frac{dT}{dy} \tag{2.57}$$

where $(N_A)_{Th}$ is the molar flux of A (kmol/m^2 s) in the Y-direction and D_{Th} is the diffusion coefficient for thermal diffusion (kmol/m s).

Eq. (2.57), with a positive value of D_{Th}, applies to the component that travels preferentially to the *low*-temperature end of the system. For the component that moves to the high-temperature end, D_{Th} is negative. In a binary mixture, the gas of higher molecular weight has the positive value of D_{Th}, and this therefore tends towards the lower-temperature end of the system.

If two vessels each containing completely mixed gas, one at temperature T_1 and the other at a temperature T_2, are connected by a lagged nonconducting pipe in which there are no turbulent eddies (such as a capillary tube), then under steady-state conditions, the rate of transfer of **A** by thermal diffusion and molecular diffusion must be equal and opposite, or

$$(N_A)_{Th} + N_A = 0 \tag{2.58}$$

N_A is given by Fick's law as

$$N_A = -D\frac{dC_A}{dy} = -DC_T\frac{dx_A}{dy} \quad \text{(Eq. 2.53)}$$

where x_A is the mole fraction of **A** and C_T is the total molar concentration at y and will not be quite constant because the temperature is varying.

Substituting Eqs (2.53), (2.57) into Eq. (2.58) gives

$$-D_{Th}\frac{1}{T}\frac{dT}{dy} - DC_T\frac{dx_A}{dy} = 0 \tag{2.59}$$

The relative magnitudes of the thermal diffusion and diffusion effects are represented by the dimensionless ratio:

$$\frac{D_{Th}}{DC_T} = K_{ABT}$$

where K_{ABT} is known as the *thermal diffusion ratio*.

Thus,

$$-K_{ABT}\frac{1}{T}\frac{dT}{dy} = \frac{dx_A}{dy} \tag{2.60}$$

If temperature gradients are small, C_T may be regarded as effectively constant. Furthermore, K_{ABT} is a function of composition, being approximately proportional to the product $x_A x_B$. It is therefore useful to work in terms of the *thermal diffusion factor* α, where

$$\alpha = \frac{K_{ABT}}{x_A(1 - x_A)}$$

Substituting for α, assumed constant, in Eq. (2.60) and integrating gives

$$\alpha \ln \frac{T_1}{T_2} = \ln\left(\frac{x_{A_2}}{1 - x_{A_2}} \frac{1 - x_{A_1}}{x_{A_1}}\right)$$

Thus,

$$\frac{x_{A_2} x_{B_1}}{x_{A_1} x_{B_2}} = \left(\frac{T_1}{T_2}\right)^\alpha \tag{2.61}$$

Eq. (2.61) gives the mole fraction of the two components **A** and **B** as a function of the absolute temperatures and the thermal diffusion factor.

Values of α taken from data in Grew and K[9] and Hirschfelder, Curtiss, and Bird[10] are given in Table 2.5.

Table 2.5 Values of thermal diffusion factor (α) for binary gas mixtures (A is the heavier component, which moves towards the cooler end)

Systems		Temperature	Mole Fraction of A	
A	B	(K)	x_A	α
D_2	H_2	288–373	0.48	0.17
He	H_2	273–760	0.50	0.15
N_2	H_2	288–373	0.50	0.34
Ar	H_2	258	0.53	0.26
O_2	H_2	90–294	0.50	0.19
CO	H_2	288–373	0.50	0.33
CO_2	H_2	288–456	0.50	0.28
C_3H_8	H_2	230–520	0.50	0.30
N_2	He	287–373	0.50	0.36
		260	0.655	0.37
Ar	He	330	0.90	0.28
		330	0.70	0.31
		330	0.50	0.37
Ne	He	205	0.46	0.31
		330	0.46	0.315
		365	0.46	0.315
O_2	N_2	293	0.50	0.018

2.2.9 Unsteady-State Mass Transfer

In many practical mass-transfer processes, unsteady-state conditions prevail. Thus, in the example given in Section 2.1, a box is divided into two compartments each containing a different gas, and the partition is removed. Molecular diffusion of the gases takes place, and concentrations and concentration gradients change with time. If a bowl of liquid evaporates into an enclosed space, the partial pressure in the gas phase progressively increases, and the concentrations and the rate of evaporation are both time-dependent.

Considering an element of gas of cross-sectional area A and of thickness δy in the direction of mass transfer in which the concentrations c_A and c_B of the components **A** and **B** are a function of both position y and time t (Fig. 2.4), then if the mass-transfer flux is composed of two components, one attributable to diffusion according to Fick's law and the other to a bulk flow velocity u_F, the fluxes of **A** and **B** at a distance y from the origin may be taken as N'_A and N'_B, respectively. These will increase to $N'_A + \left(\partial N'_A/\partial y\right)\delta y$ and $N'_B + \left(\partial N'_B/\partial y\right)\delta y$ at a distance $y+\delta y$ from the origin.

At position y, the fluxes N'_A and N'_B will be as given in Table 2.6. At a distance $y+\delta y$ from the origin, that is, at the further boundary of the element, these fluxes will increase by the amounts shown in the lower part of Table 2.6.

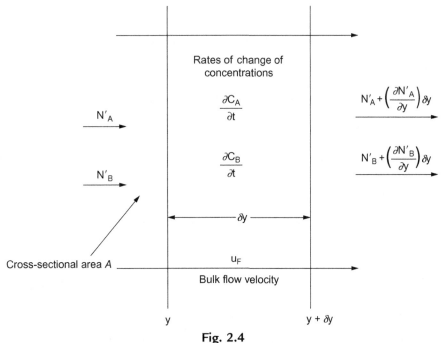

Fig. 2.4
Unsteady-state mass transfer.

Table 2.6 Fluxes of a gas mixture

	Diffusional Flux	Flux Due to Bulk Flow	Total Flux	
A	$-D\dfrac{\partial C_A}{\partial y}$	$u_F C_A$	$N'_A = -D\dfrac{\partial C_A}{\partial y} + u_F C_A$	(2.62)
B	$-D\dfrac{\partial C_B}{\partial y}$	$u_F C_B$	$N'_B = -D\dfrac{\partial C_B}{\partial y} + u_F C_B$	(2.63)
Changes in fluxes over distance δy				
A	$-D\left(\dfrac{\partial^2 C_A}{\partial y^2}\right)\delta y$	$\dfrac{\partial (u_F C_A)}{\partial y}\delta y$	$\left\{-D\dfrac{\partial^2 C_A}{\partial y^2} + \dfrac{\partial (u_F C_A)}{\partial y}\right\}\delta y$	
B	$-D\left(\dfrac{\partial^2 C_B}{\partial y^2}\right)\delta y$	$\dfrac{\partial (u_F C_B)}{\partial y}\delta y$	$\left\{-D\dfrac{\partial^2 C_B}{\partial y^2} + \dfrac{\partial (u_F C_B)}{\partial y}\right\}\delta y$	

Thus, for **A**: moles IN/unit time (at y) – moles OUT/unit time (at $y+\delta y$) = rate of change of concn. × element volume

$$\left\{-D\frac{\partial C_A}{\partial y} + u_F C_A\right\}A - \left\{-D\frac{\partial C_A}{\partial y} + u_F C_A + \frac{\partial}{\partial y}\left[-D\frac{\partial C_A}{\partial y} + u_F C_A\right]\delta y\right\}A = \frac{\partial C_A}{\partial t}(\delta y.A)$$

Simplifying

$$\frac{\partial C_A}{\partial t} = D\frac{\partial^2 C_A}{\partial y^2} - \frac{\partial (u_F C_A)}{\partial y} \tag{2.64a}$$

For component **B**,

$$\frac{\partial C_B}{\partial t} = D\frac{\partial^2 C_B}{\partial y^2} - \frac{\partial (u_F C_B)}{\partial y} \tag{2.64b}$$

and adding $\dfrac{\partial (C_A + C_B)}{\partial t} = D\dfrac{\partial^2 (C_A + C_B)}{\partial y^2} - \dfrac{\partial}{\partial y}[(C_A + C_B)u_F]$

Since, for an ideal gas, $C_A + C_B = C_F = $ constant (Eqs 2.9a–2.9c):

$$0 = 0 - \frac{\partial}{\partial y}(u_F C_T)$$

and

$$\frac{\partial u_F}{\partial y} = 0$$

where u_F is therefore independent of y.

Thus, Eq. (2.64a) can be written as

$$\frac{\partial C_A}{\partial t} = D\frac{\partial^2 C_A}{\partial y^2} - u_F\frac{\partial C_A}{\partial y} \qquad (2.65)$$

Equimolecular counterdiffusion

For equimolecular counterdiffusion, $u_F = 0$, and Eq. (2.65) simplifies to

$$\frac{\partial C_A}{\partial t} = D\frac{\partial^2 C_A}{\partial y^2} \qquad (2.66)$$

Eq. (2.66) is referred to as Fick's second law. This also applies when u_F is small, corresponding to conditions where C_A is always low. This equation can be solved for a number of important boundary conditions, and it should be compared with the corresponding equation for unsteady-state heat transfer (Eq. 1.29).

For the more general three-dimensional case where concentration gradients are changing in the x, y, and z directions, these changes must be added to give

$$\frac{\partial C_A}{\partial t} = D\left[\frac{\partial^2 C_A}{\partial x^2} + \frac{\partial^2 C_A}{\partial y^2} + \frac{\partial^2 C_A}{\partial z^2}\right] \qquad (2.67)$$

Gas absorption

In general, it is necessary to specify the physical constraints operating on the system in order to evaluate the bulk flow velocity u_F. In gas absorption, there will be no overall flux of the insoluble component **B** at the liquid interface ($y = 0$, say). In an unsteady-state process, however, where, by definition, concentrations will be changing with time throughout the system; the flux of **B** will be zero only at $y = 0$. At the interface ($y = 0$), total flux of **B** (from Eq. 2.43b) is given by

$$N'_B = -D\left(\frac{\partial C_B}{\partial y}\right)_{y=0} + u_F(C_B)_{y=0} = 0$$

or

$$u_F = \frac{D\left(\dfrac{\partial C_B}{\partial y}\right)_{y=0}}{(C_B)_{y=0}} = \frac{-D}{(C_T - C_A)_{y=0}}\left(\frac{\partial C_A}{\partial y}\right)_{y=0} \qquad (2.68)$$

Substituting in Eq. (2.65),

$$\frac{\partial C_A}{\partial t} = D\left[\frac{\partial^2 C_A}{\partial y^2} + \frac{1}{(C_T - C_A)_{y=0}}\left(\frac{\partial C_A}{\partial y}\right)_{y=0}\frac{\partial C_A}{\partial y}\right] \tag{2.69}$$

Thus, at the interface ($y=0$),

$$\left(\frac{\partial C_A}{\partial y}\right)_{y=0} = D\left[\left(\frac{\partial^2 C_A}{\partial y^2}\right)_{y=0} + \frac{1}{(C_T - C_A)_{y=0}}\left(\frac{\partial C_A}{\partial y}\right)_{y=0}^2\right] \tag{2.70}$$

This equation that is not capable of an exact analytical solution has been discussed by Arnold[11] in relation to evaporation from a free surface.

Substituting into Eq. (2.62) for N_A',

$$N_A' = -D\frac{\partial C_A}{\partial y} - \frac{DC_A}{(C_T - C_A)_{y=0}}\left(\frac{\partial C_A}{\partial y}\right)_{y=0}$$

$$= -D\left[\frac{\partial C_A}{\partial y} + \frac{C_A}{(C_T - C_A)_{y=0}}\left(\frac{\partial C_A}{\partial y}\right)_{y=0}\right] \tag{2.71}$$

2.3 Multicomponent Gas-Phase Systems

2.3.1 Molar Flux in Terms of Effective Diffusivity

For a multicomponent system, the bulk flow velocity u_F is given by

$$u_F = \frac{1}{C_T}\left(N_A' + N_B' + N_C' + \cdots\right) \tag{2.72}$$

or

$$u_F = x_A u_A + x_B u_B + x_C u_C + \cdots \tag{2.73}$$

or

$$u_F = \sum x_A u_A = \frac{1}{C_T}\sum N_A' \tag{2.74}$$

Since $N_A' = N_A + u_F C_A$ (Eq. 2.46a), then

$$N_A' = -D'\frac{dC_A}{dy} + \left[\frac{1}{C_T}\Sigma N_A'\right]C_A$$

$$= -D'C_T\frac{dx_A}{dy} + x_A\Sigma N_A' \tag{2.75}$$

where D' is the effective diffusivity for transfer of **A** in a mixture of B, C, D, and so on. For the particular case, where N'_B, N'_C, and so on are all zero,

$$N'_A = -D'C_T \frac{dx_A}{dy} + x_A \Sigma N'_A$$

or

$$N'_A = -D' \frac{C_T}{1 - x_A} \frac{dx_A}{dy} \tag{2.76}$$

A method of calculating the effective diffusivity D' in terms of each of the binary diffusivities is presented in Section 2.3.2.

2.3.2 Maxwell's Law of Diffusion

Maxwell's law for a binary system

Maxwell[12] postulated that the partial pressure gradient in the direction of diffusion for a constituent of a two-component gaseous mixture was proportional to

(a) the relative velocity of the molecules in the direction of diffusion and
(b) the product of the molar concentrations of the components.

Thus,

$$-\frac{dP_A}{dy} = FC_A C_B (u_A - u_B) \tag{2.77}$$

where u_A and u_B are the mean molecular velocities of **A** and **B**, respectively, in the direction of mass transfer and F is a coefficient.

Noting that

$$u_A = \frac{N'_A}{C_A} \tag{2.78}$$

and

$$u_B = \frac{N'_B}{C_B} \tag{2.79}$$

and using $P_A = C_A \mathbf{R}T$ (for an ideal gas) (Eq. 2.10a).

On substitution into Eq. (2.77) give

$$-\frac{dC_A}{dy} = \frac{F}{\mathbf{R}T} (N_A C_B - N_B C_A) \tag{2.80}$$

Equimolecular counterdiffusion

By definition, $N'_A = -N'_B = N_A$

Substituting in Eq. (2.80),

$$-\frac{dC_A}{dy} = \frac{FN_A}{\mathbf{R}T}(C_B + C_A) \tag{2.81}$$

or

$$N_A = -\frac{\mathbf{R}T}{FC_T}\frac{dC_A}{dy} \tag{2.82}$$

Then, by comparison with Fick's law (Eq. 2.4),

$$D = \frac{\mathbf{R}T}{FC_T} \tag{2.83}$$

or

$$F = \frac{\mathbf{R}T}{DC_T} \tag{2.84}$$

Transfer of A through stationary B

By definition, $N'_B = 0$

Thus

$$-\frac{dC_A}{dy} = \frac{F}{\mathbf{R}T}N'_A C_B \quad \text{(from Eq. 2.80)}$$

or

$$N'_A = -\frac{\mathbf{R}T}{FC_T}\frac{C_T}{C_B}\frac{dC_A}{dy} \tag{2.85}$$

Substituting from Eq. (2.83),

$$N'_A = -D\frac{C_T}{C_B}\frac{dC_A}{dy} \tag{2.86}$$

It may be noted that Eq. (2.86) is identical to Eq. (2.30) (Stefan's law), and Stefan's law can therefore also be derived from Maxwell's law of diffusion.

Maxwell's law for multicomponent mass transfer

This argument can be applied to the diffusion of a constituent of a multicomponent gas. Considering the transfer of component **A** through a stationary gas consisting of components B, C, ... if the total partial pressure gradient can be regarded as being made up of a series of terms each representing the contribution of the individual component gases, then from Eq. (2.80),

$$-\frac{dC_A}{dy} = \frac{F_{AB}N'_A C_B}{RT} + \frac{F_{AC}N'_A C_C}{RT} + \cdots$$

or

$$-\frac{dC_A}{dy} = \frac{N'_A}{RT}(F_{AB}C_B + F_{AC}C_C + \cdots) \tag{2.87}$$

From Eq. (2.84), writing

$$F_{AB} = \frac{RT}{D_{AB}C_T}, \quad \text{and so on.}$$

where D_{AB} is the diffusivity of **A** in **B** and so on.

$$-\frac{dC_A}{dy} = \frac{N'_A}{C_T}\left(\frac{C_B}{D_{AB}} + \frac{C_C}{D_{AC}} + \cdots\right) \tag{2.88}$$

$$\therefore N'_A = -\frac{C_T}{\dfrac{C_B}{D_{AB}} + \dfrac{C_C}{D_{AC}} + \cdots}\frac{dC_A}{dy}$$

$$= -\frac{1}{\dfrac{C_B}{C_T - C_A}\dfrac{1}{D_{AB}} + \dfrac{C_C}{C_T - C_A}\dfrac{1}{D_{AC}} + \cdots}\frac{C_T}{C_T - C_A}\frac{dC_A}{dy} \tag{2.89}$$

$$= -\frac{1}{\dfrac{y'_B}{D_{AB}} + \dfrac{y'_C}{D_{AC}} + \cdots}\frac{C_T}{C_T - C_A}\frac{dC_A}{dy}$$

where y'_B is the mole fraction of **B** and so on in the stationary components of the gas.

By comparing Eq. (2.89) with Stefan's law (Eq. 2.30), the effective diffusivity of **A** in the mixture (D') is given by

$$\frac{1}{D'} = -\frac{y'_B}{D_{AB}} + \frac{y'_C}{D_{AC}} + \cdots \tag{2.90}$$

Multicomponent mass transfer is discussed in more detail by Taylor and Krishna,[13] Cussler,[14] and Zielinski and Hanley.[15]

2.4 Diffusion in Liquids

Whilst the diffusion of solution in a liquid is governed by the same equations as for the gas phase, the diffusion coefficient D is about two orders of magnitude smaller for a liquid than for a gas. Furthermore, the diffusion coefficient is a much more complex function of the molecular properties.

For an ideal gas, the total molar concentration C_T is constant at a given total pressure P and temperature T. This approximation holds quite well for real gases and vapours, except at high pressures. For a liquid, however, C_T may show considerable variations as the concentrations of the components change, and in practice, the total mass concentration (density ρ of the mixture) is much more nearly constant. Thus, for a mixture of ethanol and water, for example, the mass density will range from about 790 to 1000 kg/m³, whereas the molar density will range from about 17 to 56 kmol/m³. For this reason, the diffusion equations are frequently written in the form of a mass flux J_A (mass/area × time) and the concentration gradients in terms of mass concentrations, such as c_A.

Thus, for component **A**, the mass flux is given by

$$J_A = -D\frac{dc_A}{dy} \tag{2.91}$$

$$= -D\rho\frac{d\omega_A}{dy} \tag{2.92}$$

where ρ is mass density (now taken as constant) and ω_A is the mass fraction of **A** in the liquid.

For component **B**,

$$J_B = -D\rho\frac{d\omega_B}{dy} \tag{2.93}$$

$$= D\rho\frac{d\omega_A}{dy} \quad (\text{since } \omega_A + \omega_B = 1) \tag{2.94}$$

Thus, the diffusional process in a liquid gives rise to a situation where the components are being transferred at approximately equal and opposite mass (rather than molar) rates.

Liquid-phase diffusivities are strongly dependent on the concentration of the diffusing component, which is in strong contrast to gas-phase diffusivities that are substantially independent on concentration. Values of liquid-phase diffusivities that are normally quoted apply to very dilute concentrations of the diffusing component, the only condition under which analytical solutions can be produced for the diffusion equations. For this reason, only dilute solutions are considered here, and in these circumstances, no serious error is involved in using Fick's first and second laws expressed in molar units.

The molar flux is given by

$$N_A = -D\frac{dC_A}{dy} \quad \text{(Eq. 2.4)}$$

and

$$\frac{\partial C_A}{\partial t} = D\frac{\partial^2 C_A}{\partial y^2} \quad \text{(Eq. 2.66)}$$

where D is now the liquid-phase diffusivity and C_A is the molar concentration in the liquid phase.

On integration, Eq. (2.4) becomes

$$N_A = -D\frac{C_{A_2} - C_{A_1}}{y_2 - y_1} = \frac{D}{y_2 - y_1}(C_{A_1} - C_{A_2}) \tag{2.95}$$

and $D/(y2 - y_1)$ is the liquid-phase mass-transfer coefficient.

An example of the integration of Eq. (2.66) is given in Section 2.5.3.

2.4.1 Liquid Phase Diffusivities

Values of the diffusivities of various materials in water are given in Table 2.7. Where experimental values are not available, it is necessary to use one of the predictive methods that are available.

A useful equation for the calculation of liquid-phase diffusivities of dilute solutions of nonelectrolytes has been given by Wilke and Chang.[16] This is not dimensionally consistent, and therefore, the value of the coefficient depends on the units employed. Using SI units,

$$D = \frac{1.173 \times 10^{-16}\phi_B^{1/2}M_B^{1/2}T}{\mu V_A^{0.6}} \tag{2.96}$$

where D is the diffusivity of solute **A** in solvent **B** (m²/s), ϕ_B is the association factor for the solvent (2.26 for water, 1.9 for methanol, 1.5 for ethanol, and 1.0 for unassociated solvents such as hydrocarbons and ethers), M_B is the molecular weight of the solvent, μ is the viscosity of the solution (N s/m²), T is temperature (K), and V_A is the molecular volume of the solute (m³/kmol). Values for simple molecules are given in Table 2.4. For more complex molecules, V_A is calculated by summation of the atomic volume and other contributions given in Table 2.4.

It may be noted that for water, a value of 0.0756 m³/kmol should be used.

Eq. (2.96) does not apply to either electrolytes or to concentrated solutions. Reid et al.[17] discuss diffusion in electrolytes. Little information is available on diffusivities in concentrated

Table 2.7 Diffusivities (diffusion coefficients) and Schmidt numbers, in liquids at 293 K[4]

Solute	Solvent	D (m²/s×10⁹)	Sc $(\mu/\rho D)^a$
O_2	Water	1.80	558
CO_2	Water	1.50	670
N_2O	Water	1.51	665
NH_3	Water	1.76	570
Cl_2	Water	1.22	824
Br_2	Water	1.2	840
H_2	Water	5.13	196
N_2	Water	1.64	613
HCl	Water	2.64	381
H_2S	Water	1.41	712
H_2SO_4	Water	1.73	580
HNO_3	Water	2.6	390
Acetylene	Water	1.56	645
Acetic acid	Water	0.88	1140
Methanol	Water	1.28	785
Ethanol	Water	1.00	1005
Propanol	Water	0.87	1150
Butanol	Water	0.77	1310
Allyl alcohol	Water	0.93	1080
Phenol	Water	0.84	1200
Glycerol	Water	0.72	1400
Pyrogallol	Water	0.70	1440
Hydroquinone	Water	0.77	1300
Urea	Water	1.06	946
Resorcinol	Water	0.80	1260
Urethane	Water	0.92	1090
Lactose	Water	0.43	2340
Maltose	Water	0.43	2340
Glucose	Water	0.60	1680
Mannitol	Water	0.58	1730
Raffinose	Water	0.37	2720
Sucrose	Water	0.45	2230
Sodium chloride	Water	1.35	745
Sodium hydroxide	Water	1.51	665
CO_2	Ethanol	3.4	445
Phenol	Ethanol	0.8	1900
Chloroform	Ethanol	1.23	1230
Phenol	Benzene	1.54	479
Chloroform	Benzene	2.11	350
Acetic acid	Benzene	1.92	384
Ethylene dichloride	Benzene	2.45	301

[a]Based on $\mu/\rho = 1.005 \times 10^{-6}$ m²/s for water, 7.37×10^{-7} for benzene, and 1.511×10^{-6} for ethanol, all at 293 K. The data apply only for dilute solutions.
The values are based mainly on *International Critical Tables* 5 (1928).

solutions although it appears that, for ideal mixtures, the product μD is a linear function of the molar concentration.

The calculation of liquid-phase diffusivities is discussed further in Volume 6.

2.5 Mass Transfer Across a Phase Boundary

The theoretical treatment that has been developed in Sections 2.2–2.4 relates to mass transfer within a single phase in which no discontinuities exist. In many important applications of mass transfer, however, material is transferred across a phase boundary. Thus, in distillation, a vapour and liquid are brought into contact in the fractionating column, and the more volatile material is transferred from the liquid to the vapour, whilst the less volatile constituent is transferred in the opposite direction; this is an example of equimolecular counterdiffusion. In gas absorption, the soluble gas diffuses to the surface, dissolves in the liquid, and then passes into the bulk of the liquid, and the carrier gas is not transferred. In both of these examples, one phase is a liquid and the other a gas. In liquid–liquid extraction, however, a solute is transferred from one liquid solvent to another across a phase boundary, and in the dissolution of a crystal, the solute is transferred from a solid to a liquid.

Each of these processes is characterised by a transference of material across an interface. Because no material accumulates there, the rate of transfer on each side of the interface must be the same, and therefore, the concentration gradients automatically adjust themselves so that they are proportional to the resistance to transfer in the particular phase. In addition, if there is no resistance to transfer at the interface, the concentrations on each side will be related to each other by the phase-equilibrium relationship. Whilst the existence or otherwise of a resistance to transfer at the phase boundary is the subject of conflicting views,[18] it appears likely that any resistance is not high, except in the case of crystallisation, and in the following discussion, equilibrium between the phases will be assumed to exist at the interface. Interfacial resistance may occur, however, if a surfactant is present as it may accumulate at the interface (Section 2.5.5).

The mass-transfer rate between two fluid phases will depend on the physical properties of the two phases, the concentration difference, the interfacial area, and the degree of turbulence. Mass-transfer equipment is therefore designed to give a large area of contact between the phases and to promote turbulence in each of the fluids. In the majority of plants, the two phases flow continuously in a countercurrent manner. In a steady-state process, therefore, although the composition of each element of fluid is changing as it passes through the equipment, conditions at any given point do not change with time. In most industrial equipment, the flow pattern is so complex that it is not capable of expression in mathematical terms, and the interfacial area is not known precisely.

A number of mechanisms have been suggested to represent conditions in the region of the phase boundary. The earliest of these is the *two-film* theory propounded by Whitman[19] in 1923 who suggested that the resistance to transfer in each phase could be regarded as lying in a thin film close to the interface. The transfer across these films is regarded as a steady-state process of molecular diffusion following equations of the type of Eq. (2.22). The turbulence in the bulk fluid is considered to die out at the interface of the films. In 1935, Higbie[20] suggested that the transfer process was largely attributable to fresh material being brought by the eddies to the interface, where a process of unsteady-state transfer took place for a fixed period at the freshly exposed surface. This theory is generally known as the *penetration theory*. Danckwerts[21] has since suggested a modification of this theory in which it is considered that the material brought to the surface will remain there for varying periods of time. Danckwerts also discusses the random age distribution of such elements from which the transfer is by an unsteady-state process to the second phase. Subsequently, Toor and Marchello[22] have proposed a more general theory, the *film-penetration theory*, and have shown that each of the earlier theories is a particular limiting case of their own. A number of other theoretical treatments have also been proposed, including that of Kishinevskij.[23] The two-film theory and the penetration theory will now be considered, followed by an examination of the film-penetration theory.

2.5.1 The Two-Film Theory

The two-film theory of Whitman[19] was the first serious attempt to represent conditions occurring when material is transferred from one fluid stream to another. Although it does not closely reproduce the conditions in most practical equipment, the theory gives expressions that can be applied to the experimental data that are generally available, and for that reason, it is still extensively used.

In this approach, it is assumed that turbulence dies out at the interface and that a laminar layer exists in each of the two fluids. Outside the laminar layer, turbulent eddies supplement the action caused by the random movement of the molecules, and the resistance to transfer becomes progressively smaller. For equimolecular counterdiffusion, the concentration gradient is therefore linear close to the interface and gradually becomes less at greater distances as shown in Fig. 2.5 by the full lines *ABC* and *DEF*. The basis of the theory is the assumption that the zones in which the resistance to transfer lies can be replaced by two hypothetical layers, one on each side of the interface, in which the transfer is entirely by molecular diffusion. The concentration gradient is therefore linear in each of these layers and zero outside. The broken lines *AGC* and *DHF* indicate the hypothetical concentration distributions, and the thicknesses of the two films are L_1 and L_2. Equilibrium is assumed to exist at the interface, and therefore, the relative positions of the points *C* and *D* are determined by the equilibrium relation between the phases. In Fig. 2.5, the scales are not necessarily the same on the two sides of the interface.

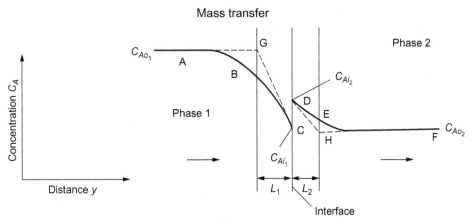

Fig. 2.5
Two-film theory.

The mass transfer is treated as a steady-state process, and therefore, the theory can be applied only if the time taken for the concentration gradients to become established is very small compared with the time of transfer or if the capacities of the films are negligible.

From Eq. (2.22), the rate of transfer per unit area in terms of the two-film theory for equimolecular counterdiffusion is given for the first phase as

$$N_A = \frac{D_1}{L_1}(C_{Ao_1} - C_{Ai_1}) = h_{D1}(C_{Ao_1} - C_{Ai_1}) \tag{2.97a}$$

where L_1 is the thickness of the film, C_{Ao1} the molar concentration outside the film, and C_{Ai1} the molar concentration at the interface.

For the second phase, with the same notation, the rate of transfer is

$$N_A = \frac{D_2}{L_2}(C_{Ai_2} - C_{Ao_2}) = h_{D2}(C_{Ai_2} - C_{Ao_2}) \tag{2.97b}$$

Because material does not accumulate at the interface, the two rates of transfer must be the same, and

$$\frac{h_{D1}}{h_{D2}} = \frac{C_{Ai_2} - C_{Ao_2}}{C_{Ao_1} - C_{Ai_1}} \tag{2.98}$$

The relation between C_{Ai1} and C_{Ai2} is determined by the phase-equilibrium relationship since the molecular layers on each side of the interface are assumed to be in equilibrium with one another. It may be noted that the ratio of the differences in concentrations is inversely proportional to the ratio of the mass-transfer coefficients. If the bulk concentrations, C_{Ao1} and C_{Ao2}, are fixed, the interface concentrations will adjust to values that satisfy Eq. (2.98). This means that, if the relative value of the coefficients changes, the interface concentrations will change too. In general, if the degree of turbulence of the fluid is increased, the effective film thicknesses will be reduced, and the mass-transfer coefficients will be correspondingly increased.

The theory is equally applicable when bulk flow occurs. In gas absorption, for example, where may be expressed the mass-transfer rate in terms of the concentration gradient in the gas phase:

$$N'_A = -D\frac{dC_A}{dy} \cdot \frac{C_T}{C_B} \quad \text{(Eq. 2.30)}$$

In this case, for a steady-state process, $(dC_A/dy)(C_T/c_B)$, as opposed to dC_A/dy, will be constant through the film, and dC_A/dy will increase as c_A decreases. Thus, lines *GC* and *DH* in Fig. 2.5 will no longer be quite straight.

2.5.2 The Penetration Theory

The penetration theory was propounded in 1935 by Higbie[20] who was investigating whether or not a resistance to transfer existed at the interface when a pure gas was absorbed in a liquid. In his experiments, a slug-like bubble of carbon dioxide was allowed rise through a vertical column of water in a 3 mm diameter glass tube. As the bubble rose, the displaced liquid ran back as a thin film between the bubble and the tube. Higbie assumed that each element of surface in this liquid was exposed to the gas for the time taken for the gas bubble to pass it, that is, for the time given by the quotient of the bubble length and its velocity. It was further supposed that during this short period, which varied between 0.01 and 0.1 s in the experiments, absorption took place as the result of unsteady-state molecular diffusion into the liquid, and for the purposes of calculation, the liquid was regarded as infinite in depth because the time of exposure was so short.

The way in which the concentration gradient builds up as a result of exposing a liquid—initially pure—to the action of a soluble gas is shown in Fig. 2.6, which is based on Higbie's calculations. The percentage saturation of the liquid is plotted against the distance from the surface for a number of exposure times in arbitrary units. Initially, only the surface layer contains solute, and the concentration changes abruptly from 100% to 0% at the surface. For progressively longer exposure times, the concentration profile develops as shown, until after an infinite time the whole of the liquid becomes saturated. The shape of the profiles is such that at any time, the effective depth of liquid that contains an appreciable concentration of solute can be specified, and hence, the theory is referred to as the *penetration theory*. If this depth of penetration is less than the total depth of liquid, no significant error is introduced by assuming that the total depth is infinite.

The work of Higbie laid the basis of the penetration theory in which it is assumed that the eddies in the fluid bring an element of fluid to the interface where it is exposed to the second

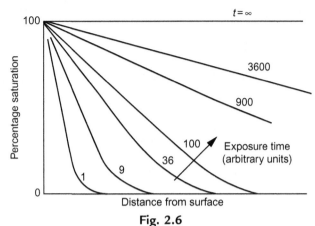

Fig. 2.6

Penetration of solute into a solvent.

phase for a definite interval of time, after which the surface element is mixed with the bulk again. Thus, fluid whose initial composition corresponds with that of the bulk fluid remote from the interface is suddenly exposed to the second phase. It is assumed that equilibrium is immediately attained by the surface layers, that a process of unsteady-state molecular diffusion then occurs, and that the element is remixed after a fixed interval of time. In the calculation, the depth of the liquid element is assumed to be infinite, and this is justifiable if the time of exposure is sufficiently short for penetration to be confined to the surface layers. Throughout, the existence of velocity gradients within the fluids is ignored, and the fluid at all depths is assumed to be moving at the same rate as the interface.

The diffusion of solute **A** away from the interface (*Y*-direction) is thus given by Eq. (2.66):

$$\frac{\partial C_A}{\partial t} = D \frac{\partial^2 C_A}{\partial y^2} \quad \text{(Eq. 2.64)}$$

for conditions of equimolecular counterdiffusion, or when the concentrations of diffusing materials are sufficiently low for the bulk flow velocity to be negligible. Because concentrations of **A** are low, there is no objection to using molar concentration for calculation of mass-transfer rates in the liquid phase (see Section 2.4).

The following boundary conditions apply:

$$
\begin{array}{lll}
t=0 & 0<y<\infty & C_A=C_{Ao} \\
t>0 & y=0 & C_A=C_{Ai} \\
t>0 & y=\infty & C_A=C_{Ao}
\end{array}
$$

where C_{Ao} is the concentration in the bulk of the phase and C_{Ai} the equilibrium value at the interface.

It is convenient to work in terms of a 'deviation' variable C' as opposed to C_A, where C' is the amount by which the concentration of **A** exceeds the initial uniform concentration C_{Ao}. This change allows some simplification of the algebra.

With the substitution,

$$C' = C_A - C_{Ao} \tag{2.99}$$

Eq. (2.66) becomes

$$\frac{\partial C'}{\partial t} = D \frac{\partial^2 C'}{\partial y^2} \tag{2.100}$$

because C_{Ao} is a constant with respect to both t and y; the boundary conditions are then

$$
\begin{array}{lll}
t = 0 & 0 < y < \infty & C' = 0 \\
t > 0 & y = 0 & C' = C'_i = C_{Ai} - C_{Ao} \\
t > 0 & y = \infty & C' = 0
\end{array}
$$

These boundary conditions are necessary and sufficient for the solution of Eq. (2.100) that is first order with respect to t and second order with respect to y.

The equation is most conveniently solved by the method of Laplace transforms and used for the solution of the unsteady-state thermal conduction problem in Chapter 1.

By definition, the Laplace transform $\overline{C'}$ of C' is given by

$$\overline{C'} = \int_0^\infty e^{-pt} C' \, dt \tag{2.101}$$

Then,

$$\overline{\frac{\partial C'}{\partial t}} = \int_0^\infty e^{-pt} \frac{\partial C'}{\partial t} dt \tag{2.102}$$

$$= [e^{-pt} C']_0^\infty + p \int_0^\infty e^{-pt} C' \, dt$$

$$= p\overline{C'} \tag{2.103}$$

Since the Laplace transform operation is independent of y,

$$\overline{\frac{\partial^2 C'}{\partial y^2}} = \frac{\partial^2 \overline{C'}}{\partial y^2} \tag{2.104}$$

Thus, taking Laplace transforms of both sides of Eq. (2.100),

$$p\overline{C'} = D\frac{\partial^2 \overline{C'}}{\partial y^2}$$

$$\therefore \frac{\partial^2 \overline{C'}}{\partial y^2} - \frac{p}{D}\overline{C'} = 0$$

Eq. (2.100) has therefore been converted from a partial differential equation in C' to an ordinary second-order linear differential equation in $\overline{C'}$.

Thus,

$$\overline{C'} = B_1 e^{\sqrt{(p/D)}y} + B_2 e^{-\sqrt{(p/D)}y} \tag{2.105}$$

When

$$y = 0, \quad C_A = C_{Ai}, \quad \text{and} \quad C' = C_{Ai} - C_{Ao} = C_i'$$

and when

$$y = \infty, \quad C_A = C_{Ao}, \quad \text{and} \quad C' = 0$$

Hence,

$$B_1 = 0$$

and

$$\overline{C'} = B_2 e^{-\sqrt{(p/D)}y} = \overline{C'}_i e^{-\sqrt{(p/D)}y} \tag{2.106}$$

Now, $B_2 = \displaystyle\int_0^\infty (C_{Ai} - C_{Ao})e^{-pt}dt$

$$= \frac{1}{p}(C_{Ai} - C_{Ao})$$

Thus,

$$\overline{C'} = \frac{1}{p}(C_{Ai} - C_{Ao})e^{-\sqrt{(p/D)}y} \tag{2.107}$$

Taking the inverse of the transform (Appendix A3, Table 12, No. 83), then

$$C' = C_A - C_{Ao} = (C_{Ai} - C_{Ao})\,\text{erfc}\left(\frac{y}{2\sqrt{Dt}}\right)$$

or

$$\frac{C_A - C_{Ao}}{C_{Ai} - C_{Ao}} = \text{erfc}\left(\frac{y}{2\sqrt{Dt}}\right) = 1 - \text{erfc}\left(\frac{y}{2\sqrt{Dt}}\right) \tag{2.108}$$

This expression gives concentration c_A as a function of position y and of time t.

erf X is known as the error function, and values are tabulated as for any other function of X; erfc X is the *complementary error function* $(1 - \text{erf } X)$.

By definition,

$$\text{erfc } X = \frac{2}{\sqrt{\pi}} \int_X^\infty e^{-x^2} dx \tag{2.109}$$

Since,

$$\int_0^\infty e^{-x^2} dx = \frac{\sqrt{\pi}}{2} \tag{2.110}$$

erfc x goes from 1 to 0 as x goes from 0 to ∞.

The concentration gradient is then obtained by differentiation of Eq. (2.108) with respect to y.

Thus,

$$\frac{1}{C_{Ai} - C_{Ao}} \frac{\partial C_A}{\partial y} = \frac{\partial}{\partial y} \left[\frac{2}{\sqrt{\pi}} \int_{(y/2\sqrt{Dt})}^\infty e^{-y^2/4Dt} d\left(\frac{y}{2\sqrt{Dt}} \right) \right]$$

$$\therefore \frac{\partial C_A}{\partial y} = -(C_{Ai} - C_{Ao}) \frac{2}{\sqrt{\pi}} \frac{1}{2\sqrt{Dt}} \left(e^{-y^2/4Dt} \right)$$

$$= -(C_{Ai} - C_{Ao}) \frac{1}{\sqrt{\pi Dt}} e^{-y^2/4Dt} \tag{2.111}$$

The mass-transfer rate at any position y at time t is given by

$$(N_A)_t = -D \frac{\partial C_A}{\partial y}$$

$$= (C_{Ai} - C_{Ao}) \sqrt{\frac{D}{\pi t}} e^{-y^2/4Dt} \tag{2.112}$$

The mass-transfer rate per unit area of surface is then given by

$$(N_A)_{t,y=0} = -D \left(\frac{\partial C_A}{\partial y} \right)_{y=0}$$

$$= (C_{Ai} - C_{Ao}) \sqrt{\frac{D}{\pi t}} \tag{2.113}$$

The point value of the mass-transfer coefficient is therefore $\sqrt{D/\pi t}$.

Regular surface renewal

It is important to note that the mass-transfer rate falls off progressively during the period of exposure, theoretically from infinity at $t=0$ to zero at $t=\infty$.

Assuming that all the surface elements are exposed for the same time t_e (Higbie's assumption), from Eq. (2.113), the moles of A (n_A) transferred at an area A in time t_e is given by

$$n_A = (C_{Ai} - C_{Ao})\sqrt{\frac{D}{\pi}}A\int_0^{t_e}\frac{dt}{\sqrt{t}}$$

$$= (C_{Ai} - C_{Ao})A\sqrt{\frac{Dt_e}{\pi}} \tag{2.114}$$

and the average rate of transfer per unit area over the exposure time t_e is given by

$$N_A = 2(C_{Ai} - C_{Ao})\sqrt{\frac{D}{\pi t_e}} \tag{2.115}$$

That is, the average rate over the time interval $t=0$ to $t=t_e$ is twice the point value at $t=t_e$.

Thus, the shorter the time of exposure, the greater is the rate of mass transfer. No precise value can be assigned to t_e in any industrial equipment, although its value will clearly become less as the degree of agitation of the fluid is increased.

If it is assumed that each element resides for the same time interval t_e in the surface, Eq. (2.115) gives the overall mean rate of transfer. It may be noted that the rate is a linear function of the driving force expressed as a concentration difference, as in the two-film theory, but that it is proportional to the diffusivity raised to the power of 0.5 instead of unity.

Eq. (2.114) forms the basis of the *laminar-jet* method of determining the molecular diffusivity of a gas in a liquid. Liquid enters the gas space from above through a sharp-edged circular hole formed in a thin horizontal plate, to give a vertical 'rod' of liquid having a flat velocity profile, which is collected in a container of slightly larger diameter than the jet. The concentration of this outlet liquid is measured in order to determine the number of moles n_A of A transferred to the laminar jet during the exposure time t_e that can be varied by altering the velocity and the length of travel of the jet. A plot of n_A versus $t_e^{1/2}$ should give a straight line, the slope of which enables the molecular diffusivity D to be calculated, since C_{Ao} is zero and C_{Ai} is the saturation concentration. The assumptions and possible sources of error in this method are discussed by Danckwerts[24]; it is important that penetration depths must be small compared with the radius of the jet.

When mass-transfer rates are very high, limitations may be placed on the rate at which a component may be transferred, by virtue of the limited frequency with which the molecules collide with the surface. For a gas, the collision rate can be calculated from the *kinetic theory*, and allowance must then be made for the fact that only a fraction of these molecules may be absorbed, with the rest being reflected. Thus, when even a pure gas is brought suddenly into contact with a fresh solvent, the initial mass-transfer rate may be controlled by the rate at which gas molecules can reach the surface, although the resistance to transfer rapidly builds up in the liquid phase to a level where this effect can be neglected. The point iswell illustrated in Example 2.4.

Example 2.4

In an experimental wetted-wall column, pure carbon dioxide is absorbed in water. The mass-transfer rate is calculated using the penetration theory, application of which is limited by the fact that the concentration should not reach more than 1% of the saturation value at a depth below the surface at which the velocity is 95% of the surface velocity. What is the maximum length of column to which the theory can be applied if the flowrate of water is 3 cm^3/s per cm of perimeter?

Viscosity of water $= 10^{-3}$ N s/m^2. Diffusivity of carbon dioxide in water $= 1.5 \times 10^{-9}$ m^2/s.

Solution
For the flow of a vertical film of fluid, the mean velocity of flow is governed by Eq. (3.87) of Vol. 1A in which sin ϕ is put equal to unity for a vertical surface:

$$u_m = \frac{\rho g s^2}{3\mu}$$

where s is the thickness of the film.

The flowrate per unit perimeter $(\rho g s^3/3 \mu) = 3 \times 10^{-4}$ m^2/sand

$$s = \left(\frac{3 \times 10^{-4} \times 10^{-3} \times 3}{1000 \times 9.81}\right)^{1/3}$$

$$= 4.51 \times 10^{-4} \, m$$

The velocity u_x at a distance y' from the vertical column wall is given by Eq. (3.85) of Vol. 1A (using y' in place of y) as

$$u_x = \frac{\rho g \left(s y' - \frac{1}{2}y'^2\right)}{\mu}$$

The free surface velocity u_s is given by substituting s for y' or

$$u_s = \frac{\rho g s^2}{2\mu}$$

Thus, $\dfrac{u_x}{u_s} = 2\left(\dfrac{y'}{s}\right) - \left(\dfrac{y'}{s}\right)^2 = 1 - \left(1 - \dfrac{y'}{s}\right)^2$

When $u_x/u_s = 0.95$, that is velocity is 95% of surface velocity, then

$$1 - \frac{y'}{s} = 0.224$$

and the distance below the free surface is $y = s - y' = 1.010 \times 10^{-4}$ m.

The relationship between concentration c_A, time, and depth is

$$\frac{C_A - C_{Ao}}{C_{Ai} - C_{Ao}} = \text{erfc}\left(\frac{y}{2\sqrt{Dt}}\right) \quad \text{(Eq. 2.108)}$$

The time at which concentration reaches 0.01 of saturation value at a depth of 1.010×10^{-4} m is given by

$$0.01 = \text{erfc}\left(\frac{1.010 \times 10^{-4}}{2\sqrt{1.5 \times 10^{-9}t}}\right)$$

Thus,

$$\text{erfc}\left(\frac{1.305}{\sqrt{t}}\right) = 0.99$$

Using tables of error functions (Appendix A3, Table 12),

$$\frac{1.305}{\sqrt{t}} = 1.822$$

and

$$t = 0.51\,\text{s}$$

The surface velocity is then

$$u_s = \frac{\rho g s^2}{2\mu} = \frac{1000 \times 9.81 \times (4.51 \times 10^{-4})^2}{2 \times 10^{-3}}$$
$$= 1\,\text{m/s}$$

and the maximum length of column is $= (1 \times 0.51) = \underline{0.51\,\text{m}}$.

Example 2.5

In a gas-liquid contactor, a pure gas is absorbed in a solvent, and the penetration theory provides a reasonable model by which to describe the transfer mechanism. As fresh solvent is exposed to the gas, the transfer rate is initially limited by the rate at which the gas molecules can reach the surface. If at 293 K and a pressure of 1 bar, the maximum possible rate of transfer of gas is 50 m³/m²s, express this as an equivalent resistance, when the gas solubility is 0.04 kmol/m³.

If the diffusivity in the liquid phase is 1.8×10^{-9} m²/s, at what time after the initial exposure will the resistance attributable to access of gas be equal to about 10% of the total resistance to transfer?

Solution

Bulk gas concentration $= \left(\frac{1}{22.4}\right)\left(\frac{273}{293}\right) = 0.0416\,\text{kmol/m}^3$

Initial mass-transfer rate $= 50 \text{ m}^3/\text{m}^2 \text{ s}$

$$= (50 \times 0.0416) = 2.08 \text{ kmol/m}^2\text{s}$$

Concentration driving force in liquid phase

$$= (0.04 - 0) = 0.04 \text{ kmol/m}^3$$

Effective mass-transfer coefficient initially

$$= \frac{2.08}{0.04} = 52.0 \text{ m/s}$$

Equivalent resistance $= 1/52.0 = 0.0192$ s/m

When this constitutes 10% of the total resistance,

liquid-phase resistance $= 0.0192 \times 9 = 0.173$ s/m

liquid-phase coefficient $= 5.78$ m/s

From Eq. (2.113), the point value of liquid-phase mass-transfer coefficient $= \sqrt{\dfrac{D}{\pi t}}$

$$= \sqrt{\frac{1.8 \times 10^{-9}}{\pi t}} = 2.394 \times 10^{-5} t^{-1/2} \text{ m/s}$$

Resistance $= 4.18 \times 10^4 \, t^{1/2}$ s/m

Thus, $4.18 \times 10^4 t^{1/2} = 0.173$, and $t = \underline{1.72 \times 10^{-11} \text{ s}}$

Thus, the limited rate of access of gas molecules is not likely to be of any significance.

Example 2.6

A deep pool of ethanol is suddenly exposed to an atmosphere of pure carbon dioxide and unsteady-state mass transfer, governed by Fick's law, and takes place for 100 s. What proportion of the absorbed carbon dioxide will have accumulated in the 1 mm layer closest to the surface in this period?

Diffusivity of carbon dioxide in ethanol $= 4 \times 10^{-9} \text{ m}^2/\text{s}$.

Solution
The accumulation in the 1 mm layer near the surface will be equal to the total amount of CO_2 entering the layer from the surface $(y=0)$ less that leaving $(y=10^{-3} \text{ m})$ in the course of 100 s.

The mass rate of transfer at any position y and time t is given by Eq. (2.112):

$$(N_A)_t = C_{Ai} \sqrt{\frac{D}{\pi t}} e^{-y^2/4Dt}$$

where N_A is expressed in moles per unit area and unit time and C_{Ao} is zero because the solvent is pure ethanol.

Considering unit area of surface, the moles transferred in time t_e at depth y is given by

$$= C_{Ai}\sqrt{\frac{D}{\pi}}\int_0^{t_e} t^{-1/2}e^{-y^2/4Dt}\,dt$$

Putting $y^2/4Dt = X^2$,

$$t^{-1/2} = \frac{2\sqrt{D}X}{y}$$

and

$$dt = \frac{y^2}{4D}\frac{-2}{X^3}\,dX$$

Thus,

$$\text{Integral} = \int_\infty^{X_e} \frac{y^2}{4D X^3}\frac{-2\,2\sqrt{D}X}{y}e^{-X^2}\,dX$$

$$= -\frac{y}{\sqrt{D}}\int_\infty^{X_e} X^{-2}e^{-X^2}\,dX$$

$$\text{Molar transfer per unit area} = C_{Ai}\sqrt{\frac{D}{\pi}}\left(\frac{-y}{\sqrt{D}}\right)\left\{\left[e^{-X^2}(-X^{-1})\right]_\infty^{X_e} - \int_\infty^{X_e}\left[-2Xe^{-X^2}(-X^{-1})\right]dX\right\}$$

$$= C_{Ai}\left(\frac{-y}{\sqrt{\pi}}\right)\left\{-X_e^{-1}e^{-X_e^2} + 2 - \int_{X_e}^\infty e^{-X^2}\,dX\right\}$$

$$= C_{Ai}\left(\frac{y}{\sqrt{\pi}}\right)\left\{\frac{2\sqrt{Dt_e}}{y}e^{-y^2/4Dt_e} - \sqrt{\pi}\,\text{erfc}\frac{y}{2\sqrt{Dt_e}}\right\}$$

$$= C_{Ai}\left\{2\sqrt{\frac{Dt_e}{\pi}}e^{-y^2/4Dt_e} - y\,\text{erfc}\frac{y}{2\sqrt{Dt_e}}\right\}$$

Putting $D = 4\times 10^{-9}$ m^2/s and $t = 100$ s:

At $y = 0$, moles transferred $= 2C_{Ai}\sqrt{\dfrac{4\times 10^{-9}\times 100}{\pi}} = 7.14\times 10^{-4}C_{Ai}$

At $y = 10^{-3}$ m, moles transferred $= C_{Ai}\{7.14\times 10^{-4}e^{-0.626} - 10^{-3}\,\text{erfc}\,0.791\}$

$$= C_{Ai}\{3.82\times 10^{-4} - 2.63\times 10^{-4}\}$$

$$= 1.19\times 10^{-4}C_{Ai}$$

The proportion of material retained in layer $= (7.14 - 1.19)/7.14 = 0.83$ or 83 per cent.

Random surface renewal

Danckwerts[21] suggested that each element of surface would not be exposed for the same time, but that a random distribution of ages would exist. It was assumed that the probability of any element of surface becoming destroyed and mixed with the bulk of the fluid was independent of thc age of the element, and on this basis, the age distribution of the surface elements was calculated using the following approach.

Supposing that the rate of production of fresh surface per unit total area of surface is s and that s is independent of the age of the element in question, the area of surface of age between t and $t+dt$ will be a function of t and may be written as f(t) dt. This will be equal to the area passing in time dt from the age range (($t-dt$) to t) to the age range (t to ($t+dt$)). Further, this in turn will be equal to the area in the age group (($t-dt$) to t), less that replaced by fresh surface in time dt, or

$$f(t)dt = f(t-dt)dt - [f(t-dt)dt]s\,dt \tag{2.116}$$

Thus,

$$\frac{f(t)-f(t-dt)}{dt} = -sf(t-dt)$$

$$\therefore f'(t) + sf(t) = 0 \quad (\text{as } dt \to 0) \tag{2.117}$$

$$\therefore e^{st}f(t) = \text{constant}$$

$$\therefore f(t) = \text{constant } e^{-st}$$

The total area of surface considered is unity, and hence,

$$\int_0^\infty f(t)dt = \text{constant} \int_0^\infty e^{-st}dt = 1 \tag{2.118}$$

$$\therefore \text{constant} \times \frac{1}{s} = 1$$

and

$$f(t) = s\, e^{-st} \tag{2.119}$$

Thus, the age distribution of the surface is of an exponential form. From Eq. (2.113), the mass-transfer rate at unit area of surface of age t is given by

$$(N_A)_t = (C_{Ai} - C_{Ao})\sqrt{\frac{D}{\pi t}} \quad (\text{Eq. 2.113})$$

Thus, the overall rate of transfer per unit area when the surface is renewed in a random manner is

$$N_A = (C_{Ai} - C_{Ao})\int_{t=0}^{t=\infty} \sqrt{\frac{D}{\pi t}}s\, e^{-st}dt$$

or

$$N_A = (C_{Ai} - C_{Ao})s\sqrt{\frac{D}{\pi}}\int_0^\infty t^{-1/2}e^{-st}dt \tag{2.120}$$

Putting $st = \beta^2$, then $s\, dt = 2\beta\, d\beta$, and

$$N_A = (C_{Ai} - C_{Ao})s\sqrt{\frac{D}{\pi}}\frac{2}{s^{1/2}}\int_0^\infty e^{-\beta^2}\, d\beta$$

Then, since the value of the integral is $\sqrt{\pi}/2$, then

$$N_A = (C_{Ai} - C_{Ao})\sqrt{Ds} \tag{2.121}$$

Eq. (2.121) might be expected to underestimate the mass-transfer rate because, in any practical equipment, there will be a finite upper limit to the age of any surface element. The proportion of the surface in the older age group is, however, very small, and the overall rate is largely unaffected. It is seen that the mass-transfer rate is again proportional to the concentration difference and to the square root of the diffusivity. The numerical value of s is difficult to estimate, although this will clearly increase as the fluid becomes more turbulent. In a packed column, s will be of the same order as the ratio of the velocity of the liquid flowing over the packing to the length of packing.

Varying interface composition

The penetration theory has been used to calculate the rate of mass transfer across an interface for conditions where the concentration C_{Ai} of solute **A** in the interfacial layers ($y = 0$) remained constant throughout the process. When there is no resistance to mass transfer in the other phase, for instance, when this consists of pure solute **A**, there will be no concentration gradient in that phase, and the composition at the interface will therefore at all times be the same as the bulk composition. Since the composition of the interfacial layers of the *penetration* phase is determined by the phase-equilibrium relationship, it, too, will remain constant, and the conditions necessary for the penetration theory to apply will hold. If, however, the other phase offers a significant resistance to transfer, this condition will not, in general, be fulfilled.

As an example, it may be supposed that in phase 1, there is a constant finite resistance to mass transfer that can in effect be represented as a resistance in a laminar film and in phase 2 the penetration model is applicable. Immediately after surface renewal has taken place, the mass-transfer resistance in phase 2 will be negligible, and therefore, the whole of the concentration driving force will lie across the film in phase 1. The interface compositions will therefore correspond to the bulk value in phase 2 (the penetration phase). As the time of exposure increases, the resistance to mass transfer in phase 2 will progressively increase, and an increasing proportion of the total driving force will lie across this phase. Thus, the interface composition, initially determined by the bulk composition in phase 2 (the penetration phase), will progressively approach the bulk composition in phase 1 as the time of exposure increases.

Because the boundary condition at $y = \infty$ ($C' = 0$) is unaltered by the fact that the concentration at the interface is a function of time, Eq. (2.106) is still applicable, although the evaluation of the constant B_2 is more complicated because $(C')_{y=0}$ is no longer constant.

$$\overline{C'} = B_2 e^{-\sqrt{(p/D)}\,y} \quad \text{(Eq. 2.106)}$$

$$\frac{d\overline{C'}}{dy} = -\sqrt{\frac{p}{D}}B_2 e^{-\sqrt{(p/D)}\,y}$$

so that

$$\left(\overline{C'}\right)_{y=0} = B_2 \tag{2.122}$$

and

$$\left(\frac{d\overline{C'}}{dy}\right)_{y=0} = -\sqrt{\frac{p}{D}}B_2 \tag{2.123}$$

In order to evaluate B_2, it is necessary to equate the mass-transfer rates on each side of the interface.

The mass-transfer rate per unit area across the film at any time t is given by

$$(N_A)_t = -\frac{D_f}{L_f}\left(C''_{Ai} - C''_{Ao}\right) \tag{2.124}$$

where D_f is the diffusivity in the film of thickness, L_f, and C''_{Ai} and C''_{Ao} are the concentrations of **A** at the interface and in the bulk.

The capacity of the film will be assumed to be small so that the holdup of solute is negligible. If Henry's law is applicable, the interface concentration in the second (penetration) phase is given by

$$C_{Ai} = \frac{1}{H}C''_{Ai} \tag{2.125}$$

where C_A is used to denote concentration in the penetration phase. Thus, by substitution, the interface composition C_{Ai} is obtained in terms of the mass-transfer rate $(N_A)_t$.

However, $(N_A)_t$ must also be given by applying Fick's law to the interfacial layers of phase 2 (the penetration phase).

Thus,

$$(N_A)_t = -D\left(\frac{\partial C_A}{\partial y}\right)_{y=0} \quad \text{(from Eq. 2.4)} \tag{2.126}$$

Combining Eqs (2.124)–(2.126),

$$(C_A)_{y=0} = C_{Ai} = \frac{C''_{Ao}}{H} + \frac{DL_f}{D_f H}\left(\frac{\partial C_A}{\partial y}\right)_{y=0}$$
(2.127)

Replacing C_A by $C' + C_{Ao}$,

$$(C')_{y=0} = \left(\frac{C''_{Ao}}{H} - C_{Ao}\right) + \frac{DL_f}{D_f H}\left(\frac{\partial C'}{\partial y}\right)_{y=0}$$

Taking Laplace transforms of each side, noting that the first term on the right-hand side is constant:

$$(\overline{C'})_{y=0} = \frac{1}{p}\left(\frac{C''_{Ao}}{H} - C_{Ao}\right) + \frac{DL_f}{D_f H}\left(\frac{d\overline{C'}}{dy}\right)_{y=0}$$
(2.128)

Substituting into Eq. (2.128) from Eqs (2.122), (2.123),

$$B_2 = \frac{1}{p}\left(\frac{C''_{Ao}}{H} - C_{Ao}\right) + \frac{DL_f}{D_f H}\left(-\sqrt{\frac{p}{D}}B_2\right)$$

or

$$B_2 = \frac{D_f}{\sqrt{D}L_f}[C''_{Ao} - HC_{Ao}]/p\left\{\frac{D_f H}{\sqrt{D}L_f} + \sqrt{p}\right\}$$
(2.129)

Substituting in Eq. (2.123),

$$\left(\frac{d\overline{C'}}{dy}\right)_{y=0} = \left(\frac{d\overline{C'}}{dy}\right)_{y=0} = -\frac{D_f}{DL_f}[C''_{Ao} - HC_{Ao}]/\sqrt{p}\left\{\frac{D_f H}{\sqrt{D}L_f} + \sqrt{p}\right\}$$
(2.130)

On inversion (see Appendix, Table 12, No. 43),

$$\left(\frac{dC'}{dy}\right)_{y=0} = -\frac{D_f}{DL_f}[C''_{Ao} - HC_{Ao}]e^{[(D_f^2 H^2)/(DL_f^2)]t}\mathrm{erfc}\sqrt{\frac{D_f^2 H^2 t}{DL_f^2}}$$
(2.131)

The mass-transfer rate at time t at the interface is then given by

$$(N_A)_t = -D\left(\frac{\partial C_A}{\partial y}\right)_{y=0} = -D\left(\frac{\partial C'}{\partial y}\right)_{y=0}$$

or

$$(N_A)_t = (C''_{Ao} - HC_{Ao})\frac{D_f}{L_f}\left(e^{[(D_f^2 H^2)/(DL_f^2)]t}\mathrm{erfc}\sqrt{\frac{D_f^2 H^2 t}{DL_f^2}}\right)$$
(2.132)

Average rates of mass transfer can be obtained, as previously, by using either the Higbie or the Danckwerts model for surface renewal.

Penetration model with laminar film at interface

Harriott[25] suggested that, as a result of the effects of interfacial tension, the layers of fluid in the immediate vicinity of the interface would frequently be unaffected by the mixing process postulated in the penetration theory. There would then be a thin laminar layer unaffected by the mixing process and offering a constant resistance to mass transfer. The overall resistance may be calculated in a manner similar to that used in the previous section where the total resistance to transfer was made up of two components—a film resistance in one phase and a penetration model resistance in the other. It is necessary in Eq. (2.132) to put the Henry's law constant equal to unity and the diffusivity D_f in the film equal to that in the remainder of the fluid D. The driving force is then $C_{Ai} - C_{Ao}$ in place of $C'_{Ao} - HC_{Ao}$, and the mass-transfer rate at time t is given for a film thickness L by

$$(N_A)_t = (C_{Ai} - C_{Ao}) \frac{D}{L} \left(e^{Dt/L^2} \operatorname{erfc} \sqrt{\frac{Dt}{L^2}} \right) \tag{2.133}$$

The average transfer rate according to the Higbie model for surface age distribution then becomes

$$N_A = 2(C_{Ai} - C_{Ao}) \sqrt{\frac{D}{\pi t_e}} \left[1 + \frac{1}{2} \sqrt{\frac{\pi L^2}{Dt_e}} \left(e^{Dt_e/L^2} \operatorname{erfc} \sqrt{\frac{Dt_e}{L^2}} - 1 \right) \right] \tag{2.134}$$

Using the Danckwerts model,

$$N_A = (C_{Ai} - C_{Ao}) \sqrt{Ds} \left(1 + \sqrt{\frac{L^2 s}{D}} \right)^{-1} \tag{2.135}$$

2.5.3 The Film-Penetration Theory

A theory that incorporates some of the principles of both the two-film theory and the penetration theory has been proposed by Toor and Marchello.[22] The whole of the resistance to transfer is regarded as lying within a laminar film at the interface, as in the two-film theory, but the mass transfer is regarded as an unsteady-state process. It is assumed that fresh surface is formed at intervals from fluid that is brought from the bulk of the fluid to the interface by the action of the eddy currents. Mass transfer then takes place as in the penetration theory, except that the resistance is confined to the finite film, and material that traverses the film is immediately completely mixed with the bulk of the fluid. For short times of exposure, when none of the diffusing material has reached the far side of the layer, the process is identical to that postulated in the penetration theory. For prolonged periods of exposure when a steady concentration gradient has developed, conditions are similar to those considered in the two-film theory.

The mass-transfer process is again governed by Eq. (2.66), but the third boundary condition is applied at $y=L$, the film thickness, and not at $y=\infty$. As before, the Laplace transform is then

$$\overline{C}' = B_1 e^{\sqrt{(p/D)}y} + B_2 e^{-\sqrt{(p/D)}y} \quad (\text{Eq.}2.105)$$

$$t>0 \quad y=0 \quad C_A = C_{Ai} \quad C' = C_{Ai} - C_{Ao} = C'_i \quad \overline{C}' = (1/p)C'_i$$
$$t>0 \quad y=L \quad C_A = C_{Ao} \quad C' = 0 \quad \overline{C}' = 0$$

Thus,

$$\frac{C'_i}{p} = B_1 + B_2$$

$$\therefore 0 = B_1 e^{\sqrt{(p/D)}L} + B_2 e^{-\sqrt{(p/D)}L}$$

$$\therefore B_1 = -B_2 e^{-2\sqrt{(p/D)}L} \quad \text{and} \quad B_2 = \frac{\overline{C}'_i}{p}\left(1 - e^{-2\sqrt{(p/D)}L}\right)^{-1}$$

$$B_1 = -\frac{C'_i}{p}e^{-2\sqrt{(p/D)}L}\left(1 - e^{-2\sqrt{(p/D)}L}\right)^{-1}$$

$$\overline{C}' = \frac{C'_i}{p}\left(1 - e^{-2\sqrt{(p/D)}L}\right)^{-1}\left(e^{-\sqrt{(p/D)}y} - e^{-\sqrt{(p/D)}(2L-y)}\right) \tag{2.136}$$

Since there is no inverse of Eq. (2.136) in its present form, it is necessary to expand using the binomial theorem. Noting that, since $2\sqrt{(p/D)}L$ is positive, $e^{-2\sqrt{(p/D)}L} < 1$, and from the binomial theorem,

$$\left(1 - e^{-2\sqrt{(p/D)}L}\right)^{-1} = \left\{1 + e^{-2\sqrt{(p/D)}L} + \cdots + e^{-2n\sqrt{(p/D)}L} + \cdots \text{to } \infty\right\}$$

$$= \sum_{n=0}^{n=\infty} e^{-2n\sqrt{(p/D)}L}$$

Substituting in Eq. (2.136),

$$\overline{C}' = \frac{C'_i}{p}\left(e^{-\sqrt{(p/D)}y} - e^{-\sqrt{(p/D)}(2L-y)}\right)\sum_{n=0}^{n=\infty} e^{-2n\sqrt{(p/D)}L}$$

$$= C'_i\left[\sum_{n=0}^{n=\infty}\frac{1}{p}e^{-\sqrt{(p/D)}(2nL+y)} - \sum_{n=0}^{n=\infty}\frac{1}{p}e^{-\sqrt{(p/D)}\{2(n+1)L-y\}}\right] \tag{2.137}$$

On inversion of Eq. (2.137),

$$\frac{C_A - C_{Ao}}{C_{Ai} - C_{Ao}} = \frac{C'}{C'_i} = \sum_{n=0}^{n=\infty} \text{erfc} \frac{(2nL+y)}{2\sqrt{Dt}} - \sum_{n=0}^{n=\infty} \text{erfc} \frac{2(n+1)L - y}{2\sqrt{Dt}} \tag{2.138}$$

Differentiating with respect to y,

$$\frac{1}{C_{Ai} - C_{Ao}} \frac{\partial C_A}{\partial y} = \sum_{n=0}^{n=\infty} -\frac{2}{\sqrt{\pi}} \frac{1}{2\sqrt{Dt}} e^{-(2nL+y)^2/(4Dt)}$$

$$- \sum_{n=0}^{n=\infty} \frac{2}{\sqrt{\pi}} \frac{1}{2\sqrt{Dt}} e^{-[2(n+1)L-y]^2/4(Dt)}$$

At the free surface, $y=0$, and

$$\frac{1}{C_{Ai} - C_{Ao}} \left(\frac{\partial C_A}{\partial y} \right)_{y=0} = -\frac{1}{\sqrt{\pi Dt}} \left(\sum_{n=0}^{n-\infty} e^{-(n^2 L^2)/(Dt)} + \sum_{n=0}^{n=\infty} e^{-[(n+1)^2 L^2]/(Dt)} \right)$$

$$= -\frac{1}{\sqrt{\pi Dt}} \left(1 + 2 \sum_{n=1}^{n=\infty} e^{-(n^2 L^2)/(Dt)} \right)$$

Now,

$$\sum_{n=0}^{n=\infty} e^{-(n^2 L^2)/Dt} = 1 + \sum_{n=0}^{n=\infty} e^{-(n^2 L^2)/Dt}$$

and

$$\sum_{n=0}^{n=\infty} e^{-[(n+1)^2 L^2]/Dt} = \sum_{n=1}^{n=\infty} e^{-(n^2 L^2)/Dt}$$

The mass-transfer rate across the interface per unit area is therefore given by

$$(N_A)_t = -D \left(\frac{\partial C_A}{\partial y} \right)_{y=0}$$

$$= (C_{Ai} - C_{Ao}) \sqrt{\frac{D}{\pi t}} \left(1 + 2 \sum_{n=1}^{n=\infty} e^{-(n^2 L^2)/(Dt)} \right) \tag{2.139}$$

Eq. (2.139) converges rapidly for high values of L^2/Dt. For low values of L^2/Dt, it is convenient to employ an alternative form by using the identity,[26] and

$$\sqrt{\frac{\alpha}{\pi}} \left(1 + 2 \sum_{n=1}^{n=\infty} e^{-n^2 \alpha} \right) \equiv 1 + 2 \sum_{n=1}^{n=\infty} e^{-(n^2 \pi^2)/\alpha} \tag{2.140}$$

Taking $\alpha = L^2/Dt$,

$$(N_A)_t = (C_{Ai} - C_{Ao})\frac{D}{L}\left(1 + 2\sum_{n=1}^{n=\infty} e^{-(n^2\pi^2 Dt)/L^2}\right) \tag{2.141}$$

It will be noted that Eqs (2.139), (2.141) become identical in form and in convergence when $L^2/Dt = \pi$:

$$\text{Then:} \quad (N_A)_t = (C_{Ai} - C_{Ao})\frac{D}{L}\left(1 + 2\sum_{n=1}^{n=\infty} e^{-n^2\pi}\right)$$

$$= (C_{Ai} - C_{Ao})\frac{D}{L}\left[1 + 2\left(e^{-\pi} + e^{-4\pi} + e^{-9\pi} + \cdots\right)\right]$$

$$= (C_{Ai} - C_{Ao})\frac{D}{L}[1 + 0.0864 + 6.92 \times 10^{-6} + 1.03 \times 10^{-12} + \cdots)$$

Thus, provided the rate of convergence is not less than that for $L^2/Dt = \pi$, all terms other than the first in the series may be neglected. Eq. (2.139) will converge more rapidly than this for $L^2/Dt > \pi$, and Eq. (2.141) will converge more rapidly for $L^2/Dt < \pi$:

Thus,

$$\pi < \frac{L^2}{Dt} < \infty \quad (N_A)_t = (C_{Ai} - C_{Ao})\sqrt{\frac{D}{\pi t}}\left(1 + 2e^{-L^2/Dt}\right) \tag{2.142}$$

$$0 < \frac{L^2}{Dt} < \pi \quad (N_A)_t = (C_{Ai} - C_{Ao})\frac{D}{L}\left(1 + 2e^{-(\pi^2 Dt)/L^2}\right) \tag{2.143}$$

It will be noted that the second terms in Eqs (2.142), (2.143) never exceeds 8.64% of the first term. Thus, with an error not exceeding 8.64%,

$$\pi < \frac{L^2}{Dt} < \infty \quad (N_A)_t = (C_{Ai} - C_{Ao})\sqrt{\frac{D}{\pi t}} \tag{2.144}$$

$$0 < \frac{L^2}{Dt} < \pi \quad (N_A)_t = (C_{Ai} - C_{Ao})\frac{D}{L} \tag{2.145}$$

The concentration profiles near an interface on the basis of

(a) the film theory (steady state),
(b) the penetration theory,
(c) the film-penetration theory

are shown in Fig. 2.7.

Thus, either the penetration theory or the film theory (Eq. 2.144 or 2.145), respectively, can be used to describe the mass-transfer process. The error will not exceed some 9% provided that the

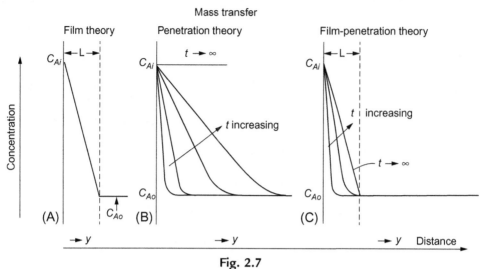

Fig. 2.7
Concentration profiles near an interface.

appropriate equation is used, Eq. (2.144) for $L^2/Dt > \pi$ and Eq. (2.145) for $L^2/Dt < \pi$. Eq. (2.145) will frequently apply quite closely in a wetted-wall column or in a packed tower with large packings. Eq. (2.144) will apply when one of the phases is dispersed in the form of droplets, as in a spray tower, or in a packed tower with small packing elements.

Eqs (2.142), (2.143) give the point value of N_A at time t. The average values N_A can then be obtained by applying the age distribution functions obtained by Higbie and by Danckwerts, respectively, as discussed Section 2.5.2.

2.5.4 Mass Transfer to a Sphere in a Homogenous Fluid

So far in this chapter, consideration has been given to transfer taking place in a single direction of a rectangular coordinate system. In many applications of mass transfer, one of the fluids is injected as approximately spherical droplets into a second immiscible fluid, and transfer of the solute occurs as the droplet passes through the continuous medium.

The case of a spherical drop of a pure liquid of radius r_0 being suddenly immersed in a gas, when the whole of the mass-transfer resistance is lying within the liquid, is now considered. It will be assumed that mass transfer is governed by Fick's law and that angular symmetry exists.

At radius r within sphere, mass-transfer rate $= -(4\pi r^2)D\left(\dfrac{\partial C_A}{\partial r}\right)$.

The change in mass-transfer rate over a distance dr

$$= \frac{\partial}{\partial r}\left(-4\pi r^2 D\frac{\partial C_A}{\partial r}\right) dr$$

Making a material balance over a shell gives

$$-\frac{\partial}{\partial r}\left(-4\pi r^2 D\frac{\partial C_A}{\partial r}\right)\,dr=\frac{\partial C_A}{\partial t}\left(4\pi r^2 dr\right)$$

$$\therefore \frac{\partial C_A}{\partial t}=D\frac{1}{r^2}\frac{\partial}{\partial r}\left(r^2\frac{\partial C_A}{\partial r}\right) \tag{2.146}$$

Eq. (2.146) may be solved by taking Laplace transforms for the boundary conditions:

$$
\begin{aligned}
&t=0 \quad 0<r<r_0 \quad C_A=0\\
&t>0 \quad r=r_0 \quad\quad C_A=C_{Ai}\ \text{(constant)}\\
&t>0 \quad r=0 \quad\quad \frac{dC_A}{dr}=0\ \text{(from symmetry)}
\end{aligned}
$$

The concentration gradient at the surface of the sphere $(r=r_0)$ is then found to be given by

$$\left(\frac{\partial C_A}{\partial r}\right)_{r=r_0}=\frac{-C_{Ai}}{r_0}+\frac{C_{Ai}}{\sqrt{\pi Dt}}\left(1+2\sum_{n=1}^{n=\infty}e^{-\left(n^2 r_0^2\right)/Dt}\right) \tag{2.147}$$

The mass-transfer rate at the surface of the sphere at time t

$$=4\pi r_0^2(-D)\left(\frac{\partial C_A}{\partial r}\right)_{r=r_0}$$

$$=4\pi r_0 D C_{Ai}\left[1-\frac{r_0}{\sqrt{\pi Dt}}\left(1+2\sum_{n=1}^{n=\infty}e^{-\left(n^2 r_0^2\right)/Dt}\right)\right] \tag{2.148}$$

The total mass transfer during the passage of a drop is therefore obtained by integration of Eq. (2.148) over the time of exposure.

2.5.5 Other Theories of Mass Transfer

Kishinevskij[23] has developed a model for mass transfer across an interface in which molecular diffusion is assumed to play no part. In this, fresh material is continuously brought to the interface as a result of turbulence within the fluid, and after exposure to the second phase, the fluid element attains equilibrium with it and then becomes mixed again with the bulk of the phase. The model thus presupposes surface renewal without penetration by diffusion, and therefore, the effect of diffusivity should not be important. No reliable experimental results are available to test the theory adequately.

2.5.6 Interfacial Turbulence

An important feature of the behaviour of the interface that has not been taken into account in the preceding treatment is the possibility of turbulence being generated by a means other than that associated with the fluid dynamics of the process. This *interfacial turbulence* may arise from local variations in the interfacial tension setup during the course of the mass-transfer process. It can occur at both gas–liquid and liquid–liquid interfaces, although the latter case has commanded the greater attention. Interfacial turbulence may give rise to violent intermittent eruptions at localised regions in the interface, as a result of which rapid mixing occurs, and the mass-transfer rate may be considerably enhanced.

The effect, which arises in cases where the interfacial tension is strongly dependent on the concentration of diffusing solute, will generally be dependent on the direction (sense) in which mass transfer is taking place.

If mass transfer causes interfacial tension to decrease, localised regions of low interfacial tension will form, and as a result, surface spreading will take place. If the gradient of interfacial tension is very high, the surface spreading will be rapid enough to give rise to intense ripples at the interface and a rapid increase in the mass-transfer rate. On the other hand, if the mass transfer gives rise to increased interfacial tension, the surface shows no tendency to spread and tends to be stable.

This phenomenon, frequently referred to as the *Marangoni effect*, explains some of the anomalously high mass-transfer rates reported in the literature.

The effect may be reduced by the introduction of surfactants that tend to concentrate at the interface where they exert a stabilising influence, although they may introduce an interface resistance and substantially reduce the mass-transfer rate. Thus, for instance, hexadecanol when added to open ponds of water will collect at the interface and substantially reduce the rate of evaporation.

Such effects are described in more detail by Sherwood, Pigford, and Wilke.[27]

2.5.7 Mass Transfer Coefficients

On the basis of each of the theories discussed, the rate of mass transfer in the absence of bulk flow is directly proportional to the driving force, expressed as a molar concentration difference, and therefore,

$$N_A = h_D(C_{Ai} - C_{Ao}) \tag{2.149}$$

where h_D is a mass-transfer coefficient (see Eq. 2.97a, 2.97b). In the two-film theory, h_D is directly proportional to the diffusivity and inversely proportional to the film thickness. According to the penetration theory, it is proportional to the square root of the diffusivity, and when all surface elements are exposed for an equal time, it is inversely proportional to the square root of time of exposure; when random surface renewal is assumed, it is proportional to the square root of the rate of renewal. In the film-penetration theory, the mass-transfer coefficient is a complex function of the diffusivity, the film thickness, and either the time of exposure or the rate of renewal of surface.

In most cases, the value of the transfer coefficient cannot be calculated from first principles, although the way in which the coefficient will vary as operating conditions are altered can frequently be predicted by using the theory that is most closely applicable to the problem in question.

The penetration and film-penetration theories have been developed for conditions of equimolecular counterdiffusion only; the equations are too complex to solve explicitly for transfer through a stationary carrier gas. For gas absorption, therefore, they apply only when the concentration of the material undergoing mass transfer is low. On the other hand, in the two-film theory, the additional contribution to the mass transfer that is caused by bulk flow is easily calculated, and h_D (Section 2.2.3) is equal to $(D/L)(C_T/C_{Bm})$ instead of D/L.

In a process where mass transfer takes place across a phase boundary, the same theoretical approach can be applied to each of the phases, though it does not follow that the same theory is best applied to both phases. For example, the film model might be applicable to one phase and the penetration model to the other. This problem is discussed in the previous section.

When the film theory is applicable to each phase (the two-film theory), the process is steady state throughout, and the interface composition does not then vary with time. For this case, the two-film coefficients can readily be combined. Because material does not accumulate at the interface, the mass-transfer rate on each side of the phase boundary will be the same, and for two phases, it follows that

$$N_A = h_{D1}(C_{Ao1} - C_{Ai1}) = h_{D2}(C_{Ai2} - C_{Ao2}) \tag{2.150}$$

If there is no resistance to transfer at the interface, C_{Ai1} and C_{Ai2} will be the corresponding values in the phase-equilibrium relationship.

Usually, the values of the concentration at the interface are not known, and the mass-transfer coefficient is considered for the overall process. Overall, transfer coefficients are then defined by

$$N_A = K_1(C_{Ao1} - C_{Ae1}) = K_2(C_{Ae2} - C_{Ao2}) \tag{2.151}$$

where C_{Ae1} is the concentration in phase 1 in equilibrium with C_{Ao2} in phase 2 and C_{Ae2} is the concentration in phase 2 in equilibrium with C_{Ao1} in phase 1. If the equilibrium relationship is linear,

$$H = \frac{C_{Ai1}}{C_{Ai2}} = \frac{C_{Ae1}}{C_{Ao2}} = \frac{C_{Ao1}}{C_{Ae2}} \tag{2.152}$$

where H is a proportionality constant.

The relationships between the various transfer coefficients are obtained as follows. From Eqs (2.150), (2.151),

$$\frac{1}{K_1} = \frac{1}{h_{D1}}\frac{C_{Ao1} - C_{Ae1}}{C_{Ao1} - C_{Ai1}} = \frac{1}{h_{D1}}\frac{C_{Ai1} - C_{Ae1}}{C_{Ao1} - C_{Ai1}} + \frac{1}{h_{D1}}\frac{C_{Ao1} - C_{Ai1}}{C_{Ao1} - C_{Ai1}}$$

But

$$\frac{1}{h_{D1}} = \frac{1}{h_{D2}}\frac{C_{Ao1} - C_{Ai1}}{C_{Ai2} - C_{Ao2}} \quad \text{(from Eq. 2.150)}$$

and hence,

$$\frac{1}{K_1} = \frac{1}{h_{D1}} + \frac{1}{h_{D2}}\left(\frac{C_{Ao1} - C_{Ai1}}{C_{Ai2} - C_{Ao2}}\right)\left(\frac{C_{Ai1} - C_{Ae1}}{C_{Ao1} - C_{Ai1}}\right)$$

From Eq. (2.152), $\dfrac{C_{Ai1} - C_{Ae1}}{C_{Ai2} - C_{Ao2}} = H$

$$\therefore \frac{1}{K_1} = \frac{1}{h_{D1}} + \frac{H}{h_{D2}} \tag{2.153}$$

Similarly,

$$\frac{1}{K_2} = \frac{1}{Hh_{D1}} + \frac{1}{h_{D2}} \tag{2.154}$$

and hence

$$\frac{1}{K_1} = \frac{H}{K_2} \tag{2.155}$$

It follows that when h_{D1} is large compared with h_{D2}, K_2, and h_{D2} are approximately equal, and, when h_{D2} is large compared with h_{D1}, K_1, and h_{D1} are almost equal.

These relations between the various coefficients are valid provided that the transfer rate is linearly related to the driving force and that the equilibrium relationship is a straight line. They are therefore applicable for the two-film theory and for any instant of time for the penetration and film-penetration theories. In general, application to time-averaged coefficients obtained from the penetration and film-penetration theories is not permissible because the condition at the interface will be time-dependent unless all of the resistance lies in one of the phases.

Example 2.8

Ammonia is absorbed at 1 bar from an ammonia-air stream by passing it up a vertical tube, down which dilute sulphuric acid is flowing. The following laboratory data are available:

Length of tube $= 825$ mm

Diameter of tube $= 15$ mm

Partial pressures of ammonia, at inlet $= 7.5$ kN/m^2; at outlet $= 2.0$ kN/m^2

Air rate $= 2 \times 10^{-5}$ kmol/s

What is the overall transfer coefficient K_G based on the gas phase?

Solution

Driving force at inlet $= 7500$ N/m^2

Driving force at outlet $= 2000$ N/m^2

Mean driving force $= \dfrac{(7500 - 2000)}{\ln(7.5/2.0)} = 4200$ N/m^2

Ammonia absorbed $= 2 \times 10^{-5} \left(\dfrac{7.5}{93.8} - \dfrac{2.0}{99.3} \right) = 1.120 \times 10^{-6}$ kmol/s

Wetted surface $= \pi \times 0.015 \times 0.825 = 0.0388$ m^2

Hence $K_G = \dfrac{(1.120 \times 10^{-6})}{(0.0388 \times 4200)} = \underline{\underline{6.87 \times 10^{-9}}}$ kmol/$\left[\text{m}^2 \text{s} \left(\text{N/m}^2 \right) \right]$.

2.5.8 Countercurrent Mass Transfer and Transfer Units

Mass-transfer processes involving two fluid streams are frequently carried out in a column; countercurrent flow is usually employed although cocurrent flow may be advantageous in some circumstances. There are two principal ways in which the two streams may be brought into contact in a continuous process so as to permit mass transfer to take place between them, and these are termed *stagewise processes* and *continuous differential contact processes*.

Stagewise Processes

In a stagewise process, the fluid streams are mixed together for a period long enough for them to come close to thermodynamic equilibrium, following which they are separated, and each phase is then passed countercurrently to the next stage where the process is repeated. An example of this type of process is the plate type of distillation column in which a liquid stream flows down the column, overflowing from each plate to the one below as shown in Fig. 2.8. The vapour passes up through the column, being dispersed into the liquid as it enters the plate, rises through the liquid, disengaging from the liquid in the vapour space above the liquid surface, and then passes upwards to repeat the process on the next plate. Mass transfer takes place as the bubbles rise through the liquid and as liquid droplets are thrown up into the vapour space above the liquid-vapour interface. On an *ideal plate*, the liquid and vapour streams leave in thermodynamic equilibrium with each another. In practice, equilibrium may not be achieved, and a *plate efficiency*, based on the compositions in either the liquid or vapour stream, is defined as the ratio of the actual change in composition to that that would have been achieved in an ideal stage. The use of stagewise processes for distillation, gas absorption, and liquid-liquid extraction is discussed in Volume 2.

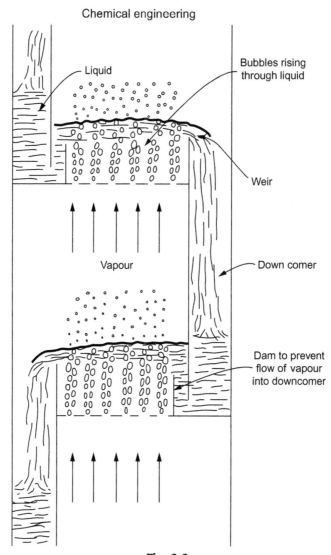

Chemical engineering

Fig. 2.8
Vapour and liquid flow in a plate column.

Continuous differential contact processes

In this process, the two streams flow countercurrently through the column and undergo a continuous change in composition. At any location is in *dynamic* rather than *thermodynamic* equilibrium. Such processes are frequently carried out in *packed columns*, in which the liquid (or one of the two liquids in the case of a liquid–liquid extraction process) wets the surface of the packing, thus increasing the interfacial area available for mass transfer and, in addition, promoting high film mass-transfer coefficients within each phase.

In a packed distillation column, the vapour stream rises against the downward flow of a liquid reflux, and a state of dynamic equilibrium is set up in a steady-state process. The more volatile constituent is transferred under the action of a concentration gradient from the liquid to the interface where it evaporates and then is transferred into the vapour stream. The less volatile component is transferred in the opposite direction, and if the molar latent heats of the components are equal, equimolecular counterdiffusion takes place.

In a packed absorption column, the flow pattern is similar to that in a packed distillation column, but the vapour stream is replaced by a mixture of carrier gas and solute gas. The solute diffuses through the gas phase to the liquid surface where it dissolves and is then transferred to the bulk of the liquid. In this case, there is no mass transfer of the carrier fluid, and the transfer rate of solute is supplemented by bulk flow.

In a liquid-liquid extraction column, the process is similar to that occurring in an absorption column except that both streams are liquids, and the lighter liquid rises through the denser one.

In distillation, equimolecular counterdiffusion takes place if the molar latent heats of the components are equal and the molar rate of flow of the two phases then remains approximately constant throughout the whole height of the column. In gas absorption, however, the mass-transfer rate is increased as a result of bulk flow, and at high concentrations of soluble gas, the molar rate of flow at the top of the column will be less than that at the bottom. At low concentrations, however, bulk flow will contribute very little to mass transfer, and in addition, flowrates will be approximately constant over the whole column.

The conditions existing in a column during the steady-state operation of a countercurrent process are shown in Fig. 2.9. The molar rates of flow of the two streams are G_1 and G_2, which will be taken as constant over the whole column. Suffixes 1 and 2 denote the two phases, and suffixes t and b relate to the top and bottom of the column.

If the height of the column is Z, its total cross-sectional area is S, and a is the interfacial area between the two phases per unit volume of column, then the rate of transfer of a component in a height dZ of column is given by

$$G_1 \frac{1}{C_T} dC_{Ao1} = h_{D1}(C_{Ai1} - C_{Ao1})Sa\,dZ \qquad (2.156)$$

$$\frac{dC_{Ao1}/dZ}{C_{Ai1} - C_{Ao1}} = \frac{h_{D1}aSC_T}{G_1} = \frac{h_{D1}aC_T}{G_1'} \qquad (2.157)$$

where G_1' is the molar rate of flow per unit cross-section of column. The exact interfacial area cannot normally be determined independently of the transfer coefficient, and therefore, values of the product $h_{D1}a$ are usually quoted for any particular system.

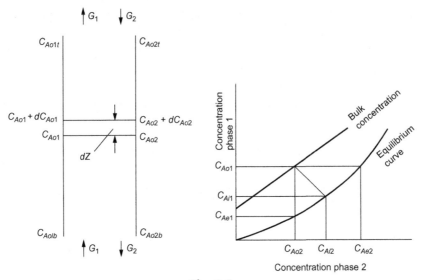

Fig. 2.9
Countercurrent mass transfer in a column.

The left-hand side of Eq. (2.157) is the rate of change of concentration with height for unit driving force and is therefore a measure of the efficiency of the column, and in this way, a high value of $(h_{D1}aC_T)/G_1'$ is associated with a high efficiency. The reciprocal of this quantity is $G_1'/(h_{D1}aC_T)$ that has linear dimensions and is known as the *height of the transfer unit* H_1 (HTU).

Rearranging Eq. (2.157) and integrating gives

$$\int_{(C_{Ao1_b})}^{(C_{Ao1_t})} \frac{dC_{Ao1}}{C_{Ai1} - C_{Ao1}} = \frac{h_{D1}aC_T}{G_1'}Z = \frac{Z}{G_1'/(h_{D1}aC_T)} = Z/H_1 \tag{2.158}$$

The right-hand side of Eq. (2.158) is the height of the column divided by the HTU, and this is known as the *number of transfer units N_1*. It is obtained by evaluating the integral on the left-hand side of the equation.

Therefore,

$$N_1 = \frac{Z}{H_1} \tag{2.159}$$

In some cases, such as the evaporation of a liquid at an approximately constant temperature or the dissolving of a highly soluble gas in a liquid, the interface concentration may be either substantially constant or negligible in comparison with that of the bulk. In such cases, the integral on the left-hand side of Eq. (2.158) may be evaluated directly to give

$$\int_{(C_{Ao1})_b}^{(C_{Ao1})_t} \frac{dC_{Ao1}}{C_{Ai1} - C_{Ao1}} = \ln \frac{(C_{Ai1} - C_{Ao1})_b}{(C_{Ai1} - C_{Ao1})_t} = \frac{Z}{\mathbf{H}_1} \tag{2.160}$$

where b and t represent values at the bottom and the top, as shown in Fig. 2.9.

Thus,

$$\frac{(C_{Ai1} - C_{Ao1})_b}{(C_{Ai1} - C_{Ao1})_t} = e^{Z/H_1} \tag{2.161}$$

Thus, H_1 is the height of column over which the *driving force changes by a factor of* e.

Eq. (2.156) can be written in terms of the film coefficient for the second phase (h_{D2}) or either of the overall transfer coefficients $(K_1$ and $K_2)$. Transfer units based on either film coefficient or overall coefficient can therefore be defined, and the following equations are analogous to Eq. (2.159):

Number of transfer units based on phase 2,

$$\mathbf{N}_2 = Z/\mathbf{H}_2 \tag{2.162}$$

Number of overall transfer units based on phase 1,

$$\mathbf{N}_{o1} = Z/\mathbf{H}_{o1} \tag{2.163}$$

Number of overall transfer units based on phase 2,

$$\mathbf{N}_{o2} = Z/\mathbf{H}_{o2} \tag{2.164}$$

Using this notation, the introduction of o into the suffix indicates an overall transfer unit.

The equations for H_1, H_2, H_{o1}, and H_{o2} are of the following form:

$$\mathbf{H}_1 = \frac{G_1'}{h_{D1}aC_T} \tag{2.165}$$

If one phase is a gas, as in gas absorption, for example, it is often more convenient to express concentrations as partial pressures in which case

$$\mathbf{H}_1 = \mathbf{H}_G = \frac{G_1'}{k_G a P} \tag{2.166}$$

where, in SI units, G_1' is expressed in $kmol/m^2$ s, k_G in $kmol/(m^2$ s $(N/m^2))$, P in N/m^2, and a in m^2/m^3.

The overall values of the HTU may be expressed in terms of the film values by using Eq. (2.153) that gives the relation between the coefficients:

$$\frac{1}{K_1} = \frac{1}{h_{D1}} + \frac{H}{h_{D2}} \quad (\text{Eq. 2.153})$$

Substituting for the coefficients in terms of the HTU,

$$\mathbf{H}_{o1}\frac{aC_T}{G_1'} = \mathbf{H}_1\frac{aC_T}{G_1'} + \frac{HaC_T}{G_2'}\mathbf{H}_2 \tag{2.167}$$

or

$$\mathbf{H}_{o1} = \mathbf{H}_1 + H\frac{G_1'}{G_2'}\mathbf{H}_2 \tag{2.168}$$

Similarly,

$$\mathbf{H}_{o2} = \mathbf{H}_2 + \frac{1}{H}\frac{G_2'}{G_1'}\mathbf{H}_1 \tag{2.169}$$

The advantage of using the transfer unit in preference to the transfer coefficient is that the former remains much more nearly constant as flow conditions are altered. This is particularly important in problems of gas absorption where the concentration of the solute gas is high and the flow pattern changes in the column because of the change in the total rate of flow of gas at different sections. In most cases, the coefficient is proportional to the flowrate raised to a power slightly less than unity, and therefore, the HTU is substantially constant.

As noted previously, for equimolecular counterdiffusion, the film transfer coefficients and hence the corresponding HTUs may be expressed in terms of the physical properties of the system and the assumed film thickness or exposure time, using the two-film, the penetration, or the film-penetration theories. For conditions where bulk flow is important, however, the transfer rate of constituent **A** is increased by the factor C_T/C_{Bm}, and the diffusion equations can be solved only on the basis of the two-film theory. In the design of equipment, it is usual to work in terms of transfer coefficients or HTUs and not to endeavour to evaluate them in terms of properties of the system.

2.6 Mass Transfer and Chemical Reaction in a Continuous Phase

In many applications of mass transfer, the solute reacts with the medium as in the case, for example, of the absorption of carbon dioxide in an alkaline solution. The mass-transfer rate then decreases in the direction of diffusion as a result of the reaction. Considering the unidirectional molecular diffusion of a component **A** through a distance δy over area **A**, then neglecting the effects of bulk flow, a material balance for an irreversible reaction of order n gives

$$\underset{\text{(at y)}}{\text{moles IN/unit time}} - \underset{\text{(at y}+\delta y)}{\text{moles OUT/unit time}} \quad = \text{rate of change of concn.} \times \text{element volume}$$

$$+ \text{reacted moles/unit volume} \times \text{element volume}$$

$$\left\{-D\frac{\partial C_A}{\partial y}\right\}A - \left\{-D\frac{\partial C_A}{\partial y} + \frac{\partial}{\partial y}\left[-D\frac{\partial C_A}{\partial y}\right]\delta y\right\}A = \frac{\partial C_A}{\partial t}(\partial y \cdot A) + \left(kC_A^n\right)(\delta y \cdot A)$$

or

$$\frac{\partial C_A}{\partial t} = D\frac{\partial^2 C_A}{\partial y^2} - kC_A^n \tag{2.170}$$

where k is the reaction rate constant. This equation has no analytical solution for the general case.

2.6.1 Steady-State Process

For a steady-state process, Eq. (2.170) becomes

$$D\frac{d^2 C_A}{dy^2} - kC_A^n = 0 \tag{2.171}$$

Eq. (2.171) may be integrated using the appropriate boundary conditions.

First-order reaction

For a first-order reaction, putting $n=1$ in Eq. (2.171), then

$$D\frac{d^2 C_A}{dy^2} - kC_A = 0 \tag{2.172}$$

The solution of Eq. (2.172) is

$$C_A = B_1' e^{\sqrt{(k/D)}y} + B_2' e^{-\sqrt{(k/D)}y} \tag{2.173}$$

B_1' and B_2' must then be evaluated using the appropriate boundary conditions.

As an example, consideration is given to the case where the fluid into which mass transfer is taking place is initially free of solute and is semiinfinite in extent. The surface concentration CA_i is taken as constant and the concentration at infinity as zero. The boundary conditions are therefore

$$y=0 \quad C_A = C_{Ai}$$
$$y=\infty \quad C_A = 0$$

Substituting these boundary conditions in Eq. (2.173) gives

$$B_1' = 0 \quad B_2' = C_{Ai}$$
$$C_A = C_{Ai}e^{-\sqrt{(k/D)}y} \tag{2.174}$$
$$\frac{dC_A}{dy} = -\sqrt{\frac{k}{D}}C_{Ai}e^{-\sqrt{(k/D)}y}$$

and

$$N_A = -D\frac{dC_A}{dy} = \sqrt{kD}C_{Ai}e^{-\sqrt{(k/D)}y} \tag{2.175}$$

At the interface, $y=0$, and

$$N_A = \sqrt{kD}C_{Ai} \tag{2.176}$$

nth-order reaction

The same boundary conditions will be used as for the first-order reaction. Eq. (2.171) may be rearranged to give

$$\frac{d^2C_A}{dy^2} - \frac{k}{D}C_A^n = 0 \tag{2.177}$$

Putting

$$\frac{dC_A}{dy} = q$$

$$\frac{d^2C_A}{dy^2} = \frac{dq}{dy} = \frac{dq}{dC_A}\frac{dC_A}{dy} = q\frac{dq}{dC_A}$$

and

$$q\frac{dq}{dC_A} - \frac{k}{D}C_A^n = 0$$

Multiplying through by dC_A and integrating,

$$\frac{q^2}{2} - \frac{1}{n+1}\frac{k}{D}C_A^{n+1} = B_3'$$

When $y=\infty$, $C_A=0$, and $\dfrac{dC_A}{dy}=q=0$ $\therefore B_3'=0$

and $\left(\dfrac{dC_A}{dy}\right)^2 = \dfrac{2}{n+1}\dfrac{k}{D}C_A^{n+1}$

Since $\dfrac{dC_A}{dy}$ is negative, the negative value of the square root will be taken to give

$$\frac{dC_A}{dy} = -\sqrt{\frac{2}{n+1}}\sqrt{\frac{k}{D}}C_A^{\frac{n+1}{2}} \tag{2.178}$$

$$N_A = -D\frac{dC_A}{dy} = \sqrt{\frac{2}{n+1}}\sqrt{kD}C_A^{\frac{n+1}{2}} \tag{2.179}$$

At the free surface, $C_A = C_{Ai}$, and

$$N_A = \sqrt{\frac{2}{n+1}} \sqrt{kD} C_{Ai}^{\frac{n+1}{2}} \tag{2.180}$$

which is identical to Eq. (2.176) for a first-order reaction when $n=1$.

Integrating Eq. (2.178) gives

$$C_A^{-\frac{n+1}{2}} dC_A = -\sqrt{\frac{2}{n+1}} \sqrt{\frac{k}{D}} dy$$

or

$$\frac{2}{1-n} C_A^{\frac{1-n}{2}} = -\sqrt{\frac{2}{n+1}} \sqrt{\frac{k}{D}} y + B_4'$$

When $y=0$, $C_A = C_{Ai}$, and $B_4' = \frac{2}{1-n} C_{Ai}^{\frac{1-n}{2}}$

and

$$C_A^{\frac{1-n}{2}} - C_{Ai}^{\frac{1-n}{2}} = (n-1) \sqrt{\frac{1}{2(n+1)}} \sqrt{\frac{k}{D}} y \tag{2.181}$$

This solution cannot be used for a first-order reaction where $n=1$ because it is then indeterminate.

Second-order reaction (n = 2)

In this case, Eq. (2.181) becomes

$$C_A^{-\frac{1}{2}} - C_{Ai}^{-\frac{1}{2}} = \sqrt{\frac{k}{6D}} y \tag{2.182}$$

and Eq. (2.180) becomes

$$N_A = \sqrt{\frac{2}{3}} \sqrt{kD} C_A^{\frac{3}{2}} \tag{2.183}$$

Example 2.9

In a gas absorption process, the solute gas **A** diffuses into a solvent liquid with which it reacts. The mass transfer is one of the steady-state unidirectional molecular diffusions, and the concentration of **A** is always sufficiently small for bulk flow to be negligible. Under these conditions, the reaction is first order with respect to the solute A.

At a depth l below the liquid surface, the concentration of **A** has fallen to one-half of the value at the surface. What is the ratio of the mass-transfer rate at this depth l to the rate at the surface? Calculate the numerical value of the ratio when $l\sqrt{k/D}=0.693$, where D is the molecular diffusivity and k the first-order rate constant.

Solution

This process is described by

$$C_A = B'_1 e^{\sqrt{(k/D)}y} + B'_2 e^{-\sqrt{(k/D)}y} \quad (\text{Eq.}\,2.173)$$

If C_{Ai} is the surface concentration ($y=0$),

$$C_{Ai} = B'_1 + B'_2$$

At $y=l$, $C_A=C_{Ai}/2$, and

$$\therefore \frac{C_{Ai}}{2} = B'_1 e^{\sqrt{(k/D)}l} + B'_2 e^{-\sqrt{(k/D)}l}$$

Solving for B'_1 and B'_2,

$$B'_1 = \frac{C_{Ai}}{2}\left(1 - 2e^{-\sqrt{(k/D)}t}\right)\left(e^{\sqrt{(k/D)}l} - e^{-\sqrt{(k/D)}l}\right)^{-1}$$

$$B'_2 = -\frac{C_{Ai}}{2}\left(1 - 2e^{\sqrt{(k/D)}t}\right)\left(e^{\sqrt{(k/D)}l} - e^{-\sqrt{(k/D)}l}\right)^{-1}$$

$$\frac{(N_A)_{y=1}}{(N_A)_{y=0}} = \frac{-D(dC_A/dy)_{y=l}}{-D(dC_A/dy)_{y=0}} = \frac{(dC_A/dy)_{y=l}}{(dC_A/dy)_{y=0}}$$

$$\frac{dC_A}{dy} = \sqrt{\frac{k}{D}}\left(B'_1 e^{\sqrt{(k/D)}y} - B'_2 e^{-\sqrt{(k/D)}y}\right)$$

$$\frac{(N_A)_{y=l}}{(N_A)_{y=0}} = \frac{B'_1 e^{\sqrt{(k/D)}l} - B'_2 e^{-\sqrt{(k/D)}l}}{B'_1 - B'_2}$$

$$= \frac{\left(1 - 2e^{-\sqrt{(k/D)}l}\right)e^{\sqrt{(k/D)}l} + \left(1 - 2e^{\sqrt{(k/D)}l}\right)e^{-\sqrt{(k/D)}l}}{\left(1 - 2e^{-\sqrt{(k/D)}l}\right) + \left(1 - 2e^{\sqrt{(k/D)}l}\right)}$$

$$= \frac{e^{\sqrt{(k/D)}l} + e^{-\sqrt{(k/D)}l} - 4}{2\left(1 - e^{-\sqrt{(k/D)}l} - e^{\sqrt{(k/D)}l}\right)}$$

When

$$l\sqrt{\tfrac{k}{D}}=0.693, \quad e^{\sqrt{(k/D)}l}=2, \quad e^{-\sqrt{(k/D)}l}=0.5$$

and

$$\frac{(N_A)_{y=1}}{(N_A)_{y=0}} = \frac{2 + \dfrac{1}{2} - 4}{2\left(1 - 2 - \dfrac{1}{2}\right)} = \underline{\underline{0.5}}$$

Example 2.10

In a steady-state process, a gas is absorbed in a liquid with which it undergoes an irreversible reaction. The mass-transfer process is governed by Fick's law, and the liquid is sufficiently deep for it to be regarded as effectively infinite in depth. On increasing the temperature, the concentration of reactant at the liquid surface C_{Ai} falls to 0.8 times its original value. The diffusivity is unchanged, but the reaction constant increases by a factor of 1.35. It is found that the mass-transfer rate at the liquid surface falls to 0.83 times its original value. What is the order of the chemical reaction?

Solution

The mass-transfer rate (moles/unit area and unit time) is given by Eq. (2.180), where denoting the original conditions by subscript 1 and the conditions at the higher temperature by subscript 2 gives

$$N_{A1} = \sqrt{\frac{2}{n+1}} \sqrt{k_1 D C_{Ai}^{\frac{n+1}{2}}} \quad \text{(Eq. 2.180)}$$

and

$$N_{A2} = 0.83 N_{A1} = \sqrt{\frac{2}{n+1}} \sqrt{1.35 k_1 D (0.8 C_{Ai})^{\frac{n+1}{2}}}$$

Substituting the numerical values gives,

$$0.83 = \sqrt{1.35} (0.8)^{\frac{n+1}{2}}$$

or

$$0.8^{\frac{n+1}{2}} = 0.714$$

Thus,

$$\frac{n+1}{2} = 1.506$$

and

$$n = 2.01$$

Thus, the reaction is of second – order.

Example 2.11

A pure gas is absorbed into a liquid with which it reacts. The concentration in the liquid is sufficiently low for the mass transfer to be covered by Fick's law, and the reaction is first order with respect to the solute gas. It may be assumed that the film theory may be applied to the liquid and that the concentration of solute gas falls from the saturation value to zero across the film. The reaction is initially carried out at 293 K. By what factor will the mass-transfer rate across the interface change, if the temperature is raised to 313 K?

The reaction rate constant at 293 K $= 2.5 \times 10^{-6} \, s^{-1}$

The energy of activation for reaction in the Arrhenius equation $=26430$ kJ/kmol $= 2.643 \times 10^7$ J/kmol

Universal gas constant, $\qquad$ **R** $= 8314$ J/kmol K

Molecular diffusivity, $\qquad$ $D = 10^{-9}$ m²/s

Film thickness, $\qquad$ $L = 10$ mm

Solubility of gas at 313 K is 80% of the solubility at 293 K.

Solution
For a first-order reaction,

$$D\frac{\partial^2 C_A}{\partial y^2} - kC_A = 0 \quad (\text{Eq.2.170})$$

Solving

$$C_A = B_1' e^{\sqrt{(k/D)}y} + B_2' e^{-\sqrt{(k/D)}y} \quad (\text{Eq.2.171})$$

When

$$y = 0, C_A = C_{AS} \text{ and when } y = L, C_A = 0.$$

Substituting

$$B_1' = -C_{AS} e^{-2L\sqrt{(k/D)}} \left(1 - e^{-2L\sqrt{(k/D)}}\right)^{-1}$$

and

$$B_2' = C_{AS} \left(1 - e^{-2L\sqrt{(k/D)}}\right)^{-1}$$

and thus,

$$\frac{C_A}{C_{AS}} = \left(e^{-\sqrt{(k/D)}y} - e^{-2L\sqrt{(k/D)}}\right)\left(1 - e^{-2L\sqrt{(k/D)}}\right)^{-1}$$

Differentiating and putting $y = 0$,

$$\frac{1}{C_{AS}}\left(\frac{dC_A}{dy}\right)_{y=0} = \frac{\left(-\sqrt{(k/D)} - \sqrt{(k/D)}e^{-2L\sqrt{(k/D)}}\right)}{\left(1 - e^{-2L}\sqrt{(k/D)}\right)}$$

The mass-transfer rate at the interface is then

$$-D\left(\frac{dC_A}{dy}\right)_{y=0} = C_{AS}\sqrt{(k/D)}\left(1 + e^{-2L\sqrt{(k/D)}}\right)\left(1 - e^{-2L\sqrt{(k/D)}}\right)^{-1}$$

When $D = 1 \times 10^{-9}$ m²/s, then at 293 K,

$$k = Ae^{-2.643 \times 10^7/(8314 \times 293)}$$

$$= 1.94 \times 10^{-5}A = 2.5 \times 10^{-6}\text{s}^{-1}$$

and $A = 0.129s^{-1}$

Since $L = 0.01$ m, then

$$2L\sqrt{(k/D)} = 2 \times 0.01(2.5 \times 10^{-6}/(1 \times 10^{-3}))^{0.5} = 1.0$$
$$e^{-2L\sqrt{(k/D)}} = e^{-1} = 0.368$$

and $C_{AS} = C_{AS1}$

The mass-transfer rate is then

$$N_{A_{293}} = C_{AS_1} \left(\frac{2.5 \times 10^{-6}}{1 \times 10^{-9}}\right)^{0.5} \frac{(1 + 0.368)}{(1 - 0.368)}$$

$$= 108.2 C_{AS_1}$$

At 313 K,

$$k = Ae^{(-2.643 \times 10^7/(8314 \times 313))} = (0.129 \times 3.37 \times 10^{-5}) = 5.0 \times 10^{-6}s^{-1}$$

$$C_{AS} = 0.8 C_{AS_1}$$

$$2L\sqrt{(k/D)} = (2 \times 0.01(5 \times 10^{-6}))/(1 \times 10^{-9})^{0.5} = 1.414$$

and $e^{-1.414} = 0.243$

The mass-transfer rate at 313 K is then

$$N_{A_{313}} = 0.8 C_{AS_1}(5 \times 10^6 \times 1 \times 10^{-9})^{0.5}(1 + 0.243)/(1 - 0.243)$$
$$= 92.9 C_{AS_1}$$

Hence, the change in the mass-transfer rate is given by the factor

$$N_{A_{313}}/N_{A_{293}} = (92.9 C_{AS_1}/108.2 C_{AS_1})$$

$$= \underline{\underline{0.86}}$$

2.6.2 Unsteady-State Process

For an unsteady-state process, Eq. (2.170) may be solved analytically only in the case of a first-order reaction ($n = 1$). In this case,

$$\frac{\partial C_A}{\partial t} = D\frac{\partial^2 C_A}{\partial y^2} - kC_A \tag{2.184}$$

The solution of this equation has been discussed by Danckwerts,[28] and here, a solution will be obtained using the Laplace transform method for a semiinfinite liquid initially free of solute. On the assumption that the liquid is in contact with pure solute gas, the concentration C_{Ai} at the liquid interface will be constant and equal to the saturation value. The boundary conditions will be those applicable to the penetration theory, that is,

$$t=0 \quad 0<y<\infty \quad C_A=0$$
$$t>0 \qquad y=0 \qquad C_A=C_{Ai}$$
$$t>0 \qquad y=\infty \qquad C_A=0$$

From Eqs (2.103), (2.104), taking Laplace transforms of both sides of Eq. (2.184) gives

$$p\overline{C}_A = D\frac{d^2\overline{C}_A}{dy^2} - k\overline{C}_A$$

or

$$\frac{d^2\overline{C}_A}{dy^2} - \frac{p+k}{D}\overline{C}_A = 0$$

Thus,

$$\overline{C}_A - B_1 e^{\sqrt{(p+k)/Dy}} + B_2 e^{-\sqrt{(p+k)/Dy}} \tag{2.185}$$

When $y=\infty$, $C_A=0$, and therefore, $\overline{C}_A=0$, from which $B_1=0$.

When $y=0$, $C_A=C_{Ai}$, and $\overline{C}_A=C_{Ai}/p$, from which $B_2=C_{Ai}/p$.

Eq. (2.185) therefore becomes

$$\overline{C}_A = \frac{C_{Ai}}{p}e^{-\sqrt{(p+k)/Dy}} \tag{2.186}$$

The mass-transfer rate N_A at the interface must be evaluated in order to obtain the rate at which gas is transferred to the liquid from the gas.

Differentiating Eq. (2.186) with respect to y gives

$$\frac{d\overline{C}_A}{dy} = \frac{\overline{dC_A}}{dy} = -\sqrt{\frac{p+k}{D}}\frac{C_{Ai}}{p}e^{-\sqrt{(p+k)/Dy}} \tag{2.187}$$

At the interface $(y=0)$,

$$\left(\frac{\overline{dC_A}}{dy}\right)_{y=0} = -\frac{C_{Ai}}{\sqrt{D}}\frac{\sqrt{p+k}}{p} \tag{2.188}$$

It is not possible to invert Eq. (2.188) directly using the transforms listed in Table 12 in Appendix A4. On putting $a=\sqrt{k}$, however, entry number 38 gives

$$\text{Inverse of } \frac{\sqrt{p}}{p-k} = \frac{1}{\sqrt{\pi t}} + \sqrt{k}e^{kt}\text{erf}\left(\sqrt{kt}\right)$$

From the *shift theorem*, if $\overline{f(t)} = \overline{f}(p)$, then

$$\overline{f(t)e^{-kt}} = \int_0^\infty f(t)e^{-kt}e^{-pt}\mathrm{d}t = \int_0^\infty f(t)e^{-(p+k)t}\mathrm{d}t = \overline{f}(p+k)$$

Thus, inverse of $\dfrac{\sqrt{p+k}}{p}$ is $\dfrac{1}{\sqrt{\pi t}}e^{-kt} + \sqrt{k}\,\mathrm{erf}\left(\sqrt{kt}\right)$.

Hence, inverting Eq. (2.188) gives

$$\left(\frac{\mathrm{d}C_A}{\mathrm{d}y}\right)_{y=0} = -\frac{C_{Ai}}{\sqrt{D}}\left\{\frac{1}{\sqrt{\pi t}}e^{-kt} + \sqrt{k}\,\mathrm{erf}\,\sqrt{kt}\right\}$$

and

$$(N_A)_t = -D\left(\frac{\mathrm{d}C_A}{\mathrm{d}y}\right)_{y=0} = C_{Ai}\sqrt{\frac{D}{k}}\left\{\sqrt{\frac{k}{\pi t}}e^{-kt} + k\,\mathrm{erf}\,\sqrt{kt}\right\} \tag{2.189}$$

Eq. (2.189) gives the instantaneous value of $(N_A)_t$ at time t.

The average value N_A for mass transfer over an exposure time t_e is given by

$$N_A = \frac{1}{t_e}\int_0^{t_e} N_A\mathrm{d}t = \frac{1}{t_e}C_{Ai}\sqrt{\frac{D}{k}}\int_0^{t_e}\left\{\sqrt{\frac{k}{\pi t}}e^{-kt} + k\,\mathrm{erf}\,\sqrt{kt}\right\}\mathrm{d}t \tag{2.190}$$

The terms in the integral of Eq. (2.190) cannot be evaluated directly. Although the integral can be rearranged to give three terms ((i), (ii), and (iii)), each of which can be integrated, by both adding and subtracting $kt\sqrt{(k/\pi t)}e^{-kt}$ and splitting the first term into two parts. This gives

$$N_A = \frac{1}{t_e}C_{Ai}\sqrt{\frac{D}{k}}\int_0^{t_e}\left\{\underbrace{\left(k\,\mathrm{erf}\,\sqrt{kt} + kt\sqrt{\frac{k}{\pi t}}e^{-kt}\right)}_{\text{(i)}}\right.$$

$$\left. + \underbrace{\left(-kt\sqrt{\frac{k}{\pi t}}e^{-kt} + \frac{1}{2}\sqrt{\frac{k}{\pi t}}e^{-kt}\right)}_{\text{(ii)}} + \underbrace{\frac{1}{2}\sqrt{\frac{k}{\pi t}}e^{-kt}}_{\text{(iii)}}\right\}\mathrm{d}t$$

Considering each of the above terms ((i), (ii), (iii)) in turn,

(i) Noting that $\dfrac{\mathrm{d}}{\mathrm{d}t}\left(k\,\mathrm{erf}\,\sqrt{kt}\right) = k\dfrac{\mathrm{d}}{\mathrm{d}t}\left\{\dfrac{2}{\sqrt{\pi}}\displaystyle\int_0^{\sqrt{kt}} e^{-kt}\mathrm{d}\left(\sqrt{kt}\right)\right\}$

$$= k\frac{\mathrm{d}}{\mathrm{d}t}\left\{\frac{2}{\sqrt{\pi}}\int_0^t e^{-kt}\left(\frac{1}{2}\sqrt{\frac{k}{t}}\right)\mathrm{d}t\right\} = k\sqrt{\frac{k}{\pi t}}e^{-kt}$$

$$\int_0^{t_e}\left(k\,\mathrm{erf}\,\sqrt{kt} + kt\sqrt{\frac{k}{\pi t}}e^{-kt}\right)\mathrm{d}t = k\int_0^{t_e}\left(\frac{\mathrm{d}}{\mathrm{d}t}\left(t\,\mathrm{erf}\,\sqrt{kt}\right)\right)\mathrm{d}t = kt_e\,\mathrm{erf}\,\sqrt{kt_e}$$

(ii) $$\int_0^{t_e} \left(\sqrt{\frac{kt}{\pi}}(-ke^{-kt}) + \frac{1}{2}\sqrt{\frac{k}{\pi t}}e^{-kt} \right) dt = \int_0^{t_e} \left(\frac{d}{dt}\left(\sqrt{\frac{kt}{\pi}}e^{-kt} \right) \right) dt = \sqrt{\frac{kt_e}{\pi}}e^{-kt_e}$$

(iii) $$\int_0^{t_e} \left(\frac{1}{2}\sqrt{\frac{k}{\pi t}}e^{-kt} \right) dt = \int_0^{t_e} \left(\frac{1}{2}\sqrt{\frac{k}{\pi t}}e^{-kt} \right)\left(2\sqrt{\frac{t}{k}}d\sqrt{kt} \right)$$

$$= \frac{1}{2}\cdot\frac{2}{\sqrt{\pi}}\int_0^{t_e} e^{-kt}d\sqrt{kt} = \frac{1}{2}\mathrm{erf}\sqrt{kt_e}$$

Thus, $N_A = \dfrac{1}{t_e}C_{Ai}\sqrt{\dfrac{D}{k}}\left\{ kt_e\,\mathrm{erf}\sqrt{kt_e} + \sqrt{\dfrac{kt_e}{\pi}}e^{-kt_e} + \dfrac{1}{2}\mathrm{erf}\sqrt{kt_e} \right\}$

$$= C_{Ai}\sqrt{\frac{D}{k}}\left\{ \left(k + \frac{1}{2t_e} \right)\mathrm{erf}\sqrt{kt_e} + \sqrt{\frac{k}{\pi t_e}}e^{-kt_e} \right\} \qquad (2.191)$$

Thus, the mass-transfer coefficient, enhanced by chemical reaction, h'_D, is given by

$$h'_D = \frac{N_A}{C_{Ai}} = \sqrt{\frac{D}{k}}\left\{ \left(k + \frac{1}{2t_e} \right)\mathrm{erf}\sqrt{kt_e} + \sqrt{\frac{k}{\pi t_e}}e^{-kt_e} \right\} \qquad (2.192)$$

Two special cases are now considered:

(1)   *When the reaction rate is very low* and $k \to 0$.
Using the Taylor series, for small values of $\sqrt{kt_e}$, and therefore neglecting higher powers of $\sqrt{kt_e}$,

$$\mathrm{erf}\sqrt{kt_e} = \left(\mathrm{erf}\sqrt{kt_e} \right)_{\sqrt{kt_e} \to 0} + \left[\sqrt{kt_e}\left\{ \frac{d}{d\sqrt{kt}}\left(\mathrm{erf}\sqrt{kt} \right) \right\} \right]_{\sqrt{kt} \to 0}$$

Then,

$$\lim \left[\left(k + \frac{1}{2t_e} \right)\mathrm{erf}\sqrt{kt_e} \right]_{k \to 0} = \frac{1}{2t_e}\left[\left(\sqrt{kt_e} \right)\frac{d}{d(\sqrt{kt})}\left\{ \frac{2}{\sqrt{\pi}}\int_0^{t_e} e^{-kt}d\left(\sqrt{kt} \right) \right\} \right]_{k \to 0}$$

$$= \frac{1}{2t_e}\left[\frac{2}{\sqrt{\pi}}e^{-kt}\left(\sqrt{kt_e} \right) \right]_{k \to 0}$$

$$= \sqrt{\frac{k}{\pi t_e}}$$

Also,

$$\lim \left[\sqrt{\frac{k}{\pi t_e}}e^{-kt_e} \right]_{k \to 0} = \sqrt{\frac{k}{\pi t_e}}$$

Substituting in Eq. (2.192) gives

$$h'_D \rightarrow \sqrt{\frac{D}{k}}\left(2\sqrt{\frac{k}{\pi t_e}}\right) = 2\sqrt{\frac{D}{\pi t_e}} \tag{2.193}$$

as for mass transfer without chemical reaction (Eq. 2.115).

(2) *When the reaction rate is very high,* erf $\sqrt{kt_e} \rightarrow 1$, $k \gg (1/t_e)$, and from Eq. (2.192),

$$h'_D = \sqrt{\frac{D}{k}}\{[k \times 1] + 0\} = \sqrt{Dk} \tag{2.194}$$

and is independent of exposure time.

2.7 Mass Transfer and Chemical Reaction in a Catalyst Pellet

When an irreversible chemical reaction is carried out in a packed or fluidised bed composed of catalyst particles, the overall reaction rate is influenced by

(i) the chemical kinetics,
(ii) mass-transfer resistance within the pores of the catalyst particles, and
(iii) resistance to mass transfer of the reactant to the outer surface of the particles.

In general, the concentration of the reactant will decrease from C_{Ao} in the bulk of the fluid to C_{Ai} at the surface of the particle, to give a concentration driving force of $(C_{Ao} - C_{Ai})$. Thus, within the pellet, the concentration will fall progressively from C_{Ai} with distance from the surface. This presupposes that no distinct adsorbed phase is formed in the pores. In this section, the combined effects of mass transfer and chemical reaction within the particle are considered, and the effects of external mass transfer are discussed in Section 2.9.4.

In the absence of a resistance to mass transfer, the concentration of reactant will be uniform throughout the whole volume of the particle and equal to that at its surface (C_{Ai}).

The reaction rate per unit volume of particle $\mathcal{R}'_{vn}$ for an nth-order reaction is then given by

$$\mathcal{R}'_{vn} = kC_{Ai}^n \tag{2.195}$$

When the mass-transfer resistance within the particle is significant, a concentration gradient of reactant is established within the particle, with the concentration, and hence the reaction rate, decreasing progressively with distance from the particle surface. The overall reaction rate is therefore less than that given by Eq. (2.195).

The internal structure of the catalyst particle is often of a complex labyrinth-like nature, with interconnected pores of a multiplicity of shapes and sizes. In some cases, the pore size may be less than the mean free path of the molecules, and both molecular and Knudsen diffusion may occur simultaneously. Furthermore, the average length of the diffusion path will be extended as a result of the tortuosity of the channels. In view of the difficulty of precisely defining the pore structure, the particle is assumed to be pseudohomogeneous in composition, and the diffusion process is characterised by an effective diffusivity D_e (Eq. 2.8).

The ratio of the overall rate of reaction to that that would be achieved in the absence of a mass-transfer resistance is referred to as the *effectiveness factor η*. Scott and Dullien[29] describe an apparatus incorporating a diffusion cell in which the effective diffusivity D_e of a gas in a porous medium may be measured. This approach allows for the combined effects of molecular and Knudsen diffusion and takes into account the effect of the complex structure of the porous solid and the influence of tortuosity that affects the path length to be traversed by the molecules.

The effectiveness factor depends not only on the reaction rate constant and the effective diffusivity but also on the size and shape of the catalyst pellets. In the following analysis, detailed consideration is given to particles of two regular shapes:

(i) *Flat platelets* in which the mass-transfer process can be regarded as one-dimensional, with mass transfer taking place perpendicular to the faces of the platelets. Furthermore, the platelets will be assumed to be sufficiently thin for deviations from unidirectional transfer due to end effects to be negligible.

(ii) *Spherical particles* in which mass transfer takes place only in the radial direction and in which the area available for mass transfer decreases towards the centre of the particle.

Some consideration is also given to particles of other shapes (e.g. long, thin cylinders), but the mathematics becomes complex, and no detailed analysis will be given. It will be shown, however, that the effectiveness factor does not critically depend on the shape of the particles, provided that their characteristic length is defined in an appropriate way. Some comparison is made between calculated results and experimental measurements with particles of frequently ill-defined shapes.

The treatment here is restricted to first-order irreversible reactions under steady-state conditions. Higher order reactions are considered by Aris.[30]

2.7.1 Flat Platelets

In a thin flat platelet, the mass-transfer process is symmetrical about the centre plane, and it is necessary to consider only one-half of the particle. Furthermore, again from considerations of symmetry, the concentration gradient and mass-transfer rate at the centre plane will be zero. The governing equation for the steady-state process involving a first-order reaction is obtained by substituting D_e for D in Eq. (2.172):

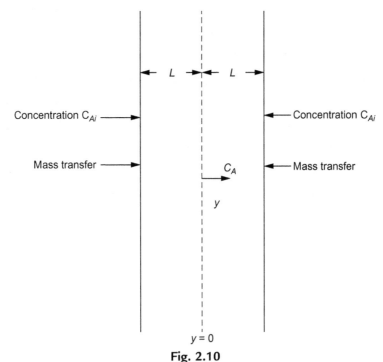

Fig. 2.10
Mass transfer and reaction in a platelet.

$$\frac{d^2 C_A}{dy^2} - \frac{k}{D_e} C_A = 0 \tag{2.196}$$

The process taking place only in the right-hand half of the pellet illustrated in Fig. 2.10 is considered. If the plate has a total thickness $2L$ and the centre plane is taken as the origin ($y=0$), then the following boundary conditions will apply:

$$y=0 \qquad \frac{dC_A}{dy}=0 \qquad \text{(from symmetry)}$$

$$y=L \qquad C_A = C_{Ai} \qquad \text{(the concentration at the surface of the particle)}$$

The general solution of Eq. (2.196) is

$$C_A = B_1' e^{\sqrt{\frac{k}{D_e}}y} + B_2' e^{-\sqrt{\frac{k}{D_e}}y} \tag{2.197}$$

Putting $\dfrac{k}{D_e} = \lambda^2$,

$$C_A = B_1' e^{\lambda y} + B_2' e^{-\lambda y}$$

and

$$\frac{dC_A}{dy} = \lambda \left(B_1' e^{\lambda y} - B_2' e^{-\lambda y} \right)$$

When $y = 0$, $\dfrac{dC_A}{dy} = 0$, and $B_1' = B_2'$.

And $C_A = B_2' \left(e^{\lambda y} + e^{-\lambda y} \right)$

When $y = L$,

$$C_A = C_{Ai}, \text{ and } B_2' = C_{Ai} \left(e^{\lambda L} + e^{-\lambda L} \right)^{-1}$$

and

$$\frac{C_A}{C_{Ai}} = \frac{e^{\lambda y} + e^{-\lambda y}}{e^{\lambda L} + e^{-\lambda L}} \tag{2.198}$$

The concentration gradient at a distance y from the surface is therefore given by

$$\frac{1}{C_{Ai}} \frac{dC_A}{dy} = \frac{\lambda \left(e^{\lambda y} - e^{-\lambda y} \right)}{e^{\lambda L} + e^{-\lambda L}}$$

The mass-transfer rate per unit area at the surface of the particle is then

$$(N_A)_{y=L} = -D_e \left(\frac{dC_A}{dy} \right)_{y=L} = -C_{Ai} D_e \lambda \frac{e^{\lambda L} - e^{-\lambda L}}{e^{\lambda L} + e^{-\lambda L}}$$

or

$$(N_A)_{y=L} = -C_{Ai} \sqrt{kD_e} \tanh \lambda L \tag{2.199}$$

The quantity $L\sqrt{\frac{k}{D_e}} = \lambda L$ is known as the *Thiele modulus*, ϕ.[31] The negative sign indicates that the transfer is in the direction of y negative, that is, towards the centre of the pellet. The rate of transfer of **A** (moles/unit area and unit time) from the external fluid to the surface of the half pellet (and therefore, in a steady-state process, the rate at which it is reacting) is therefore

$$(-N_A)_{y=L} = C_{Ai} \sqrt{kD_e} \tanh \lambda L \tag{2.200}$$

If there were no resistance to mass transfer, the concentration of **A** would be equal to C_{Ai} everywhere in the pellet, and the reaction rate per unit area in the half pellet (of volume per unit area equal to L) would be given by

$$\mathfrak{R}_p = kC_{Ai}L \tag{2.201}$$

The effectiveness factor η is equal to the ratio of the rates given by Eqs (2.200), (2.201), or

$$\eta = \frac{1}{\phi} \tanh \phi \tag{2.202}$$

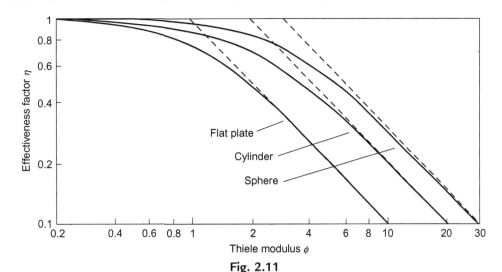

Fig. 2.11

Effectiveness factor η as a function of Thiele modulus ϕ ($\phi=\lambda L$ for platelet, $\phi=\lambda r_c$ for cylinder, and $\phi=\lambda r_o$ for sphere).

A logarithmic plot of the dimensionless quantities ϕ and η is shown in Fig. 2.11. It may be seen that η approaches unity at low values of ϕ and becomes proportional to ϕ^{-1} at high values of ϕ.

Three regions can be distinguished:

(i) $\phi <$ approximately 0.3 tanh $\phi \to \phi$, $\eta \to 1$ In this region, the mass-transfer effects are small, and the rate is determined almost entirely by the reaction kinetics.

(ii) $\phi >$ approximately 3 tanh $\phi \to 1$ $\eta \to \phi^{-1}$
Since $\phi = \sqrt{\frac{k}{D_e}}L$, this corresponds to a region with high values of k (implying high reaction rates), coupled with low values of D_e and high values of L (implying a high mass-transfer resistance). In this region, mass-transfer considerations are therefore of dominant importance, and reaction tends to be confined to a thin region close to the particle surface.

(iii) $0.3 < \phi < 3$
This is a transitional region in which reaction kinetics and mass-transfer resistance both affect the overall reaction rate.

2.7.2 Spherical Pellets

The basic differential equation for mass transfer accompanied by an nth-order chemical reaction in a spherical particle is obtained by taking a material balance over a spherical shell of inner radius r and outer radius $r+\delta r$, as shown in Fig. 2.12.

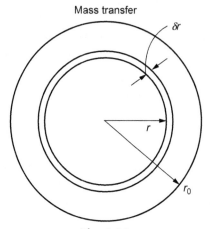

Fig. 2.12
Mass transfer and chemical reaction in a spherical particle.

At radius r, the mass-transfer rate (moles per unit time) is given by

$$-D_e \frac{\partial C_A}{\partial r} 4\pi r^2$$

The corresponding mass-transfer rate at radius $r+\delta r$ is

$$-D_e \frac{\partial C_A}{\partial r} 4\pi r^2 + \frac{\partial}{\partial r}\left\{-D_e \frac{\partial C_A}{\partial r} 4\pi r^2\right\}\delta r$$

The change in mass-transfer rate over the distance δr is therefore

$$4\pi D_e \frac{\partial}{\partial r}\left\{r^2 \frac{\partial C_A}{\partial r}\right\}\delta r$$

The rate of removal of **A** (moles/unit time) by chemical reaction is

$$kC_A^n \left(4\pi r^2 \delta r\right)$$

The rate of accumulation of **A** (moles/unit time) is given by

$$\frac{\partial C_A}{\partial t}\left(4\pi r^2 \delta r\right)$$

A material balance gives

$$4\pi D_e \frac{\partial}{\partial r}\left\{r^2 \frac{\partial C_A}{\partial r}\right\}\delta r = kC_A^n \left(4\pi r^2 \delta r\right) + \frac{\partial C_A}{\partial t}\left(4\pi r^2 \delta r\right)$$

that is,

$$\frac{\partial C_A}{\partial t} = D_e \frac{1}{r^2}\frac{\partial}{\partial r}\left(r^2 \frac{\partial C_A}{\partial r}\right) - kC_A^n \tag{2.203}$$

It may be noted that, in the absence of a chemical reaction, Eq. (2.203) reduces to Eq. (2.146). For a steady-state process, $\partial C_A/\partial t = 0$, and for a first-order reaction, $n = 1$. Thus,

$$\frac{d}{dr}\left(r^2\frac{dC_A}{dr}\right) = \frac{k}{D_e}r^2C_A = \lambda^2 r^2 C_A$$

where $\lambda^2 = k/D_e$.

Thus,

$$r^2\frac{d^2C_A}{dr^2} + 2r\frac{dC_A}{dr} = \frac{k}{D_e}r^2C_A = \lambda^2 r^2 C_A \tag{2.204}$$

or

$$\frac{d^2C_A}{dr^2} + \frac{2}{r}\frac{dC_A}{dr} - \lambda^2 C_A = 0 \tag{2.205}$$

Eq. (2.205) may be solved by putting $C_A = \dfrac{\psi}{r}$ to give

$$r^2\frac{d^2\left(\frac{\psi}{r}\right)}{dr^2} + 2r\frac{d\left(\frac{\psi}{r}\right)}{dr} = \lambda^2 r\psi \tag{2.206}$$

Since

$$\frac{d\left(\frac{\psi}{r}\right)}{dr} = \frac{1}{r}\frac{d\psi}{dr} - \frac{1}{r^2}\psi$$

and

$$\frac{d^2\left(\frac{\psi}{r}\right)}{dr^2} = \frac{1}{r}\frac{d^2\psi}{dr^2} - \frac{1}{r^2}\frac{d\psi}{dr} - \frac{1}{r^2}\frac{d\psi}{dr} + \frac{2}{r^3}\psi$$

Substituting in Eq. (2.206) gives

$$r\frac{d^2\psi}{dr^2} - 2\frac{d\psi}{dr} + \frac{2}{r}\psi + 2\frac{d\psi}{dr} - \frac{2}{r}\psi = \lambda^2 r$$

or

$$\frac{d^2\psi}{dr^2} - \lambda^2\psi = 0$$

Solving

$$\psi = B_1'' e^{\lambda r} + B_2'' e^{-\lambda r}$$

or

$$C_A = B_1'' \frac{1}{r} e^{\lambda r} + B_2'' \frac{1}{r} e^{-\lambda r}$$

and

$$\frac{dC_A}{dr} = B_1'' \left(\frac{1}{r} \lambda e^{\lambda r} - \frac{1}{r^2} e^{\lambda r} \right) + B_2'' \left(\frac{1}{r} (-\lambda) e^{-\lambda r} - \frac{1}{r^2} e^{-\lambda r} \right)$$

The boundary conditions are now considered. When $r = 0$, then from symmetry,

$$\frac{dC_A}{dr} = 0$$

giving

$$0 = B_1'' \left(-\frac{1}{r^2} e^0 \right) + B_2'' \left(-\frac{1}{r^2} e^0 \right) \text{ since } \left(\frac{1}{r^2} >> \frac{1}{r} \right)_{r \to 0}$$

Thus, $B_2'' = -B_1''$

When $r = r_0$,

$$C_A = C_{Ai}$$

and

$$C_{Ai} = B_1'' \frac{1}{r_0} e^{\lambda r_0} - B_1'' \frac{1}{r_0} e^{-\lambda r_0}$$

and

$$B_1'' = C_{Ai} r_0 \left(e^{\lambda r_0} - e^{-\lambda r_0} \right)^{-1}$$

Thus,

$$\frac{C_A}{C_{Ai}} = \frac{r_0}{r} \frac{e^{\lambda r} - e^{-\lambda r}}{e^{\lambda r_0} - e^{-\lambda r_0}}$$

and

$$\frac{C_A}{C_{Ai}} = \frac{r_0}{r} \frac{\sinh \lambda r}{\sinh \lambda r_0} \tag{2.207}$$

Differentiating with respect to r to obtain the concentration gradient,

$$\frac{1}{C_{Ai}} \frac{dC_A}{dr} = \frac{r_0}{\sinh \lambda r_0} \left\{ \frac{1}{r} \lambda \cosh \lambda r - \frac{1}{r^2} \sinh \lambda r \right\}$$

At the surface,

$$\left(\frac{dC_A}{dr}\right)_{r=r_0} = \frac{C_{Ai}r_0}{\sinh \lambda r_0}\left\{\frac{1}{r_0}\lambda \cosh \lambda r_0 - \frac{1}{r_0^2}\sinh \lambda r_0\right\}$$

$$= \frac{C_{Ai}}{r_0}\{\lambda r_0 \coth \lambda r_0 - 1\}$$

Thus, the mass transfer at the outer surface of the pellet (moles/unit time) is

$$-D_e\left(\frac{dC_A}{dr}\right)_{r=r_0}\left(4\pi r_0^2\right) = -C_{Ai}D_e 4\pi r_0\{\lambda r_0 \coth \lambda r_0 - 1\} \tag{2.208}$$

in the direction of r positive (away from the centre of the particle) and

$$C_{Ai}D_e 4\pi r_0\{\lambda r_0 \coth \lambda r_0 - 1\} \tag{2.209}$$

towards the centre of the pellet.

With no resistance to mass transfer, the concentration is C_{Ai} throughout the whole spherical pellet, and the reaction rate, which must be equal to the mass-transfer rate in a steady-state process, is

$$= \frac{4}{3}\pi r_0^3 k C_{Ai} \tag{2.210}$$

The effectiveness factor η is obtained by dividing Eq. (2.209) by Eq. (2.210) to give

$$\eta = \frac{C_{Ai}D_e 4\pi r_0\{\lambda r_0 \coth \lambda r_0 - 1\}}{\frac{4}{3}\pi r_0^3 k C_{Ai}}$$

$$= 3\frac{D_e}{k}\frac{1}{r_0^2}\{\lambda r_0 \coth \lambda r_0 - 1\}$$

$$= \frac{3}{r_0^2 \lambda^2}\{\lambda r_0 \coth \lambda r_0 - 1\} \tag{2.211}$$

Thus,

$$\eta = \frac{3}{\phi_0}\coth \phi_0 - \frac{3}{\phi_0^2} = \frac{3}{\phi_0}\left(\coth \phi_0 - \frac{1}{\phi_0}\right) \tag{2.212}$$

where $\phi_0 = \lambda r_0$ is the *Thiele modulus* for the spherical particle.

It is useful to redefine the characteristic linear dimension L of the spherical particle as its volume per unit surface area. This is, in effect, consistent with the definition of L adopted for the platelet where L is half its thickness. Then, for the sphere,

$$L = \frac{(4/3)\pi r_0^3}{4\pi r_0^2} = \frac{r_0}{3} \tag{2.213}$$

The Thiele modulus ϕ_L then becomes

$$\phi_L = \lambda L = \sqrt{\frac{k}{D_e}} L \tag{2.214}$$

Substituting for r_0 in Eq. (2.215), the effectiveness factor η may be written as

$$\eta = \frac{3}{9\lambda^2 L^2} \{3\lambda L \coth 3\lambda L - 1\}$$

$$= \frac{1}{\phi_L} \left(\coth 3\phi_L - \frac{1}{3\phi_L} \right) \tag{2.215}$$

A logarithmic plot of η versus ϕ_L in Fig. 2.13 shows that, using this definition of L, the curves for the slab or platelet and the spherical particle come very close together.

2.7.3 Other Particle Shapes

The relationship between effectiveness factor η and Thiele modulus ϕ_L may be calculated for several other regular shapes of particles, where again the characteristic dimension of the particle is defined as the ratio of its volume to its surface area. It is found that the results all fall quite closely together, irrespective of the particle shape. For a long cylindrical particle with a

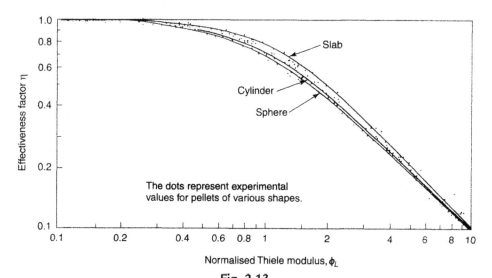

Fig. 2.13

Effectiveness factors η as a function of the normalised Thiele modulus $\phi_L = \lambda \frac{V_p}{A_p} = \lambda L$ for a first-order reaction.

high length to diameter ratio, the effects of diffusion at the end faces may be neglected, and the mass-transfer process may be regarded as taking place solely in the radial direction. The equation for mass transfer and chemical reaction in the cylinder may be derived in a manner analogous to that used for Eq. (2.205) for the sphere:

$$\frac{d^2 C_A}{dr^2} + \frac{1}{r}\frac{dC_A}{dr} - \lambda^2 C_A = 0 \tag{2.216}$$

The solution of this equation is in the form of a Bessel function.[32] Again, the characteristic length of the cylinder may be defined as the ratio of its volume to its surface area; in this case, $L = r_c/2$. It may be seen in Fig. 2.13 that, when the effectiveness factor η is plotted against the normalised Thiele modulus, the curve for the cylinder lies between the curves for the slab and the sphere. Furthermore, for these three particles, the effectiveness factor is not critically dependent on shape.

The results of investigations with particles of a variety of shapes, mainly irregular ones, have been reported by Rester and Aris,[33] and the results are shown as data points in Fig. 2.13. This provides additional evidence that the $\eta - \phi_L$ plot is not particularly sensitive to the shape of the particles, and a single curve can be used for most practical applications, particularly at high values of the Thiele modulus where reaction is confined to a thin region close to the surface whose curvature is then unimportant.

Example 2.12

Estimate the Thiele modulus and the effectiveness factor for a reactor in which the catalyst particles are

> Thin rectangular platelets, the ends of which are sealed so that mass transfer is unidirectional and perpendicular to the surface of the particle. The total thickness of the particles is 8 mm. Spherical particles is 10 mm in diameter.

The first-order rate constant is 5×10^{-4} s^{-1}, and the effective diffusivity of the reactants in the pores of the particles is 2×10^{-9} m^2/s.

Solution

For the reacting system, $\lambda = \sqrt{\dfrac{k}{D_e}} = \sqrt{\dfrac{5 \times 10^{-4}}{2 \times 10^{-9}}} = 500\,\text{m}^{-1}$.

(i) For the platelet of thickness 8 mm, $L = \left(\frac{1}{2} \times 8 \times 10^{-3}\right) = 0.004\,\text{m}^{-1}$, and the Thiele modulus $\phi = \lambda L = (500 \times 0.004) = 2.0$.

From Eq. (2.202), the effectiveness factor η is given by

$$\eta = \frac{1}{\phi}\tanh\phi = \frac{1}{2}\tanh 2 = \underline{\underline{0.482}}$$

(ii) For the sphere of diameter 10 mm, $r_0 = 0.005$ m^{-1}, and the Thiele modulus $\phi_0 = \lambda r_0 = (500 \times 0.005) = 2.5$.

From Eq. (2.212), the effectiveness factor η is given by

$$\eta = \frac{3}{\phi_0}\left(\coth\phi_0 - \frac{1}{\phi_0}\right) = \frac{3}{2.5}\left(\coth 2.5 - \frac{1}{2.5}\right) = \underline{\underline{0.736}}$$

Alternatively,

$$\phi_L = \lambda\frac{r_0}{3} = (500 \times 0.005)/3 = 0.833$$

and

$$\eta = \frac{1}{\phi_L}\left(\coth 3\phi_L - \frac{1}{3\phi_L}\right) = \frac{1}{0.833}\left(\coth(3 \times 0.833) - \frac{1}{3 \times 0.833}\right) \quad \text{(from Eq. 2.215)}$$
$$= \underline{\underline{0.736}}$$

Example 2.13

A first-order chemical reaction takes place in a reactor in which the catalyst pellets are platelets of thickness 5 mm. The effective diffusivity D_e for the reactants in the catalyst particle is 10^{-5} m²/s, and the first-order rate constant k is 14.4 s^{-1}.

Calculate

the effectiveness factor η and
the concentration of reactant at a position halfway between the centre and the outside of the pellet (i.e. at a position one quarter of the way across the particle from the outside).

Solution

For the particle, the Thiele modulus $\phi = \sqrt{\frac{k}{D_e}}L = \lambda L$

$$= \sqrt{\frac{14.4}{10^{-5}}} \times (2.5 \times 10^{-3}) = 3$$

From Eq. (2.202), the effectiveness factor, $\eta = \frac{1}{\phi}\tanh\phi$

$$= \frac{1}{3}\tanh 3 = 0.332$$

Thus, $\eta\lambda \approx 1$, corresponding to the region where mass-transfer effects dominate.

The concentration profile is given by Eq. (2.198) as

$$\frac{C_A}{C_{Ai}} = \frac{\cosh\lambda y}{\cosh\lambda L}$$

The concentration c_A at $y = 1.25 \times 10^{-3}$ m

$$= 0.015\frac{\cosh 1.5}{\cosh 3} = \underline{\underline{0.035\,\text{kmol/m}^3}}$$

2.7.4 Mass Transfer and Chemical Reaction With a Mass Transfer Resistance External to the Pellet

When the resistance to mass transfer to the external surface of the pellet is significant compared with that within the particle, part of the concentration driving force is required to overcome this external resistance, and the concentration of reacting material at the surface of the pellet C_{Ai} is less than that in the bulk of the fluid phase C_{Ao}. In Sections 2.7.1–2.7.3, the effect of mass-transfer resistance within a porous particle is expressed as an effectiveness factor, by which the reaction rate within the particle is reduced as a result of this resistance. The reaction rate per unit volume of particle (moles/unit volume and unit time) for a first-order reaction is given by

$$\mathfrak{R}_v = \eta k C_{Ai} \tag{2.217}$$

For a particle of volume V_p, the reaction rate per unit time for the particle is given by

$$\mathfrak{R}_p = \eta k V_p C_{Ai} \tag{2.218}$$

When there is an external mass-transfer resistance, the value of C_{Ai} (the concentration at the surface of the particle) is less than that in the bulk of the fluid (C_{Ao}) and will not be known. However, if the value of the external mass-transfer coefficient is known, the mass-transfer rate from the bulk of the fluid to the particle may be expressed as

$$\mathfrak{R}_p = h_D A_p (C_{Ao} - C_{Ai}) \tag{2.219}$$

giving

$$C_{Ai} = C_{Ao} - \frac{\mathfrak{R}_p}{h_D A_p} \tag{2.220}$$

Substituting in Eq. (2.219),

$$\mathfrak{R}_p = \frac{k C_{Ao}}{\dfrac{1}{\eta V_p} + \dfrac{k}{h_D A_p}} \tag{2.221}$$

Then, dividing by V_p gives the reaction rate $\mathfrak{R}_v$ (moles per unit volume of particle in unit time) as

$$\mathfrak{R}_v = \frac{k C_{Ao}}{\dfrac{1}{\eta} + \dfrac{k}{h_D} L} \tag{2.222}$$

where $L = V_p/A_p$, the particle length term used in the normalised Thiele modulus. Information on external mass-transfer coefficients for single spherical particle and particles in fixed beds is given in Section 2.9.4, and particles in fluidised beds are discussed in Volume 2.

Example 2.14

A hydrocarbon is cracked using a silica-alumina catalyst in the form of spherical pellets of mean diameter 2.0 mm. When the reactant concentration is 0.011 kmol/m^3, the reaction rate is 8.2×10^{-2} kmol/(m^3 catalyst) s. If the reaction is of first order and the effective diffusivity D_e is 7.5×10^{-8} m^2/s, calculate the value of the effectiveness factor η. It may be assumed that the effect of mass-transfer resistance in the fluid external to the particles may be neglected.

Solution

Since the value of the first-order rate constant is not given, λ and ϕ_L cannot be calculated directly. The reaction rate per unit volume of catalyst $\mathfrak{R}_v = \eta k C_{Ai}$ (Eq. 2.217)

and the Thiele modulus $\phi_L = \sqrt{\dfrac{k}{D_e}} L = \dfrac{\sqrt{k}}{\sqrt{7.5 \times 10^{-8}}} \dfrac{1 \times 10^{-3}}{3}$

$$= 1.217\sqrt{k} \tag{i}$$

The effectiveness factor η is given by Eq. (2.215) as

$$\eta = \frac{1}{\phi_L}\left(\coth 3\phi_L - \frac{1}{3\varphi_L}\right)$$

If $\phi_L > 3$,

$$\eta \approx \varphi_L^{-1} \tag{ii}$$

It is assumed that the reactor is operating in this regime and the assumption is then checked.

Substituting numerical values in Eq. (2.217),

$$8.2 \times 10^{-2} = \eta k(0.011) \tag{iii}$$

From Eqs (i), (ii),

$$\eta = \phi_L^{-1} = 0.822\frac{1}{\sqrt{k}}$$

From Eq. (iii),

$$8.2 \times 10^{-2} = 0.822\frac{1}{\sqrt{k}}k(0.011)$$

and $k = 82.2\,\text{s}^{-1}$

From equation (i), $\phi_L = 11.04$ and $\underline{\eta = 0.0906}$

This result may be checked by using Eq. (2.215):

$$\eta = \frac{1}{11.04}\left(\coth 33.18 - \frac{1}{3 \times 11.04}\right)$$
$$= \underline{\underline{0.0878}}$$

This value is sufficiently close for practical purposes to the value of 0.0906, calculated previously. If necessary, a second iteration may be carried out.

2.8 Taylor-Aris Dispersion

Taylor-Aris dispersion is the phenomenon that occurs when mass transfer of a given species occurs in laminar flow in tubes and channels. Such a configuration is of relevance in applications such as transport and distribution of injected drugs, transport in microfluidic devices (to be discussed in the following chapter), and also in the measurement of solute diffusion coefficients in small capillaries. The phenomenon discussed in this section is most pertinent under laminar flow conditions, and hence, it is assumed that laminar flow exists in a tube or rectangular channel. It is well known that the laminar flow velocity profile under steady, fully developed conditions is a parabolic function of the radial coordinate (for tubes) and wall-normal coordinate (for channels). The velocity is zero at the rigid walls of the tube and increases in a parabolic manner to reach a maximum at the centreline of the tube/channel. Now, let us imagine introducing a thin element of a dye (i.e. solute) at a given time instant for a short time duration. This protocol of dye injection is usually idealised as a 'pulse' input in reaction engineering and process dynamics. The solute particles present at various radial locations in the tube will move with different velocities in the absence of diffusion. Without diffusion, an initial disc (spanning the cross-section of the tube) will expand as a paraboloid, whose extent will increase in time. However, molecular diffusion of the solute molecules redistributes the solute molecules in both the direction along the flow (i.e. axial direction) and in the direction perpendicular to the flow (i.e. radial direction). The interplay between the nonuniform velocity distribution in the tube and the molecular diffusion of solute molecules results in a particular mechanism of dispersion referred to as the 'Taylor-Aris' dispersion. The approximate analysis of Taylor[34–36] resulted in an expression for mass flux along the tube axis, which is given by the sum of convective flux and the diffusive flux:

$$J = UC_m + \mathcal{D}_{eff} \frac{\partial C_m}{\partial z} \tag{2.223}$$

where J is the mass flux along the flow direction; U is the average velocity of the laminar flow; C_m is the average concentration of the solute over the tube cross-section; z is the flow direction; and $\mathcal{D}_{eff}$ is the effective diffusion coefficient, also referred to as the dispersion coefficient. For the case of Poiseuille flow in a tube, Taylor's analysis yielded an expression for $\mathcal{D}_{eff}$ as

$$\mathcal{D}_{eff} = \mathcal{D} + \frac{R^2 U^2}{48 \mathcal{D}} \tag{2.224}$$

where $\mathcal{D}$ is the coefficient of molecular diffusion and R is the tube radius. The above relation also suggests a method for measurement of $\mathcal{D}$ from an experimental measurement of the effective diffusivity $\mathcal{D}_{eff}$.[37] In this section, we will provide a brief description of the analysis that yields the above result. A more detailed and rigorous analysis of the problem is given in Leal.[38]

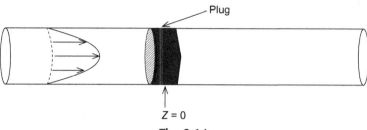

Fig. 2.14

Schematic representation of the initial condition in Taylor dispersion. A thin slice of a concentrated solute is introduced at $z=0$ at time $t=0$.

We consider a tube in which there is a laminar flow of a Newtonian fluid (Fig. 2.14). At time $t=0$, we introduce a thin 'slice' of a coloured dye (that spans the entire cross-section of the tube) at some axial location $z=0$. The concentration of the solute is assumed to be uniform across the cross-section initially. Our goal here is to derive an expression for the evolution of the cross-sectional averaged concentration profile in the tube at later times. The nonuniform velocity profile convects this initial pulse at different speeds at different radial locations in the tube, with the maximum convection happening at the centre of the tube and nearly zero convection near the tube walls. This initial pulse will thus be deformed by the flow and will extend in the z-direction. The governing equation for the nondimensional concentration $\theta = (C(r,z)/C_0)$, where C_0 is the concentration of the solute in the initial pulse, is given by the convective-diffusion equation:

$$\frac{\partial \theta}{\partial t} + 2U\left(1 - \frac{r^2}{a^2}\right)\frac{\partial \theta}{\partial z} = \mathcal{D}\left[\frac{1}{r}\frac{\partial}{\partial r}\left(r\frac{\partial \theta}{\partial r}\right) + \frac{\partial^2 \theta}{\partial z^2}\right] \tag{2.225}$$

The boundary conditions are that at the centreline $r=0$, the concentration θ is finite, and at the tube wall $r=R$, the concentration satisfies the no-flux condition $\dfrac{\partial \theta}{\partial r}=0$. By using R to nondimensionalise lengths and R/U to nondimensionalise time, we obtain the following nondimensional governing equation for θ:

$$\frac{\partial \theta}{\partial t^*} + 2\left(1 - r^{*2}\right)\frac{\partial \theta}{\partial z^*} = \frac{1}{Pe}\left[\frac{1}{r^*}\frac{\partial}{\partial r^*}\left(r^*\frac{\partial \theta}{\partial r^*}\right) + \frac{\partial^2 \theta}{\partial z^{*2}}\right] \tag{2.226}$$

where r^* and z^*, respectively, denote the nondimensional radial and axial coordinates and $Pe = UR/\mathcal{D}$ is the (nondimensional) Peclet number. In Taylor's analysis, the Peclet number is considered to be very large, that is, $Pe \gg 1$. For $Pe \gg 1$, the transport is dominated by convection, and the initial flat pulse will spread along the flow direction, and the concentration in the z direction will become parabolic and will continue to grow in time. At sufficiently later time, the longitudinal extent of the spread will be much larger compared with the tube radius R. Thus, the choice of R as the length scale in the z direction is not appropriate at late times and is valid only at early times, that is, for $t < R/U$ or $t^* < 1$. This implies that the

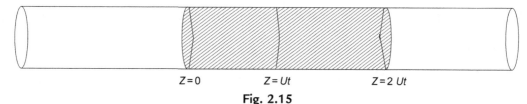

Z = 0 Z = Ut Z = 2 Ut

Fig. 2.15

Schematic representation of spread of the introduced dye at later times. The peak of the concentration lies at $z = Ut$, and the spread of the solute is about this moving peak along the axial direction.

scale R used for nondimensionalising lengths in the z direction may not be appropriate at longer times. Following the work of Taylor,[34] we make a change to a moving coordinate system that translated with the mean velocity U of the flow. Then, the dimensional convective-diffusion equation in the new frame becomes (with $\tilde{z} = z - Ut$) (Fig. 2.15)

$$\frac{D\theta}{Dt} + U\left(1 - 2\frac{r^2}{R^2}\right)\frac{\partial\theta}{\partial\tilde{z}} = \mathcal{D}\left[\frac{1}{r}\frac{\partial}{\partial r}\left(r\frac{\partial\theta}{\partial r}\right) + \frac{\partial^2\theta}{\partial\tilde{z}^2}\right] \tag{2.227}$$

In order to describe the phenomena at late times, we need to reconsider our nondimensionalisation scheme above and introduce a new length scale l_c and a new timescale l_c/U. We define our nondimensional coordinates as $r^* = r/R$, $z^* = z/l_c$, and $t^* = t/(l_c/U)$. The new choice of scales yields the following nondimensional governing equation:

$$Pe\frac{R}{l_c}\left(\frac{D\theta}{Dt^*} + \left(1 - 2r^{*2}\right)\frac{\partial\theta}{\partial\tilde{z}^*}\right) = \left[\frac{1}{r^*}\frac{\partial}{\partial r^*}\left(r^*\frac{\partial\theta}{\partial r^*}\right) + \frac{R^2}{l_c^2}\frac{\partial^2\theta}{\partial\tilde{z}^{*2}}\right], \tag{2.228}$$

The above nondimensionalised equation indicates that the term governing the diffusion of the solute in the axial direction is multiplied by R^2/l_c^2 and is negligible when $l_c \gg R$. At long times after the introduction of the pulse, we are mainly concerned about the evolution of the concentration in the axial direction, and hence, it is useful to define a cross-sectional averaged concentration $\langle\theta\rangle$ as follows:

$$\theta = \langle\theta\rangle + \theta' \tag{2.229}$$

where θ' is the deviation from the cross-sectional average and

$$\langle\theta\rangle \equiv \frac{1}{2\pi}\int_0^{2\pi}\int_0^1 \theta r^* dr^* d\phi, \tag{2.230}$$

where ϕ is the polar angle in the cylindrical coordinate system. The decomposition of θ (Eq. 2.229) is substituted in (2.228), and the resulting equation is averaged over the cross-section to yield

$$Pe\frac{RD\langle\theta\rangle}{l_c Dt^*} = \frac{R^2}{l_c^2}\frac{\partial^2\langle\theta\rangle}{\partial\tilde{z}^{*2}} - Pe\frac{R}{l_c}\left\langle\left(1 - 2r^{*2}\right)\frac{\partial\theta'}{\partial\tilde{z}^*}\right\rangle \tag{2.231}$$

When $Pe \gg 1$, we may neglect the axial conduction term in the above equation, to give

$$\frac{D\langle\theta\rangle}{Dt^*} = -\left\langle u'\frac{\partial\theta'}{\partial\tilde{z}^*}\right\rangle, \tag{2.232}$$

where $u' = (u - \langle u\rangle) = 2(1 - r^{*2}) - 1$. Taylor showed that the right side of the above equation is equal to $\frac{\partial^2\langle\theta\rangle}{\partial\tilde{z}^2}$, and thus, the previous equation reduces to a dimensional form as

$$\frac{D\langle\theta\rangle}{Dt^*} = \mathcal{D}\frac{Pe^2}{48}\frac{\partial^2\theta}{\partial\tilde{z}^{*2}}, \tag{2.233}$$

that when converted to the stationary frame of reference becomes (in dimensional form)

$$\frac{\partial\langle\theta\rangle}{\partial t} + U\frac{\partial\langle\theta\rangle}{\partial z} = \mathcal{D}\frac{Pe^2}{48}\frac{\partial^2\theta}{\partial z^2}. \tag{2.234}$$

This is a convective-diffusion equation in the z direction, and the effective diffusivity can be inferred from the above equation as

$$\mathcal{D}_{eff} = \mathcal{D}(R^2 U^2)/(48\mathcal{D}^2). \tag{2.235}$$

This result shows, perhaps counter to intuition, that as molecular diffusivity $\mathcal{D}$ is increased, the dispersion coefficient decreases. However, due to the axial convection, at the centreline, the concentration is the highest, but the radial gradients are also the highest near the tip of the parabola, which leads to rapid diffusion of the concentration in the radial direction, and the solute concentration is convected downstream at a lower rate. Another inference that is counter to intuition is that the spreading front moves at the average velocity of the flow, instead of the maximum velocity at the centreline. Thus, the effective dispersion in the axial direction actually decreases with increase in molecular diffusivity of the solute species. The solution of (2.233) yields a concentration profile that is a Gaussian in the flow (z) direction, in the frame of reference that moves with the average fluid velocity.

The results of this analysis are very relevant in the quantitative understanding of dispersion in flow through tubes and in the inference of molecular diffusivity of a species from the experimental measurement of its dispersion coefficient.

2.9 Practical Studies of Mass Transfer

The principal applications of mass transfer are in the fields of distillation, gas absorption, and other separation processes involving mass transfer, which are discussed in Volume 2. In particular, mass-transfer coefficients and heights of transfer units in distillation and in gas absorption are discussed in Volume 2. In this section, an account is given of some of the experimental studies of mass transfer in equipment of simple geometry, in order to provide a historical perspective.

2.9.1 The j-Factor of Chilton and Colburn for Flow in Tubes

Heat transfer

Because the mechanisms governing mass transfer are similar to those involved in both heat transfer by conduction and convection and in momentum transfer (fluid flow), quantitative relations exist between the three processes, and these are discussed in Chapter 4. There is generally more published information available on heat transfer than on mass transfer, and these relationships often therefore provide a useful means of estimating mass-transfer coefficients.

Results of experimental studies of heat transfer may be conveniently represented by means of the *j*-factor method developed by Colburn[39] and by Chilton and Colburn[40] for representing data on heat transfer between a turbulent fluid and the wall of a pipe.

$$Nu = 0.023 Re^{0.8} Pr^{0.33} \quad \text{(Eq. 9.64)}$$

where the viscosity is measured at the mean film temperature and Nu, Re, and Pr denote the Nusselt, Reynolds, and Prandtl numbers, respectively.

If both sides of the equation are divided by the product $Re\, Pr$,

$$St = 0.023 Re^{-0.2} Pr^{-0.67} \tag{2.236}$$

where $St (= h/C_p \rho u)$ is the Stanton number.

Eq. (2.236) may be rearranged to give

$$j_h = St\, Pr^{0.67} = 0.023 Re^{-0.2} \tag{2.237}$$

The left-hand side of Eq. (2.237) is referred to as the *j*-factor for heat transfer (j_h). Chilton and Colburn found that a plot of j_h against Re gave approximately the same curve as the friction chart (ϕ vs. Re) for turbulent flow of a fluid in a pipe.

The right-hand side of Eq. (2.237) gives numerical values that are very close to those obtained from the Blasius equation for the friction factor ϕ for the turbulent flow of a fluid through a smooth pipe at Reynolds numbers up to about 10^6.

$$\phi = 0.0396 Re^{-0.25} \quad \text{(Eq. 3.11 of Vol. 1A)}$$

Re	3×10^3	10^4	3×10^4	10^5	3×10^5	10^6
ϕ (Eq. 3.11 of Vol. 1A)	0.0054	0.0040	0.0030	0.0022	0.0017	0.0013
ϕ (Eq. 2.237)		0.0046	0.0037	0.0029	0.0023	0.0019	0.0015

Mass transfer

Several workers have measured the rate of transfer from a liquid flowing down the inside wall of a tube to a gas passing countercurrently upwards. Gilliland and Sherwood[41] vaporised a number of liquids including water, toluene, aniline and propyl, amyl, and butyl alcohols into an air stream flowing up the tube. A small tube, diameter $d = 25$ mm and length $= 450$ mm, was used fitted with calming sections at the top and bottom, and the pressure was varied from 14 to about 300 kN/m^2.

The experimental results were correlated using an equation of the form:

$$\frac{h_D}{u}\frac{C_{Bm}}{C_T}\left(\frac{\mu}{\rho D}\right)^{0.56} = 0.023 Re^{-0.17} \tag{2.238}$$

The index of the Schmidt group Sc is less than the value of 0.67 for the Prandtl group for heat transfer, but the range of values of Sc used was very small.

There has for long been uncertainty concerning the appropriate value to be used for the exponent of the Schmidt number in Eq. (2.238). Sherwood et al.[27] have analysed experimental results obtained by a number of workers for heat transfer to the walls of a tube for a wide range of gases and liquids (including water, organic liquids, oils, and molten salts) and offered a logarithmic plot of the Stanton number ($St = h/C_p\rho u$) against Prandtl number (Pr) for a constant Reynolds number of 10,000. Superimposed on the graph are results for mass transfer obtained from experiments on the dissolution of solute from the walls of tubes composed of solid organics into liquids and on the evaporation of liquid films from the walls of tubes to turbulent air streams using a wetted-wall column, again all at a Reynolds number of 10,000: these results were plotted as Stanton number for mass transfer (h_D/u) against Schmidt number (Sc). There is very close agreement between the results for heat transfer and for mass transfer, with a line slope of about -0.67 giving a satisfactory correlation of the results. The range of values of Prandtl and Schmidt numbers was from about 0.4 to 10,000. This established that the exponents for both the Prandtl and Schmidt numbers in the j-factors should be the same, namely, 0.67. These conclusions are consistent with the experimental results of Linton and Sherwood[42] who measured the rates of dissolution of cast tubes of benzoic acid, cinnamic acid, and β-naphthol into water, giving Schmidt numbers in the range 1000–3000.

In defining a j-factor (j_d) for mass transfer, there is therefore good experimental evidence for modifying the exponent of the Schmidt number in Gilliland and Sherwood's correlation (Eq. 2.238). Furthermore, there is no very strong case for maintaining the small differences in the exponent of Reynolds number. On this basis, the j-factor for mass transfer may be defined as follows:

$$j_d = \frac{h_D}{u}\frac{C_{Bm}}{C_T} Sc^{0.67} = 0.023 Re^{-0.2} \tag{2.239}$$

The term C_{Bm}/C_T (the ratio of the logarithmic mean concentration of the insoluble component to the total concentration) is introduced because $h_D(C_{Bm}/C_T)$ is less dependent than h_D on the concentrations of the components. This reflects the fact that the analogy between momentum, heat, and mass transfer relates only to that part of the mass transfer that is *not* associated with the *bulk flow* mechanism; this is a fraction C_{Bm}/C_T of the total mass transfer. For equimolecular counterdiffusion, as in binary distillation when the molar latent heats of the components are equal, the term C_{Bm}/C_T is omitted as there is no bulk flow contributing to the mass transfer.

Sherwood and Pigford[7] have shown that if the data of Gilliland and Sherwood[41] and others[40,43,44] are plotted with the Schmidt group raised to this power of 0.67, as shown in Fig. 2.16, a reasonably good correlation is obtained. Although the points are rather more scattered than with heat transfer, it is reasonable to assume that both j_d and j_h are approximately equal to ϕ. Eqs (2.237), (2.239) apply in the absence of ripples that can be responsible for a very much increased rate of mass transfer. The constant of 0.023 in the equations will then have a higher value.

By equating j_h and j_d (Eqs 2.237, 2.239), the mass-transfer coefficient may be expressed in terms of the heat-transfer coefficient, giving

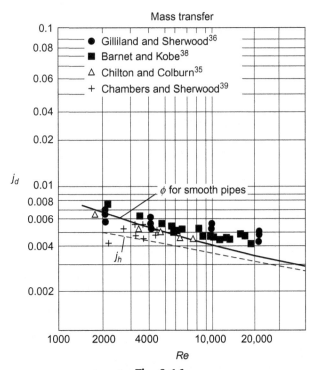

Fig. 2.16

Mass transfer in wetted-wall columns.

$$h_D = \left(\frac{h}{C_p \rho}\right)\left(\frac{C_T}{C_{Bm}}\right)\left(\frac{\text{Pr}}{\text{Sc}}\right)^{0.67} \tag{2.240}$$

2.9.2 Mass Transfer at Plane Surfaces

Many of the earlier studies of mass transfer involved measuring the rate of vaporisation of liquids by passing a turbulent air stream over a liquid surface. In addition, some investigations have been carried out in the absence of air flow, under what have been termed still air conditions. Most of these experiments have been carried out in some form of wind tunnel where the rate of flow of air and its temperature and humidity could be controlled and measured. In these experiments, it was found to be important to keep the surface of the liquid level with the rim of the pan in order to avoid the generation of eddies at the leading edge.

Hinchley and Himus[45] measured the rate of evaporation from heated rectangular pans fitted flush with the floor of a wind tunnel (0.46 m wide by 0.23 m high) and showed that the rate of vaporisation was proportional to the difference between the saturation vapour pressure of the water P_s and the partial pressure of water in the air P_w. The results for the mass rate of evaporation W were represented by an empirical equation of the form:

$$W = \text{constant}\,(P_s - P_w) \tag{2.241}$$

where the constant varies with the geometry of the pan and the air velocity.

This early work showed that the driving force in the process was the pressure difference $(P_s - P_w)$. Systematic work in more elaborate equipment by Powell and Griffiths[46] Wade,[47] and Pasquill[48] then followed. Wade, who vaporised a variety of organic liquids, including acetone, benzene, and trichloroethylene at atmospheric pressure, used a small pan 88 mm square in a wind tunnel. Powell and Griffiths stretched canvas sheeting over rectangular pans and, by keeping the canvas wet at all times, ensured that it behaved as a free water surface. In all of these experiments, the rate of vaporisation showed a similar form of dependence on the partial pressure difference and the rate of flow of the air stream. Powell and Griffiths found that the vaporisation rate per unit area decreased in the downwind direction. For rectangular pans, of length L, the vaporisation rate was proportional to $L^{0.77}$. This can be explained in terms of the thickening of the boundary layer (see Chapter 3) and the increase in the partial pressure of vapour in the air stream arising from the evaporation at upstream positions.

In these experiments, it might be anticipated that, with high concentrations of vapour in the air, the rate of evaporation would no longer be linearly related to the partial pressure difference because of the contribution of bulk flow to the mass-transfer process (Section 2.2.3), although there is no evidence of this even at mole fractions of vapour at the surface as high as 0.5. Possibly, the experimental measurements were not sufficiently sensitive to detect this effect.

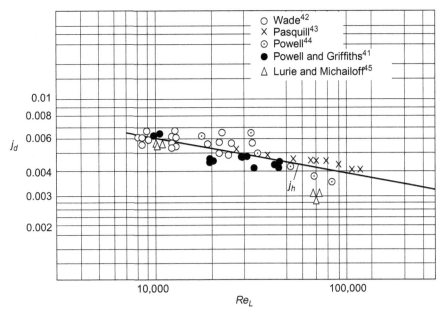

Fig. 2.17

Evaporation from plane surfaces.

Sherwood and Pigford[7] have plotted the results of several workers[46–50] in terms of the Reynolds number Re_L, using the length L of the pan as the characteristic linear dimension. Fig. 2.17, taken from this work, shows j_d plotted against Re_L for a number of liquids evaporating into an air stream. Although the individual points show some scatter, j_d is seen to follow the same general trend as j_h in this work. The Schmidt number was varied over such a small range that the correlation was not significantly poorer if it was omitted from the correlation.

Maisel and Sherwood[51] also carried out experiments in a wind tunnel in which water was evaporated from a wet porous surface preceded by a dry surface of length L_0. Thus, a velocity boundary layer had become established in the air before it came into contact with the evaporating surface. The results were correlated by

$$j_d = 0.0415 Re_L^{-0.2} \left[1 - \left(\frac{L_0}{L} \right)^{0.8} \right]^{-0.11} \tag{2.242}$$

where L is the total length of the surface (dry + wet).

2.9.3 Effect of Surface Roughness and Form Drag

The results discussed in Section 2.9.2 give reasonably good support to the treatment of heat, mass, and momentum transfer by the *j*-factor method, although it is important to remember that, in all the cases considered, the drag is almost entirely in the form of skin friction (i.e. viscous drag at the surface). As soon as an attempt is made to apply the relation to cases where form drag (i.e. additional drag caused by the eddies set up as a result of the fluid impinging on an obstruction) is important, such as beds of granular solids or evaporation from cylinders or spheres, the *j*-factor and the friction factor are found no longer to be equal. This problem receives further consideration in Volume 2. Sherwood[52,53] carried out experiments where the form drag was large compared with the skin friction, as calculated approximately by subtracting the form drag from the total drag force. In this way, reasonable agreement between the corresponding value of the friction factor ϕ and j_h and j_d was obtained.

Gamson et al.[54] have successfully used the *j*-factor method to correlate their experimental results for heat and mass transfer between a bed of granular solids and a gas stream.

Pratt[55] has examined the effect of using artificially roughened surfaces and of introducing 'turbulence promoters', which increase the amount of form drag. It was found that the values of ϕ and the heat- and mass-transfer coefficients were a minimum for smooth surfaces and all three quantities increased as the surface roughness was increased. ϕ increased far more rapidly than either of the other two quantities, however, and the heat- and mass-transfer coefficients were found to reach a limiting value, whereas ϕ could be increased almost indefinitely. Pratt has suggested that these limiting values are reached when the velocity gradient at the surface corresponds with that in the turbulent part of the fluid, that is, at a condition where the buffer layer ceases to exist (Chapter 3).

2.9.4 Mass Transfer From a Fluid to the Surface of Particles

It is necessary to calculate mass-transfer coefficients between a fluid and the surface of a particle in a number of important cases, including

  (i)   gas absorption in a spray tower,
 (ii)   evaporation of moisture from the surface of droplets in a spray tower,
(iii) reactions between a fluid and dispersed liquid droplets or solid particles as, for instance, in a combustion process where the oxygen in the air must gain access to the external surfaces, and
(iv) catalytic reactions involving porous particles where the reactant must be transferred to the outer surface of the particle before it can diffuse into the pores and make contact with the active sites on the catalyst.

There have been comparatively few experimental studies in this area, and the results of different workers do not always show a high degree of consistency. Frequently, estimates of mass-transfer coefficients have been made by applying the analogy between heat transfer and mass transfer and thereby utilising the larger body of information that is available on heat transfer.

Interest extends from transfer to single particles to systems in which the particles are in the form of fixed or fluidised beds. The only case for which there is a rigorous analytical solution is that for heat by conduction and mass transfer by diffusion to a sphere.

Mass transfer to single particles

Mass transfer from a single spherical drop to still air is controlled by molecular diffusion, and at low concentrations when bulk flow is negligible, the problem is analogous to that of heat transfer by conduction from a sphere, which is considered in Chapter 1, Section 1.3.4. Thus, for steady-state radial diffusion into a large expanse of stationary fluid in which the partial pressure falls off to zero over an infinite distance, the equation for mass transfer will take the same form as that for heat transfer (Eq. 1.26):

$$Sh' = \frac{h_D d'}{D} = 2 \tag{2.243}$$

where Sh' is the Sherwood number that, for mass transfer, is the counterpart of the Nusselt number $Nu'(= hd'/k)$ for heat transfer to a sphere. This value of 2 for the Sherwood number is the theoretical minimum in any continuous medium and is increased if the concentration difference occurs over a finite, as opposed to an infinite, distance and if there is turbulence in the fluid.

For conditions of forced convection, Frössling[56] studied the evaporation of drops of nitrobenzene, aniline, and water and of spheres of naphthalene, into an air stream. The drops were mainly small and of the order of 1 mm diameter. Powell[49] measured the evaporation of water from the surfaces of wet spheres up to 150 mm diameter and from spheres of ice.

The experimental results of Frössling may be represented by the equation:

$$Sh' = \frac{h_D d'}{D} = 2.0\left(1 + 0.276 Re'^{0.5} Sc^{0.33}\right) \tag{2.244}$$

Sherwood and Pigford[7] found that the effect of the Schmidt group was also influenced by the Reynolds group and that the available data were fairly well correlated as shown in Fig. 2.18, in which $(h_D d')/D$ is plotted against $Re' Sc^{0.61}$.

Garner and Keey[57,58] dissolved pelleted spheres of organic acids in water in a low-speed water tunnel at particle Reynolds numbers between 2.3 and 255 and compared their results with other data available at Reynolds numbers up to 900. Natural convection was found to exert some influence at Reynolds numbers up to 750. At Reynolds numbers greater than 250, the results are correlated by Eq. (2.243):

$$Sh' = 0.94 Re'^{0.5} Sc^{0.33} \tag{2.245}$$

Mass transfer under conditions of natural convection was also investigated.

MASS TRANSFER

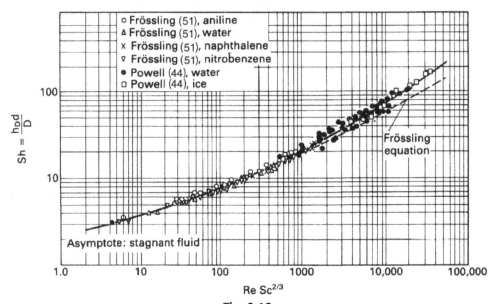

Fig. 2.18

Mass transfer to single spheres.

Ranz and Marshall[59] have carried out a comprehensive study of the evaporation of liquid drops and confirm that Eq. (2.244) correlates the results of a number of workers. A value of 0.3 in place of 0.276 for the coefficient is suggested, although the spread of the experimental results is such that the difference is not statistically significant.

Rowe et al.[60] have reviewed the literature on heat and mass transfer between spherical particles and fluid. For heat transfer, their results that are discussed in Chapter 1, Section 1.4.6, are generally well represented by Eq. (1.100):

$$Nu' = 2 + \beta'' Re'^{0.5} Pr^{0.33} \quad (0.4 < \beta'' < 0.8) \tag{2.246}$$

For mass transfer,

$$Sh' = \alpha' + \beta' Re'^{0.5} Sc^{0.33} \quad (0.3 < \beta' < 1.0) \tag{2.247}$$

The constant α appears to be a function of the Grashof number but approaches a value of about 2 as the Grashof number approaches zero.

In an experimental investigation,[60] they confirmed that Eqs (2.246), (2.247) can be used to represent the results obtained for transfer from both air and water to spheres. The constants β', β'' varied from 0.68 to 0.79.

There is therefore broad agreement between the results of Frössling,[56] Ranz and Marshall,[59] and Rowe et al.[60] The variations in the values of the coefficient are an indication of the degree of reproducibility of the experimental results. However, Brian and Hayes[61] who carried a numerical solution of the equations for heat and mass transfer suggest that, at high values of $Re'Sc^{0.33}$, these equations tend to underestimate the value of the transfer coefficient, and an equation that can be expressed in the following form is proposed:

$$Sh' = \left[4.0 + 1.21(Re'Sc)^{0.67}\right]^{0.5} \tag{2.248}$$

Mass transfer to particles in a fixed or fluidised bed

Experimental results for fixed packed beds are very sensitive to the structure of the bed, which may be strongly influenced by its method of formation. Gupta and Thodos[62] have studied both heat transfer and mass transfer in fixed beds and have shown that the results for both processes may be correlated by similar equations based on *j*-factors (see Section 2.9.1). Rearrangement of the terms in the mass-transfer equation permits the results for the Sherwood number (Sh') to be expressed as a function of the Reynolds (Re_c) and Schmidt numbers (Sc):

$$Sh' = 2.06\frac{1}{e}Re_c'^{0.425}Sc^{0.33} \tag{2.249}$$

where e is the voidage of the bed and Re_c' is the particle Reynolds number incorporating the superficial velocity of the fluid (u_c).

Kramers[63] carried out experiments on heat transfer to particles in a fixed bed and has expressed his results in the form of a relation between the Nusselt, Prandtl, and Reynolds numbers. This symequation may be rewritten to apply to mass transfer, by using the analogy between the two processes, giving

$$Sh' = 2.0 + 1.3Sc^{0.15} + 0.66Sc^{0.31}Re_c'^{0.5} \tag{2.250}$$

In selecting the most appropriate equation for any particular operation, it is recommended that the original references be checked to ascertain in which study the experimental conditions were closest.

Both heat transfer and mass transfer between a fluid and particles in a fluidised bed are discussed in Volume 2. The results are sensitive to the quality of fluidisation and particularly to the uniformity of distribution of the particles in the fluid. In most cases, it is found that the same correlations for both heat transfer and mass transfer are applicable to fixed and fluidised beds.

2.10 Nomenclature

		Units in SI System	Dimensions in M, N, L, T, θ
A_p	External surface area of particle	m^2	L^2
a	Interfacial area per unit volume	m^2/m^3	L^{-1}
B_1, B_2	Integration constants	kmol s/m^3	$NL^{-3}T$
B_1', B_2'	Integration constants	kmol/m^3	NL^{-3}
C	Molar concentration	kmol/m^3	NL^{-3}
C_A, C_B	Molar concentration of A, B	kmol/m^3	NL^{-3}
C_p	Specific heat at constant pressure	J/kg K	$L^2T^{-2}\theta^{-1}$
C_T	Total molar concentration	kmol/m^3	NL^{-3}
C_{Bm}	Logarithmic mean value of C_B	kmol/m^3	NL^{-3}
C'	$C_A - C_{A0}$	kmol/m^3	NL^{-3}
C_A'', C_B''	Molar concentration of A, B in film	kmol/m^3	NL^{-3}
$\overline{C'}$	Laplace transform of C'	kmol s/m^3	$NL^{-3}T$
c	Mass concentration	kg/m^3	ML^{-3}
D	Diffusivity	m^2/s	L^2T^{-1}
D_e	Effective diffusivity within catalyst particle	m^2/s	L^2T^{-1}
D_f	Diffusivity of fluid in film	m^2/s	L^2T^{-1}
D_{Kn}	Knudsen diffusivity	m^2/s	L^2T^{-1}
D_L	Liquid-phase diffusivity	m^2/s	L^2T^{-1}
D_{AB}	Diffusivity of **A** in **B**	m^2/s	L^2T^{-1}
D_{Th}	Coefficient of thermal diffusion	kmol/ms	$NL^{-1}T^{-1}$
D_{BA}	Diffusivity of **B** in **A**	m^2/s	L^2T^{-1}
D'	Effective diffusivity in multicomponent system	m^2/s	L^2T^{-1}
d	Pipe diameter	m	L
d'	Diameter of sphere	m	L
e	Bed voidage	–	–
E_D	Eddy diffusivity	m^2/s	L^2T^{-1}
F	Coefficient in Maxwell's law of diffusion	m^3/kmols	$N^{-1}L^3T^{-1}$
f	Ratio—N_B'/N_A'	–	–
G	Molar flow of stream	kmol/s	NT^{-1}
G'	Molar flow per unit area	kmol/m^2s	$NL^{-2}T^{-1}$
	Ratio of equilibrium values of concentrations in two phases	–	–
H	Height of transfer unit	m	L

h	Heat transfer coefficient	W/m^2 K	MT$^{-3}\theta^{-1}$
h_D	Mass-transfer coefficient	m/s	LT^{-1}
h_D'	Mass-transfer coefficient enhanced by chemical reaction	m/s	LT^{-1}
H	Henry's law constant, C_{Ai}''/C_{Ai} (Eq. 2.125)	–	–
j_d	j-factor for mass transfer	–	–
j_h	j-factor for heat transfer	–	–
K	Overall mass-transfer coefficient	m/s	LT^{-1}
J	Flux as mass per unit area and unit time	kg/m^{2}s	ML^{-2}T^{-1}
K_{ABT}	Ratio of transport rate by thermal diffusion to that by Fick's law	–	–
k_G	Mass-transfer coefficient for transfer through stationary fluid	kmol s/kg m	NM^{-1}L^{-1}T
k_G'	Mass-transfer coefficient for equimolecular counterdiffusion	kmol s/kg m	NM^{-1}L^{-1}T
k_x	Mass-transfer coefficient (mole fraction driving force)	kmol/m^{2}s	NL^{-2}T^{-1}
k	Reaction rate constant first-order reaction	s^{-1}	T^{-1}
	nth-order reaction	kmol^{1-n}m^{3n-3} s^{-1}	N^{1-n} L^{3n-3} T^{-1}
L	Length of surface, or film thickness, or half-thickness of platelet or V_p/A_p	m	L
L_f	Thickness of film	m	L
L_0	Length of surface, unheated	m	L
M	Molecular weight (relative molecular mass)	kg/kmol	MN^{-1}
N	Molar rate of diffusion per unit area (average value)	kmol/m^{2}s	NL^{-2}T^{-1}
$(N)_t$	Molar rate of diffusion per unit area at time t	kmol/m^{2}s	NL^{-2}T^{-1}
N'	Total molar rate of transfer per unit area	kmol/m^{2}s	NL^{-2}T^{-1}
$(N)_{Th}$	Molar flux due to thermal diffusion	kmol/m^{2}s	NL^{-2}T^{-1}
N	Number of transfer units	–	–
n	Number of moles of gas	kmol	N
n	Order of reaction or number of term in series	–	–

P	Total pressure	N/m^2	ML^{-1}T^{-2}
P_A, P_B	Partial pressure of A, B	N/m^2	ML^{-1}T^{-2}
P_s	Vapour pressure of water	N/m^2	ML^{-1}T^{-2}
P_w	Partial pressure of water in gas stream	N/m^2	ML^{-1}T^{-2}
P_{Bm}	Logarithmic mean value of P_B	N/m^2	ML^{-1}T^{-2}
p	Parameter in Laplace transform	s^{-1}	T^{-1}
q	Concentration gradient dC_A/dy	kmol/m^4	NL^{-4}
R	Shear stress acting on surface	N/m^2	ML^{-1}T^{-2}
R	Universal gas constant	8314 J/kmol K	MN^{-1}L^{2}T$^{-2}\theta^{-1}$
r	Radius within sphere or cylinder	m	L
r_c	External radius of cylinder	m	L
r_0	External radius of sphere	m	L
$\mathfrak{R}_p$	Reaction rate per particle for first-order reaction	kmol/s	N T^{-1}
$\mathfrak{R}_v$	Reaction rate per unit volume of particle for first-order reaction	kmol/m^{3}s	N L^{-3}T^{-1}
$\mathfrak{R}'_{vn}$	Reaction rate per unit volume of particle for nth-order reaction (no mass-transfer resistance)	kmol/m^{3}s	N L^{-3}T^{-1}
S	Cross-sectional area of flow	m^2	L^2
s	Rate of production of fresh surface per unit area	s^{-1}	T^{-1}
T	Absolute temperature	K	θ
t	Time	s	T
t_e	Time of exposure of surface element	s	T
u	Mean velocity	m/s	LT^{-1}
u_c	Superficial velocity (volumetric flowrate/total area)	m/s	LT^{-1}
u_A, u_B	Mean molecular velocity in direction of transfer	m/s	LT^{-1}
u_{DA}, u_{DB}	Diffusional velocity of transfer	m/s	LT^{-1}
u_F	Velocity due to bulk flow	m/s	LT^{-1}
u_m	Molecular velocity	m/s	LT^{-1}
u_s	Stream velocity	m/s	LT^{-1}
V	Volume	m^3	L^3
V_p	Volume of catalyst particle	m^3	L^3
V	Molecular volume	m^3/kmol	N^{-1}L^3
V$_o$	Correction term in Eq. (2.96)	m^3/kmol	N^{-1}L^3
W	Mass rate of evaporation	kg/s	MT^{-1}

x	Distance from leading edge of surface or in X-direction	m	L
	or mole fraction	–	–
y	Distance from surface or in direction of diffusion	m	L
y'	Mol fraction in stationary gas	–	–
Z	Height of column	m	L
z	Distance in Z-direction	m	L
α	Thermal diffusion factor	–	–
α'	Term in Eq. (2.247)	–	–
β', β''	Coefficient in Eqs (2.247), (2.246)	–	–
η	Effectiveness factor	–	–
λ	$\sqrt{(k/D_e)}$	s^{-1}	L^{-1}
λ_m	Mean free path of molecules	m	L
μ	Viscosity of fluid	N s/m^2	$\text{ML}^{-1}\text{T}^{-1}$
ρ	Density of fluid	kg/m^3	ML^{-3}
ϕ	Friction factor $(R/\rho u_2)$	–	–
ϕ	Thiele modulus based on L, rc, or r0	–	–
ϕ_L	Thiele modulus based on length term $L = Vp/Ap$	–	–
Ψ	rCA	kmol/m^2	NL^{-2}
ω	Mass fraction	–	–
N_u	Nusselt number hd/k	–	—
N_u'	Nusselt number for sphere hd'/k	–	—
P_r	Prandtl number $Cp\mu/k$	–	–
R_e	Reynolds number $ud\,\rho/\mu$	–	–
R_e'	Reynolds number for sphere $ud'\,\rho/\mu$	–	–
Re_c'	Reynolds number for particle in fixed or fluidised bed $ucd'\,\rho/\mu$	–	–
Re_L	Reynolds number for flat plate $usL\rho/\mu$	–	–
Sc	Schmidt number $\mu/\rho D$	–	–
Sh'	Sherwood number for sphere hDd'/D	–	–
St	Stanton number $h/Cp\rho u$	–	–

* Dimensions depend on order of reaction.

Suffixes		–	–
0	Value in bulk of phase	–	–
1	Phase 1	–	–
2	Phase 2	–	–
A	Component **A**	–	–

B	Component **B**	–	–
AB	Of **A** in **B**	–	–
b	Bottom of column	–	–
e	Value in equilibrium with bulk of other phase	–	–
G	Gas phase	–	–
i	Interface value	–	–
L	Liquid phase	–	–
o	Overall value (for height and number of transfer units)	–	–
	or value in bulk of phase	–	–
t	Top of column	–	–

References

1. Fick A. Ueber diffusion. *Ann Phys* 1855;**94**:59.
2. Present RD. *Kinetic theory of gases*. New York: McGraw-Hill; 1958.
3. Stefan J. *Wiener Akad Wissensch* 1873;**68**:385; 1879;**79**:169; 1889;**98**:1418; Ann Physik 1890;**41**:723. Versuche über die Verdampfung.
4. Perry RH, Green DW, editors. *Perry's chemical engineers' handbook*. 6th ed. New York: McGraw-Hill; 1984.
5. Winkelmann A. Ueber die Diffusion von Gasen und Dämpfen. *Ann Phys* 1884;**22**(1):152.
6. Gilliland ER. Diffusion coefficients in gaseous systems. *Ind Eng Chem* 1934;**26**:681.
7. Sherwood TK, Pigford RL. *Absorption and extraction*. New York: McGraw-Hill; 1952.
8. Fuller EN, Schettler PD, Giddings JC. A new method for prediction of binary gas-phase diffusion coefficients. *Ind Eng Chem* 1966;**58**:19.
9. Grew KE, Ibbs TL. *Thermal diffusion in gases*. Cambridge: Cambridge University Press; 1952.
10. Hirschfelder JO, Curtiss CF, Bird RB. *Molecular theory of gases and liquids*. New York: Wiley; 1954. p. 584–5.
11. Arnold JH. Studies in diffusion. III. Steady-state vaporization and absorption. *Am Inst Chem Eng J* 1944;**40**:361.
12. Maxwell JC. The dynamical theory of gases. *Philos Trans R Soc* 1867;**157**:49.
13. Taylor R, Krishna R. *Multicomponent mass transfer*. New York: Wiley; 1993.
14. Cussler EL. *Multicomponent diffusion*. Amsterdam: Elsevier; 1976.
15. Zielinski JM, Hanley BF. Practical friction-based approach to modeling multicomponent diffusion. *AICHE J* 1999;**45**:1.
16. Wilke CR, Chang P. Correlation of diffusion coefficients in dilute solutions. *AICHE J* 1955;**1**:264.
17. Reid RC, Prausnitz JM, Sherwood TK. *The properties of gases and liquids*. New York: McGraw-Hill; 1977.
18. Goodridge F, Bricknell DJ. Interfacial resistance in the carbon dioxide-water system. *Trans Inst Chem Eng* 1962;**40**:54.
19. Whitman WG. The two-film theory of absorption. *Metall Chem Eng* 1923;**29**:147.
20. Higbie R. The rate of absorption of pure gas into a still liquid during short periods of exposure. *Am Inst Chem Eng J* 1935;**31**:365.
21. Danckwerts PV. Significance of liquid film coefficients in gas absorption. *Ind Eng Chem* 1951;**43**:1460.
22. Toor HL, Marchello JM. Film-penetration model for mass and heat transfer. *AICHE J* 1958;**4**:97.

23. Kishinevskij MK. The kinetics of absorption under intense mixing. *Zhur Priklad Khim* 1951;**24**:542. *J Appl Chem USSR* **24**:593.
24. Danckwerts PV. *Gas–liquid reactions*. New York: McGraw-Hill; 1970.
25. Harriott P. A random eddy modification of the penetration theory. *Chem Eng Sci* 1962;**17**:149.
26. Dwight HB. *Tables of integrals and other mathematical data*. New York: MacMillan; 1957.
27. Sherwood TK, Pigford RL, Wilke CR. *Mass transfer*. New York: McGraw-Hill; 1975.
28. Danckwerts PV. Absorption by simultaneous diffusion and chemical reaction. *Trans Faraday Soc* 1950;**46**:300.
29. Scott DS, Dullien FAL. Diffusion of ideal gases in capillaries and porous solids. *AICHE J* 1962;**8**:113.
30. Aris R. *The mathematical theory of diffusion and reaction in permeable catalysts*. London: Oxford University Press; 1975.
31. Thiele EW. Relation between catalyst activity and size of particle. *Ind Eng Chem* 1939;**24**:916.
32. Mickley HS, Sherwood TK, Reed CE. *Applied mathematics in chemical engineering*. 2nd ed. New York: McGraw-Hill; 1957.
33. Rester S, Aris R. Communications on the theory of diffusion and reaction. II. The effect of shape on the effectiveness factor. *Chem Eng Sci* 1969;**24**:793.
34. Taylor GI. Dispersion of soluble matter in solvent flowing slowly through a tube. *Proc R Soc Lond Ser A* 1953;**219**:186.
35. Taylor GI. Conditions under which dispersion of a solute in a stream of solvent can be used to measure molecular diffusion. *Proc R Soc Lond Ser A* 1954;**225**:473.
36. Aris R. On the dispersion of a solute in a fluid flowing through a tube. *Proc R Soc Lond Ser A* 1956;**235**:67.
37. Bello MS, Rezzonico R, Righetti PG. Use of Taylor-Aris dispersion for measurement of a solute diffusion coefficient in thin capillaries. *Science* 1994;**266**:773–6.
38. Leal LG. *Advanced transport phenomena: fluid mechanics and convective transport processes*. Cambridge: Cambridge University Press; 2007.
39. Colburn AP. A method of correlating forced convection heat transfer data and a comparison with fluid friction. *Am Inst Chem Eng J* 1933;**29**:174.
40. Chilton TH, Colburn AP. Mass transfer (absorption) coefficients—production from data on heat transfer and fluid friction. *Ind Eng Chem* 1934;**26**:1183.
41. Gilliland ER, Sherwood TK. Diffusion of vapours into air streams. *Ind Eng Chem* 1934;**26**:516.
42. Linton WH, Sherwood TK. Mass transfer from solid shapes to water in streamline and turbulent flow. *Chem Eng Prog* 1950;**46**:258.
43. Barnet WI, Kobe KA. Heat and vapour transfer in a wetted-wall column. *Ind Eng Chem* 1941;**33**:436.
44. Chambers FS, Sherwood TK. Absorption of nitrogen dioxide by aqueous solutions. *Ind Eng Chem* 1937;**29**:579. *Trans Am Inst Chem Eng* 1937;**33**:579.
45. Hinchley JW, Himus GW. Evaporation in currents of air. *Trans Inst Chem Eng* 1924;**2**:57.
46. Powell RW, Griffiths E. The evaporation of water from plane and cylindrical surfaces. *Trans Inst Chem Eng* 1935;**13**:175.
47. Wade SH. Evaporation of liquids in currents of air. *Trans Inst Chem Eng* 1942;**20**:1.
48. Pasquill F. Evaporation from a plane free liquid surface into a turbulent air stream. *Proc R Soc Lond A* 1943;**182**:75.
49. Powell RW. Further experiments on the evaporation of water from saturated surfaces. *Trans Inst Chem Eng* 1940;**18**:36.
50. Lurie M, Michailoff M. Evaporation from free water surfaces. *Ind Eng Chem* 1936;**28**:345.
51. Maisel DS, Sherwood TK. Evaporation of liquids into turbulent gas streams. *Chem Eng Prog* 1950;**46**:131.
52. Sherwood TK. Mass transfer and friction in turbulent flow. *Am Inst Chem Eng J* 1940;**36**:817.
53. Sherwood TK. Heat transfer, mass transfer, and fluid friction. *Ind Eng Chem* 1950;**42**:2077.
54. Gamson BW, Thodos G, Hougen OA. Heat mass and momentum transfer in the flow of gases through granular solids. *Am Inst Chem Eng J* 1943;**39**:1.
55. Pratt HRC. The application of turbulent flow theory to transfer processes in tubes containing turbulence promoters and packings. *Trans Inst Chem Eng* 1950;**28**:77.

56. Frössling N. Über die Verdunstung fallender Tropfen. *Gerlands Beitr Geophys* 1938;**52**:170.
57. Garner FH, Keey RB. Mass transfer from single solid spheres. I. Transfer at low Reynolds numbers. *Chem Eng Sci* 1958;**9**:119.
58. Garner FH, Keey RB. Mass transfer from single solid spheres. II. Transfer in free convection. *Chem Eng Sci* 1959;**9**:218.
59. Ranz WE, Marshall WR. Evaporation from drops part II. *Chem Eng Prog* 1952;**48**:173.
60. Rowe PN, Claxton KT, Lewis JB. Heat and mass transfer from a single sphere in an extensive flowing fluid. *Trans Inst Chem Eng* 1965;**41**:T14.
61. Brian PLT, Hayes HB. Effects of transpiration and changing diameter on heat and mass transfer. *AICHE J* 1969;**15**:419.
62. Gupta AS, Thodos G. Direct analogy between heat and mass transfer in beds of spheres. *AICHE J* 1963;**9**:751.
63. Kramers H. Heat transfer from spheres to flowing media. *Physica* 1946;**12**:61.

Further Reading

1. Bennett CO, Myers JE. *Momentum, heat, and mass transfer*. 3rd ed. New York: McGraw-Hill; 1983.
2. Bird RB, Stewart WE, Lightfoot EN. *Transport phenomena*. New York: Wiley; 1960.
3. Cussler EL. *Diffusion. Mass transfer in fluid systems*. 2nd ed. Cambridge: Cambridge University Press; 1997.
4. Eckert ERG, Drake RM. *Analysis of heat and mass transfer*. New York: McGraw-Hill; 1972.
5. Edwards DA, Brenner H, Wasan DT. *Interface transport processes and rheology*. Oxford: Butterworth-Heinemann; 1991.
6. Hewitt GF, Shires GL, Polezhaev YV. *Encyclopedia of heat and mass transfer*. Boca Raton, NY: CRC Press; 1997.
7. Sherwood TK. A review of the development of mass transfer theory. *Chem Eng Educ* 1974;**Fall**:204.
8. Thomas JM, Thomas WJ. *Practice and principles of homogeneous catalysis*. VCH; 1977.
9. Treybal RE. *Mass-transfer operations*. 3rd ed. New York: McGraw-Hill; 1980.

PART 3

Momentum, Heat and Mass Transfer

CHAPTER 3

The Boundary Layer

3.1 Introduction

When a fluid flows over a solid surface, that part of the stream that is close to the surface suffers a significant retardation owing to the no-slip boundary condition at the fluid-solid boundary, and thus a velocity profile develops in the fluid. The velocity gradients are steepest close to the surface and become progressively smaller with distance from the surface. Although theoretically there is no outer limit at which the velocity gradient becomes zero, it is convenient to divide the flow into two parts for practical purposes:

(1) A *boundary layer* close to the surface in which the velocity increases from zero at the surface itself to a near-constant stream velocity at its outer boundary
(2) A region outside the boundary layer in which the velocity gradient in a direction perpendicular to the surface is negligibly small and in which the velocity is everywhere equal to the free stream velocity

The thickness of the boundary layer may be arbitrarily defined as the distance from the surface at which the velocity reaches some proportion (such as 0.9, 0.99, and 0.999) of the undisturbed stream velocity. Alternatively, it may be possible to approximate to the velocity profile by means of an equation that is then used to give the distance from the surface at which the velocity gradient is zero and the velocity is equal to the stream velocity. Difficulties arise in comparing the thicknesses obtained using these various definitions, because velocity is changing so slowly with distance that a small difference in the criterion used for the selection of velocity will account for a very large difference in the corresponding thickness of the boundary layer.

The flow conditions in the boundary layer are of considerable interest to chemical engineers because these influence not only the drag effect of the fluid on the surface but also the heat- or mass-transfer rates where a temperature or a concentration gradient exists.

It is convenient first to consider the flow over a thin plate inserted parallel to the flow of a fluid with a constant stream velocity u_s. It will be assumed that the plate is sufficiently wide for conditions to be constant across any finite width w of the plate that is being considered. Furthermore, the extent of the fluid in a direction perpendicular to the surface is considered as sufficiently large for the velocity of the fluid remote from the surface to be unaffected and

Coulson and Richardson's Chemical Engineering. https://doi.org/10.1016/B978-0-08-102550-5.00003-1

377

to remain constant at the stream velocity u_s. Whilst part of the fluid flows on one side of the flat plate and part on the other, the flow on only one side is considered.

On the assumption that there is no slip at the surface (discussed in Section 3.3), the fluid velocity at all points on the surface, where $y=0$, will be zero. At some position, a distance x from the leading edge, the velocity will increase from zero at the surface to approach the stream velocity u_s asymptotically. At the leading edge, that is where $x=0$, the fluid will have been influenced by the surface for only an infinitesimal time, and therefore, only the molecular layer of fluid at the surface will have been retarded. At progressively greater distances (x) along the surface, the fluid will have been retarded for a greater time, and the effects will be felt to greater depths in the fluid. Thus, the thickness (δ) of the boundary layer will increase, starting from a zero value at the leading edge. Furthermore, the velocity gradient at the surface $(du_x/dy)_{y=0}$ (where u_x is the velocity in the X-direction at a distance y from the surface) will become less because the velocity will change by the same amount (from $u_x=0$ at $y=0$ to $u_x=u_s$ at $y=\delta$) over a greater distance. The development of the boundary layer is illustrated in Fig. 3.1.

Near the leading edge of the surface where the boundary layer thickness is small, the flow will be streamline, or laminar, and the shear stresses will arise solely from viscous shear effects. When the boundary layer thickness exceeds a critical value, however, the laminar flow ceases to be stable, and turbulence sets in. The transition from streamline to turbulent flow is not as sharply defined as in pipe flow (discussed in Chapter 3 of Vol. 1A) and may be appreciably influenced by small disturbances generated at the leading edge of the surface and by roughness and imperfections of the surface in itself, all of which lead to an early development of turbulence. However, for equivalent conditions, the important flow parameter is the Reynolds number Re_δ $(=u_s\delta\rho/\mu)$. Because δ can be expressed as a function of x, the distance from the leading edge of the surface, the usual criterion is taken as the value of the Reynolds number Re_x $(=u_sx\rho/\mu)$. If the location of the transition point is at a distance x_c from the leading edge, then $Re_{x_c}=u_sx_c\rho/\mu$ is of the order of 10^5.

The transition from streamline to turbulent flow in the neighbourhood of a surface has been studied by Brown[1] with the aid of a three-dimensional smoke tunnel with transparent faces. Photographs were taken using short-interval flash lamps giving effective exposure times

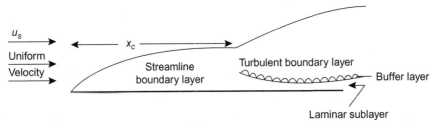

Fig. 3.1
Development of the boundary layer.

Fig. 3.2
Formation of vortices in flow over an aerofoil.

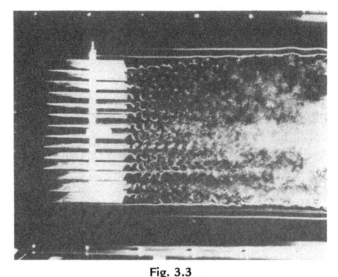

Fig. 3.3
Wake produced by cascade of flat plates with sharpened leading edges and blunt trailing edges.

of about 20 µs. Fig. 3.2 illustrates the flow over an aerofoil and the manner in which vortices are created. Fig. 3.3 shows the wake of a cascade of flat plates with sharpened leading edges and blunt trailing edges.

When the flow in the boundary layer is turbulent, streamline flow persists in a thin region close to the surface called the *laminar sublayer*. This region is of particular importance because, in heat or mass transfer, it is where the greater part of the resistance to transfer lies. High heat- and mass-transfer rates therefore depend on the laminar sublayer being thin. Separating the laminar sublayer from the turbulent part of the boundary layer is the *buffer layer* (Fig. 3.1), in which the contributions of the viscous effects and of the turbulent eddies are of comparable magnitudes. This phenomenon is discussed in more detail in Chapter 4, although the general picture is as follows.

The average size of the eddies, or turbulent circulating currents, becomes progressively smaller as the surface is approached. The eddies are responsible for transference of momentum from the faster-moving fluid remote from the surface to slower-moving fluid near the surface. In this way, momentum is transferred by a progression of eddies of diminishing dimensions down to about 1 mm, at which point they tend to die out and viscous forces (due to transfer by molecular-scale movement) predominate near the surface. Although the laminar sublayer is taken to be the region where eddies are completely absent, there is evidence that occasionally eddies do penetrate right up to the surface. The situation is therefore highly complex, and any attempt to model the process involves simplifying assumptions.

The following treatment, based on the simplified approach suggested by Prandtl,[2] involves the following three assumptions:

(1)   That the flow may be considered essentially as unidirectional (X-direction) and that the effects of velocity components perpendicular to the surface *within* the boundary layer may be neglected (i.e., $u_y \ll u_x$). This condition will not be met at very low Reynolds numbers where the boundary layer thickens rapidly.

(2)   That the existence of the buffer layer may be neglected and that in turbulent flow, the boundary layer may be considered as consisting of a turbulent region adjacent to a laminar sublayer that separates it from the surface.

(3)   That the stream velocity does not change in the direction of flow. On this basis, from Bernoulli's theorem, the pressure then does not change (i.e., $\partial P/\partial x = 0$). In practice, $\partial P/\partial x$ may be positive or negative. If positive, a greater retardation of the fluid will result, and the boundary layer will thicken more rapidly. If $\partial P/\partial x$ is negative, the converse will be true.

For flow against a pressure gradient ($\partial P/\partial x$ positive in the direction of flow), the combined force due to pressure gradient and friction may be sufficient to bring the fluid completely to rest and to cause some backflow close to the surface. When this occurs, the fluid velocity will be zero, not only at the surface but also at a second position a small distance away. In these circumstances, the boundary layer is said to *separate*, and circulating currents are set up as shown in Fig. 3.4. An example of this phenomenon is described in Chapter 3 of Vol. 1A in which flow relative to a cylinder or sphere is considered.

The procedure adopted here consists of taking a momentum balance on an element of fluid. The resulting *Momentum Equation* involves no assumptions concerning the nature of the flow. However, it includes an integral, the evaluation of which requires a knowledge of the velocity profile $u_x = f(y)$. At this stage, assumptions must be made concerning the nature of the flow in order to obtain realistic expressions for the velocity profile.

3.2 The Momentum Equation

It will be assumed that a fluid of density ρ and viscosity μ flows over a plane surface and the velocity of flow outside the boundary layer is u_s. A boundary layer of thickness δ forms near the surface, and at a distance y from the surface, the velocity of the fluid is reduced to a value u_x.

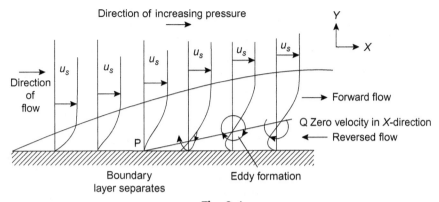

Fig. 3.4
Boundary layer separation.

The equilibrium is considered of an element of fluid bounded by the planes 1-2 and 3-4 at distances x and $x+dx$, respectively, from the leading edge; the element is of length l in the direction of flow and is of depth w in the direction perpendicular to the plane 1-2-3-4. The distance l is greater than the boundary layer thickness δ (Fig. 3.5), and conditions are constant over the width w. The velocities and forces in the X-direction are now considered.

At plane 1-2, mass rate of flow through a strip of thickness dy at distance y from the surface

$$= \rho u_x w\, dy$$

The total flow through plane 1-2

$$= w \int_0^l \rho u_x dy \tag{3.1}$$

The rate of transfer of momentum through the elementary strip

$$= \rho u_x w\, dy\, u_x = w\rho u_x^2\, dy$$

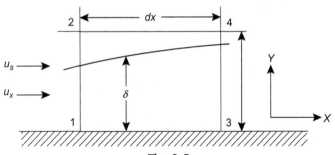

Fig. 3.5
Element of boundary layer.

The total rate of transfer of momentum through plane 1-2

$$M_i = w \int_0^l \rho u_x^2 dy \tag{3.2}$$

In passing from plane 1-2 to plane 3-4, the mass flow changes by

$$w \frac{\partial}{\partial x} \left(\int_0^l \rho u_x dy \right) dx \tag{3.3}$$

and the momentum flux changes by

$$M_{ii} - M_i = w \frac{\partial}{\partial x} \left(\int_0^l \rho u_x^2 dy \right) dx \tag{3.4}$$

where M_{ii} is the momentum flux across the plane 3-4.

A mass flow of fluid equal to the difference between the flows at planes 3-4 and 1-2 (Eq. 3.3) must therefore occur through plane 2-4, as it is assumed that there is uniformity over the width of the element.

Since plane 2-4 lies outside the boundary layer, the fluid crossing this plane must have a velocity u_s in the X-direction. Because the fluid in the boundary layer is being retarded, there will be a smaller flow at plane 3-4 than at 1-2, and hence, the flow through plane 2-4 is outwards, and fluid leaves the element of volume.

Thus, the rate of transfer of momentum through plane 2-4 out of the element is

$$M_{iii} = -wu_s \frac{\partial}{\partial x} \left(\int_0^l \rho u_x dy \right) dx \tag{3.5}$$

It will be noted that the derivative is negative, which indicates a positive outflow of momentum from the element.

3.2.1 Steady-State Momentum Balance Over the Element 1-2-3-4

The terms that must be considered in the momentum balance for the X-direction are as follows:

   (i)   The momentum flux M_i through plane 1-2 *into* the element
  (ii)   The momentum flux M_{ii} through plane 3-4 *out of* the element
 (iii)   The momentum flux M_{iii} through plane 3-4 *out of* the element

Thus, the net momentum flux out of the element M_{ex} is given by

$$\underbrace{w \frac{\partial}{\partial x} \left(\int_0^l \rho u_x^2 dy \right) dx}_{M_{ii} - M_i} + \underbrace{\left\{ -wu_s \frac{\partial}{\partial x} \left(\int_0^l \rho u_x dy \right) dx \right\}}_{M_{iii}}$$

Then, since u_s is assumed not to vary with x,

$$M_{ex} = \left\{ -w\frac{\partial}{\partial x}\left(\int_0^l \rho u_x(u_s - u_x)\,dy \right)dx \right\} \tag{3.6}$$

The net rate of change of momentum in the X-direction on the element must be equal to the momentum added from outside, through plane 2-4, together with the net force acting on it.

The forces in the X-direction acting on the element of fluid are the following:

(1) A shear force resulting from the shear stress R_0 acting at the surface. This is a retarding force, and therefore, R_0 is negative.
(2) The force produced as a result of any difference in pressure dP between the planes 3-4 and 1-2. However, if the velocity u_s outside the boundary layer remains constant, from Bernoulli's theorem, there can be no pressure gradient in the X-direction and $\partial P/\partial x = 0$.

Thus, the net force acting F is just the retarding force attributable to the shear stress at the surface only and thus,

$$F = R_0 w\,dx \tag{3.7}$$

Equating the net momentum flux out of the element to the net retarding force (Eqs 3.6, 3.7) and simplifying give

$$\frac{\partial}{\partial x}\int_0^l \rho(u_s - u_x)u_x\,dy = -R_0 \tag{3.8}$$

This expression, known as the *momentum equation*, may be integrated provided that the relation between u_x and y is known.

If the velocity of the main stream remains constant at u_s and the density ρ may be taken as constant, Eq. (3.8) then becomes

$$\rho\frac{\partial}{\partial x}\int_0^l (u_s - u_x)u_x\,dy = -R_0 \tag{3.9}$$

It may be noted that no assumptions have been made concerning the nature of the flow within the boundary layer and therefore this relation is applicable to both the streamline and the turbulent regions. The relation between u_x and y is derived for streamline and turbulent flow over a plane surface, and the integral in Eq. (3.9) is evaluated.

3.3 The Streamline Portion of the Boundary Layer

In the streamline boundary layer, the only forces acting within the fluid are pure viscous forces, and no transfer of momentum takes place by eddy motion.

Assuming that the relation between u_x and y can be expressed approximately by

$$u_x = u_0 + ay + by^2 + cy^3 \qquad (3.10)$$

The coefficients a, b, c, and u_0 may be evaluated because the boundary conditions that the relation must satisfy are known, as shown in Fig. 3.6.

In fluid dynamics, it is generally assumed that the velocity of flow at a solid boundary, such as a pipe wall, is zero. This is referred to as the *no-slip condition*. If the fluid 'wets' the surface, this assumption can be justified in physical terms since the molecules are small compared with the irregularities on the surface of even the smoothest pipe, and therefore, a fluid layer is effectively held stationary at the wall. All gases and the majority of liquids have a sufficiently low contact angle for this condition to be met.

If the fluid does not wet the wall, the no-slip condition no longer applies, and the pressure gradient at a given flowrate will be lower. This effect is particularly important with the flow of molten and in some microfluidic applications, as discussed later in Chapter 6 of this volume. However, in most applications in chemical engineering, wall slip is not a significant phenomenon and the no-slip boundary condition indeed describes the flow phenomena very well.

It is assumed here that the fluid in contact with the surface is at rest and therefore u_0 must be zero. Furthermore, all the fluid close to the surface is moving at very low velocity, and therefore, any changes in its momentum as it flows parallel to the surface must be extremely small. Consequently, the net shear force acting on any element of fluid near the surface is negligible, the retarding force at its lower boundary being balanced by the accelerating force at its upper boundary. Thus, the shear stress R_0 in the fluid near the surface must approach a constant value.

Since $R_0 = -\mu(\partial u_x/\partial y)_{y=0}$, $\partial u_x/\partial y$ must also be constant at small values of y and

$$\left(\frac{\partial^2 u_x}{\partial y^2}\right)_{y=0} = 0$$

At the distant edge of the boundary layer, it is assumed that the velocity just equals the main stream velocity and that there is no discontinuity in the velocity profile.

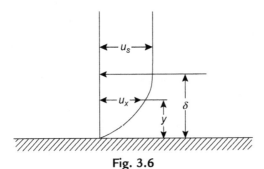

Fig. 3.6
Velocity distribution in streamline boundary layer.

Thus, when $y = \delta$,

$$u_x = u_s \text{ and } \frac{\partial u_x}{\partial y} = 0$$

Now, with $u_0 = 0$, Eq. (3.10) becomes

$$u_x = ay + by^2 + cy^3$$
$$\frac{\partial u_x}{\partial y} = a + 2by + 3cy^3$$

and

$$\frac{\partial^2 u_x}{\partial y^2} = 2b + 6cy$$

At $y = 0$,

$$\frac{\partial^2 u_x}{\partial y^2} = 0$$

Thus,

$$b = 0$$

At $y = \delta$,

$$u_x = a\delta + c\delta^3 = u_s$$

and

$$\frac{\partial u_x}{\partial y} = a + 3c\delta^2 = 0$$

Thus,

$$a = -3c\delta^2$$

Hence,

$$c = -\frac{u_s}{2\delta^3} \text{ and } a = \frac{3u_s}{2\delta}$$

The equation for the velocity profile is therefore

$$u_x = \frac{3u_s}{2}\frac{y}{\delta} - \frac{u_s}{2}\left(\frac{y}{\delta}\right)^3 \tag{3.11}$$

$$\frac{u_x}{u_s} = \frac{3}{2}\left(\frac{y}{\delta}\right) - \frac{1}{2}\left(\frac{y}{\delta}\right)^3 \tag{3.12}$$

Eq. (3.12) corresponds closely to experimentally determined velocity profiles in a laminar boundary layer.

This relation applies over the range $0 < y < \delta$.

When $y > \delta$, then,

$$u_x = u_s \tag{3.13}$$

The integral in the momentum equation (3.9) can now be evaluated for the streamline boundary layer by considering the ranges $0 < y < \delta$ and $\delta < y < l$ separately.

Thus,

$$
\begin{aligned}
\int_0^l (u_s - u_x) u_x \mathrm{d}y &= \int_0^\delta u_s^2 \left(1 - \frac{3y}{2\delta} + \frac{y^3}{2\delta^3}\right)\left(\frac{3y}{2\delta} - \frac{y^3}{2\delta^3}\right) \mathrm{d}y + \int_\delta^l (u_s - u_s) u_s \mathrm{d}y \\
&= u_s^2 \int_0^\delta \left(\frac{3y}{2\delta} - \frac{9y^2}{4\delta^2} - \frac{1}{2}\frac{y^3}{\delta^3} + \frac{3}{2}\frac{y^4}{\delta^4} - \frac{1}{4}\frac{y^6}{\delta^6}\right) \mathrm{d}y \\
&= u_s^2 \delta \left(\frac{3}{4} - \frac{3}{4} - \frac{1}{8} + \frac{3}{10} - \frac{1}{28}\right) \\
&= \frac{39}{280} \delta u_s^2
\end{aligned}
\tag{3.14}
$$

In addition,

$$R_0 = -\mu \left(\frac{\partial u_x}{\partial y}\right)_{y=0} = -\frac{3}{2}\mu \frac{u_s}{\delta} \tag{3.15}$$

Substitution from Eqs (3.14), (3.15) in Eq. (3.9) gives

$$\rho \frac{\partial}{\partial x}\left(\frac{39}{280}\delta u_s^2\right) = \frac{3}{2}\mu \frac{u_s}{\delta}$$

$$\therefore \quad \delta \mathrm{d}\delta = \left(\frac{140}{13}\right)\frac{\mu}{\rho u_s}\frac{1}{u_s}\mathrm{d}x$$

$$\therefore \quad \frac{\delta^2}{2} = \left(\frac{140}{13}\right)\left(\frac{\mu x}{\rho u_s}\right) \quad (\text{since } \delta = 0 \text{ when } x = 0) \tag{3.16}$$

Thus,

$$\delta = 4.64 \sqrt{\frac{\mu x}{\rho u_s}}$$

and

$$\frac{\delta}{x} = 4.64 \sqrt{\frac{\mu}{x \rho u_s}} = 4.64 Re_x^{-1/2} \tag{3.17}$$

The rate of thickening of the boundary layer is then obtained by differentiating Eq. (3.17):

$$\frac{d\delta}{dx} = 2.32 Re_x^{-1/2} \tag{3.18}$$

This relation for the thickness of the boundary layer has been obtained on the assumption that the velocity profile can be described by a polynomial of the form of Eq. (3.10) and that the main stream velocity is reached at a distance δ from the surface, whereas, in fact, the stream velocity is approached asymptotically. Although Eq. (3.11) gives the velocity u_x accurately as a function of y, it does not provide a means of calculating accurately the distance from the surface at which u_x has a particular value when u_x is near u_s, because $\partial u_x/\partial y$ is then small. The thickness of the boundary layer as calculated is therefore a function of the particular approximate relation that is taken to represent the velocity profile. This difficulty can be overcome by introducing a new concept, the *displacement thickness δ^**.

When a viscous fluid flows over a surface, it is retarded, and the overall flowrate is therefore reduced. A nonviscous fluid, however, would not be retarded, and therefore, a boundary layer would not form. The displacement thickness δ^* is defined as the distance the surface would have to be moved in the Y-direction in order to obtain the same rate of flow with this nonviscous fluid as would be obtained for the viscous fluid with the surface retained at $x=0$.

The mass rate of flow of a frictionless fluid between $y=\delta^*$ and $y=\infty$

$$= \rho \int_{\delta^*}^{\infty} u_s dy$$

The mass rate of flow of the real fluid between $y=0$ and $y=\infty$

$$= \rho \int_{0}^{\infty} u_x dy$$

Then, by definition of the displacement thickness,

$$\rho \int_{\delta^*}^{\infty} u_s dy = \rho \int_{0}^{\infty} u_x dy$$

or

$$\int_{0}^{\infty} u_s dy - \int_{0}^{\delta^*} u_s dy = \int_{0}^{\infty} u_x dy$$

and

$$\delta^* = \int_{0}^{\infty} \left(1 - \frac{u_x}{u_s} \right) dy \tag{3.19}$$

Using Eq. (3.12) to give the velocity profile,

$$\delta^* = \int_0^\delta \left(1 - \frac{3y}{2\delta} + \frac{1}{2}\frac{y^3}{\delta^3}\right) dy$$

since Eq. (3.12) applies only over the limits, $0 < y < \delta$, and outside this region, $u_x = u_s$, and the integral is zero.

Then,

$$\delta^* = \delta\left(1 - \frac{3}{4} + \frac{1}{8}\right)$$

and

$$\frac{\delta^*}{\delta} = 0.375 \tag{3.20}$$

3.3.1 Shear Stress at the Surface

The shear stress in the fluid at the surface is given by

$$R_0 = -\mu\left(\frac{\partial u_x}{\partial y}\right)_{y=0}$$

$$= -\frac{3}{2}\mu\frac{u_s}{\delta} \quad \text{(from Eq. 3.15)}$$

$$= -\frac{3}{2}\mu u_s \frac{1}{x}\frac{1}{4.65}\sqrt{\frac{x\rho u_s}{\mu}}$$

$$= -0.323\rho u_s^2\sqrt{\frac{\mu}{x\rho u_s}} = -0.323\rho u_s^2 Re_x^{-1/2}$$

The shear stress R acting on the surface itself is equal and opposite to the shear stress on the fluid at the surface, that is, $R = -R_0$.

$$\frac{R}{\rho u_x^2} = 0.323 Re_x^{-1/2} \tag{3.21}$$

Eq. (3.21) gives the point values of R and $R/\rho u_s^2$ at $x = x$. In order to calculate the total frictional force acting at the surface, it is necessary to multiply the average value of R between $x = 0$ and $x = x$ by the area of the surface.

The average value of $R/\rho u_s^2$ denoted by the symbol $(R/\rho u_s^2)_m$ is then given by

$$\left(\frac{R}{\rho u_s^2}\right)_m x = \int_0^x \frac{R}{\rho u_s^2}\,\mathrm{d}x$$

$$= \int_0^x 0.323\sqrt{\frac{\mu}{x\rho u_s}}\,\mathrm{d}x \quad \text{(from Eq. 3.21)}$$

$$= 0.646x\sqrt{\frac{\mu}{x\rho u_s}} \tag{3.22}$$

$$\left(\frac{R}{\rho u_s^2}\right)_m = 0.646\sqrt{\frac{\mu}{x\rho u_s}}$$

$$= 0.646Re_x^{-0.5} \approx 0.65Re_x^{-0.5}$$

3.4 The Turbulent Boundary Layer

3.4.1 The Turbulent Portion

Eq. (3.12) does not fit velocity profiles measured in a turbulent boundary layer, and an alternative approach must be used. In the simplified treatment of the flow conditions within the turbulent boundary layer, the existence of the buffer layer, shown in Fig. 3.1, is neglected, and it is assumed that the boundary layer consists of a laminar sublayer, in which momentum transfer is by molecular motion alone, outside which there is a turbulent region in which transfer is effected entirely by eddy motion (Fig. 3.7). The approach is based on the assumption that the shear stress at a plane surface can be calculated from the simple power law developed by Blasius, already referred to in Chapter 3 of Vol. 1A.

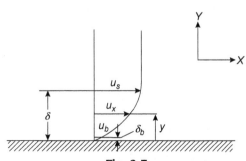

Fig. 3.7
Turbulent boundary layer.

Blasius[3] has given the following approximate expression for the shear stress at a plane smooth surface over which a fluid is flowing with a velocity u_s, for conditions where $Re_x < 10^7$:

$$\frac{R}{\rho u_s^2} = 0.0228 \left(\frac{\mu}{u_s \delta \rho}\right)^{0.25} \tag{3.23}$$

Thus, the shear stress is expressed as a function of the boundary layer thickness δ, and it is therefore implicitly assumed that a certain velocity profile exists in the fluid. As a first assumption, it may be assumed that a simple power relation exists between the velocity and the distance from the surface in the boundary layer or

$$\frac{u_x}{u_s} = \left(\frac{y}{\delta}\right)^f \tag{3.24}$$

Hence $R = 0.0228 \rho u_s^2 \left(\frac{\mu}{u_s \delta \rho}\right)^{0.25}$

$$= 0.0228 \rho^{0.75} \mu^{0.25} \delta^{-0.25} u_s^{1.75}$$

$$= 0.0228 \rho^{0.75} \mu^{0.25} \delta^{-0.25} u_x^{1.75} \left(\frac{\delta}{y}\right)^{1.75f} \quad \text{(from Eq. 3.24)} \tag{3.25}$$

$$= 0.0228 \rho^{0.75} \mu^{0.25} u_x^{1.75} y^{-1.75f} \delta^{1.75f - 0.25}$$

If the velocity profile is the same for all stream velocities, the shear stress must be defined by specifying the velocity u_x at any distance y from the surface. The boundary layer thickness, determined by the velocity profile, is then no longer an independent variable so that the index of δ in Eq. (3.25) must be zero or

$$1.75f - 0.25 = 0$$

and

$$f = \frac{1}{7}$$

Thus,

$$\frac{u_x}{u_s} = \left(\frac{y}{\delta}\right)^{1/7} \tag{3.26}$$

Eq. (3.26) is sometimes known as the *Prandtl one-seventh power law*.

Differentiating Eq. (3.26) with respect to y gives

$$\frac{\partial u_x}{\partial y} = \frac{1}{7} u_s \delta^{-1/7} y^{-6/7} = \left(\frac{1}{7}\right)\left(\frac{u_s}{y}\right)\left(\frac{y}{\delta}\right)^{1/7} \tag{3.27}$$

This relation is not completely satisfactory in that it gives an infinite velocity gradient at the surface, where the laminar sublayer exists, and a finite velocity gradient at the outer edge of the boundary layer. This is in contradiction to the conditions that must exist in the stream. However, little error is introduced by using this relation for the whole of the boundary layer in the momentum equation because, firstly, both the velocities and hence the momentum fluxes near the surface are very low and, secondly, it gives the correct value of the velocity at the edge of the boundary layer. Accepting Eq. (3.26) for the limits $0 < y < \delta$, the integral in Eq. (3.9) becomes

$$\int_0^l (u_s - u_x)u_x dy = u_s^2 \left\{ \int_0^\delta \left[1 - \left(\frac{y}{\delta}\right)^{1/7} \right] \left(\frac{y}{\delta}\right)^{1/7} dy \right\} + \int_\delta^l (u_s - u_s)u_s dy$$

$$= u_s^2 \int_0^\delta \left[\left(\frac{y}{\delta}\right)^{1/7} - \left(\frac{y}{\delta}\right)^{2/7} \right] dy$$

$$= u_s^2 \delta \left(\frac{7}{8} - \frac{7}{9}\right)$$

$$= \frac{7}{72} u_s^2 \delta$$

(3.28)

From the Blasius equation,

$$-R_0 = R = 0.228\rho u_s^2 \left(\frac{\mu}{u_s \delta \rho}\right)^{1/4} \quad \text{(from Eq. 3.23)}$$

Substituting from Eqs (3.23), (3.28) in Eq. (3.9) gives

$$\rho \frac{\partial}{\partial x}\left[\frac{7}{72}u_s^2 \delta\right] = 0.0228\rho u_s^2 \left(\frac{\mu}{u_s \delta \rho}\right)^{1/4}$$

$$\delta^{1/4} d\delta = 0.235 \left(\frac{\mu}{u_s \rho}\right)^{1/4} dx$$

and

$$\frac{4}{5}\delta^{5/4} = 0.235x \left(\frac{\mu}{u_s \rho}\right)^{1/4} + \text{constant}$$

Putting the constant equal to zero, implies that $\delta = 0$ when $x = 0$, that is that the turbulent boundary layer extends to the leading edge of the surface. An error is introduced by this assumption, but it is found to be small except where the surface is only slightly longer than the critical distance x_c for the laminar-turbulent transition.

Thus,

$$\delta = 0.376x^{0.8}\left(\frac{\mu}{u_s\rho}\right)^{0.2}$$

$$= 0.376x\left(\frac{\mu}{u_s\rho x}\right)^{0.2} \tag{3.29}$$

or

$$\frac{\delta}{x} = 0.376Re_x^{-0.2} \tag{3.30}$$

The rate of thickening of the boundary layer is obtained by differentiating Eq. (3.30) to give

$$\frac{d\delta}{dx} = 0.301Re_x^{-0.2} \tag{3.31}$$

The displacement thickness δ^* is given by Eq. (3.19):

$$\delta^* = \int_0^\infty \left(1 - \frac{u_x}{u_s}\right)dy \quad (\text{Eq. } 3.19)$$

$$= \int_0^\delta \left(1 - \left(\frac{y}{\delta}\right)^{1/7}\right)dy \left(\text{since } 1 - \frac{u_x}{u_x} = 0 \text{ when } y > \delta\right) \tag{3.32}$$

$$= \frac{1}{8}\delta$$

As noted previously, δ^* is independent of the particular approximation used for the velocity profile.

It is of interest to compare the rates of thickening of the streamline and turbulent boundary layers at the transition point. Taking a typical value of $Re_{xc} = 10^5$, then,

For the streamline boundary layer, from Eq. (3.18), $\frac{d\delta}{dx} = 0.0073$

For the turbulent boundary layer, from Eq. (3.31), $\frac{d\delta}{dx} = 0.0301$

Thus, the turbulent boundary layer is thickening at about four times the rate of the streamline boundary layer at the transition point.

3.4.2 The Laminar Sublayer

If at a distance x from the leading edge the laminar sublayer is of thickness δ_b and the total thickness of the boundary layer is δ, the properties of the laminar sublayer can be found by equating the shear stress at the surface as given by the Blasius equation (3.23) to that obtained from the velocity gradient near the surface.

It has been noted that the shear stress and hence the velocity gradient are almost constant near the surface. Since the laminar sublayer is very thin, the velocity gradient within it may therefore be taken as constant.

Thus, the shear stress in the fluid at the surface,

$$R_0 = -\mu \left(\frac{\partial u_x}{\partial y} \right)_{y=0} = -\mu \frac{u_x}{y}, \quad \text{where } y < \delta_b$$

Equating this to the value obtained from Eq. (3.23) gives

$$0.0228 \rho u_s^2 \left(\frac{\mu}{u_s \delta \rho} \right)^{1/4} = \mu \frac{u_x}{y}$$

and

$$u_x = 0.0228 \rho u_s^2 \frac{1}{\mu} \left(\frac{\mu}{u_s \delta \rho} \right)^{1/4} y$$

If the velocity at the edge of the laminar sublayer is u_b, that is, if $u_x = u_b$ when $y = \delta_b$,

$$u_b = 0.0228 \rho u_s^2 \frac{1}{\mu} \left(\frac{\mu}{u_s \delta \rho} \right)^{1/4} \delta_b$$

$$= 0.0228 \frac{\rho u_s^2}{\mu} \frac{\mu}{u_s \delta \rho} \delta_b \left(\frac{\mu}{u_s \delta \rho} \right)^{-3/4}$$

Thus,

$$\frac{\delta_b}{\delta} = \frac{1}{0.0228} \left(\frac{u_b}{u_s} \right) \left(\frac{\mu}{u_s \delta \rho} \right)^{3/4} \tag{3.33}$$

The velocity at the inner edge of the turbulent region must also be given by the equation for the velocity distribution in the turbulent region.

Hence,

$$\left(\frac{\delta_b}{\delta} \right)^{1/7} = \frac{u_b}{u_s} \quad \text{(from Eq. 3.26)}$$

Thus,

$$\left(\frac{u_b}{u_s} \right)7 = \frac{1}{0.0228} \left(\frac{u_b}{u_s} \right) \left(\frac{\mu}{u_s \delta \rho} \right)^{3/4} \quad \text{(from Eq. 3.33)}$$

or

$$\frac{u_b}{u_s} = 1.87 \left(\frac{\mu}{u_s \delta \rho}\right)^{1/8}$$

$$= 1.87 Re_\delta^{-1/8}$$

(3.34)

Since

$$\delta = 0.376 x^{0.8} \left(\frac{\mu}{u_s \rho}\right)^{0.2} \quad \text{(Eq. 3.29)}$$

$$\frac{u_b}{u_s} = 1.87 \left(\frac{u_s \rho}{\mu} 0.376 \frac{x^{0.8} \mu^{0.2}}{u_s^{0.2} \rho^{0.2}}\right)^{-1/8}$$

$$= \frac{1.87}{0.376^{1/8}} \left(\frac{u_s^{0.8} x^{0.8} \rho^{0.8}}{\mu^{0.8}}\right)^{-1/8}$$

(3.35)

$$= 2.11 Re_x^{-0.1} \approx 2.1 Re_x^{-0.1}$$

The thickness of the laminar sublayer is given by

$$\frac{\delta_b}{\delta} = \left(\frac{u_b}{u_s}\right)^7 = \frac{190}{Re_x^{0.7}} \quad \text{(from Eqs 3.26, 3.35)}$$

or

$$\frac{\delta_b}{x} = \frac{190}{Re_x^{0.7}} \frac{0.376}{Re_x^{-0.9}} \quad \text{(from Eq. 3.30)}$$

$$= 71.5 Re_x^{-0.9}$$

(3.36)

Thus, $\delta_b \propto x^{0.1}$; that is, δ_b increases very slowly as x increases. Further, $\delta_b \propto u_s^{-0.9}$ and therefore decreases rapidly as the velocity is increased, and heat- and mass-transfer coefficients are therefore considerably influenced by the velocity.

The shear stress at the surface, at a distance x from the leading edge, is given by

$$R_0 = -\mu \frac{u_b}{\delta_b}$$

Since $R_0 = -R$, then

$$R = \mu 2.11 u_s Re_x^{-0.1} \frac{1}{x} \frac{1}{71.5} Re_x^{0.9} \quad \text{(from Eqs 3.35, 3.36)}$$

$$= 0.0296 Re_x^{0.8} \frac{\mu u_s}{x}$$

(3.37)

$$= 0.0296 \rho u_s^2 Re_x^{-0.2} \approx 0.03 \rho u_s^2 Re_x^{-0.2}$$

or

$$\frac{R}{\rho u_s^2} = 0.0296 Re_x^{-0.2} \tag{3.38}$$

or approximately

$$\frac{R}{\rho u_s^2} = 0.03 Re_x^{-0.2} \tag{3.39}$$

The mean value of $R/\rho u_s^2$ over the range $x=0$ to $x=x$ is given by

$$\left(\frac{R}{\rho u_s^2}\right)_m x = \int_0^x \left(\frac{R}{\rho u_s^2}\right) dx$$

$$= \int_0^x 0.0296 \left(\frac{\mu}{u_s x \rho}\right)^{0.2} dx$$

$$= 0.0296 \left(\frac{\mu}{u_s x \rho}\right)^{0.2} \frac{x}{0.8}$$

or

$$\left(\frac{R}{\rho u_s^2}\right)_m = 0.037 Re_x^{-0.2} \tag{3.40}$$

The total shear force acting on the surface is found by adding the forces acting in the streamline $(x < x_c)$ and turbulent $(x > x_c)$ regions. This can be done provided the critical value Re_{xc}, is known.

In the streamline region,

$$\left(\frac{R}{\rho u_s^2}\right)_m = 0.646 Re_x^{-0.5} \quad \text{(Eq. 3.22)}$$

In the turbulent region,

$$\left(\frac{R}{\rho u_s^2}\right)_m = 0.037 Re_x^{-0.2} \quad \text{(Eq. 3.40)}$$

In calculating the mean value of $(R/\rho u_s^2)_m$ in the turbulent region, it was assumed that the turbulent boundary layer extended to the leading edge. A more accurate value for the mean value of $(R/\rho u_s^2)_m$ over the whole surface can be obtained by using the expression for streamline conditions over the range from $x=0$ to $x=x_c$ (where x_c is the critical distance from the leading edge) and the expression for turbulent conditions in the range $x=x_c$ *to* $x=x$.

Thus,

$$
\left(\frac{R}{\rho u_s^2}\right)_m = \frac{1}{x}\left(0.646 Re_{x_c}^{-0.5} x_c + 0.037 Re_x^{-0.2} x - 0.037 Re_{x_c}^{-0.2} x_c\right)
$$

$$
= 0.646 Re_{x_c}^{-0.5}\frac{Re_{x_c}}{Re_x} + 0.037 Re_x^{-0.2} - 0.037 Re_{x_c}^{-0.2}\frac{Re_{x_c}}{Re_x} \tag{3.41}
$$

$$
= 0.037 Re_x^{-0.2} + Re_x^{-1}\left(0.646 Re_{x_c}^{0.5} - 0.037 Re_{x_c}^{0.8}\right)
$$

Example 3.1

Water flows at a velocity of 1 m/s over a plane surface 0.6 m wide and 1 m long. Calculate the total drag force acting on the surface if the transition from streamline to turbulent flow in the boundary layer occurs when the Reynolds group $Re_{xc} = 10^5$.

Solution

Taking $\mu = 1$ mN s/m$^2 = 10^{-3}$ N s/m^2, at the far end of the surface, $Re_x = (1 \times 1 \times 10^3)/10^{-3} = 10^6$ mean value of $R/\rho u_y^2$ from Eq. (3.41)

$$
= 0.037\left(10^6\right)^{-0.2} + \left(10^6\right)^{-1}\left[0.646\left(10^5\right)^{0.5} - 0.037\left(10^5\right)^{0.8}\right]
$$

$$
= 0.00214
$$

$$
\text{Total drag force} = \frac{R}{\rho u_s^2}\left(\rho u_s^2\right) \times (\text{area of surface})
$$

$$
= \left(0.00214 \times 1000 \times 1^2 \times 1 \times 0.6\right)
$$

$$
= \underline{1.28\text{N}}
$$

Example 3.2

Calculate the thickness of the boundary layer at a distance of 150 mm from the leading edge of a surface over which oil, of viscosity 0.05 N s/m^2 and density 1000 kg/m^3, flows with a velocity of 0.3 m/s. What is the displacement thickness of the boundary layer?

Solution

$$
Re_x = (0.150 \times 0.3 \times 1000/0.05) = 900
$$

For streamline flow,

$$
\frac{\delta}{x} = \frac{4.64}{Re_x^{0.5}} \quad (\text{from Eq. 3.17})
$$

$$
= \frac{4.64}{900^{0.5}} = 0.1545
$$

Hence,

$$\delta = (0.1545 \times 0.150) = 0.0232\,\mathrm{m} = \underline{23.2\,\mathrm{mm}}$$

and from Eq. (3.20), the displacement thickness $\delta^* = (0.375 \times 23.2) = \underline{8.7\,\mathrm{mm}}$.

3.5 Boundary Layer Theory Applied to Pipe Flow

3.5.1 Entry Conditions

When a fluid flowing with a uniform velocity enters a pipe, a boundary layer forms at the walls and gradually thickens with distance from the entry point. Since the fluid in the boundary layer is retarded and the total flow remains constant, the fluid in the central stream is accelerated. At a certain distance from the inlet, the boundary layers, which have formed in contact with the walls, join at the axis of the pipe and, from that point onwards, occupy the whole cross-section, and consequently remain of a constant thickness. *Fully developed flow* then exists. If the boundary layers are still streamline when fully developed flow commences, the flow in the pipe remains streamline. On the other hand, if the boundary layers are already turbulent, turbulent flow will persist, as shown in Fig. 3.8.

An approximate experimental expression for the inlet length L_e for laminar flow is

$$\frac{L_e}{d} = 0.0575 Re \qquad (3.42)$$

where d is the diameter of the pipe and Re is the Reynolds group with respect to pipe diameter and based on the mean velocity of flow in the pipe. This expression is only approximate and is inaccurate for Reynolds numbers in the region of 2500 because the boundary layer thickness increases very rapidly in this region. An average value of L_e at a Reynolds number of 2500 is about 100 d. The inlet length is somewhat arbitrary as steady conditions in the pipe

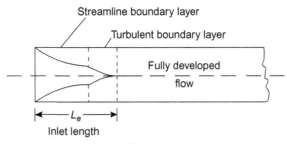

Fig. 3.8
Conditions at entry to pipe.

are approached asymptotically, the boundary layer thickness being a function of the assumed velocity profile.

At the inlet to the pipe, the velocity across the whole section is constant. The velocity at the pipe axis will progressively increase in the direction of flow and reach a maximum value when the boundary layers join. Beyond this point, the velocity profile and the velocity at the axis will not change. Since the fluid at the axis has been accelerated, its kinetic energy per unit mass will increase, and therefore, there must be a corresponding fall in its pressure energy.

Under streamline conditions, the velocity at the axis u_s will increase from a value u at the inlet to a value $2u$ where fully developed flow exists, as shown in Fig. 3.9, because the mean velocity of flow u in the pipe is half of the axial velocity, from Eq. (3.36) of Vol. 1A.

Thus, the kinetic energy per unit mass of the fluid at the axis inlet $= \frac{1}{2}u^2$
The corresponding kinetic energy at the end of the inlet length $= \frac{1}{2}(2u)^2 = 2u^2$.
The increase in the kinetic energy per unit mass $= \frac{3}{2}u^2$.
Thus, the fall in pressure due to the increase of velocity of the fluid $= \frac{3}{2}\rho u^2$.

If the flow in the pipe is turbulent, the velocity at the axis increases from u to only about $u/0.817$, as given by Eq. (3.63) of Vol. 1A.

Under these conditions, the fall in pressure

$$
\begin{aligned}
&= \frac{1}{2}\rho u^2 \left(\frac{1}{0.817^2} - 1 \right) \\
&\approx \frac{1}{4}\rho u^2
\end{aligned}
\tag{3.43}
$$

If the fluid enters the pipe from a duct of larger cross-section, the existence of a radial velocity component gives rise to the formation of a *vena contracta* near the entry to the pipe, but this has been neglected here.

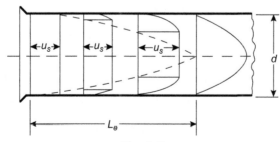

Fig. 3.9
Development of the laminar velocity profile at the entry to a pipe.

3.5.2 Application of the Boundary-Layer Theory

The velocity distribution and frictional resistance have been calculated from purely theoretical considerations for the streamline flow of a fluid in a pipe. The boundary layer theory can now be applied in order to calculate, approximately, the conditions when the fluid is turbulent. For this purpose, it is assumed that the boundary layer expressions may be applied to flow over a cylindrical surface and that the flow conditions in the region of fully developed flow are the same as those when the boundary layers first join. The thickness of the boundary layer is thus taken to be equal to the radius of the pipe, and the velocity at the outer edge of the boundary layer is assumed to be the velocity at the axis. Such assumptions are valid very close to the walls, although significant errors will arise near the centre of the pipe.

The velocity of the fluid may be assumed to obey the Prandtl one-seventh power law, given by Eq. (3.26). If the boundary layer thickness δ is replaced by the pipe radius r, this is then given by

$$\frac{u_x}{u_s} = \left(\frac{y}{r}\right)^{1/7} \tag{3.44}$$

The relation between the mean velocity and the velocity at the axis is derived using this expression in Chapter 3 of Vol. 1A. There, the mean velocity u is shown to be 0.82 times the velocity u_s at the axis, although in this calculation the thickness of the laminar sublayer was neglected and the Prandtl velocity distribution assumed to apply over the whole cross-section. The result therefore is strictly applicable only at very high Reynolds numbers where the thickness of the laminar sublayer is very small. At lower Reynolds numbers, the mean velocity will be rather less than 0.82 times the velocity at the axis.

The expressions for the shear stress at the walls, the thickness of the laminar sublayer, and the velocity at the outer edge of the laminar sublayer may be applied to the turbulent flow of a fluid in a pipe. It is convenient to express these relations in terms of the mean velocity in the pipe, the pipe diameter, and the Reynolds group with respect to the mean velocity and diameter.

The shear stress at the walls is given by the Blasius equation (3.23) as

$$\frac{R}{\rho u_s^2} = 0.0228 \left(\frac{\mu}{u_s r \rho}\right)^{1/4}$$

Writing $u = 0.817 u_s$ and $d = 2r$,

$$\frac{R}{\rho u^2} = 0.0386 \left(\frac{\mu}{u d \rho}\right)^{1/4} = 0.0386 Re^{-1/4} \tag{3.45}$$

This equation is more usually written

$$\frac{R}{\rho u^2} = 0.0396 Re^{-1/4} \quad (\text{see Eq. 3.11}) \tag{3.46}$$

The discrepancy between the coefficients in Eqs (3.45), (3.46) is attributable to the fact that the effect of the curvature of the pipe wall has not been taken into account in applying the equation for flow over a plane surface to flow through a pipe. In addition, it takes no account of the existence of the laminar sublayer at the walls.

Eq. (3.46) is applicable for Reynolds numbers up to 10^5.

The velocity at the edge of the laminar sublayer is given by

$$\frac{u_b}{u_s} = 1.87 \left(\frac{\mu}{u_s r \rho} \right)^{1/8} \quad (\text{Eq. 3.34})$$

that becomes

$$\frac{u_b}{u} = 2.49 \left(\frac{\mu}{u d \rho} \right)^{1/8} \tag{3.47}$$

$$= 2.49 Re^{-1/8}$$

and

$$\frac{u_b}{u_s} = 2.0 Re^{-1/8} \tag{3.48}$$

The thickness of the laminar sublayer is given by

$$\frac{\delta_b}{r} = \left(\frac{u_b}{u_s} \right)^7 \quad (\text{from Eq. 3.26})$$

$$= (1.87)^7 \left(\frac{\mu}{u_s r \rho} \right)^{7/8} \quad (\text{from Eq. 3.34})$$

Thus,

$$\frac{\delta_b}{d} = 62 \left(\frac{\mu}{u d \rho} \right)^{7/8} \tag{3.49}$$

$$= 62 Re^{-7/8}$$

The thickness of the laminar sublayer is therefore almost inversely proportional to the Reynolds number and hence to the velocity.

Example 3.3

Calculate the thickness of the laminar sublayer when benzene flows through a pipe 50 mm in diameter at 2 L/s. What is the velocity of the benzene at the edge of the laminar sublayer? Assume that fully developed flow exists within the pipe and that for benzene, $\rho = 870 \text{ kg/m}^3$ and $\mu = 0.7 \text{ mN s/m}^2$.

Solution

$$\text{The mass flowrate of benzene} = \left(2 \times 10^{-3} \times 870\right)$$

$$= 1.74 \text{ kg/s}$$

Thus,

$$\text{Reynolds number} = \frac{4G}{\mu \pi D} = \frac{4 \times 1.74}{0.7 \times 10^{-3} \pi \times 0.050} \quad \text{(Eq. 4.52)}$$

$$= 63{,}290$$

From Eq. (3.49),

$$\frac{\delta_b}{d} = 62 Re^{-7/8}$$

$$\delta_b = \frac{(62 \times 0.050)}{63{,}290^{7/8}}$$

$$= 1.95 \times 10^{-4} \text{ m}$$

or

$$\underline{0.195 \text{ mm}}$$

$$\text{The mean velocity} = \frac{1.74}{\left(870 \times (\pi/4)0.050^2\right)}$$

$$= 1.018 \text{ m/s}$$

From Eq. (3.47),

$$\frac{u_b}{u} = \frac{2.49}{Re^{1/8}}$$

from which

$$u_b = \frac{(2.49 \times 1.018)}{63{,}290^{1/8}}$$

$$= \underline{\underline{0.637 \text{ m/s}}}$$

3.6 The Boundary Layer for Heat Transfer

3.6.1 Introduction

Where a fluid flows over a surface that is at a different temperature, heat transfer occurs, and a temperature profile is established in the vicinity of the surface. A number of possible conditions may be considered. At the outset, the heat-transfer rate may be sufficient to change the temperature of the fluid stream significantly, or it may remain at a substantially constant temperature. Furthermore, a variety of conditions may apply at the surface. Thus, the surface may be maintained at a constant temperature, particularly if it is in good thermal conduct with a heat source or sink of high thermal capacity. Alternatively, the heat flux at the surface may be maintained constant, or conditions may be intermediate between the constant temperature and the constant heat flux conditions. In general, there is likely to be a far greater variety of conditions as compared with those in the momentum-transfer problem previously considered. Temperature gradients are likely to be highest in the vicinity of the surface, and it is useful to develop the concept of a *thermal boundary layer*, analogous to the velocity boundary layer already considered, within which the whole of the temperature gradient may be regarded as existing.

Thus, a velocity boundary layer and a thermal boundary layer may develop simultaneously. If the physical properties of the fluid do not change significantly over the temperature range to which the fluid is subjected, the velocity boundary layer will not be affected by the heat-transfer process. If physical properties are altered, there will be an interactive effect between the momentum- and heat-transfer processes, leading to a comparatively complex situation in which numerical methods of solution will be necessary.

In general, the thermal boundary layer will not correspond with the velocity boundary layer. In the following treatment, the simplest noninteracting case is considered with physical properties assumed to be constant. The stream temperature is taken as constant (θ_s). In the first case, the wall temperature is also taken as a constant, and then by choosing the temperature scale so that the wall temperature is zero, the boundary conditions are similar to those for momentum transfer.

It will be shown that the momentum and thermal boundary layers coincide only if the Prandtl number is unity, implying equal values for the kinematic viscosity (μ/ρ) and the thermal diffusivity ($D_H = k/C_p\rho$).

The condition of constant heat flux at the surface, as opposed to constant surface temperature, is then considered in a later section.

3.6.2 The Heat Balance

The procedure here is similar to that adopted previously. A heat balance, as opposed to a momentum balance, is taken over an element that extends beyond the limits of both the velocity and thermal boundary layers. In this way, any fluid entering or leaving the element through the

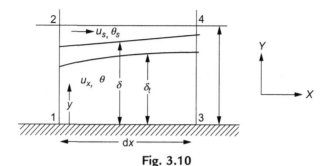

Fig. 3.10
The thermal boundary layer.

face distant from the surface is at the stream velocity u_s and stream temperature θ_s. A heat balance is made therefore on the element shown in Fig. 3.10 in which the length l is greater than the velocity boundary layer thickness δ and the thermal boundary layer thickness δ_t.

The rate of heat transfer through an element of width w of the plane 1-2 of thickness dy at a distance y from the surface is

$$= C_p \rho \theta u_x w \, dy \tag{3.50}$$

where C_p is the specific heat capacity of the fluid at constant pressure, ρ is the density of the fluid, and θ and u_x are the temperature and velocity at a distance y from the surface.

The total rate of transfer of heat through the plane 1-2 is then

$$= C_p \rho w \int_0^l \theta u_x \, dy \tag{3.51}$$

Assuming that the physical properties of the fluid are independent of temperature. In the distance dx, this heat flow changes by an amount given by

$$C_p \rho w \frac{\partial}{\partial x} \left(\int_0^l \theta u_x \, dy \right) dx \tag{3.52}$$

It is shown in Section 3.2 that there is a mass rate of flow of fluid through plane 2-4, out of the element equal to $\rho w (\partial/\partial x) \left(\int_0^l u_x \, dy \right) dx$.

Since the plane 2-4 lies outside the boundary layers, the heat leaving the element through the plane as a result of this flow is

$$C_p \rho \theta_s w \frac{\partial}{\partial x} \left(\int_0^l u_x \, dy \right) dx \tag{3.53}$$

where θ_s is the temperature outside the thermal boundary layer.

The heat transferred by thermal conduction into the element through plane 1-3 is

$$= -kw\mathrm{d}x\left(\frac{\partial\theta}{\partial y}\right)_{y=0} \tag{3.54}$$

If the temperature θ_s of the main stream is unchanged, a heat balance on the element gives

$$C_p\rho w\left(\frac{\partial}{\partial x}\int_0^l \theta u_x \mathrm{d}y\right)\mathrm{d}x = C_p\rho\theta_s w\left(\frac{\partial}{\partial x}\int_0^l u_x \mathrm{d}y\right)\mathrm{d}x - k\left(\frac{\partial\theta}{\partial y}\right)_{y=0} w\mathrm{d}x$$

or

$$\frac{\partial}{\partial x}\int_0^l u_x(\theta_s - \theta)\mathrm{d}y = D_H\left(\frac{\partial\theta}{\partial y}\right)_{y=0} \tag{3.55}$$

where D_H $(=k/C_p\rho)$ is the thermal diffusivity of the fluid.

The relations between u_x and y have already been obtained for both streamline and turbulent flow. A relation between θ and y for streamline conditions in the boundary layer is now derived, although it is not possible to define the conditions in the turbulent boundary layer sufficiently precisely to derive a similar expression for that case.

3.6.3 Heat Transfer for Streamline Flow Over a Plane Surface—Constant Surface Temperature

The flow of fluid over a plane surface, heated at distances greater than x_0 from the leading edge, is now considered. As shown in Fig. 3.11, the velocity boundary layer starts at the leading edge and the thermal boundary layer at a distance x_0 from it. If the temperature of the heated portion of the plate remains constant, this may be taken as the datum temperature. It is assumed that the temperature at a distance y from the surface may be represented by a polynomial of the form:

$$\theta = a_0 y + b_0 y^2 + c_0 y^3 \tag{3.56}$$

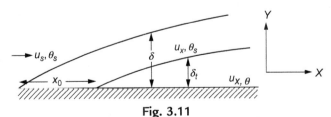

Fig. 3.11
Thermal boundary layer—streamline flow.

If the fluid layer in contact with the surface is assumed to be at rest, any heat flow in the vicinity of the surface must be by pure thermal conduction. Thus, the heat transferred per unit area and unit time q_0 is given by

$$q_0 = -k \left(\frac{\partial \theta}{\partial y} \right)_{y=0}$$

If the temperature of the fluid element in contact with the wall is to remain constant, the heat-transfer rate into and out of the element must be the same or

$$\left(\frac{\partial \theta}{\partial y} \right)_{y=0} = \text{a constant} \quad \text{and} \quad \left(\frac{\partial^2 \theta}{\partial y^2} \right)_{y=0} = 0$$

At the outer edge of the thermal boundary layer, the temperature is θ_s and the temperature gradient $(\partial \theta / \partial y) = 0$ if there is to be no discontinuity in the temperature profile.

Thus, the conditions for the thermal boundary layer, with respect to temperature, are the same as those for the velocity boundary layer with respect to velocity. Then, if the thickness of the thermal boundary layer is δ_t, the temperature distribution is given by

$$\frac{\theta}{\theta_s} = \frac{3}{2} \left(\frac{y}{\delta_t} \right) - \frac{1}{2} \left(\frac{y}{\delta_t} \right)^3 \tag{3.57}$$

that may be compared with Eq. (3.12), and

$$\left(\frac{\partial \theta}{\partial y} \right)_{y=0} = \frac{3 \theta_s}{2 \delta_t} \tag{3.58}$$

It is assumed that the velocity boundary layer is everywhere thicker than the thermal boundary layer so that $\delta > \delta_t$ (Fig. 3.11). Thus, the velocity distribution everywhere within the thermal boundary layer is given by Eq. (3.12). The implications of this assumption are discussed later.

The integral in Eq. (3.55) clearly has a finite value within the thermal boundary layer, although it is zero outside it. When the expression for the temperature distribution in the boundary layer is inserted, the upper limit of integration must be altered from l to δ_t.

Thus,

$$\int_0^l (\theta_s - \theta) u_x \mathrm{d}y = \theta_s u_s \int_0^{\delta_t} \left[1 - \frac{3}{2} \frac{y}{\delta_t} + \frac{1}{2} \left(\frac{y}{\delta_t} \right)^3 \right] \left[\frac{3y}{2\delta} - \frac{1}{2} \left(\frac{y}{\delta} \right)^3 \right] \mathrm{d}y$$

$$= \theta_s u_s \left[\frac{3 \delta_t^2}{4 \delta} - \frac{3 \delta_t^2}{4 \delta} - \frac{1 \delta_t^4}{8 \delta^3} + \frac{3}{20} \left(\frac{\delta_t^2}{\delta} + \frac{\delta_t^4}{\delta^3} \right) - \frac{1}{28} \frac{\delta_t^4}{\delta^3} \right] \tag{3.59}$$

$$= \theta_s u_s \left(\frac{3}{20} \frac{\delta_t^2}{\delta} - \frac{3}{280} \frac{\delta_t^4}{\delta^3} \right)$$

$$= \theta_s u_s \delta \left(\frac{3}{20} \sigma^2 - \frac{3}{280} \sigma^4 \right)$$

where $\sigma = \delta_t/\delta$.

Since $\delta_t < \delta$, the second term is small compared with the first and

$$\int_0^l (\theta_s - \theta)u_x dy \approx \frac{3}{20}\theta_s u_s \delta\sigma^2 \tag{3.60}$$

Substituting from Eqs (3.58), (3.60) in Eq. (3.55) gives

$$\frac{\partial}{\partial x}\left(\frac{3}{20}\theta_s u_s \delta\sigma^2\right) = D_H\frac{3\theta_s}{2\delta_t} = D_H\frac{3\theta_s}{2\delta\sigma}$$

$$\therefore \quad \frac{1}{10}u_s\delta\sigma\frac{\partial}{\partial x}(\delta\sigma^2) = D_H$$

$$\therefore \quad \frac{1}{10}u_s\left(\delta\sigma^3\frac{\partial\delta}{\partial x} + 2\delta^2\sigma^2\frac{\partial\sigma}{\partial x}\right) = D_H \tag{3.61}$$

It has already been shown that

$$\delta^2 = \frac{280\mu x}{13\rho u_s} = 21.5\frac{\mu x}{\rho u_s} \quad \text{(Eq. 3.16)}$$

and hence,

$$\delta\frac{\partial\delta}{\partial x} = \frac{140}{13}\frac{\mu}{\rho u_s}$$

Substituting in Eq. (3.61) and

$$\frac{u_s}{10\rho u_s}\frac{\mu}{\left(\frac{140}{13}\sigma^3 + \frac{560}{13}x\sigma^2\frac{\partial\sigma}{\partial x}\right)} = D_H$$

$$\therefore \quad \frac{14}{13}\frac{\mu}{\rho D_H}\left(\sigma^3 + 4x\sigma^2\frac{\partial\sigma}{\partial x}\right) = 1$$

and

$$\sigma^3 + \frac{4x}{3}\frac{\partial\sigma^3}{\partial x} = \frac{13}{14}Pr^{-1}$$

where the Prandtl number $Pr = C_p\mu/k = \mu/\rho D_H$.

$$\therefore \quad \frac{3}{4}x^{-1}\sigma^3 + \frac{\partial\sigma^3}{\partial x} = \frac{13}{14}Pr^{-1}\frac{3}{4}x^{-1}$$

and

$$\frac{3}{4}x^{-1/4}\sigma^3 + x^{3/4}\frac{\partial\sigma^3}{\partial x} = \frac{13}{14}Pr - 1\frac{3}{4}x^{-1/4}$$

Integrating,

$$x^{3/4}\sigma^3 = \frac{13}{14}Pr^{-1}x^{3/4} + \text{constant}$$

or

$$\sigma^3 = \frac{13}{14}Pr^{-1} + \text{constant}\,x^{-3/4}$$

When $x = x_0$,

$$\sigma = 0$$

so that

$$\text{constant} = -\frac{13}{14}Pr^{-1}x_0^{3/4}$$

Hence,

$$\sigma^3 = \frac{13}{14}Pr^{-1}\left[1 - \left(\frac{x_0}{x}\right)^{3/4}\right]$$

and

$$\sigma = 0.976Pr^{-1/3}\left[1 - \left(\frac{x_0}{x}\right)^{3/4}\right]^{1/3} \tag{3.62}$$

If the whole length of the plate is heated, $x_0 = 0$, and

$$\sigma = 0.976Pr^{-1/3} \tag{3.63}$$

In this derivation, it has been assumed that $\sigma < 1$.

For all liquids other than molten metals, $Pr > 1$, and hence, from Eq. (3.63), $\sigma < 1$.

For gases, $Pr \neq 0.6$, so that $\sigma \neq 1.18$.

Thus, only a small error is introduced when this expression is applied to gases. The only serious deviations occur for molten metals, which have very low Prandtl numbers.

If h is the heat-transfer coefficient, then

$$q_0 = -h\theta_s$$

and

$$-h\theta_s = -k\left(\frac{\partial\theta}{\partial y}\right)_{y=0}$$

or from Eq. (3.58),

$$h = \frac{k}{\theta_s} \frac{3\theta_s}{2} \frac{1}{\delta_t}$$

$$= \frac{3}{2} \frac{k}{\delta_t} = \frac{3}{2} \frac{k}{\delta\sigma}$$

(3.64)

Substituting for δ from Eq. (3.17) and σ from Eq. (3.62) gives

$$h = \frac{3k}{2} \frac{1}{4.64} \sqrt{\frac{\rho u_s}{\mu x}} \frac{Pr^{1/3}}{0.976 \left[1 - (x_0/x)^{3/4}\right]^{1/3}}$$

or

$$\frac{hx}{k} = 0.332 Pr^{1/3} Re_x^{1/2} \frac{1}{\left[1 - (x_0/x)^{3/4}\right]^{1/3}}$$

(3.65)

If the surface is heated over its entire length, so that $x_0 = 0$, then

$$Nu_x = \frac{hx}{k} = 0.332 Pr^{1/3} Re_x^{1/2}$$

(3.66)

It is seen from Eq. (3.66) that the heat-transfer coefficient theoretically has an infinite value at the leading edge, where the thickness of the thermal boundary layer is zero, and that it decreases progressively as the boundary layer thickens. Eq. (3.66) gives the point value of the heat-transfer coefficient at a distance x from the leading edge. The mean value between $x = 0$ and $x = x$ is given by

$$h_m = x^{-1} \int_0^x h \, dx$$

$$= x^{-1} \int_0^x \psi x^{-1/2} dx$$

(3.67)

where ψ is not a function of x.

Thus,

$$h_m = x^{-1} \left[2\psi x^{1/2}\right]_0^x = 2h$$

(3.68)

The mean value of the heat-transfer coefficient between $x = 0$ and $x = x$ is equal to twice the point value at $x = x$. The mean value of the Nusselt group is given by

$$(Nu_x)_m = 0.664 Pr^{1/3} Re_x^{1/2}$$

(3.69)

3.6.4 Heat Transfer for Streamline Flow Over a Plane Surface—Constant Surface Heat Flux

Another important case is where the heat flux, as opposed to the temperature at the surface, is constant; this may occur where the surface is electrically heated. Then, the temperature difference $|\theta_s - \theta_0|$ will increase in the direction of flow (X-direction) as the value of the heat-transfer coefficient decreases due to the thickening of the thermal boundary layer. The equation for the temperature profile in the boundary layer becomes

$$\frac{\theta - \theta_0}{\theta_s - \theta_0} = \frac{3}{2}\left(\frac{y}{\delta_t}\right) - \frac{1}{2}\left(\frac{y}{\delta_t}\right)^3 \tag{3.70}$$

from Eq. (3.57), and the temperature gradient at the walls is given by

$$\left(\frac{\partial \theta}{\partial y}\right)_{y=0} = \frac{3(\theta_0 - \theta_0)}{2\delta_t} \tag{3.71}$$

from Eq. (3.58).

The value of the integral in the energy balance (Eq. 3.55) is again given by Eq. (3.60) (substituting $(\theta_s - \theta_0)$ for θ_s). The heat flux q_0 at the surface is now constant, and the right-hand side of Eq. (3.55) may be expressed as $(-q_0/C_p\rho)$. Thus, for constant surface heat flux, Eq. (3.55) becomes

$$\frac{\partial}{\partial x}\left(\frac{3}{20}(\theta_s - \theta_0)u_s\delta\sigma^2\right) = -\frac{q_0}{C_p\rho} \tag{3.72}$$

Eq. (3.72) cannot be integrated directly, however, because the temperature driving force $(\theta_s - \theta_0)$ is not known as a function of location x on the plate. The solution of Eq. (3.72) involves a quite complex procedure that is given by Kays and Crawford[4] and takes the following form:

$$\frac{hx}{k} = Nu_x = 0.453Pr^{1/3}Re_x^{1/2} \tag{3.73}$$

By comparing Eqs (3.61), (3.66), it is seen that the local Nusselt number and the heat-transfer coefficient are both some 36% higher for a constant surface heat flux as compared with a constant surface temperature.

The average value of the Nusselt group $(Nu_x)_m$ is obtained by integrating over the range $x = 0$ to $x = x$, giving

$$(Nu_x)_m = 0.906Pr^{1/3}Re_x^{1/2} \tag{3.74}$$

3.7 The Boundary Layer for Mass Transfer

If a concentration gradient exists within a fluid flowing over a surface, mass transfer will take place, and the whole of the resistance to transfer can be regarded as lying within a *diffusion boundary layer* in the vicinity of the surface. If the concentration gradients and hence the mass-transfer rates are small, variations in physical properties may be neglected, and it can be shown that the velocity and thermal boundary layers are unaffected.[5] For low concentrations of the diffusing component, the effects of bulk flow will be small, and the mass balance equation for component A is

$$\frac{\partial}{\partial x}\int_0^l (C_{As} - C_A)u_x dy = D\left(\frac{\partial C_A}{\partial y}\right)_{y=0} \tag{3.75}$$

where C_A and C_{As} are the molar concentrations of **A** at a distance y from the surface and outside the boundary layer, respectively, and l is a distance at right angles to the surface that is greater than the thickness of any of the boundary layers. Eq. (3.70) is obtained in exactly the same manner as Eq. (3.55) for heat transfer.

Again, the form of the concentration profile in the diffusion boundary layer depends on the conditions that are assumed to exist at the surface and in the fluid stream. For the conditions corresponding to those used in consideration of the thermal boundary layer, that is, constant concentrations both in the stream outside the boundary layer and at the surface, the concentration profile is of similar form to that given by Eq. (3.70):

$$\frac{C_A - C_{A0}}{C_{As} - C_{A0}} = \frac{3}{2}\left(\frac{y}{\delta_D}\right) - \frac{1}{2}\left(\frac{y}{\delta_D}\right)^3 \tag{3.76}$$

where δ_D is the thickness of the concentration boundary layer, C_A is the concentration of A at $y=y$, C_{A0} is the concentration of A at the surface $(y=0)$, and C_{As} is the concentration of A outside the boundary layer.

Substituting from Eq. (3.76) to evaluate the integral in Eq. (3.75), assuming that mass transfer takes place over the whole length of the surface $(x_0=0)$, by analogy with Eq. (3.63) gives

$$\frac{\delta_D}{\delta} \approx 0.976 Sc^{-1/3} \tag{3.77}$$

where $Sc = \mu/\rho D$ is the Schmidt number. Eq. (3.77) is applicable provided that $Sc > 1$. If Sc is only slightly less than 1, a negligible error is introduced, and it is therefore applicable to most mixtures of gases, as seen in Table 2.2 (Chapter 2). The arguments are identical to those relating to the validity of Eq. (3.63) for heat transfer.

The point values of the Sherwood number Sh_x and mass-transfer coefficient h_D are then given by

$$Sh_x = \frac{h_D x}{D} = 0331 Sc^{1/3} Re_x^{1/2} \tag{3.78}$$

The mean value of the coefficient between $x=0$ and $x=x$ is then given by

$$(Sh_x)m = 0.662 Sc^{1/3} Re_x^{1/2} \tag{3.79}$$

In Eqs (3.78), (3.79), Sh_x and $(Sh_x)_m$ represent the point and mean values, respectively, of the Sherwood numbers.

3.8 Nomenclature

		Units in SI System	Dimensions in M, N, L, T, θ
a	Coefficient of y	K/m	$L^{-1}\theta$
a_0	Coefficient of y	s^{-1}	B^{-1}
b	Coefficient of y^2	$m^{-1} s^{-1}$	$L^{-1}T^{-1}$
b_0	Coefficient of y^2	K/m^2	$L^{-2}\theta$
C_a	Molar concentration of A	$kmol/m^3$	NL^{-3}
C_{A0}	Molar concentration of A at surface $(y=0)$	$kmol/m^3$	NL^{-3}
C_{As}	Molar concentration of A outside boundary layer	$kmol/m^3$	NL^{-3}
C_p	Specific heat at constant pressure	J/kg K	$L^2T^{-2}\theta^{-1}$
c	Coefficient of y^3	$m^{-2} s^{-1}$	$L^{-2}T^{-1}$
c_0	Coefficient of y^3	K/m^3	$L^{-3}\theta$
D	Molecular diffusivity	m^2/s	L^2T^{-1}
D_H	Thermal diffusivity	m^2/s	L^2T^{-1}
d	Pipe diameter	m	L
F	Retarding force	N	MLT^{-2}
f	Index	–	–
h	Heat-transfer coefficient	W/m^2 K	$MT^{-3}\theta^{-1}$
h_D	Mass-transfer coefficient	$kmol/((m^2)(s)(kmol/m^3))$	LT^{-1}
h_m	Mean value of heat-transfer coefficient	W/m^2 K	$MT^{-3}\theta^{-1}$
k	Thermal conductivity	W/m K	$MLT^{-3}\theta^{-1}$
Le	Inlet length of pipe	m	L
l	Thickness of element of fluid	m	L
M	Momentum flux	N	MLT^{-2}

P	Total pressure	N/m^2	$\mathbf{ML^{-1}T^{-2}}$
q_0	Rate of transfer of heat per unit area at walls	W/m^2	$\mathbf{MT^{-3}}$
R	Shear stress acting on surface	N/m^2	$\mathbf{ML^{-1}T^{-2}}$
R_0	Shear stress acting on fluid at surface	N/m^2	$\mathbf{ML^{-1}T^{-2}}$
r	Radius of pipe	m	$\mathbf{L}$
t	Time	s	$\mathbf{T}$
u	Mean velocity	m/s	$\mathbf{LT^{-1}}$
u_b	Velocity at edge of laminar sublayer	m/s	$\mathbf{LT^{-1}}$
u_0	Velocity of fluid at surface	m/s	$\mathbf{LT^{-1}}$
u_s	Velocity of fluid outside boundary layer or at pipe axis	m/s	$\mathbf{LT^{-1}}$
u_x	Velocity in X-direction at $y=y$	m/s	$\mathbf{LT^{-1}}$
w	Width of surface	m	$\mathbf{L}$
x	Distance from leading edge of surface in X-direction	m	$\mathbf{L}$
x_c	Value of x at which flow becomes turbulent	m	$\mathbf{L}$
x_0	Unheated length of surface	m	$\mathbf{L}$
y	Distance from surface	m	$\mathbf{L}$
δ	Thickness of boundary layer	m	$\mathbf{L}$
δ_b	Thickness of laminar sublayer	m	$\mathbf{L}$
δ_D	Diffusion boundary layer thickness	m	$\mathbf{L}$
δ_t	Thickness of thermal boundary layer	m	$\mathbf{L}$
δ^*	Displacement thickness of boundary layer	m	$\mathbf{L}$
θ	Temperature at $y=y$	K	$\boldsymbol{\theta}$
θ_s	Temperature outside boundary layer or at pipe axis	K	$\boldsymbol{\theta}$
μ	Viscosity of fluid	N s/m^2	$\mathbf{ML^{-1}T^{-1}}$
ρ	Density of fluid	kg/m^3	$\mathbf{ML^3}$
σ	Ratio of δ_t to δ	–	–
Nu_x	Nusselt number hx/k	–	–
Re	Reynolds number $ud\,\rho/\mu$	–	–
Re_x	Reynolds number $u_s x\rho/\mu$	–	–
Re_{xc}	Reynolds number $u_s x_c\rho/\mu$	–	–
Re_δ	Reynolds number $u_s\delta\rho/\mu$	–	–
Pr	Prandtl number $C_p\mu/k$	–	–
Sc	Schmidt number $\mu/\rho D$	–	–
Sh_x	Sherwood number $h_D x/D$	–	–

References

1. Brown FNM. The organized boundary layer, In: *Proceedings of the Midwestern conference on fluid mechanics (Sept. 1959)*; 1959. p. 331.
2. Prandtl L. *The essentials of fluid dynamics*. New York: Hafner; 1949.
3. Blasius H. Das Ähnlichkeitsgesetz bei Reibungsvorgängen in Flüssigkeiten. *Forsch Ver deut Ing* 1913;**131**.
4. Kays WM, Crawford ME. *Convective heat and mass transfer*. 3rd ed. New York: McGraw-Hill; 1998.
5. Eckert ERG, Drake Jr RM. *Analysis of heat and mass transfer*. New York: McGraw-Hill; 1972.

Further Reading

1. Schlichting H. *Boundary layer theory* [trans. by Kestin J]. 6th ed. New York: McGraw-Hill; 1968.
2. White FM. *Viscous fluid flow*. New York: McGraw-Hill; 1974.

CHAPTER 4

Quantitative Relations Between Transfer Processes

4.1 Introduction

In the previous chapters, the stresses arising from relative motion within a fluid, the transfer of heat by conduction and convection, and the mechanism of mass transfer are all discussed. These three major processes of momentum, heat, and mass transfer have, however, been regarded as independent problems.

In most of the unit operations encountered in the chemical and process industries, one or more of the processes of momentum, heat, and mass transfer are involved. Thus, in the flow of a fluid under adiabatic conditions through a bed of granular particles, a pressure gradient is set up in the direction of flow, and a velocity gradient develops approximately perpendicularly to the direction of motion in each fluid stream; momentum transfer then takes place between the fluid elements that are moving at different velocities. If there is a temperature difference between the fluid and the pipe wall or the particles, heat transfer will take place as well, and the convective component of the heat transfer will be directly affected by the flow pattern of the fluid. Here, then, is an example of a process of simultaneous momentum and heat transfer in which the same fundamental mechanism is affecting both processes. Fractional distillation and gas absorption are frequently carried out in a packed column in which the gas or vapour stream rises countercurrently to a liquid. The function of the packing in this case is to provide a large interfacial area between the phases and to promote turbulence within the fluids. In a very turbulent fluid, the rates of transfer per unit area of both momentum and mass are high, and as the pressure drop rises, the rates of transfer of both momentum and mass increase together. In some cases, momentum, heat, and mass transfer all occur simultaneously, for example, in a water-cooling tower (see Chapter 5), where the transfer of sensible heat and evaporation both take place from the surface of the water droplets. It will now be shown not only that the process of momentum, heat, and mass transfer are physically related but also that quantitative relations between them can be developed.

Another form of interaction between the transfer processes is responsible for the phenomenon of *thermal diffusion* in which a component in a mixture moves under the action of a temperature

Coulson and Richardson's Chemical Engineering. https://doi.org/10.1016/B978-0-08-102550-5.00004-3

gradient. Although there are important applications of thermal diffusion, the magnitude of the effect is usually small relative to that arising from concentration gradients.

When a fluid is flowing under streamline conditions over a surface, a forward component of velocity is superimposed on the random distribution of velocities of the molecules, and movement at right angles to the surface occurs solely as a result of the random motion of the molecules. Thus, if two adjacent layers of fluid are moving at different velocities, there will be a tendency for the faster-moving layer to be retarded and the slower-moving layer to be accelerated by virtue of the continuous passage of molecules in each direction. There will therefore be a net transfer of momentum from the fast-moving to the slow-moving stream. Similarly, the molecular motion will tend to reduce any temperature gradient or any concentration gradient if the fluid consists of a mixture of two or more components. At the boundary, the effects of the molecular transfer are balanced by the drag forces at the surface.

If the motion of the fluid is turbulent, the transfer of fluid by eddy motion is superimposed on the molecular transfer process. In this case, the rate of transfer to the surface will be a function of the degree of turbulence. When the fluid is highly turbulent, the rate of transfer by molecular motion will be negligible compared with that by eddy motion. For small degrees of turbulence, the two may be of the same order.

It was shown in the previous chapter that when a fluid flows under turbulent conditions over a surface, the flow can conveniently be divided into three regions:

(1)   At the surface, the laminar sublayer, in which the only motion at right angles to the surface is due to molecular diffusion
(2)   Next, the buffer layer, in which molecular diffusion and eddy motion are of comparable magnitude
(3)   Finally, over the greater part of the fluid, the turbulent region, in which eddy motion is large compared with molecular diffusion

In addition to momentum, both heat and mass can be transferred either by molecular diffusion alone or by molecular diffusion combined with eddy diffusion. Because the effects of eddy diffusion are generally far greater than those of the molecular diffusion, the main resistance to transfer will lie in the regions where only molecular diffusion is occurring. Thus, the main resistance to the flow of heat or mass to a surface lies within the laminar sublayer. It is shown in Chapter 3 that the thickness of the laminar sublayer is almost inversely proportional to the Reynolds number for fully developed turbulent flow in a pipe. Thus, the heat- and mass-transfer coefficients are much higher at high Reynolds numbers.

There are strict limitations to the application of the analogy between momentum transfer on the one hand and heat and mass transfer on the other. Firstly, it must be borne in mind that momentum is a vectorial quantity, whereas heat and mass are scalar quantities. Secondly, the quantitative relations apply only to that part of the momentum transfer that arises from *skin friction.* If *form drag* is increased, there is little corresponding increase in the rates at which heat transfer and mass transfer will take place.

Skin friction is the drag force arising from shear stress attributable to the viscous force in the laminar region in the neighbourhood of a surface. *Form drag* is the inertial component arising from vortex formation arising from the presence of an obstruction of flow by, for instance, a baffle or a roughness element on the surface of the pipe. Thus, in the design of contacting devices such as column packings, it is important that they are so shaped that the greater part of the pressure drop is attributable to skin friction rather than to form drag.

4.2 Transfer by Molecular Diffusion

4.2.1 Momentum Transfer

When the flow characteristics of the fluid are *Newtonian*, the shear stress R_y in a fluid is proportional to the velocity gradient and to the viscosity.

$$\text{Thus, for constant density}: \quad R_y = -\mu \frac{\mathrm{d}u_x}{\mathrm{d}y} = -\frac{\mu}{\rho} \frac{\mathrm{d}(\rho u_x)}{\mathrm{d}y} \quad \text{(cf. Eq. 3.3)} \qquad (4.1)$$

where u_x is the velocity of the fluid parallel to the surface at a distance y from it.

The shear stress R_y within the fluid, at a distance y from the boundary surface, is a measure of the rate of transfer of momentum per unit area at right angles to the surface.

Since (ρu_x) is the momentum per unit volume of the fluid, the rate of transfer of momentum per unit area is proportional to the gradient in the Y-direction of the momentum per unit volume. The negative sign indicates that momentum is transferred from the fast-moving to the slow-moving fluid and the shear stress acts in such a direction as to oppose the motion of the fluid.

4.2.2 Heat Transfer

From the definition of thermal conductivity, the heat transferred per unit time through unit area at a distance y from the surface is given by

$$q_y = -k \frac{\mathrm{d}\theta}{\mathrm{d}y} = -\left(\frac{k}{C_p \rho}\right) \frac{\mathrm{d}(C_p \rho \theta)}{\mathrm{d}y} \quad \text{(cf. Eq. 9.11)} \qquad (4.2)$$

where C_p is the specific heat of the fluid at constant pressure, θ the temperature, and k the thermal conductivity. C_p and ρ are both assumed to be constant.

The term $(C_p \rho \theta)$ represents the heat content per unit volume of fluid, and therefore, the flow of heat is proportional to the gradient in the Y-direction of the heat content per unit volume. The proportionality constant $k/C_p \rho$ is called the thermal diffusivity D_H.

4.2.3 Mass Transfer

It is shown in Chapter 2, from Fick's law of diffusion, that the rate of diffusion of a constituent A in a mixture is proportional to its concentration gradient.

Thus, from Eq. (2.4),

$$N_A = -D\frac{dC_A}{dy} \tag{4.3}$$

where N_A is the molar rate of diffusion of constituent A per unit area, C_A the molar concentration of constituent A, and D the diffusivity.

The essential similarity between the three processes is that the rates of transfer of momentum, heat, and mass are all proportional to the concentration gradients of these quantities. In the case of gases, the proportionality constants μ/ρ, D_H, and D, all of which have the dimensions length^2/time, all have a physical significance. For liquids, the constants cannot be interpreted in a similar manner. The viscosity, thermal conductivity, and diffusivity of a gas will now be considered.

4.2.4 Viscosity

Consider the flow of a gas parallel to a solid surface and the movement of molecules at right angles to this direction through a plane a–a of unit area, parallel to the surface and sufficiently close to it to be within the laminar sublayer (Fig. 4.1). During an interval of time dt, molecules with an average velocity i_1u_m in the Y-direction will pass through the plane (where u_m is the root-mean-square velocity and i_1 is some fraction of it, depending on the actual distribution of velocities).

If all these molecules can be considered as having the same component of velocity in the Y-direction, molecules from a volume i_1u_m dt will pass through the plane in time dt.

If $\mathbf{N}$ is the numerical concentration of molecules close to the surface, the number of molecules passing $= i_1u_m\mathbf{N}$ dt.

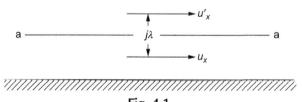

Fig. 4.1
Transfer of momentum near a surface.

Thus, the rate of passage of molecules $= i_1 u_m \mathbf{N}$.

These molecules have a mean velocity u_x (say) in the X-direction.

Thus, the rate at which momentum is transferred across the plane away from the surface

$$= i_1 \mathbf{N} u_m m u_x$$

where m is the mass of each molecule.

By similar reasoning, there must be an equivalent stream of molecules also passing through the plane in the opposite direction; otherwise, there would be a resultant flow perpendicular to the surface.

If this other stream of molecules has originated at a distance $j\lambda$ from the previous ones and the mean component of their velocities in the X-direction is u'_x (where λ is the mean free path of the molecules and j is some fraction of the order of unity), then:

The net rate of transfer of momentum away from the surface

$$= i_1 \mathbf{N} u_m m \left(u_x - u'_x \right)$$

The gradient of the velocity with respect to the Y-direction

$$= \frac{du_x}{dy} = \frac{\left(u'_x - u_x \right)}{j\lambda}$$

since λ is small.

Thus, the rate of transfer of momentum per unit area can be written as

$$R_y = -i_1 \mathbf{N} u_m m j\lambda \frac{du_x}{dy}$$

$$= -i_1 j \rho u_m \lambda \frac{du_x}{dy}$$

(4.4)

(since $\mathbf{N}m = \rho$, the density of the fluid).

But

$$R_y = -\mu \frac{du_x}{dy} \quad \text{(from Eq. 4.1)}$$

$\therefore$

$$\frac{\mu}{\rho} = i_1 j u_m \lambda$$

(4.5)

The value of the product $i_1 j$ has been variously given by different workers; from statistical treatment of the velocities of the molecules,[1] a value of 0.5 will be taken.

Thus,

$$\frac{\mu}{\rho} = \frac{1}{2} u_m \lambda \tag{4.6}$$

It is now possible to give a physical interpretation to the Reynolds number:

$$Re = \frac{ud\rho}{\mu} = ud\frac{2}{u_m \lambda} = 2\frac{u}{u_m}\frac{d}{\lambda} \tag{4.7}$$

or Re is proportional to the product of the ratio of the flow velocity to the molecular velocity and the ratio of the characteristic linear dimension of the system to the mean free path of the molecules.

From the kinetic theory,[1] $u_m = \sqrt{(8\mathbf{R}T/\pi M)}$ and is independent of pressure, and $\rho\lambda$ is a constant.

Thus, the viscosity of a gas would be expected to be a function of temperature but not of pressure.

4.2.5 Thermal Conductivity

Considering now the case where there is a temperature gradient in the Y-direction, the rate of passage of molecules through the unit plane a–a $= i_2 u_m \mathbf{N}$ (where i_2 is some fraction of the order of unity). If the temperature difference between two planes situated a distance $j\lambda$ apart is $(\theta - \theta')$, the net heat transferred as one molecule passes in one direction and another molecule passes in the opposite direction is $c_m(\theta - \theta')$, where c_m is the heat capacity per molecule.

The net rate of heat transfer per unit area $= i_2 u_m \mathbf{N} c_m(\theta - \theta')$.

The temperature gradient $d\theta/dy = (\theta' - \theta)/j\lambda$ since λ is small.

Thus, the net rate of heat transfer per unit area

$$q = -i_2 j u_m \mathbf{N} c_m \lambda \frac{d\theta}{dy}$$

$$= -i_2 j u_m C_v \rho \lambda \frac{d\theta}{dy} \tag{4.8}$$

since $\mathbf{N}c_m = \rho C_v$, the specific heat per unit volume of fluid,

and

$$q = -k\frac{d\theta}{dy} \quad \text{(from Eq. 4.2)}$$

Thus, the thermal diffusivity

$$\frac{k}{C_p\rho} = i_2 j u_m \lambda \frac{C_v}{C_p} \qquad (4.9)$$

From statistical calculations,[1] the value of $i_2 j$ has been given as $(9\gamma - 5)/8$ (where $\gamma = C_p/C_v$, the ratio of the specific heat at constant pressure to the specific heat at constant volume).

Thus,

$$\frac{k}{C_p\rho} = u_m \lambda \frac{9\gamma - 5}{8\gamma}. \qquad (4.10)$$

The Prandtl number Pr is defined as the ratio of the kinematic viscosity to the thermal diffusivity.

Thus,

$$Pr = \frac{\mu/\rho}{k/C_p\rho} = \frac{C_p\mu}{k} = \frac{\frac{1}{2}u_m\lambda}{u_m\lambda(9\gamma - 5)/8\gamma}$$

$$= \frac{4\gamma}{9\gamma - 5} \qquad (4.11)$$

Values of Pr calculated from Eq. (4.11) are in close agreement with practical figures.

4.2.6 Diffusivity

Considering the diffusion, in the Y-direction, of one constituent **A** of a mixture across the plane a–a, if the numerical concentration is $\mathbf{N}_A$ on one side of the plane and $\mathbf{N}'_A$ on the other side at a distance of $j\lambda$, the net rate of passage of molecules per unit area

$$= i_3 u_m \left(\mathbf{N}_A - \mathbf{N}'_A\right)$$

where i_3 is an appropriate fraction of the order of unity.

The rate of mass transfer per unit area

$$= i_3 u_m \left(\mathbf{N}_A - \mathbf{N}'_A\right)m$$

The concentration gradient of **A** in the Y-direction

$$= \frac{\mathrm{d}C_A}{\mathrm{d}y} = \frac{\left(\mathbf{N}'_A - \mathbf{N}_A\right)m}{j\gamma}$$

Thus, the rate of mass transfer per unit area

$$= -i_3j\lambda u_m \frac{dC_A}{dy} \tag{4.12}$$

$$= -D\frac{dC_A}{dy} \quad \text{(from Eq.4.3)}$$

Thus,

$$D = i_3ju_m\lambda \tag{4.13}$$

There is, however, no satisfactory evaluation of the product i_3j.

The ratio of the kinematic viscosity to the diffusivity is the Schmidt number, Sc, where

$$Sc = \frac{(\mu/\rho)}{D} = \frac{\mu}{\rho D} \tag{4.14}$$

It is thus seen that the kinematic viscosity, the thermal diffusivity, and the diffusivity for mass transfer are all proportional to the product of the mean free path and the root-mean-square velocity of the molecules and that the expressions for the transfer of momentum, heat, and mass are of the same form.

For liquids, the same qualitative forms of relationships exist, but it is not possible to express the physical properties of the liquids in terms of molecular velocities and distances.

4.3 Eddy Transfer

In the previous section, the molecular basis for the processes of momentum transfer, heat transfer, and mass transfer has been discussed. It has been shown that, in a fluid in which there is a momentum gradient, a temperature gradient, or a concentration gradient, the consequential momentum-, heat-, and mass-transfer processes arise as a result of the random motion of the molecules. For an ideal gas, the kinetic theory of gases is applicable, and the physical properties μ/ρ, $k/C_p\rho$, and D, which determine the transfer rates, are all seen to be proportional to the product of a molecular velocity and the mean free path of the molecules.

A fluid in turbulent flow is characterised by the presence of circulating or eddy currents, and these are responsible for fluid mixing that, in turn, gives rise to momentum, heat, or mass transfer when there is an appropriate gradient of the 'property' in question. The following simplified analysis of the transport processes in a turbulent fluid is based on the work and ideas of Prandtl. By analogy with kinetic theory, it is suggested that the relationship between transfer rate and driving force should depend on quantities termed the *eddy kinematic viscosity E*, the *eddy thermal diffusivity E_H*, and the *eddy diffusivity E_D* analogous to μ/ρ, $k/C_p\rho$, and D for molecular transport. Extending the analogy further, E, E_H, and E_D might be expected to be proportional to the product of a velocity term and a length term, each of which is a characteristic of the eddies in the fluid, whereas μ/ρ, $k/C_p\rho$, and D are all physical properties of the fluid, and for a material of given composition at a specified temperature and pressure that have unique values, the eddy terms E, E_H, and E_D all depend on the intensity of the eddies. In general, therefore, they are a function of the flow pattern and vary from point to point within the fluid.

In Chapter 3, the concept of a *boundary layer* is discussed. It is suggested that, when a fluid is in turbulent flow over a surface, the eddy currents tend to die out in the region very close to the surface, giving rise to a *laminar sublayer* in which E, E_H, and E_D are all very small. With increasing distance from the surface, these quantities become progressively greater, rising from zero in the laminar sublayer to values considerably in excess of μ/ρ, $k/C_p\rho$, and D in regions remote from the surface. Immediately, outside the laminar sublayer is a buffer zone in which the molecular and eddy terms are of comparable magnitudes. At its outer edge, the eddy terms have become much larger than the molecular terms, and the latter can then be neglected—in what can now be regarded as the fully turbulent region.

4.3.1 The Nature of Turbulent Flow

In turbulent flow, there is a complex interconnected series of circulating or eddy currents in the fluid, generally increasing in scale and intensity with the increase of distance from any boundary surface. If, for steady-state turbulent flow, the velocity is measured at any fixed point in the fluid, both its magnitude and direction will be found to vary in a random manner with time. This is because a random velocity component, attributable to the circulation of the fluid in the eddies, is superimposed on the steady-state mean velocity. No net motion arises from the eddies, and therefore, their time average in any direction must be zero. The instantaneous magnitude and direction of velocity at any point is therefore the vector sum of the steady and fluctuating components.

If the magnitude of the fluctuating velocity component is the same in each of the three principal directions, the flow is termed *isotropic*. If they are different, the flow is said to be *anisotropic*. Thus, if the root-mean-square values of the random velocity components in the X, Y, and Z directions are, respectively, $\sqrt{\overline{u_{Ex}^2}}$, $\sqrt{\overline{u_{Ey}^2}}$, and $\sqrt{\overline{u_{Ez}^2}}$, then for isotropic turbulence

$$\sqrt{\overline{u_{Ex}^2}} = \sqrt{\overline{u_{Ey}^2}} = \sqrt{\overline{u_{Ez}^2}} \tag{4.15}$$

There are two principal characteristics of turbulence. One is the *scale* that is a measure of the mean size of the eddies, and the other is the *intensity* that is a function of the circulation velocity $\left(\sqrt{\overline{u_E^2}}\right)$ within the eddies. Both the scale and the intensity increase as the distance from a solid boundary becomes greater. During turbulent flow in a pipe, momentum is transferred from large eddies in the central core through successively smaller eddies as the walls are approached. Eventually, when the laminar sublayer is reached, the eddies die out completely. However, the laminar sublayer should not be regarded as a completely discrete region, because there is evidence that from time to time eddies do penetrate and occasionally completely disrupt it.

The intensity of turbulence I is defined as the ratio of the mean value of the fluctuating component of velocity to the steady-state velocity. For flow in the X-direction parallel to a surface, this may be written as

$$I = \frac{\sqrt{\frac{1}{3}\left(\overline{u_{Ex}^2} + \overline{u_{Ey}^2} + \overline{u_{Ez}^2}\right)}}{u_x} \tag{4.16}$$

For isotropic turbulence, from Eq. (4.15), this becomes

$$I = \frac{\sqrt{\overline{u_E^2}}}{u_x} \tag{4.17}$$

The intensity of turbulence will vary with the geometry of the flow system. Typically, for a fluid flowing over a plane surface or through a pipe, it may have a value of between 0.005 and 0.02. In the presence of packings and turbulence-promoting grids, very much higher values (0.05–0.1) are common.

The scale of turbulence is given approximately by the diameter of the eddy or by the distance between the centres of successive eddies. The scale of turbulence is related to the dimensions of the system through which the fluid is flowing. The size of largest eddies is clearly limited by the diameter of the pipe or duct. As the wall is approached, their average size becomes less, and momentum transfer takes place by interchange through a succession of eddies of progressively smaller size (down to about 1 mm) until they finally die out as the laminar sublayer is approached near the walls.

An idea of the scale of turbulence can be obtained by measuring instantaneous values of velocities at two different points within the fluid and examining how the correlation coefficient for the two sets of values changes as the distance between the points is increased.

When these are close together, most of the simultaneously measured velocities will relate to fluid in the same eddy, and the correlation coefficient will be high. When the points are further apart, the correlation coefficient will fall because in an appreciable number of the pairs of measurements the two velocities will relate to different eddies. Thus, the distance apart of the measuring stations at which the correlation coefficient becomes very poor is a measure of scale of turbulence. Frequently, different scales of turbulence can be present simultaneously. Thus, when a fluid in a tube flows past an obstacle or suspended particle, eddies may form in the wake of the particles, and their size will be of the same order as the size of the particle; in addition, there will be larger eddies limited in size only by the diameter of the pipe.

4.3.2 Mixing Length and Eddy Kinematic Viscosity

Prandtl[2,3] and Taylor[4] both developed the concept of a *mixing length* as a measure of the distance that an element of fluid must travel before it loses its original identity and becomes fully assimilated by the fluid in its new position. Its magnitude will be of the same order as the scale of turbulence or the eddy size. The mixing length is analogous in concept to the *mean free path* of gas molecules that, according to the kinetic theory, is the mean distance a molecule travels before it collides with another molecule and loses its original identity.

In turbulent flow over a surface, a velocity gradient and hence a momentum gradient exist within the fluid. Any random movement perpendicular to the surface gives rise to a momentum transfer. Elements of fluid with high velocities are brought from remote regions towards the surface and change places with slower-moving fluid elements. This mechanism is essentially similar to that involved in the random movement of molecules in a gas. It is therefore suggested that an *eddy kinematic viscosity E* for eddy transport may be defined that is analogous to the kinematic viscosity μ/ρ for molecular transport. Then, for isotropic turbulence,

$$E \propto \lambda_E u_E \qquad (4.18)$$

where λ_E is the mixing length and u_E is some measure of the linear velocity of the fluid in the eddies.

On this basis, the momentum transfer rate per unit area in a direction perpendicular to the surface at some position y is given by

$$R_y = -E \frac{d(\rho u_x)}{dy} \qquad (4.19)$$

For constant density,

$$R_y = -E\rho \frac{du_x}{dy} \qquad (4.20)$$

Prandtl has suggested that u_E is likely to increase as both the mixing length λ_E and the modulus of the velocity gradient $|du_x/dy|$ increase. The simplest form of relation between the three quantities is that

$$u_E \propto \lambda_E \left| \frac{du_x}{dy} \right| \qquad (4.21)$$

This is tantamount to saying that the velocity change over a distance equal to the mixing length approximates to the eddy velocity. This cannot be established theoretically but is probably a reasonable assumption.

Combining Eqs (4.18) and (4.21) gives

$$E \propto \lambda_E \left\{ \lambda_E \left| \frac{du_x}{dy} \right| \right\} \tag{4.22}$$

Arbitrarily putting the proportionality constant equal to unity, then

$$E = \lambda_E^2 \left| \frac{du_x}{dy} \right| \tag{4.23}$$

Eq. (4.23) implies a small change in the definition of λ_E.

In the neighbourhood of a surface, the velocity gradient will be positive, and the modulus sign in Eq. (4.23) may be dropped. On substitution into Eq. (4.20),

$$R_y = -\rho \lambda_E^2 \left(\frac{du_x}{dy} \right)^2 \tag{4.24}$$

or

$$\sqrt{\frac{-R_y}{\rho}} = \lambda_E \frac{du_x}{dy} \tag{4.25}$$

In Eqs (4.19) and (4.20), R_y represents the momentum transferred per unit area and unit time. This momentum transfer *tends* to accelerate the slower-moving fluid close to the surface and to retard the faster-moving fluid situated at a distance from the surface. It gives rise to a stress R_y at a distance y from the surface since, from Newton's law of motion, force equals the rate of change of momentum. Such stresses, caused by the random motion in the eddies, are sometimes referred to as *Reynolds stresses*.

The problem can also be approached in a slightly different manner. In Fig. 4.2, the velocity profile is shown near a surface. At point 1, the velocity is u_x, and at point 2, the velocity is u'_x. For an eddy velocity u_{Ey} in the direction perpendicular to the surface, the fluid is transported away from the surface at a mass rate per unit area equal to $u_{Ey}\rho$; this fluid must be replaced by an equal mass of fluid that is transferred in the opposite direction. The momentum transferred *away* from the surface per unit and unit time is given by

$$R_y = \rho u_{Ey} (u_x - u'_x)$$

If the distance between the two locations is approximately equal to the mixing length λ_E and if the velocity gradient is nearly constant over that distance,

$$\frac{u'_x - u_x}{\lambda_E} \approx \frac{du_x}{dy}$$

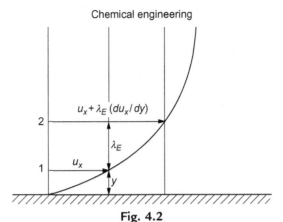

Fig. 4.2
Prandtl mixing length.

Again, assuming that

$$u'_x - u_x \approx u_{Ey}$$

then

$$R = -\rho \lambda_E^2 \left(\frac{du_x}{dy}\right)^2 \quad \text{(Eq. 12.24)}$$

It is assumed throughout that no mixing takes place with the intervening fluid when an eddy transports fluid elements over a distance equal to the mixing length.

Close to a surface $R_y \rightarrow R_0$, the value at the surface.

The shear stress R acting on the surface must be equal and opposite to that in the fluid in contact with the surface, that is, $R = -R_0$ and

$\therefore$

$$\sqrt{\frac{R}{\rho}} = \lambda_E \frac{du_x}{dy} \tag{4.26}$$

$\sqrt{R/\rho}$ is known as the *shearing stress velocity* or *friction velocity* and is usually denoted by u^*.

In steady-state flow over a plane surface or close to the wall for flow in a pipe, u^* is constant, and Eq. (4.26) can be integrated provided that the relation between λ_E and y is known. λ_E will increase with y, and if a linear relation is assumed, then

$$\lambda_E = Ky \tag{4.27}$$

This is the simplest possible form of relation; its use is justified only if it leads to results that are in conformity with experimental results for velocity profiles.

Then,

$$u^* = Ky\frac{du_x}{dy} \tag{4.28}$$

On integration,

$$\frac{u_x}{u^*} = \frac{1}{K}\ln y + B$$

where B is a constant

or

$$\frac{u_x}{u^*} = \frac{1}{K}\ln\frac{yu^*\rho}{\mu} + B' \tag{4.29}$$

Since $(\mu^*\rho/\mu)$ is constant, B' will also be constant.

Writing the dimensionless velocity term $u_x/u^* = u^+$ and the dimensionless derivative of y $(yu^*\rho/\mu) = y^+$, then

$$u^+ = \frac{1}{K}\ln y^+ + B' \tag{4.30}$$

If Eq. (4.29) is applied to the outer edge of the boundary layer when $y = \delta$ (boundary layer thickness) and $u_x = u_s$ (the stream velocity), then

$$\frac{u_s}{u^*} = \frac{1}{K}\ln\frac{\delta u^*\rho}{\mu} + B' \tag{4.31}$$

Subtracting Eq. (4.29) from Eq. (4.31),

$$\frac{u_s - u_x}{u^*} = \frac{1}{K}\ln\frac{\delta}{y} \tag{4.32}$$

Using experimental results for the flow of fluids over both smooth and rough surfaces, Nikuradse[5,6] found K to have a value of 0.4.

Thus,

$$\frac{u_s - u_x}{u^*} = 2.5\ln\frac{\delta}{y} \tag{4.33}$$

For fully developed flow in a pipe, $\delta = r$, and u_s is the velocity at the axis, and then,

$$\frac{u_s - u_x}{u^*} = 2.5\ln\frac{r}{y} \tag{4.34}$$

Eq. (4.34) is known as the *velocity-defect law* (Fig. 4.3).

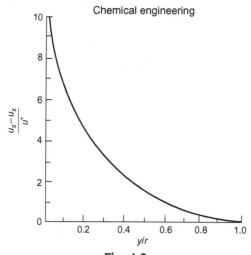

Fig. 4.3
Velocity-defect law.

The application to pipe flow is not strictly valid because $u^* \left(= \sqrt{R/\rho}\right)$ is constant only in regions close to the wall. However, Eq. (4.34) appears to give a reasonable approximation to velocity profiles for turbulent flow, except near the pipe axis. The errors in this region can be seen from the fact that on the differentiation of Eq. (4.34) and putting $y=r$, the velocity gradient on the centre line is $2.5u^*/r$ instead of zero.

Inserting $K=0.4$ in Eq. (4.27) gives the relation between mixing length (λ_E) and distance (y) from the surface:

$$\frac{\lambda_E}{y}=0.4 \qquad\qquad (4.35)$$

Eq. (4.35) applies only in those regions where eddy transfer dominates, that is, outside both the laminar sublayer and the buffer layer (see below).

4.4 Universal Velocity Profile

For fully developed turbulent flow in a pipe, the whole of the flow may be regarded as lying within the boundary layer. The cross-section can then conveniently be divided into three regions:

(a) The *turbulent core* in which the contribution of eddy transport is so much greater than that of molecular transport that the latter can be neglected

(b) The *buffer layer* in which the two mechanisms are of comparable magnitude

(c)   The *laminar sublayer* in which turbulent eddies have effectively died out so that only molecular transport need be considered

It is now possible to consider each of these regions in turn and to develop a series of equations to represent the velocity over the whole cross-section of a pipe. Together, they constitute the *universal velocity profile*.

4.4.1 The Turbulent Core

Eq. (4.30) applies in the turbulent core, except near the axis of the pipe where the shear stress is markedly different from that at the walls. Inserting the value of 0.4 for K,

$$u^+ = 2.5 \ \ln y^+ + B' \tag{4.36}$$

Plotting experimental data on velocity profiles as u^+ against log y^+ (as in Fig. 4.4) gives a series of parallel straight lines of slope 2.5 and with intercepts at $\ln y^+ = 0$ varying with the relative roughness of the surface (e/d). For smooth surfaces ($e/d = 0$), $B' = 5.5$. B' becomes progressively smaller as the relative roughness increases.

Thus, for a smooth pipe,

$$u^+ = 2.5 \ \ln y^+ + 5.5 \tag{4.37}$$

and for a rough pipe,

$$u^+ = 2.5 \ \ln y^+ + B' \tag{4.38}$$

where B' is a function of e/d and is less than 5.5.

Eqs (4.37) and (4.38) correlate experimental data well for values of y^+ exceeding 30.

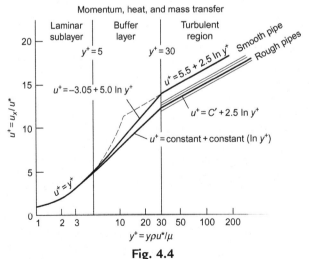

Fig. 4.4
The universal velocity profile.

4.4.2 The Laminar Sublayer

In the laminar sublayer, turbulence has died out, and momentum transfer is attributable solely to viscous shear. Because the layer is thin, the velocity gradient is approximately linear and equal to u_b/δ_b where u_b is the velocity at the outer edge of a laminar sublayer of thickness δ_b (see Chapter 3):

$$R = \mu\frac{u_b}{\delta_b} = \mu\frac{u_x}{y}$$

Then,

$$u^{*2} = \frac{R}{\rho} = \frac{\mu u_x}{\rho y}$$

$\therefore$

$$\frac{u_x}{u^*} = \frac{yu^*\rho}{\mu} \tag{4.39}$$

or

$$u^+ = y^+ \tag{4.40}$$

This relationship holds reasonably well for values of y^+ up to about 5, and it applies to both rough and smooth surfaces.

4.4.3 The Buffer Layer

The buffer layer covers the intermediate range $5 < y^+ < 30$. A straight line may be drawn to connect the curve for the laminar sublayer (Eq. 4.40) at $y^+ = 5$ with the line for the turbulent zone for flow over a smooth surface at $y^+ = 30$ (Eq. 4.37) (see Fig. 4.4). The data for the intermediate region are well correlated by this line whose equation must be of the form

$$u^+ = a \ln y^+ + a' \tag{4.41}$$

The line passes through the points ($y^+ = 5$ and $u^+ = 5$) and ($y^+ = 30$ and $u^+ = 2.5 \ln 30 + 5.5$), and therefore, a and a' may be evaluated to give

$$u^+ = 5.0 \ln y^+ - 3.05 \tag{4.42}$$

An equation, similar in form to Eq. (4.42), will be applicable for rough surfaces, but the values of the two constants will be different.

4.4.4 Velocity Profile for All Regions

For a smooth pipe, therefore, the complete universal velocity profile is given by

$$0 < y^+ < 5 \quad u^+ = y^+ \quad \text{(Eq. 12.40)}$$

$$5 < y^+ < 30 \quad u^+ = 5.0 \ \ln y^+ - 3.05 \quad \text{(Eq. 12.42)}$$

$$y^+ > 30 \quad u^+ = 2.5 \ \ln y^+ + 5.5 \quad \text{(Eq. 12.37)}$$

A simplified form of velocity profile is obtained by neglecting the existence of the buffer layer and assuming that there is a sudden transition from the laminar sublayer to an eddy-dominated turbulent regime. The transition will occur at the point of intersection of the curves (shown by broken lines in Fig. 4.4) representing Eqs (4.40) and (4.37). Solving the equations simultaneously gives $y^+ = 11.6$ as the point of intersection.

For flow in a pipe, the dimensionless distance y^+ from the walls and the corresponding velocity u^+ may be expressed in terms of three dimensionless quantities: the pipe Reynolds number $ud\,\rho/\mu$, the pipe friction factor ϕ $(=R/\rho u^2)$, and the ratio y/d or y/r.

Since

$$u^* = \sqrt{\frac{R}{\rho}} = \sqrt{\frac{R}{\rho u^2}} u = \phi^{1/2} u$$

$$u^+ = \frac{u_x}{u^*} = \frac{u_x}{u} \phi^{-1/2} \tag{4.43}$$

$$y^+ = \frac{y u^* \rho}{\mu} = \frac{y u^*}{d\,u} \frac{ud\rho}{\mu} = \frac{y}{d} \phi^{1/2} Re \tag{4.44}$$

4.4.5 Velocity Gradients

By differentiation of Eqs (4.40), (4.42), and (4.37), respectively, the corresponding values of the gradient du^+/dy^+ are obtained:

$$0 < y^+ < 5 \qquad \frac{du^+}{dy^+} = 1 \tag{4.45}$$

$$5 < y^+ < 30 \qquad \frac{du^+}{dy^+} = \frac{5.0}{y^+} \tag{4.46}$$

$$y^+ > 30 \qquad \frac{du^+}{dy^+} = \frac{2.5}{y^+} \tag{4.47}$$

Eqs (4.45) and (4.47) are applicable to both rough and smooth surfaces; Eq. (4.46) is valid only for a smooth surface $(e/d. \rightarrow 0)$.

Velocity gradients are directly related to du^+/dy^+,

since

$$\frac{du_x}{dy} = \frac{du^+}{dy^+}\frac{\rho u^{*2}}{\mu} \tag{4.48}$$

Thus,

$$0 < y^+ < 5 \quad \frac{du_x}{dy} = \frac{\rho u^{*2}}{\mu} = \frac{R}{\mu} \tag{4.49}$$

$$\begin{array}{c} 5 < y^+ < 30 \\ \text{(smooth surfaces)} \end{array} \quad \frac{du_x}{dy} = \frac{5.0\rho u^{*2}}{y^+\mu} = \frac{5.0R}{y^+\mu} \tag{4.50a}$$

$$\frac{du_x}{dy} = 5.0\frac{u^*}{y} = 5.0\phi^{1/2}\frac{u}{y} \tag{4.50b}$$

$$y^+ > 30 \quad \frac{du_x}{dy} = \frac{2.5\rho u^{*2}}{y^+\mu} = \frac{2.5R}{y^+\mu} \tag{4.51a}$$

$$\frac{du_x}{dy} = 2.5\frac{u^*}{y} = 2.5\phi^{1/2}\frac{u}{y} \tag{4.51b}$$

4.4.6 Laminar Sublayer and Buffer Layer Thicknesses

On the basis of the universal velocity profile, the laminar sublayer extends from $y^+ = 0$ to $y^+ = 5$ and the buffer layer from $y^+ = 5$ to $y^+ = 30$.

From the definition of y^+,

$$y = y^+\frac{\mu}{u^*\rho}$$

$$= y^+\frac{\mu}{ud\rho}\frac{u}{u^*}d \tag{4.52}$$

Thus,

$$\frac{y}{d} = Re^{-1}\left(\frac{R}{\rho u^2}\right)^{-1/2}y^+ \tag{4.53}$$

$$= Re^{-1}\phi^{-1/2}y^+ \tag{4.53a}$$

Putting $y^+ = 5$, the laminar sublayer thickness (δ_b) is given by

$$\frac{\delta_b}{d} = 5Re^{-1}\phi^{-1/2} \tag{4.54}$$

The buffer layer extends to $y^+ = 30$, where

$$\frac{y}{\delta} = 30Re^{-1}\phi^{-1/2} \tag{4.55}$$

If the buffer layer is neglected, it has been shown (Section 4.4.4) that the laminar sublayer will extend to $y^+ = 11.6$ giving

$$\frac{\delta_b}{d} = 11.6Re^{-1}\phi^{-1/2} \tag{4.56}$$

Using the Blasius equation (Eq. 3.46) to give an approximate value for $R/\rho u^2$ for a smooth pipe,

$$\phi = 0.0396Re^{-1/4} \quad \text{(from Eq. 3.46)}$$

$$\frac{\delta_b}{d} = 58Re^{-7/8} \tag{4.57}$$

The equation should be compared with Eq. (3.49) obtained using Prandtl's simplified approach to boundary layer theory that also disregards the existence of the buffer layer:

$$\frac{\delta_b}{d} = 62Re^{-7/8} \quad \text{(Eq. 3.49)}$$

Similarly, the velocity u_b at the edge of the laminar sublayer is given by

$$u^+ = \frac{u_b}{u^*} = 11.6 \tag{4.58}$$

(since $u^+ = y^+$ in the laminar sublayer).

Thus, using the Blasius equation,

$$\frac{u_b}{u} = 11.6\phi^{1/2} \tag{4.59}$$

Substituting for ϕ in terms of Re,

$$\frac{u_b}{u} = 2.32Re^{-1/8} \tag{4.60}$$

Again, Eq. (4.60) can be compared with Eq. (3.40), derived on the basis of the Prandtl approach:

$$\frac{u_b}{u} = 2.49Re^{-1/8} \quad \text{(Eq. 3.47)}$$

4.4.7 Variation of Eddy Kinematic Viscosity

Since the buffer layer is very close to the wall, R_y can be replaced by R_0,

and

$$R_0 = -(\mu + E\rho)\frac{du_x}{dy} \tag{4.61}$$

Thus,

$$\frac{-R_0}{\rho} = u^{*2} = \left(\frac{\mu}{\rho} + E\right)\frac{du_x}{dy} \tag{4.62}$$

For $5 < y^+ < 30$, substituting for du_x/dy from Eq. (4.50a) for a smooth surface,

$$u^{*2} = \left(\frac{\mu}{\rho} + E\right)\frac{5\rho u^{*2}}{\mu y^+}$$

giving

$$E = \frac{\mu}{\rho}\left(\frac{y^+}{5} - 1\right) \tag{4.63}$$

Thus, E varies from zero at $y^+ = 5$ to $(5\mu/\rho)$ at $y^+ = 30$.

For values y^+ greater than 30, μ/ρ is usually neglected in comparison with E. The error is greatest at $y^+ = 30$ where $E/(\mu/\rho) = 5$, but it rapidly becomes smaller at larger distances from the surface.

Thus, for $y^+ > 30$, from Eq. (4.62),

$$u^{*2} = E\frac{du_x}{dy} \tag{4.64}$$

Substituting for du_x/dy from Eq. (4.51a),

$$u^{*2} = E\frac{2.5\rho u^{*2}}{\mu y^+}$$

Thus,

$$E = 0.4y^+\frac{\mu}{\rho} \tag{4.65a}$$

$$= 0.4u^*y. \tag{4.65b}$$

It should be noted that Eq. (4.65a) gives $E = 12\ (\mu/\rho)$ at $y^+ = 30$, compared with $5\ (\mu/\rho)$ from Eq. (4.63). This arises because of the discontinuity in the universal velocity profile at $y^+ = 30$.

4.4.8 Approximate Form of Velocity Profile in Turbulent Region

A simple approximate form of the relation between u^+ and y^+ for the turbulent flow of a fluid in a pipe of circular cross-section may be obtained using the Prandtl one-seventh power law and the Blasius equation for a smooth surface. These two equations have been shown (Section 3.4) to be mutually consistent.

The Prandtl one-seventh power law gives

$$\frac{u_x}{u_{CL}} = \left(\frac{y}{r}\right)^{1/7} \quad \text{(Eq. 3.59 of Vol. 1A)}$$

Then,

$$u^+ = \frac{u_x}{u^*} = \frac{u_{CL}}{u^*}\left(\frac{y}{r}\right)^{1/7} = \frac{u_{CL}}{r^{1/7}}\frac{1}{u^*}\left(\frac{y^+\mu}{\rho u^*}\right)^{1/7} \tag{4.66}$$

The Blasius relation between friction factor and Reynolds number for turbulent flow is

$$\phi = \frac{R}{\rho u^2} = 0.0396 Re^{-1/4} \quad \text{(Eq. 3.11)}$$

Thus,

$$\frac{R}{\rho} = u^{*2} = 0.0396\rho^{-1/4}\mu^{1/4}(2r)^{-1/4}u^{1.75}$$

Again, from the Prandtl one-seventh power law,

$$\frac{u}{u_{CL}} = \frac{49}{60} \quad \text{(Eq. 3.63)}$$

Thus,

$$\frac{u_{CL}}{r^{1/7}} = (0.0396)^{-1/1.75}2^{1/7}\rho^{1/7}\mu^{-1/7}u^{*8/7}\frac{60}{49}$$
$$= 8.56\rho^{1/7}\mu^{-1/7}u^{*8/7}$$

Substituting in Eq. (4.66),

$$u^+ = \left(8.56\rho^{1/7}\mu^{-1/7}u^{*8/7}\right)\frac{1}{u^*}\left(\frac{\mu}{\rho u^*}\right)^{1/7}y^{+1/7}$$

that is,

$$u^+ = 8.56 y^{+1/7} \tag{4.67}$$

In Table 4.1, the values of u^+ calculated from Eq. (4.67) are compared with those given by the universal velocity profile (Eqs 4.37, 4.40, and 4.42). It will be seen that there is almost

Table 4.1 Comparison of values of u+ calculated from Eq. (4.67) with those given by the universal velocity profile

y^+	u^+ from Eq. (4.67)	u^+ from UVP	Percent difference (based on column 3)
10	11.89	8.46	+40.5
15	12.60	10.49	+20.1
20	13.13	11.93	+10.1
25	13.56	14.65	−7.4
30	13.92	14.00	−0.0
100	16.53	17.01	−2.9
300	19.33	19.76	−2.2
1000	22.96	22.76	+0.0
2000	25.35	24.50	+3.5
3000	26.87	25.52	+5.3
4000	27.99	26.23	+6.7
5000	28.90	26.79	+7.9
10,000	31.91	28.52	+11.9
100,000	44.34	34.28	+29.3

exact correspondence at $y^+ = 1000$ and differences are less than 6% in the range $30 < y^+ < 3000$.

4.4.9 Effect of Curvature of Pipe Wall on Shear Stress

Close to the wall of a pipe, the effect of the curvature of the wall has been neglected, and the shear stress in the fluid has been taken to be independent of the distance from the wall. However, this assumption is not justified near the centre of the pipe.

As shown in Chapter 3 of Vol. 1A, the shear stress varies linearly over the cross-section rising from zero at the axis of the pipe to a maximum value at the walls.

A force balance taken over the whole cross-section gives

$$-\frac{dP}{dx}\pi r^2 = -R_0 2\pi r$$

Thus,

$$-R_0 = -\frac{dP}{dx}\frac{r}{2} \tag{4.68}$$

Taking a similar force balance over the central core of fluid lying at distances greater than y from the wall, that is, for a plug of radius $(r-y)$,

$$-R_y = -\frac{dP}{dx}\frac{r-y}{2} \tag{4.69}$$

Thus,

$$\frac{R_y}{R_0} = 1 - \frac{y}{r} \tag{4.70}$$

At radius $(r-y)$, Eq. (4.61) becomes

$$R_y = -(\mu + E\rho)\frac{du_x}{dy} \tag{4.71}$$

Neglecting μ/ρ compared with E and substituting for R_y from Eq. (4.70), then

$$\frac{-R_0}{\rho}\left(1 - \frac{\gamma}{r}\right) = E\frac{du_x}{dy}$$

This leads to

$$E = 0.4y + \frac{\mu}{\rho}\left(1 - \frac{y}{r}\right) \tag{4.72a}$$

$$= 0.4u^*y\left(1 - \frac{y}{r}\right) \tag{4.72b}$$

$$= 0.4\phi^{1/2}uy\left(1 - \frac{y}{r}\right) \tag{4.72c}$$

4.5 Friction Factor for a Smooth Pipe

Eq. (4.37) can be used in order to calculate the friction factor $\phi = R/\rho u^2$ for the turbulent flow of fluid in a pipe. It is first necessary to obtain an expression for the mean velocity u of the fluid from the relation:

$$u = \frac{\int_0^r [2\pi(r - y)dy u_x]}{\pi r^2}$$

$$= 2\int_0^1 u_x\left(1 - \frac{y}{r}\right)d\left(\frac{y}{r}\right) \tag{4.73}$$

The velocity at the pipe axis u_s is obtained by putting $y = r$ into Eq. (4.37).

Thus,

$$u_s = u^*\left(2.5 \ln\frac{r\rho u^*}{\mu} + 5.5\right) \tag{4.74}$$

This is not strictly justified because Eq. (4.74) gives a finite, instead of zero, velocity gradient when applied at the centre of the pipe.

Substituting for u_x in Eq. (4.73) from Eq. (4.34),

$$u = 2\int_0^1 \left(u_s - 2.5u^* \ln\frac{r}{y}\right)\left(1 - \frac{y}{r}\right)d\left(\frac{y}{r}\right)$$

$$\frac{u}{u_s} = 2\int_0^1 \left(1 + 2.5\frac{u^*}{u_s} \ln\frac{y}{r}\right)\left(1 - \frac{y}{r}\right)d\left(\frac{y}{r}\right)$$

$$= 2\left[\frac{y}{r} - \frac{1}{2}\left(\frac{y}{r}\right)^2\right]_0^1 + 5.0\frac{u^*}{u_s}\left\{\left[\left(\ln\frac{y}{r}\right)\left[\frac{y}{r} - \frac{1}{2}\left(\frac{y}{r}\right)^2\right]\right]_0^1 \right.$$

$$\left. - \int_0^1 \left(\frac{y}{r}\right)^{-1}\left[\frac{y}{r} - \frac{1}{2}\left(\frac{y}{r}\right)^2\right]d\left(\frac{y}{r}\right)\right\}$$

$$= 1 + 5.0\frac{u^*}{u_s}\left\{0 - \left[\left(\frac{y}{r}\right) - \frac{1}{4}\left(\frac{y}{r}\right)^2\right]_0^1\right\}$$

$$= 1 + 5\frac{u^*}{u_s}\left(-\frac{3}{4}\right)$$

$$= 1 - 3.75\frac{u^*}{u_s} \tag{4.75}$$

Substituting into Eq. (4.74),

$$u + 3.75u^* = u^*\left\{2.5 \ln\left[\left(\frac{d\rho u}{\mu}\right)\left(\frac{r}{d}\right)\left(\frac{u^*}{u}\right)\right] + 5.5\right\}$$

$$\frac{u}{u^*} = 2.5 \ln\left\{(Re)\frac{u^*}{u}\right\}$$

Now,

$$\phi = \frac{R}{\rho u^2} = \left(\frac{u^*}{u}\right)^2$$

$\therefore$

$$\phi^{-1/2} = 2.5 \ln\left[(Re)\phi^{1/2}\right] \tag{4.76}$$

The experimental results for ϕ as a function of Re closely follow Eq. (4.76) modified by a correction term of 0.3 to give

$$\phi^{-1/2} = 2.5 \ln \left[(Re)\phi^{1/2} \right] + 0.3 \qquad (4.77)$$

The correction is largely associated with the errors involved in using Eq. (4.74) at the pipe axis.

Eq. (4.77) is identical to Eq. (3.12) of Vol. 1A.

Example 4.1

Air flows through a smooth circular duct of internal diameter of 250 mm at an average velocity of 15 m/s. Calculate the fluid velocity at points 50 and 5 mm from the wall. What will be the thickness of the laminar sublayer if this extends to $u^+ = y^+ = 5$? The density and viscosity of air may be taken as 1.10 kg/m^3 and 20×10^{-6} N s/m^2, respectively.

Solution

Reynolds number,

$$Re = \frac{(0.250 \times 15 \times 1.10)}{(20 \times 10^{-6})} = 2.06 \times 10^5$$

Hence, from Fig. 3.7 of Vol. 1A,

$$\frac{R}{\rho u^2} = 0.0018$$

$$u_s = \frac{u}{0.817} = \left(\frac{15}{0.817} \right) = 18.4 \text{m/s}$$

$$u^* = u\sqrt{\frac{R}{\rho u^2}} = 15\sqrt{0.0018} = 0.636 \text{m/s}$$

At 50 mm from the wall,

$$\frac{y}{r} = \left(\frac{0.050}{0.125} \right) = 0.40$$

Hence, from Eq. (4.34),

$$u_x = u_s + 2.5u^* \ln \left(\frac{y}{r} \right)$$

$$= 18.4 + (2.5 \times 0.636 \ln 0.4)$$

$$= \underline{\underline{16.9 \text{m/s}}}$$

At 5 mm from the wall, $y/r = 0.005/0.125 = 0.04$.

Hence,

$$u_x = 18.4 + 2.5 \times 0.636 \ln 0.04$$

$$= \underline{\underline{13.3 \text{m/s}}}$$

The thickness of the laminar sublayer is given by Eq. (4.54):

$$\delta_b = \frac{5d}{Re\sqrt{(R/\rho u^2)}}$$
$$= \frac{(5 \times 0.250)}{\left(2.06 \times 10^5 \sqrt{(0.0018)}\right)}$$
$$= 1.43 \times 10^{-4}\,\text{m}$$

or

$$\underline{0.143\,\text{mm}}$$

4.6 Effect of Surface Roughness on Shear Stress

Experiments have been carried out on artificially roughened surfaces in order to determine the effect of obstructions of various heights.[5,6] Experimentally, it has been shown that the shear force is not affected by the presence of an obstruction of height e unless

$$\frac{u_e e \rho}{\mu} > 40 \tag{4.78}$$

where u_e is the velocity of the fluid at a distance e above the surface.

If the obstruction lies entirely within the laminar sublayer, the velocity u_e is given by

$$R = \mu \left(\frac{du_x}{dy}\right)_{y=0}$$
$$= \mu \left(\frac{u_e}{e}\right), \quad \text{approximately}$$

The shearing stress velocity is given by

$$u^* = \sqrt{\frac{R}{\rho}} = \sqrt{\frac{\mu u_e}{pe}}$$

so that

$$u_e = \frac{pe}{\mu} u^{*2}$$

Thus,

$$\frac{u_e e \rho}{\mu} = \frac{e\rho}{\mu} u^{*2} \frac{e\rho}{\mu} = \left(\frac{e\rho u^*}{\mu}\right)^2 \tag{4.79}$$

$e\rho u^*/\mu$ is known as the *roughness Reynolds number*, Re_r.

For the flow of a fluid in a pipe,

$$u^* = \sqrt{\frac{R}{\rho}} = u\phi^{1/2}$$

where u is the mean velocity over the whole cross-section. Re_r will now be expressed in terms of the three dimensionless groups used in the friction chart.

Thus,

$$Re_r = \frac{e\rho u^*}{\mu}$$

$$= \left(\frac{d\rho u}{\mu}\right)\left(\frac{e}{d}\right)\phi^{1/2}$$

$$= Re\left(\frac{e}{d}\right)\phi^{1/2} \tag{4.80}$$

The shear stress should then be unaffected by the obstruction if $Re_r < \sqrt{40}$, that is, if $Re_r < 6.5$.

If the surface has a number of closely spaced obstructions, however, all of the same height e, the shear stress is affected when $Re_r >$ about 3. If the obstructions are of varying heights, with e as the arithmetic mean, the shear stress is increased if $Re_r >$ about 0.3, because the effect of one relatively large obstruction is greater than that of several small ones.

For hydrodynamically smooth pipes, through which fluid is flowing under turbulent conditions, the shear stress is given approximately by the Blasius equation:

$$\phi = \frac{R}{\rho u^2} \propto Re^{-1/4} \quad \text{(from Eq. 11.46)}$$

so that

$$R \propto u^{1.75}$$

and is independent of the roughness.

For smooth pipes, the frictional drag at the surface is known as *skin friction*. With rough pipes, however, an additional drag known as *form drag* results from the eddy currents caused by the impact of the fluid on the obstructions, and when the surface is very rough, it becomes large compared with the skin friction. Since form drag involves the dissipation of kinetic energy, the losses are proportional to the square of the velocity of the fluid, so that $R \propto u^2$. This applies when $Re_r > 50$.

Thus, when

$$Re_r < 0.3, \quad R \propto u^{1.75}$$

when

$$Re_r > 50, \quad R \propto u^2$$

and when

$$0.3 < Re_r < 50, \quad R \propto u^w \text{ where } 1.75 < w < 2.$$

When the thickness of the laminar sublayer is large compared with the height of the obstructions, the pipe behaves as a smooth pipe (when $e < \delta_b/3$). Since the thickness of the laminar sublayer decreases as the Reynolds number is increased, a surface that is hydrodynamically smooth at low Reynolds numbers may behave as a rough surface at higher values. This explains the shapes of the curves obtained for ϕ plotted against Reynolds number (Fig. 3.7 of Vol. 1A). The curves, for all but the roughest of pipes, follow the curve for the smooth pipe at low Reynolds numbers and then diverge at higher values. The greater the roughness of the surface, the lower is the Reynolds number at which the curve starts to diverge. At high Reynolds numbers, the curves for rough pipes become parallel to the Reynolds number axis, indicating that skin friction is negligible and $R \propto u^2$. Under these conditions, the shear stress can be calculated from Eq. (3.14) of Vol. 1A.

Nikuradse's data[5,6] for rough pipes give

$$u^+ = 2.5 \ \ln\left(\frac{y}{e}\right) + 8.5 \tag{4.81}$$

4.7 Simultaneous Momentum, Heat and Mass Transfer

It has been seen that when there is a velocity gradient in a fluid, the turbulent eddies are responsible for transferring momentum from the regions of high velocity to those of low velocity; this gives rise to shear stresses within the field. It has been suggested that the eddy kinematic viscosity E can be written as the product of an eddy velocity u_E and a mixing length λ_E giving

$$E \propto \lambda_E u_E \quad \text{(Eq. 12.18)}$$

If u_E is expressed as the product of the mixing length and the modulus of velocity gradient and if the proportionality constant is equal to unity,

$$E = \lambda_E^2 \left|\frac{du_x}{dy}\right| \quad \text{(Eq. 12.23)}$$

If there is a temperature gradient within the fluid, the eddies will be responsible for heat transfer, and an eddy thermal diffusivity E_H may be defined in a similar way. It is suggested that, since the mechanism of transfer of heat by eddies is essentially the same as that for the transfer of momentum, E_H is related to mixing length and velocity gradient in a similar manner.

Thus,

$$E_H = \lambda_E^2 \left| \frac{du_x}{dy} \right| \tag{4.82}$$

On a similar basis, an eddy diffusivity for mass transfer E_D can be defined for systems in which concentration gradients exist as

$$E_D = \lambda_E^2 \left| \frac{du_x}{dy} \right| \tag{4.83}$$

E/E_H is termed the *turbulent Prandtl number* and E/E_D the *turbulent Schmidt number*.

E, E_H, and E_D are all nearly equal, and therefore, both the dimensionless numbers are approximately equal to unity.

Thus, for the eddy transfer of heat,

$$q_y = -E_H \frac{d(C_p \rho \theta)}{dy} = -\lambda_E^2 \left| \frac{du_x}{dy} \right| \frac{d(C_p \rho \theta)}{dy} \tag{4.84}$$

and similarly for mass transfer,

$$N_A = -E_D \frac{dC_A}{dy} = -\lambda_E^2 \left| \frac{du_x}{dy} \right| \frac{dC_A}{dy} \tag{4.85}$$

In the neighbourhood of a surface, du_x/dy will be positive, and thus,

$$R_y = -\lambda_E^2 \frac{du_x}{dy} \frac{d(\rho u_x)}{dy} \tag{4.86}$$

$$q_y = -\lambda_E^2 \frac{du_x}{dy} \frac{d(C_p \rho \theta)}{dy} \tag{4.87}$$

and

$$N_A = -\lambda_E^2 \frac{du_x}{dy} \frac{dC_A}{dy} \tag{4.88}$$

For conditions of constant density, Eq. (4.25) gives

$$\sqrt{\frac{-R_y}{\rho}} = \lambda_E \frac{du_x}{dy} \quad \text{(Eq. 12.25)}$$

When molecular and eddy transport both contribute significantly, it may be assumed as a first approximation that their effects are additive. Then,

$$R_y = -\left(\frac{\mu}{\rho} + E\right)\frac{d(\rho u_x)}{dy} = -\mu\frac{du_x}{dy} - \lambda_E^2\rho\left(\frac{du_x}{dy}\right)^2 \tag{4.89}$$

$$q_y = -\left(\frac{k}{C_p\rho} + E_H\right)\frac{d(C_p\rho\theta)}{dy} = -k\frac{d\theta}{dy} - \lambda_E^2\rho C_p\left(\frac{du_x}{dy}\right)\left(\frac{d\theta}{dy}\right) \tag{4.90}$$

$$N_A = -(D + E_D)\frac{dC_A}{dy} = -D\frac{dC_A}{dy} - \lambda_E^2\left(\frac{du_x}{dy}\right)\left(\frac{dC_A}{dy}\right) \tag{4.91}$$

and

$$\frac{\lambda_E}{y} \approx 0.4 \quad (\text{Eq. 4.35})$$

Whereas the kinematic viscosity μ/ρ, the thermal diffusivity $k/C_p\rho$, and the diffusivity D are physical properties of the system and can therefore be taken as constant provided that physical conditions do not vary appreciably, the eddy coefficients E, E_H, and E_D will be affected by the flow pattern and will vary throughout the fluid. Each of the eddy coefficients is proportional to the square of the mixing length. The mixing length will normally increase with distance from a surface, and the eddy coefficients will therefore increase rapidly with position.

The relations are summarised in Table 4.2.

Before the equations given in columns 2 and 3 of Table 4.2 can be integrated, it is necessary to know how E, E_H, and E_D vary with position. In Section 4.4, an estimate has been made of how E varies with the dimensionless distance $y^+(=yu^*\rho/\mu)$ from the surface, by using the concept of the universal velocity profile for smooth surfaces.

For molecular transport alone (the laminar sublayer), $y^+ < 5$:

$$E \to 0$$

For combined molecular and eddy transport (the buffer zone), $5 < y < 30$:

Table 4.2 Relations between physical properties

	Molecular processes only	Molecular and eddy transfer together	Eddy transfer predominating
Momentum transfer	$R_y = -\frac{\mu}{\rho}\frac{d(\rho u_x)}{dy}$	$R_y = -\left(\frac{\mu}{\rho} + E\right)\frac{d(\rho u_x)}{dy}$	$R_y = -E\frac{d(\rho u_x)}{dy}$
Heat transfer	$q_y = -\frac{k}{C_p\rho}\frac{d(C_p\rho\theta)}{dy}$	$q_y = -\left(\frac{k}{C_p\rho} + E_H\right)\frac{d(C_p\rho\theta)}{dy}$	$q_y = -E_H\frac{d(C_p\rho\theta)}{dy}$
Mass transfer	$N_A = -D\frac{dC_A}{dy}$	$N_A = -(D + E_D)\frac{dC_A}{dy}$	$N_A = -E_D\frac{dC_A}{dy}$
		where $E \approx E_D \approx \lambda_E^2\left\|\frac{du_x}{dy}\right\|$ and $\lambda_E \approx 0.4y$	

$$E = \frac{\mu}{\rho}\left(\frac{y^+}{5} - 1\right) \quad \text{(Eq. 4.63)}$$

For the region where the eddy mechanism predominates, $y^+ > 30$:

$$E = \frac{\mu}{\rho} \cdot \frac{y^+}{2.5} \quad \text{(from Eq. 4.65a)}$$

Then, using the approximation $E = E_H = E_D$, these values may be inserted in the equations in Table 4.2. However, the limits of the buffer zone ($5 < y^+ < 30$) may be affected because of differences in the thicknesses of the boundary layers for momentum, heat, and mass transfer (Chapter 3).

4.7.1 Mass Transfer

In the buffer zone, $5 < y^+ < 30$, and

$$N_A = -(D + E_D)\frac{dC_A}{dy}$$

$$= -\left[D + \frac{\mu}{\rho}\left(\frac{y^+}{5} - 1\right)\right]\frac{dC_A}{dy^+} \cdot \frac{u^*\rho}{\mu}$$

$$= -u^*\left[Sc^{-1} + \frac{y^+}{5} - 1\right]\frac{dC_A}{dy^+}$$

Thus,

$$N_A \int_{y_1^+}^{y_2^+} \frac{dy^+}{\frac{y^+}{5} + \left(Sc^{-1} - 1\right)} = -u^* \int_{C_{A1}}^{C_{A2}} dC_A$$

giving

$$N_A = \frac{u^*(C_{A1} - C_{A2})}{5\ln\left[\frac{y_2^+ + 5(Sc^{-1} - 1)}{y_1^+ + 5(Sc^{-1} - 1)}\right]} \tag{4.92}$$

In the fully turbulent region, $y^+ > 30$, and

$$N_A = -E_D\frac{dC_A}{dy}$$

$$-\frac{\mu}{\rho}\frac{y^+}{2.5}\frac{dC_A}{dy^+} \cdot \frac{u^*\rho}{\mu}$$

$$= -u^*\frac{y^+}{2.5}\frac{dC_A}{dy^+}$$

giving

$$N_A = \frac{u^*(C_{A1} - C_{A2})}{2.5 \ln \frac{y_2^+}{y_1^+}}$$ (4.93)

4.7.2 Heat Transfer

The corresponding equations for heat transfer can be obtained in an exactly analogous manner.

In the buffer zone, $5 < y^+ < 30$, and

$$q = \frac{C_p \rho u^*(T_1 - T_2)}{5 \ln \left[\frac{y_2^+ + 5(Pr^{-1} - 1)}{y_1^+ + 5(Pr^{-1} - 1)}\right]}$$ (4.94)

In the fully turbulent region, $y^+ > 30$:

$$q = \frac{C_p \rho u^*(T_1 - T_2)}{2.5 \ln \frac{y_2^+}{y_1^+}}$$ (4.95)

4.8 Reynolds Analogy

4.8.1 Simple Form of Analogy Between Momentum, Heat and Mass Transfer

The simple concept of the Reynolds analogy was first suggested by Reynolds[7] to relate heat-transfer rates to shear stress, but it is also applicable to mass transfer. It is assumed that the elements of fluid are brought from remote regions to the surface by the action of the turbulent eddies; the elements do not undergo any mixing with the intermediate fluid through which they pass, and they instantaneously reach equilibrium on contact with the interfacial layers. An equal volume of fluid is, at the same time, displaced in the reverse direction. Thus, in a flowing fluid, there is a transference of momentum and a simultaneous transfer of heat if there is a temperature gradient and of mass if there is a concentration gradient. The turbulent fluid is assumed to have direct access to the surface, and the existence of a buffer layer and laminar sublayer is neglected. Modification of the model has been made by Taylor[4] and Prandtl[8,9] to take account of the laminar sublayer. Subsequently, the effect of the buffer layer has been incorporated by applying the *universal velocity profile*.

Consider the equilibrium set up when an element of fluid moves from a region at high temperature, lying outside the boundary layer, to a solid surface at a lower temperature if no mixing with the intermediate fluid takes place. Turbulence is therefore assumed to persist right

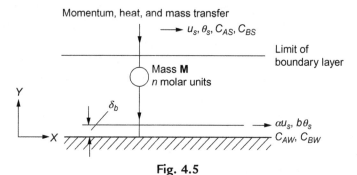

Fig. 4.5

The Reynolds analogy—momentum, heat, and mass transfer.

up to the surface. The relationship between the rates of transfer of momentum and heat can then be deduced as follows (Fig. 4.5).

Consider the fluid to be flowing in a direction parallel to the surface (X-direction) and for momentum and heat transfer to be taking place in a direction at right angles to the surface (Y-direction positive away from surface).

Suppose a mass M of fluid situated at a distance from the surface to be moving with a velocity u_s in the X-direction. If this element moves to the surface where the velocity is zero, it will give up its momentum Mu_s in time t, say. If the temperature difference between the element and the surface is θ_s and C_p is the specific heat of the fluid, the heat transferred to the surface will be $M C_p \theta_s$. If the surface is of area A, the rate of heat transfer is given by

$$\frac{MC_p\theta_s}{At} = -q_0 \tag{4.96}$$

where $-q_0$ is the heat transferred to the surface per unit area per unit time (NB—the negative sign has been introduced as the positive direction is away from the surface).

If the shear stress at the surface is R_0, it will equal the rate of change in momentum per unit area.

Thus,

$$\frac{Mu_s}{At} = -R_0. \tag{4.97}$$

A similar argument can be applied to the mass-transfer process when a concentration gradient exists. Thus, if the molar concentration of **A** remote from the surface is c_{As} and at the surface it is c_{Aw}, the moles of A transferred to the surface will be $(M/\rho)(c_{As}-c_{Aw})$, if the density (ρ) can be assumed to be constant over the range of concentrations encountered. Thus, the moles of A transferred to the surface per unit area and unit time $(-N_A)_{y=0}$ is given by

$$\frac{1}{At}\frac{M}{\rho}(c_{As} - c_{Aw}) = (-N_A)_{y=0} \tag{4.98}$$

Dividing Eq. (4.96) by (4.97),

$$\frac{C_p \theta_s}{u_s} = \frac{-q_0}{-R_0} \tag{4.99}$$

or

$$\frac{-R_0}{u_s} = \frac{-q_0}{C_p \theta_s} \tag{4.100}$$

Again, dividing Eq. (4.98) by Eq. (4.97),

$$\frac{C_{As} - C_{Aw}}{\rho u_s} = \frac{(-N_A)_{y=0}}{-R_0}$$

or

$$\frac{-R_0}{\rho u_s} = \frac{(-N_A)_{y=0}}{C_{As} - C_{Aw}} \tag{4.101}$$

Now, R_0 (the shear stress in the fluid at the surface) is equal and opposite to R, the shear stress acting on the surface; $-q_0/\theta_s$ is by definition the heat-transfer coefficient at the surface (h); and $(-N_A)_{y=0}/(C_{As} - C_{Aw})$ is the mass-transfer coefficient (h_D). Then, dividing both sides of Eq. (4.100) by ρu_s and of Eq. (4.101) by u_s to make them dimensionless,

$$\frac{R}{\rho u_s^2} = \frac{h}{C_p \rho u_s} = St \tag{4.102}$$

where St denotes the Stanton number ($h/C_p \rho u_s$)

and

$$\frac{R}{\rho u_s^2} = \frac{h_D}{u_s} \tag{4.103}$$

h_D/u_s is sometimes referred to as the Stanton number for mass transfer.

Thus,

$$\frac{h_D}{u_s} = \frac{h}{C_p \rho u_s} \tag{4.104}$$

or

$$h_D = \frac{h}{C_p \rho} \tag{4.105}$$

Eq. (4.105) is often referred to as the *Lewis relation*. It provides an approximate method for evaluating a mass-transfer coefficient if the heat-transfer coefficient is known. The assumption that the turbulent eddies can penetrate right up to the surface is justified however only in special circumstances and the problem is considered further in the next section.

For flow in a smooth pipe, the friction factor for turbulent flow is given approximately by the Blasius equation and is proportional to the Reynolds number (and hence the velocity) raised to a power of $-1/4$. From Eqs (4.102) and (4.103), therefore, the heat- and mass-transfer coefficients are both proportional to $u_s^{07.5}$.

The application of the analogies to the problems of heat and mass transfer to plane surfaces and to pipe walls for fully developed flow is discussed later.

Example 4.2

Air at 330 K, flowing at 10 m/s, enters a pipe of inner diameter of 25 mm, maintained at 415 K. The drop of static pressure along the pipe is 80 N/m^2 per metre length. Using the Reynolds analogy between heat transfer and fluid friction, estimate the air temperature 0.6 m along the pipe.

Solution

From Eqs (4.102) and (3.18), of Vol. 1A

$$-\Delta P = 4\left(R/\rho u^2\right)(l/d)\rho u^2 = 4\left(h/C_p\rho u\right)(l/d)\rho u^2$$

(using the mean pipeline velocity u in the Reynolds analogy).

Then, in SI units,

$$-\Delta P/l = 80 = 4\left[h/\left(C_p \times 10\right)\right](1/0.025) \times 10^2$$

and

$$h = 0.05C_p \ \text{W/m}^2\text{K}$$

In passing through a length dL of pipe, the air temperature rises from T to $T+dT$.

The heat taken up per unit time by the air is given by

$$dQ = \left(\rho u C_p\right)\left(\pi d^2/4\right)dT$$

The density of air at 1 bar and 330

$$K = (MP)/(RT) = \frac{29 \times 10^5}{8314 \times 330} = 1.057 \, \text{kg/m}^3$$

Thus,

$$dQ = \left(1.057 \times 10 \times C_p\right)\left[\pi(0.025)^2/4\right]dT$$
$$= 0.0052C_p dT \ \text{W} \tag{i}$$

The heat transferred through the pipe wall is also given by

$$dQ = h(\pi d\,dL)(415 - T)$$

$$= (0.05C_p)(\pi \times 0.025 \, dL)(415 - T)$$

$$= 0.039C_p(415 - T)dL \quad W \tag{ii}$$

Equating (i) and (ii),

$$\int_{330}^{T_0} \frac{dT}{415 - T} = 0.75 \int_0^{0.6} dL$$

giving

$$\ln[85/(415 - T_0)] = 0.45$$

and

$$85/(415 - T_0) = e^{0.45} = 1.57$$

and

$$\underline{\underline{T_0 = 360 \, K}}$$

4.8.2 Mass Transfer With Bulk Flow

When the mass-transfer process deviates significantly from equimolecular counterdiffusion, allowance must be made for the fact that there may be a very large difference in the molar rates of transfer of the two components. Thus, in a gas absorption process, there will be no transfer of the insoluble component **B** across the interface, and only the soluble component **A** will be transferred. This problem will now be considered in relation to the Reynolds analogy. However, it gives manageable results only if physical properties such as density are taken as constant and therefore results should be applied with care.

Consider the movement of an element of fluid consisting of n molar units of a mixture of two constituents A and B from a region outside the boundary layer, where the molecular concentrations are c_{As} and c_{Bs}, to the surface where the corresponding concentrations are c_{Aw} and c_{Bw}. The total molar concentration is everywhere c_T. The transfer is effected in a time t and takes place at an area A of surface.

There is no net transference of the component B. When n molar units of material are transferred from outside the boundary layer to the surface,

$$\text{Transfer of } \mathbf{A} \text{ towards surface} = n\frac{C_{As}}{C_T}$$

$$\text{Transfer of } \mathbf{B} \text{ towards surface} = n\frac{C_{Bs}}{CT}$$

In this case, the molar rate of transfer of **B** away from the surface is equal to the transfer towards the surface:

$$\text{Transfer of } \textbf{B} \text{ away from surface} = n\frac{C_{Bs}}{C_T}$$

Associated transfer of **A** away from surface

$$= n\frac{C_{Bs}}{C_T}\frac{C_{Aw}}{C_{Bw}}$$

Thus, the net transfer of A towards the surface is given by

$$-N_A' At = n\left(\frac{C_{As}}{C_T} - \frac{C_{Bs}}{C_T}\frac{C_{Aw}}{C_{Bw}}\right)$$

$$= n\left(\frac{C_T C_{As} - C_{Aw}C_{As} - C_{Aw}C_T + C_{Aw}C_{As}}{C_T(C_T - C_{Aw})}\right)$$

$$= n\frac{(C_{As} - C_{Aw})}{C_{Bw}}$$

It is assumed that the total molar concentration is everywhere constant. Thus, the rate of transfer per unit area and unit time is given by

$$-N_A' = \frac{n(C_{As} - C_{Aw})}{C_{Bw}At} \tag{4.106}$$

The net transfer of momentum per unit time

$$= -R_0 A = \frac{n\rho u_s}{C_T t} \tag{4.107}$$

where ρ is taken as the mean mass density of the fluid.

$$\therefore$$

$$-R_0 = \frac{n\rho u_s}{C_T tA} \tag{4.108}$$

Dividing Eqs (4.106) and (4.108) gives

$$\frac{N_A'}{R_0} = \frac{C_{As} - C_{Aw}}{\rho u_s}\frac{C_T}{C_{Bw}} \tag{4.109}$$

Writing $R_0 = -R$ and defining the mass-transfer coefficient by the relation, then

$$\frac{-N_A'}{C_{As} - C_{Aw}} - h_D \tag{4.110}$$

$$\frac{h_D}{u_s}\frac{C_{Bw}}{C_T}=\frac{R}{\rho u_s^2}$$ (4.111)

Thus, there is a direct proportionality between the momentum transfer and that portion of the mass transfer that is *not* attributable to bulk flow.

For heat transfer,

$$\frac{R}{\rho u_s^2}=\frac{h}{C_p\rho u_s}$$ (Eq. 4.102)

Thus, for simultaneous heat transfer and mass transfer with bulk flow giving rise to no net transfer of component B, combination of Eqs (4.111) and (4.102) gives

$$\frac{h_D}{u_s}\frac{C_{Bw}}{C_T}=\frac{h}{C_p\rho u_s}$$

or

$$h_D\frac{C_{Bw}}{C_T}=\frac{h}{C_p\rho}$$ (4.112)

Eq. (4.112) is the form of the *Lewis relation* that is applicable to mass transfer with bulk flow.

4.8.3 Taylor–Prandtl Modification of Reynolds Analogy for Heat Transfer and Mass Transfer

The original Reynolds analogy involves a number of simplifying assumptions that are justifiable only in a limited range of conditions. Thus, it was assumed that fluid was transferred from outside the boundary layer to the surface without mixing with the intervening fluid, it was brought to rest at the surface, and thermal equilibrium was established. Various modifications have been made to this simple theory to take account of the existence of the laminar sublayer and the buffer layer close to the surface.

Taylor[4] and Prandtl[8,9] allowed for the existence of the laminar sublayer but ignored the existence of the buffer layer in their treatment and assumed that the simple Reynolds analogy was applicable to the transfer of heat and momentum from the main stream to the edge of the laminar sublayer of thickness δ_b. Transfer through the laminar sublayer was then presumed to be attributable solely to molecular motion.

If au_s and $b\theta_s$ are the velocity and temperature, respectively, at the edge of the laminar sublayer (see Fig. 4.5), applying the Reynolds analogy (Eq. 4.99) for transfer across the turbulent region is given by

$$\frac{-q_0}{-R_0} = \frac{C_p(\theta_s - b\theta_s)}{u_s - \alpha u_s}$$

(4.113)

The rate of transfer of heat by conduction through the laminar sublayer from a surface of area A is given by

$$-q_0 A = \frac{kb\theta_s A}{\delta_b}$$

(4.114)

The rate of transfer of momentum is equal to the shearing force, and therefore,

$$-R_0 A = \frac{\mu \alpha u_s A}{\delta_b} = RA$$

(4.115)

Dividing Eqs (4.114) and (4.115) gives

$$\frac{-q_0}{-R_0} = \frac{kb\theta_s}{\mu \alpha u_s}$$

(4.116)

Thus, from Eq. (4.113) and (4.116),

$$\frac{kb\theta_s}{\mu \alpha u_s} = \frac{C_p(1-b)\theta_s}{(1-\alpha)u_s}$$

$\therefore$

$$\frac{Pr(1-b)}{b} = \frac{(1-\alpha)}{\alpha}$$

$\therefore$

$$\frac{b}{\alpha} = \frac{1}{\alpha + (1-\alpha)Pr^{-1}}$$

Substituting in Eq. (4.116),

$$\frac{-q_0}{-R_0} = \frac{C_p\theta_s}{u_s}\frac{1}{1+\alpha(Pr-1)} = \frac{h\theta_s}{R}$$

or

$$St = \frac{h}{C_p \rho u_s} = \frac{R/\rho u_s^2}{1+\alpha(Pr-1)}$$

(4.117)

The quantity α, which is the ratio of the velocity at the edge of the laminar sublayer to the stream velocity, was evaluated in Chapter 3 in terms of the Reynolds number for flow over the surface. For flow over a plane surface, from Chapter 3,

$$\alpha = 2.1 Re_x^{-0.1} \quad \text{(Eq. 3.35)}$$

where Re_x is the Reynolds number $u_s x\rho/\mu$, x being the distance from the leading edge of the surface.

For flow through a pipe of diameter d (Chapter 3),

$$\alpha = 2.0 Re^{-1/8} \quad (\text{Eq. 3.48})$$

where Re is the Reynolds number $ud\rho/\mu$.

Alternatively, from Eq. (4.59),

$$\alpha = \frac{u_b}{u} = 11.6\phi^{1/2} \approx 11.6\left(\frac{R}{\rho u_s^2}\right)^{1/2}$$

For mass transfer to a surface, a similar relation to Eq. (4.117) can be derived for equimolecular counterdiffusion except that the Prandtl number is replaced by the Schmidt number. It follows that

$$\frac{h_D}{u_s} = \frac{R/\rho u_s^2}{1 + \alpha(Sc - 1)} \tag{4.118}$$

where Sc is the Schmidt number $(\mu/\rho D)$. It is possible also to derive an expression to take account of bulk flow, but many simplifying assumptions must be made, and the final result is not very useful.

It is thus seen that by taking account of the existence of the laminar sublayer, correction factors are introduced into the simple Reynolds analogy.

For heat transfer, the factor is $(1 + \alpha (Pr - 1))$, and for mass transfer, it is $(1 + \alpha (Sc - 1))$.

There are two sets of conditions under which the correction factor approaches unity:

(i) For gases, both the Prandtl and Schmidt groups are approximately unity, and therefore, the simple Reynolds analogy is closely followed. Furthermore, the Lewis relation that is based on the simple analogy would be expected to hold closely for gases. The Lewis relation will also hold for any system for which the Prandtl and Schmidt numbers are equal. In this case, the correction factors for heat and mass transfer will be approximately equal.

(ii) When the fluid is highly turbulent, the laminar sublayer will become very thin, and the velocity at the edge of the laminar sublayer will be small. In these circumstances again, the correction factor will approach unity.

Eqs (4.117) and (4.118) provide a means of expressing the mass-transfer coefficient in terms of the heat-transfer coefficient.

From Eq. (4.117),

$$\frac{R}{\rho u_s^2} = \frac{h}{C_p \rho u_s}[1 + \alpha(Pr - 1)] \tag{4.119}$$

From Eq. (4.118),

$$\frac{R}{\rho u_s^2} = \frac{h_D}{u_s}[1 + \alpha(Sc - 1)] \tag{4.120}$$

Thus,

$$h_D = \frac{h}{C_p\rho}\frac{1 + \alpha(Pr - 1)}{1 + \alpha(Sc - 1)} \tag{4.121}$$

Eq. (4.121) is a modified form of the *Lewis relation*, which takes into account the resistance to heat and mass transfer in the laminar sublayer.

4.8.4 Use of Universal Velocity Profile in Reynolds Analogy

In the Taylor–Prandtl modification of the theory of heat transfer to a turbulent fluid, it was assumed that the heat passed directly from the turbulent fluid to the laminar sublayer and the existence of the buffer layer was neglected. It was therefore possible to apply the simple theory for the boundary layer in order to calculate the heat transfer. In most cases, the results so obtained are sufficiently accurate, but errors become significant when the relations are used to calculate heat transfer to liquids of high viscosities. A more accurate expression can be obtained if the temperature difference across the buffer layer is taken into account. The exact conditions in the buffer layer are difficult to define, and any mathematical treatment of the problem involves a number of assumptions. However, the conditions close to the surface over which fluid is flowing can be calculated approximately using the *universal velocity profile*.[10]

The method is based on the calculation of the total temperature difference between the fluid and the surface, by adding the components attributable to the laminar sublayer, the buffer layer, and the turbulent region. In the steady state, the heat flux (q_0) normal to the surface will be constant if the effects of curvature are neglected.

4.8.4.1 Laminar sublayer (0<y+<5)

Since the laminar sublayer is thin, the temperature gradient may be assumed to be approximately linear (see also Section 3.6.1).

Thus,

$$q_0 = -k\frac{\theta_5}{y_5} \tag{4.122}$$

where y_5 and θ_5 are, respectively, the values of y and θ at $y^+ = 5$. As before, the temperature scale is so chosen that the surface temperature is zero.

By definition,

$$y^+ = \frac{yu^*\rho}{\mu}$$

Thus,

$$y_5 = \frac{5\mu}{u^*\rho}$$

Substituting in Eq. (4.122),

$$\theta_5 = \frac{-q_0}{k}\frac{5\mu}{u^*\rho} \tag{4.123}$$

4.8.4.2 Buffer layer (5 < y+ < 30)

It has been shown that it is reasonable to assume that the eddy kinematic viscosity E and the eddy thermal diffusivity E_H are equal. The variation of E through the buffer zone is given by

$$E = \frac{\mu}{\rho}\left(\frac{y^+}{5} - 1\right) \quad \text{(Eq. 12.67)}$$

The heat-transfer rate in the buffer zone is given by

$$q_0 = -\left(k + E_H C_p\rho\right)\frac{d\theta}{dy} \quad \text{(from Eq. 4.94)}$$

Substituting from Eq. (4.63) and putting $E_H = E$,

$$-q_0 = \left[k + C_p\rho\frac{\mu}{\rho}\left(\frac{y^+}{5} - 1\right)\right]\frac{d\theta}{dy} \tag{4.124}$$

From the definition of y^+,

$$\frac{d\theta}{dy} = \frac{u^*\rho}{\mu}\frac{d\theta}{dy^+} \tag{4.125}$$

Substituting from Eq. (4.125) into Eq. (4.124),

$$-q_0 = \rho u^*\left[\frac{k}{\mu} + C_p\left(\frac{y^+}{5} - 1\right)\right]\frac{d\theta}{dy^+}$$

$$= \frac{C_p\rho u^*}{5}\left[\frac{5k}{C_p\mu} - 5 + y^+\right]\frac{d\theta}{dy^+}$$

∴

$$\frac{d\theta}{dy^+} = \frac{-5q_0}{C_p\rho u^*[5(Pr^{-1}-1)+y^+]} \tag{4.126}$$

Integrating between the limits of $y^+=5$ and $y^+=30$,

$$\theta_{30} - \theta_5 = \frac{-5q_0}{C_p\rho u^*} \ln(5Pr+1). \tag{4.127}$$

4.8.4.3 Turbulent zone (y+>30)

In this region, the Reynolds analogy can be applied. Eq. (4.113) becomes

$$\frac{-q_0}{-R_0} = \frac{C_p(\theta_s - \theta_{30})}{u_s - u_{30}}$$

The velocity at $y^+=30$ for a smooth surface is given by

$$\frac{u_x}{u^*} = u^+ = 5.0 \ln y^+ - 3.05 \quad \text{(Eq. 12.42)}$$

$\therefore$

$$u_{30} = u^*(5.0 \ln 30 - 3.05)$$

and

$$\theta_s - \theta_{30} = \frac{-q_0}{C_p\rho u^{*2}}(u_s - 5.0u^* \ln 30 + 3.05u^*) \tag{4.128}$$

The overall temperature difference θ_s is obtained by the addition of Eqs (4.123), (4.127), and (4.128):

$$\text{Thus } \theta_s = \frac{-q_0}{C_p\rho u^{*2}}\left[\frac{5C_p\mu}{k}u^* + 5u^* \ln(5Pr+1) + u_s - 5.0u^* \ln 30 + 3.05u^*\right]$$

$$= \frac{-q_0}{C_p\rho u^{*2}}\left[5Pru^* + 5u^* \ln\left(\frac{Pr}{6} + \frac{1}{30}\right) + u_s + 3.05u^*\right] \tag{4.129}$$

that is,

$$\frac{u^{*2}}{u_s^2} = \frac{-q_0}{C_p\rho u_s\theta_s}\left\{1 + 5\frac{u^*}{u_s}\left[Pr + \ln\left(\frac{Pr}{6} + \frac{1}{30}\right) + \ln 5 - 1\right]\right\} \quad (\text{since } 5 \ln 5 = 8.05)$$

and

$$\frac{R}{\rho u_s^2} = \frac{h}{C_p\rho u_s}\left\{1 + 5\sqrt{\frac{R}{\rho u_s^2}}\left[(Pr-1) + \ln\left(\frac{5}{6}Pr + \frac{1}{6}\right)\right]\right\} \tag{4.130}$$

where

$$\frac{h}{C_p \rho u_s} = St \text{ (the Stanton number)}$$

A similar expression can also be derived for mass transfer in the absence of bulk flow:

$$\frac{R}{\rho u_s^2} = \frac{h_D}{u_s}\left\{1 + 5\sqrt{\frac{R}{\rho u_s^2}}\left[(Sc - 1) + \ln\left(\frac{5}{6}Sc + \frac{1}{6}\right)\right]\right\} \qquad (4.131)$$

4.8.5 Flow Over a Plane Surface

The simple Reynolds analogy gives a relation between the friction factor $R/\rho u_s^2$ and the Stanton number for heat transfer:

$$\frac{R}{\rho u_s^2} = \frac{h}{C_p \rho u_s} \quad \text{(Eq. 4.102)}$$

This equation can be used for calculating the point value of the heat-transfer coefficient by substituting for $R/\rho u_s^2$ in terms of the Reynolds group Re_x using Eq. (3.39):

$$\frac{R}{\rho u_s^2} = 0.03 Re_x^{-0.2} \quad \text{(Eq. 3.39)}$$

and

$$St = \frac{h}{C_p \rho u_s} = 0.03 Re_x^{-0.2} \qquad (4.132)$$

Eq. (4.132) gives the point value of the heat-transfer coefficient. If the whole surface is effective for heat transfer, the mean value is given by

$$St = \frac{h}{C_p \rho u_s} = \frac{1}{x}\int_0^x 0.03 Re_x^{-0.2} dx$$

$$= 0.037 Re_x^{-0.2} \qquad (4.133)$$

These equations take no account of the existence of the laminar sublayer and therefore give unduly high values for the transfer coefficient, especially with liquids. The effect of the laminar sublayer is allowed for by using the Taylor–Prandtl modification:

$$St = \frac{h}{C_p \rho u_s} = \frac{R/\rho u_s^2}{1 + \alpha(Pr - 1)} \quad \text{(Eq. 4.117)}$$

where

$$\alpha = \frac{u_b}{u_s} = 2.1 Re_x^{-0.1} \quad \text{(Eq. 11.35)}$$

Thus,

$$St = Nu_x Re_x^{-1} Pr^{-1} = \frac{0.03 Re_x^{-0.2}}{1 + 2.1 Re_x^{-0.1}(Pr - 1)} \tag{4.134}$$

This expression will give the point value of the Stanton number and hence of the heat-transfer coefficient. The mean value over the whole surface is obtained by integration. No general expression for the mean coefficient can be obtained, and a graphical or numerical integration must be carried out after the insertion of the appropriate values of the constants.

Similarly, substitution may be made from Eqs (3.39) to (4.130) to give the point values of the Stanton number and the heat-transfer coefficient; thus,

$$St = \frac{h}{C_p \rho u_s} = \frac{0.03 Re_x^{-0.2}}{1 + 0.87 Re_x^{-0.1} \left[(Pr - 1) + \ln\left(\frac{5}{6} Pr + \frac{1}{6} \right) \right]} \tag{4.135}$$

Mean values may be obtained by graphical or numerical integration.

The same procedure may be used for obtaining relationships for mass-transfer coefficients, for equimolecular counterdiffusion or where the concentration of the nondiffusing constituent is small:

$$\frac{R}{\rho u_s^2} = \frac{h_D}{u_s} \quad \text{(Eq. 4.103)}$$

For flow over a plane surface, substitution from Eq. (3.32) gives

$$\frac{h_D}{u_s} = 0.03 Re_x^{-0.2} \tag{4.136}$$

Eq. (4.136) gives the point value of h_D. The mean value over the surface is obtained in the same manner as Eq. (4.133) as

$$\frac{h_D}{u_s} = 0.037 Re_x^{-0.2} \tag{4.137}$$

For mass transfer through a stationary second component,

$$\frac{R}{\rho u_s^2} = \frac{h_D}{u_s} \frac{C_{Bw}}{C_T} \quad \text{(Eq. 12.111)}$$

The correction factor C_{Bw}/C_T must then be introduced into Eqs (4.136) and (4.137).

The above equations are applicable only when the Schmidt number Sc is very close to unity or where the velocity of flow is so high that the resistance of the laminar sublayer is small. The resistance of the laminar sublayer can be taken into account, however, for equimolecular counterdiffusion or for low concentration gradients by using Eq. (4.118):

$$\frac{h_D}{u_s} = \frac{R/\rho u_s^2}{1 + \alpha(Sc - 1)} \quad \text{(Eq. 12.118)}$$

Substitution for $R/\rho u_s^2$ and α using Eqs (3.39) and (3.35) gives

$$\frac{h_D}{u_s} = Sh_x Re_x^{-1} Sc^{-1} = \frac{0.03 Re_x^{-0.2}}{1 + 2.1 Re_x^{-0.1}(Sc - 1)} \quad \text{(4.138)}$$

4.8.6 Flow in a Pipe

For the inlet length of a pipe in which the boundary layers are forming, the equations in the previous section will give an approximate value for the heat-transfer coefficient. It should be remembered, however, that the flow in the boundary layer at the entrance to the pipe may be streamline and the point of transition to turbulent flow is not easily defined. The results therefore are, at best, approximate.

In fully developed flow, Eqs (4.102) and (4.117) can be used, but it is preferable to work in terms of the mean velocity of flow and the ordinary pipe Reynolds number Re. Furthermore, the heat-transfer coefficient is generally expressed in terms of a driving force equal to the difference between the bulk fluid temperature and the wall temperature. If the fluid is highly turbulent, however, the bulk temperature will be quite close to the temperature θ_s at the axis.

Into Eqs (4.99) and (4.117), Eq. (3.63) of Vol. 1A may then be substituted:

$$u = 0.817 u_s \quad \text{(Eq. 3.63 of Vol. 1A)}$$

$$\frac{u_b}{u} = 2.49 Re^{-1/8} \quad \text{(Eq. 3.47)}$$

$$\frac{R}{\rho u^2} = 0.396 Re^{-1/4} \quad \text{(Eq. 3.46)}$$

Firstly, using the simple Reynolds analogy (Eq. 4.102),

$$\frac{h}{C_p \rho u} = \left(\frac{h}{C_p \rho u_s}\right)\left(\frac{u_s}{u}\right)$$

$$= \left(\frac{R}{\rho u_s^2}\right)\left(\frac{u_s}{u}\right)$$

$$= \left(\frac{R}{\rho u^2}\right)\left(\frac{u}{u_s}\right)$$

$$= 0.032 Re^{-1/4} \tag{4.139}$$

Then, using the Taylor–Prandtl modification (Eq. 4.117),

$$\frac{h}{C_p \rho u} = \frac{0.032 Re^{-1/4}}{1 + (u_b/u)(u/u_s)(Pr - 1)}$$

$$= \frac{0.032 Re^{-1/4}}{1 + 2.0 Re^{-1/8}(Pr - 1)} \tag{4.140}$$

Finally, using Eq. (4.135),

$$St = \frac{h}{C_p \rho u} = \frac{0.817(R/\rho u^2)}{1 + 0.817\sqrt{(R/\rho u^2)}5\left[(Pr - 1) + \ln\left(\frac{5}{6}Pr + \frac{1}{6}\right)\right]}$$

$$= \frac{0.032 Re^{1/4}}{1 + 0.817 Re^{-1/8}\left[(Pr - 1) + \ln\left(\frac{5}{6}Pr + \frac{1}{6}\right)\right]} \tag{4.141}$$

For mass transfer, the equations corresponding to Eqs (4.137) and (4.138) are obtained in the same way as the analogous heat-transfer equations.

Thus, using the simple Reynolds analogy for equimolecular counterdiffusion,

$$\frac{h_D}{u} = 0.032 Re^{-1/4} \tag{4.142}$$

and for diffusion through a stationary gas,

$$\left(\frac{h_D}{u}\right)\left(\frac{C_{Bw}}{C_T}\right) = 0.032 Re^{-1/4} \tag{4.143}$$

Using the Taylor–Prandtl form for equimolecular counterdiffusion or low concentration gradients,

$$\frac{h_D}{u} = \frac{0.032 Re^{-1/4}}{1 + 2.0 Re^{-1/8}(Sc - 1)} \tag{4.144}$$

Example 4.3

Water flows at 0.50 m/s through a 20 mm tube lined with β-naphthol. What is the mass-transfer coefficient if the Schmidt number is 2330?

Solution

Reynolds number, $Re = (0.020 \times 0.50 \times 1000/1 \times 10^{-3}) = 10{,}000$

From Eq. (4.144),

$$\frac{h_D}{u} = 0.032 Re^{-1/4} \left[1 + 2.0(Sc - 1)Re^{-1/8}\right]^{-1}$$

$\therefore$

$$h_D = 0.032 \times 0.50 \times 0.1[1 + 2 \times 2329 \times 0.316]^{-1}$$
$$= \underline{\underline{1.085 \times 10^{-6}\,\text{kmol/m}^2\,\text{s}\,(\text{kmol/m}^3)}} = \underline{\underline{1.085 \times 10^{-6}\,\text{m/s}}}$$

Example 4.4

Calculate the rise in temperature of water that is passed at 3.5 m/s through a smooth 25 mm diameter pipe, 6 m long. The water enters at 300 K, and the tube wall may be assumed constant at 330 K. The following methods may be used:

(a) The simple Reynolds analogy (Eq. 4.139)
(b) The Taylor–Prandtl modification (Eq. 4.140)
(c) The universal velocity profile (Eq. 4.141)
(d) $Nu = 0.023 Re^{0.8} Pr^{0.33}$ (Eq. 1.64)

Solution
Taking the fluid properties at 310 K and assuming that fully developed flow exists, an approximate solution will be obtained neglecting the variation of properties with temperature:

$$Re = \frac{0.025 \times 3.5 \times 1000}{0.7 \times 10^{-3}} = 1.25 \times 10^5$$

$$Pr = \frac{4.18 \times 10^3 \times 0.7 \times 10^{-3}}{0.65} = 4.50$$

(a) *Reynolds analogy*

$$\frac{h}{C_p \rho u} = 0.032 Re^{-0.25} \quad \text{(Eq. 12.139)}$$

$$h = \left[4.18 \times 1000 \times 1000 \times 3.5 \times 0.032 \left(1.25 \times 10^5\right)^{-0.25}\right]$$

$$= 24{,}902\,\text{W/m}^2\,\text{K} \quad \text{or} \quad 24.9\,\text{kW/m}^2\,\text{K}$$

Heat transferred per unit time in length dL of pipe $= h\pi\,0.025\,\text{d}L\,(330 - \theta)$ kW, where θ is the temperature at a distance L m from the inlet.

Rate of increase of heat content of fluid $= \left(\frac{\pi}{4}0.025^2 \times 3.5 \times 1000 \times 4.18\right)$ dθ kW

The outlet temperature θ' is then given by

$$\int_{300}^{\theta'} \frac{\text{d}\theta}{(330 - \theta)} = 0.0109 h \int_0^6 \text{d}L$$

where h is in kW/m^2 K.

$\therefore$

$$\log_{10}(330 - \theta') = \log_{10} 30 - \left(\frac{0.0654h}{2.303}\right) = 1.477 - 0.0283h$$

In this case,

$$h = 24.9 \, \text{kW/m}^2 \, \text{K}$$

$\therefore$

$$\log_{10}(330 - \theta') = (1.477 - 0.75) = 0.772$$

and

$$\underline{\underline{\theta' = 324.1 \, \text{K}}}$$

(b) *Taylor–Prandtl equation*

$$\frac{h}{C_p \rho u} = 0.032 Re^{-1/4} \left[1 + 2.0 Re^{-1/8}(Pr - 1)\right]^{-1} \quad (\text{Eq. 4.140})$$

$\therefore$

$$h = \frac{24.9}{(1 + 2.0 \times 3.5/4.34)}$$
$$= 9.53 \, \text{kW/m}^2 \, \text{K}$$

and

$$\log_{10}(330 - \theta') = 1.477 - (0.0283 \times 9.53) = 1.207$$
$$\underline{\underline{\theta' = 313.9 \, \text{K}}}$$

(c) *Universal velocity profile equation*

$$\frac{h}{C_p \rho u} = 0.032 Re^{-1/4} \left\{1 + 0.82 Re^{-1/8}[(Pr - 1) + \ln(0.83 Pr + 0.17)]\right\}^{-1} \quad (\text{Eq. 4.141})$$

$$= \frac{24.9}{1 + (0.82/4.34)(3.5 + 2.303 \times 0.591)}$$
$$= 12.98 \, \text{kW/m}^2 \, \text{K}$$

$\therefore$

$$\log_{10}(330 - \theta') = 1.477 - (0.0283 \times 12.98) = 1.110$$

and

$$\theta' = 317.1\,\text{K}$$

(d) $Nu = 0.023 Re^{0.8} Pr^{0.33}$

$$h = \frac{0.023 \times 0.65}{0.0250} (1.25 \times 10^5)^{0.8} (4.50)^{0.33} \quad \text{(Eq. 9.64)}$$

$$= 0.596 \times 1.195 \times 10^4 \times 1.64$$

$$= 1.168 \times 10^4 \, \text{W/m}^2\,\text{K or } 11.68\,\text{kW/m}^2\,\text{K}$$

and

$$\log_{10}(330 - \theta') = 1.477 - (0.0283 \times 11.68) = 1.147$$

$$\theta' = 316.0\,\text{K}$$

Comparing the results:

Method	h (kW/m² K)	θ' (K)
(a)	24.9	324.1
(b)	9.5	313.9
(c)	13.0	317.1
(d)	11.7	316.0

It is seen that the simple Reynolds analogy is far from accurate in calculating heat transfer to a liquid.

Example 4.5

The tube in Example 4.3 is maintained at 350 K, and air is passed through it at 3.5 m/s, the initial temperature of the air being 290 K. What is the outlet temperature of the air for the four cases used in Example 4.3?

Solution
Taking the physical properties of air at 310 K and assuming that fully developed flow exists in the pipe, then

$$Re = 0.0250 \times 3.5 \times \frac{(29/22.4)(273/310)}{0.018 \times 10^{-3}} = 5535$$

$$Pr = \frac{1.003 \times 1000 \times 0.018 \times 10^{-3}}{0.024} = 0.75$$

The heat-transfer coefficients and final temperatures are then calculated as in Example 4.3 to give the following:

Method	h (W/m^2 K)	θ (K)
(a)	15.5	348.1
(b)	18.3	349.0
(c)	17.9	348.9
(d)	21.2	349.4

In this case, the result obtained using the Reynolds analogy agrees much more closely with the other three methods.

4.9 Nomenclature

		Units in SI system	Dimensions in M, N, L, T, θ
A	Area of surface	m^2	L^2
a	Constant in Eq. (4.41)	—	—
a'	Constant in Eq. (4.41)	—	—
B	Integration constant	m/s	LT^{-1}
B'	B/u^*	—	—
b	Ratio of θ_b to θ_s	—	—
C	Molar concentration	kmol/m^3	NL^{-3}
C_A, C_B	Molar concentration of **A**, **B**	kmol/m^3	NL^{-3}
C_{As}, C_{Bs}	Molar concentration of **A**, **B** outside boundary layer	kmol/m^3	NL^{-3}
C_{Aw}, C_{Bw},	Molar concentration of **A**, **B** at wall	kmol/m^3	NL^{-3}
C_p	Specific heat at constant pressure	J/kg K	$L^2T^{-2}\theta^{-1}$
C_T	Total molar concentration	kmol/m^3	NL^{-3}
C_v	Specific heat at constant volume	J/kg K	$L^2T^{-2}\theta^{-1}$
c_m	Heat capacity of one molecule	J/K	$ML^2T^{-2}\theta^{-1}$
D	Diffusivity	m^2/s	L^2T^{-1}
D_H	Thermal diffusivity	m^2/s	L^2T^{-1}
d	Pipe or particle diameter	m	L
E	Eddy kinematic viscosity	m^2/s	L^2T^{-1}
E_D	Eddy diffusivity	m^2/s	L^2T^{-1}
E_H	Eddy thermal diffusivity	m^2/s	L^2T^{-1}
e	Surface roughness	m	L
h	Heat-transfer coefficient	W/m^2 K	$MT^{-3}\theta^{-1}$
h_D	Mass-transfer coefficient	m/s	LT^{-1}
I	Intensity of turbulence	—	—
i	Fraction of root-mean-square velocity of molecules	—	—
j	Fraction of mean free path of molecules	—	—

K	Ratio of mixing length to distance from surface	—	—
k	Thermal conductivity	W/m K	$\mathbf{MLT^{-3}\theta^{-1}}$
M	Molecular weight	kg/kmol	$\mathbf{MN^{-1}}$
$\mathbf{M}$	Mass of fluid	kg	$\mathbf{M}$
m	Mass of gas molecule	kg	$\mathbf{M}$
N	Molar rate of diffusion per unit area	kmol/m^2 s	$\mathbf{NL^{-2}T^{-1}}$
N'	Total molar rate of transfer per unit area	kmol/m^2 s	$\mathbf{NL^{-2}T^{-1}}$
$\mathbf{N}$	Number of molecules per unit volume at $y=y$	m^{-3}	$\mathbf{L^{-3}}$
$\mathbf{N'}$	Number of molecules per unit volume $y=y+j\lambda$	m^{-3}	$\mathbf{L^{-3}}$
n	Number of molar units	kmol	$\mathbf{N}$
q_0	Rate of transfer of heat per unit area at walls	W/m^2	$\mathbf{MT^{-3}}$
q_y	Rate of transfer of heat per unit area at $y=y$	W/m^2	$\mathbf{MT^{-3}}$
R	Shear stress acting on surface	N/m^2	$\mathbf{ML^{-1}T^{-2}}$
R_0	Shear stress acting on fluid at surface	N/m^2	$\mathbf{ML^{-1}T^{-2}}$
R_y	Shear stress in fluid at $y=y$	N/m^2	$\mathbf{ML^{-1}T^{-2}}$
$\mathbf{R}$	Universal gas constant	8314 J/ kmol K	$\mathbf{MN^{-1}L^2T^{-2}\theta^{-1}}$
r	Radius of pipe	m	$\mathbf{L}$
T	Absolute temperature	K	$\boldsymbol{\theta}$
t	Time	s	$\mathbf{T}$
u	Mean velocity	m/s	$\mathbf{LT^{-1}}$
u_b	Velocity at edge of laminar sublayer	m/s	$\mathbf{LT^{-1}}$
u_E	Mean velocity in eddy	m/s	$\mathbf{LT^{-1}}$
u_{Ex}	Mean component of eddy velocity in X-direction	m/s	$\mathbf{LT^{-1}}$
u_{Ey}	Mean component of eddy velocity in Y-direction	m/s	$\mathbf{LT^{-1}}$
u_{Ez}	Mean component of eddy velocity in Z-direction	m/s	$\mathbf{LT^{-1}}$
u_e	Velocity at distance e from surface	m/s	$\mathbf{LT^{-1}}$
u_m	Root-mean-square velocity of molecules	m/s	$\mathbf{LT^{-1}}$
u_s	Velocity of fluid outside boundary layer or at pipe axis	m/s	$\mathbf{LT^{-1}}$
u_x	Velocity in X-direction at $y=y$	m/s	$\mathbf{LT^{-1}}$
u'_x	Velocity in X-direction at $y=y+j\lambda$ or $y+\lambda_E$	m/s	$\mathbf{LT^{-1}}$

u^+	Ratio of u_y to u^*	—	—
u^*	Shearing stress velocity, $\sqrt{R/\rho}$	m/s	$\mathbf{LT^{-1}}$
w	Exponent of velocity	—	—
y	Distance from surface	m	$\mathbf{L}$
y^+	Ratio of y to $\mu/\rho u^*$	—	—
α	Ratio of u_b to u_s	—	—
γ	Ratio of C_p to C_v	—	—
δ	Thickness of boundary layer	m	$\mathbf{L}$
δ_b	Thickness of laminar sublayer	m	$\mathbf{L}$
λ	Mean free path of molecules	m	$\mathbf{L}$
λ_E	Mixing length	m	$\mathbf{L}$
μ	Viscosity of fluid	N s/m^2	$\mathbf{ML^{-1}T^{-1}}$
Φ	Friction factor $R/\rho u^2$	—	—
ρ	Density of fluid	kg/m^3	$\mathbf{ML^{-3}}$
θ	Temperature at $y=y$	K	$\boldsymbol{\theta}$
θ_s	Temperature outside boundary layer or at pipe axis	K	$\boldsymbol{\theta}$
θ'	Temperature at $y=y+j\lambda$	K	$\boldsymbol{\theta}$
Nu	Nusselt number hd/k	—	—
Nu_x	Nusselt number hx/k	—	—
Re	Reynolds number $ud\rho/\mu$	—	—
Re_r	Roughness Reynolds number $u^*e\rho/\mu$	—	—
Re_x	Reynolds number $u_s x\rho/\mu$	—	—
Pr	Prandtl number $C_p\mu/k$	—	—
Sc	Schmidt number $\mu/\rho D$	—	—
Sh_x	Sherwood number h_D/D	—	—
St	Stanton number $h/C_p\rho u$ or $h/C_p\rho u_s$	—	—

Subscripts
A, B   For component **A, B**

References

1. Jeans JH. *Kinetic theory of gases.* Cambridge: Cambridge University Press; 1940.
2. Prandtl L. Untersuchungen zur ausgebildeten Turbulenz. *Z Angew Math Mech* 1925;**5**:136.
3. Prandtl L. Neuere Ergebnisse der Turbulenzforschung. *Z Ver Deut Ing* 1933;**77**:105.
4. Taylor GI. Conditions at the surface of a hot body exposed to the wind. *NACA Rep Mem* 1916;**272**:423.
5. Nikuradse J. Gesetzmässigkeiten der turbulenten Strömung in glatten Röhren. *Forsch Ver Deut Ing* 1932;**356**.
6. Nikuradse J. Strömungsgesetze in rauhen Röhren. *Forsch Ver Deut Ing* 1933;**361**.
7. Reynolds O. On the extent and action of the heating surface for steam boilers. *Proc Manchester Lit Phil Soc* 1874;**14**:7.

8. Prandtl L. Eine Beziehung zwischen Wärmeaustausch und Strömungswiderstand der Flüssigkeiten. *Physik Z* 1910;**11**:1072.

9. Prandtl L. Bemerkung über den Wärmeübergang im Röhr. *Physik Z* 1928;**29**:487.

10. Martinelli RC. Heat transfer to molten metals. *Trans Am Soc Mech Eng* 1947;**69**:947.

Further Reading

1. Bennett CO, Myers JE. *Momentum, heat and mass transfer*. 3rd ed. New York: McGraw-Hill; 1983.

2. Brodkey RS, Hershey HC. *Transport phenomena*. New York: McGraw-Hill; 1988.

3. Cussler EL. *Diffusion Mass transfer in fluid systems*. 2nd ed. Cambridge: Cambridge University Press; 1997.

4. Hinze JO. *Turbulence*. 2nd ed. New York: McGraw-Hill; 1975.

5. Middleman S. *An introduction to mass and heat transfer*. New York: Wiley; 1997.

CHAPTER 5

Applications in Humidification and Water Cooling

5.1 Introduction

In the processing of materials, it is often necessary either to increase the amount of vapour present in a gas stream, an operation known as *humidification*, or to reduce the vapour present, a process referred to as *dehumidification.* In humidification, the vapour content may be increased by passing the gas over a liquid that then evaporates into the gas stream. This transfer into the main stream takes place by diffusion, and at the interface, simultaneous heat and mass transfer take place according to the relations considered in previous chapters. In the reverse operation, that is, dehumidification, partial condensation must be effected and the condensed vapour removed.

The most widespread application of humidification and dehumidification involves the air–water system, and a discussion of this system forms the greater part of the present chapter. Although the drying of wet solids is an example of a humidification operation, the reduction of the moisture content of the solids is the main objective, and the humidification of the airstream is a secondary effect. Much of the present chapter is, however, of vital significance in any drying operation. Air-conditioning and gas drying also involve humidification and dehumidification operations. For example, moisture must be removed from wet chlorine so that the gas can be handled in steel equipment that otherwise would be severely corroded. Similarly, the gases used in the manufacture of sulphuric acid must be dried or dehumidified before entering the converters, and this is achieved by passing the gas through a dehydrating agent such as sulphuric acid, in essence an absorption operation, or by an alternative dehumidification process discussed later.

In order that hot condenser water may be reused in a plant, it is normally cooled by contact with an airstream. The equipment usually takes the form of a tower in which the hot water is run in at the top and allowed to flow downwards over a packing against a countercurrent flow of air that enters at the bottom of the cooling tower. The design of such towers forms an important part of the present chapter, though at the outset it is necessary to consider basic definitions of the various quantities involved in humidification, in particular *wet-bulb* and *adiabatic saturation temperatures*, and the way in which humidity data are presented on charts and graphs. While the present discussion is devoted to the very important air–water system, which is in some ways unique, the same principles may be applied to other liquids and gases, and this topic is covered in a final section.

5.2 Humidification Terms

5.2.1 Definitions

The more important terms used in relation to humidification are defined as follows:

Humidity (H)	Mass of vapour associated with unit mass of dry gas
Humidity of saturated gas (H_0)	Humidity of the gas when it is saturated with vapour at a given temperature
Percentage humidity	$100 \, (H/H_0)$
Humid heat (s)	Heat required to raise unit mass of dry gas and its associated vapour through unit temperature difference at constant pressure or $s = C_a + HC_w$ where C_a and C_w are the specific heat capacities of the gas and the vapour, respectively. (For the air–water system, the humid heat is approximately $s = 1.00 + 1.9H$ kJ/kg K)
Humid volume	Volume occupied by unit mass of dry gas and its associated vapour
Saturated volume	Humid volume of saturated gas
Dew point	Temperature at which the gas is saturated with vapour. As a gas is cooled, the dew point is the temperature at which condensation will first occur
Percentage relative humidity	$\left(\dfrac{\text{Partial pressure of vapourin gas}}{\text{Partial pressure of vapour in saturated gas}}\right) \times 100$

The above nomenclature conforms with the recommendations of BS1339,[1] although there are some ambiguities in the standard.

The relationship between the partial pressure of the vapour and the humidity of a gas may be derived as follows. In unit volume of gas,

$$\text{Mass of vapour} = \frac{P_w M_w}{RT}$$

and

$$\text{Mass of noncondensable gas} = \frac{(P - P_w)M_A}{RT}$$

The humidity is therefore given by

$$H = \frac{P_w}{P - P_w}\left(\frac{M_w}{M_A}\right)$$ (5.1)

and the humidity of the saturated gas is

$$H_0 = \frac{P_{w0}}{P - P_{w0}}\left(\frac{M_w}{M_A}\right)$$ (5.2)

where P_w is the partial pressure of vapour in the gas, P_{w0} the partial pressure of vapour in the saturated gas at the same temperature, M_A the mean molecular weight of the dry gas, M_w the molecular mass of the vapour, P the total pressure, **R** the gas constant (8314 J/kmol K in SI units), and T the absolute temperature.

For the air–water system, P_w is frequently small compared with P, and hence substituting for the molecular masses,

$$H = \frac{18}{29}\left(\frac{P_w}{P}\right)$$

The relationship between the percentage humidity of a gas and the percentage relative humidity may be derived as follows:

The percentage humidity, by definition $= 100H/H_0$.

Substituting from Eqs (5.1) and (5.2) and simplifying,

$$\begin{aligned}
\text{Percentage humidity} &= \left(\frac{P - P_{w0}}{P - P_w}\right) \cdot \left(\frac{P_w}{P_{w0}}\right) \times 100 \\
&= \frac{(P - P_{w0})}{(P - P_w)} \times (\text{percentage relative humidity})
\end{aligned}$$ (5.3)

When $(P - P_{w0})/(P - P_w) \approx 1$, the percentage relative humidity and the percentage humidity are equal. This condition is approached when the partial pressure of the vapour is only a small proportion of the total pressure or when the gas is almost saturated, that is, as $P_w \rightarrow P_{w0}$.

Example 5.1

In a process in which it is used as a solvent, benzene is evaporated into dry nitrogen. At 297 K and 101.3 kN/m^2, the resulting mixture has a percentage relative humidity of 60. It is required to recover 80% of the benzene present by cooling to 283 K and compressing to a suitable pressure. What should this pressure be? The vapour pressure of benzene is 12.2 kN/m^2 at 297 K and 6.0 kN/m^2 at 283 K.

Solution

From the definition of percentage relative humidity (RH),

$$P_w = P_{w0}\left(\frac{RH}{100}\right)$$

$$\text{At } 297\text{K}, \quad P_w = (12.2 \times 1000) \times \left(\frac{60}{100}\right) = 7320\,\text{N/m}^2$$

In the benzene–nitrogen mixture,

$$\text{Mass of benzene} = \frac{P_w M_w}{RT} = \frac{(7320 \times 78)}{8314 \times 297} = 0.231\,\text{kg}$$

$$\text{Mass of nitrogen} = \frac{(P - P_w)M_A}{RT} = \frac{[(101.3 - 732) \times 1000 \times 28]}{(8314 \times 297)} = 1.066\,\text{kg}$$

Hence, the humidity is

$$H = \left(\frac{0.231}{1.066}\right) = 0.217\,\text{kg/kg}$$

In order to recover 80% of the benzene, the humidity must be reduced to 20% of the initial value. As the vapour will be in contact with liquid benzene, the nitrogen will be saturated with benzene vapour, and hence at 283 K,

$$H_0 = \frac{(0.217 \times 20)}{100} = 0.0433\,\text{kg/kg}$$

Thus, in Eq. (5.2),

$$0.0433 = \left(\frac{6000}{P - 6000}\right)\left(\frac{78}{28}\right)$$

from which

$$P = 3.92 \times 10^5\,\text{N/m}^2 = 392\,\text{kN/m}^2$$

Example 5.2

In a vessel at 101.3 kN/m^2 and 300 K, the percentage relative humidity of the water vapour in the air is 25. If the partial pressure of water vapour when air is saturated with vapour at 300 K is 3.6 kN/m^2, calculate.

(a) the partial pressure of the water vapour in the vessel,
(b) the specific volumes of the air and water vapour,
(c) the humidity of the air and humid volume,
(d) the percentage humidity.

Solution

(a) From the definition of percentage relative humidity,

$$P_w = P_{w0}\frac{RH}{100} = 3600 \times \left(\frac{25}{100}\right) = 900\,\text{N/m}^2 = 0.9\,\text{kN/m}^2$$

(b) In 1 m^3 of air,

$$\text{Mass of water vapour} = \frac{(900 \times 18)}{8314 \times 300} = 0.0065 \, \text{kg}$$

$$\text{Mass of air} = \frac{[(101.3 - 0.9) \times 1000 \times 29]}{(8314 \times 300)} = 1.167 \, \text{kg}$$

Hence, specific volume of water vapour at $0.9 \, \text{kN/m}^2 = \left(\frac{1}{0.0065}\right) = 154 \, \text{m}^3/\text{kg}$.

specific volume of air at $100.4 \, \text{kN/m}^2 = \left(\frac{1}{1.167}\right) = 0.857 \, \text{m}^3/\text{kg}$

(c) Humidity,

$$H = \left(\frac{0.0065}{1.1673}\right) = 0.0056 \, \text{kg/kg}$$

(Using the approximate relationship,

$$H = \frac{(18 \times 900)}{(29 \times 101.3 \times 1000)} = 0.0055 \, \text{kg/kg})$$

∴ humid volume, volume of 1 kg air + associated vapour = specific volume of air at 100.4 kN/m^2

$$= 0.857 \, \text{m}^3/\text{kg}$$

(d) From Eq. (5.3),

$$\text{Percentage humidity} = \frac{[(101.3 - 3.6) \times 1000]}{[(101.3 - 0.9) \times 1000]} \times 25$$

$$= 24.3$$

5.2.2 Wet-Bulb Temperature

When a stream of unsaturated gas is passed over the surface of a liquid, the humidity of the gas is increased due to evaporation of the liquid. The temperature of the liquid falls below that of the gas, and heat is transferred from the gas to the liquid. At equilibrium, the rate of heat transfer from the gas just balances that required to vaporise the liquid, and the liquid is said to be at the *wet-bulb temperature*. The rate at which this temperature is reached depends on the initial temperatures and the rate of flow of gas past the liquid surface. With a small area of contact between the gas and the liquid and a high gas flow rate, the temperature and the humidity of the gas stream remain virtually unchanged.

The rate of transfer of heat from the gas to the liquid can be written as

$$Q = hA(\theta - \theta_w) \tag{5.4}$$

where Q is the heat flow, h the coefficient of heat transfer, A the area for transfer, and θ and θ_w are the temperatures of the gas and liquid phases.

The liquid evaporating into the gas is transferred by diffusion from the interface to the gas stream as a result of a concentration difference $(c_0 - c)$, where c_0 is the concentration of the vapour at the surface (mass per unit volume) and c is the concentration in the gas stream. The rate of evaporation is then given by

$$W = h_D A(c_0 - c) = h_D A \frac{M_w}{RT}(P_{w0} - P_w) \tag{5.5}$$

where h_D is the coefficient of mass transfer.

The partial pressures of the vapour, P_w and P_{w0}, may be expressed in terms of the corresponding humidities H and H_w by Eqs (5.1) and (5.2).

If P_w and P_{w0} are small compared with P, $(P - P_w)$ and $(P - P_{w0})$ may be replaced by a mean partial pressure of the gas P_A and

$$W = h_{DA} A \frac{(H_w - H)M_w}{RT} \cdot \left(P_A \frac{M_A}{M_w} \right) \tag{5.6}$$

$$= h_D A \rho_A (H_w - H)$$

where ρ_A is the density of the gas at the partial pressure P_A.

The heat transfer required to maintain this rate of evaporation is

$$Q = h_D A \rho_A (H_w - H)\lambda \tag{5.7}$$

where λ is the latent heat of vaporisation of the liquid.

Thus, equating Eqs (5.4) and (5.7),

$$(H - H_w) = -\frac{h}{h_D \rho_A \lambda}(\theta - \theta_w) \tag{5.8}$$

Both h and h_D are dependent on the equivalent gas-film thickness, and thus, any decrease in the thickness, as a result of increasing the gas velocity, for example, increases both h and h_D. At normal temperatures, (h/h_D) is virtually independent of the gas velocity provided this is greater than about 5 m/s. Under these conditions, heat transfer by convection from the gas stream is large compared with that from the surroundings by radiation and conduction.

The wet-bulb temperature θ_w depends only on the temperature and the humidity of the gas, and values normally quoted are determined for comparatively high gas velocities, such that the condition of the gas does not change appreciably as a result of being brought into contact with the liquid and the ratio (h/h_D) has reached a constant value. For the air–water system, the ratio $(h/h_D \rho_A)$ is about 1.0 kJ/kg K and varies from 1.5 to 2.0 kJ/kg K for organic liquids.

Example 5.3

Moist air at 310 K has a wet-bulb temperature of 300 K. If the latent heat of vaporisation of water at 300 K is 2440 kJ/kg, estimate the humidity of the air and the percentage relative humidity. The total pressure is 105 kN/m^2, and the vapour pressure of water vapour at 300 K is 3.60 and 6.33 kN/m^2 at 310 K.

Solution

The humidity of air saturated at the wet-bulb temperature is given by.

$$H_w = \frac{P_{w0}}{P0P_{w0}} \frac{M_w}{M_A} \quad \text{(Eq. 13.2)}$$

$$= \left(\frac{3.6}{105.0 - 3.6}\right)\left(\frac{18}{29}\right) = 0.0220 \, \text{kg/kg}$$

Therefore, taking $(h/h_D\rho_A)$ as 1.0 kJ/kg K, in Eq. (5.8),

$$(0.0220 - H) = \left(\frac{1.0}{2440}\right)(310 - 300)$$

$$H = \underline{\underline{0.018 \text{ kg/kg}}}$$

or

$$\text{at } 310\,\text{K}, \quad P_{w0} = 6.33\,\text{kN/m}^2$$

In Eq. (5.2),

$$0.0780 = \frac{18P_w}{(105.0 - P_w)29}$$

$$\therefore \ P_w = 2.959\,\text{kN/m}^2$$

and the percentage relative humidity

$$= \frac{(100 \times 2.959)}{6.33} = \underline{\underline{46.7\%}}$$

5.2.3 Adiabatic Saturation Temperature

In the system just considered, neither the humidity nor the temperature of the gas is appreciably changed. If the gas is passed over the liquid at such a rate that the time of contact is sufficient for equilibrium to be established, the gas will become saturated, and both phases will be brought to the same temperature. In a thermally insulated system, the total sensible heat falls by an amount equal to the latent heat of the liquid evaporated. As a result of continued passage of the gas, the temperature of the liquid gradually approaches an equilibrium value that is known as the *adiabatic saturation temperature*.

These conditions are achieved in an infinitely tall thermally insulated humidification column through which gas of a given initial temperature and humidity flows countercurrently to the liquid under conditions where the gas is completely saturated at the top of the column. If the liquid is continuously circulated around the column and if any fresh liquid that is added is at the same temperature as the circulating liquid, the temperature of the liquid at the top and bottom of the column and of the gas at the top approaches the adiabatic saturation temperature. Temperature and humidity differences are a maximum at the bottom and zero at the top, and therefore, the rates of transfer of heat and mass decrease progressively from the bottom to the top of the tower. This is illustrated in Fig. 5.1.

Making a heat balance over the column, it is seen that the heat of vaporisation of the liquid must come from the sensible heat in the gas. The temperature of the gas falls from θ to the adiabatic saturation temperature θ_s, and its humidity increases from H to H_s (the saturation value at θ_s). Then, working on the basis of unit mass of dry gas,or

$$(\theta - \theta_s)s = (H_s - H)\lambda$$
$$(H - H_s) = -\frac{s}{\lambda}(\theta - \theta_s)$$

(5.9)

where s is the humid heat of the gas and λ the latent heat of vaporisation at θ_s. s is almost constant for small changes in H.

Eq. (5.9) indicates an approximately linear relationship between humidity and temperature for all mixtures of gas and vapour having the same adiabatic saturation temperature θ_s. A curve of humidity versus temperature for gases with a given adiabatic saturation temperature is known as an *adiabatic cooling line*. For a range of adiabatic saturation temperatures, a family of curves, approximating to straight lines of slopes equal to $-(s/\lambda)$, is obtained. These lines are not exactly straight and parallel because of variations in λ and s.

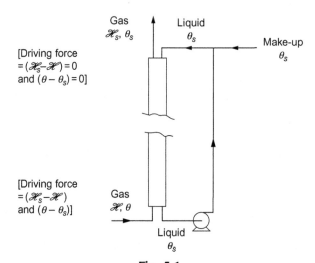

Fig. 5.1

Adiabatic saturation temperature θ_s.

Comparing Eqs (5.8) and (5.9), it is seen that the adiabatic saturation temperature is equal to the wet-bulb temperature when $s = h/h_D\rho_A$. This is the case for most water vapour systems and accurately so when $H = 0.047$. The ratio $(h/h_D\rho_A s) = b$ is sometimes known as the *psychrometric ratio*, and as indicated, b is approximately unity for the air–water system. For most systems involving air and an organic liquid, $b = 1.3$–2.5 and the wet-bulb temperature is higher than the adiabatic saturation temperature. This was confirmed in 1932 by Sherwood and Comings[2] who worked with water, ethanol, n-propanol, n-butanol, benzene, toluene, carbon tetrachloride, and n-propyl acetate and found that the wet-bulb temperature was always higher than the adiabatic saturation temperature except in the case of water.

In Chapter 4, it is shown that when the Schmidt and Prandtl numbers for a mixture of gas and vapour are approximately equal to unity, the *Lewis relation* applies or

$$h_D = \frac{h}{C_p\rho} \quad \text{(Eq.4.105)}$$

where C_p and ρ are the mean specific heat and density of the vapour phase.

Therefore,

$$\frac{h}{h_D\rho_A} = \frac{C_p\rho}{\rho_A} \tag{5.10}$$

Where the humidity is relatively low, $C_p \approx s$ and $\rho \approx \rho_A$, and hence,

$$s \approx \frac{h}{h_D\rho_A} \tag{5.11}$$

For systems containing vapour other than that of water, s is only approximately equal to $h/h_D\rho_A$, and the difference between the two quantities may be as high as 50%.

If an unsaturated gas is brought into contact with a liquid that is at the adiabatic saturation temperature of the gas, a simultaneous transfer of heat and mass takes place. The temperature of the gas falls, and its humidity increases (Fig. 5.2). The temperature of the liquid at any instant tends to change and approaches the wet-bulb temperature corresponding to the particular condition of the gas at that moment. For a liquid other than water, the adiabatic saturation temperature is less than the wet-bulb temperature, and therefore in the initial stages, the temperature of the liquid rises. As the gas becomes humidified, however, its wet-bulb temperature falls and consequently the temperature to which the liquid is tending decreases as evaporation takes place. In due course, therefore, a point is reached where the liquid actually reaches the wet-bulb temperature of the gas in contact with it. It does not remain at this temperature, however, because the gas is not then completely saturated, and further humidification is accompanied by a continued lowering of the wet-bulb temperature. The temperature of the liquid therefore starts to fall and continues to fall until the gas is completely saturated. The liquid and gas are then both at the adiabatic saturation temperature.

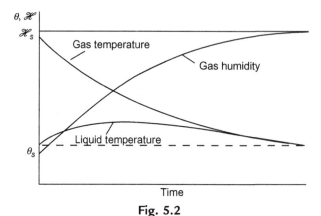

Fig. 5.2
Saturation of gas with liquid other than water at the adiabatic saturation temperature.

The air–water system is unique, however, in that the Lewis relation holds quite accurately, so that the adiabatic saturation temperature is the same as the wet-bulb temperature. If, therefore, an unsaturated gas is brought into contact with water at the adiabatic saturation temperature of the gas, there is no tendency for the temperature of the water to change, and it remains in a condition of dynamic equilibrium through the whole of the humidification process (Fig. 5.3). In this case, the adiabatic cooling line represents the conditions of gases of constant wet-bulb temperatures and constant adiabatic saturation temperatures. The change in the condition of a gas as it is humidified with water vapour is therefore represented by the adiabatic cooling line, and the intermediate conditions of the gas during the process are readily obtained. This is particularly useful because only partial humidification is normally obtained in practice.

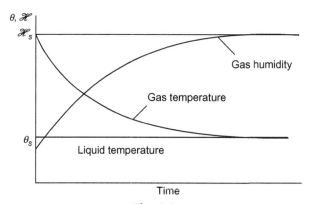

Fig. 5.3
Saturation of air with water at adiabatic saturation temperature.

5.3 Humidity Data for the Air–Water System

To facilitate calculations, various properties of the air–water system are plotted on a *psychrometric* or *humidity chart*. Such a chart is based on either the temperature or the enthalpy of the gas. The temperature-humidity chart is the more commonly used though the enthalpy–humidity chart is particularly useful for determining the effect of mixing two gases or of mixing a gas and a liquid. Each chart refers to a particular total pressure of the system. A humidity-temperature chart for the air–water system at atmospheric pressure, based on the original chart by Grosvenor,[3] is given in Fig. 5.4, and the corresponding humidity-enthalpy chart is given in Fig. 5.5.

5.3.1 Temperature–Humidity Chart

In Fig. 5.4, it will be seen that the following quantities are plotted against temperature:

(i)   The *humidity H* for various values of the percentage humidity.
For saturated gas,

$$H_0 = \frac{P_{w0}}{P - P_{w0}} \left(\frac{M_w}{M_A}\right) \quad \text{(Eq.13.2)}$$

From Eq. (5.1) for a gas with a humidity less than the saturation value,

$$H = \frac{P_w}{P - P_w} \left(\frac{M_w}{M_A}\right) = H_0 \frac{P_w}{P_{w0}} \frac{P_w - P_{w0}}{P - P_w} \tag{5.12}$$

(ii) *The specific volume of dry gas.* This is a linear function of temperature.
(iii) *The saturated volume.* This increases more rapidly with temperature than the specific volume of dry gas because both the quantity and the specific volume of vapour increase with temperature. At a given temperature, the humid volume varies linearly with humidity, and hence, the humid volume of unsaturated gas can be found by interpolation.
(iv) The latent heat of vaporisation.

In addition, the *humid heat* is plotted as the abscissa in Fig. 5.4 with the humidity as the ordinate.

Adiabatic cooling lines are included in the diagram, and as already discussed, these have a slope of $-(s/\lambda)$, and they are slightly curved since s is a function of H. On the chart, they appear as straight lines, however, since the inclination of the axis has been correspondingly adjusted. Each adiabatic cooling line represents the composition of all gases whose adiabatic saturation temperature is given by its point of intersection with the 100% humidity curve. For the air–water system, the adiabatic cooling lines represent conditions of constant wet-bulb

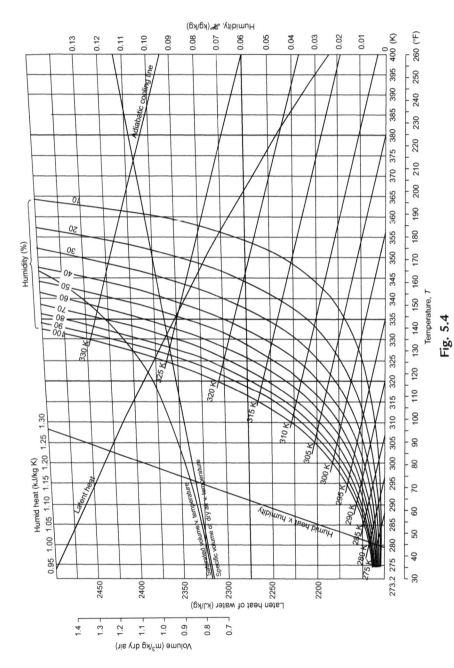

Fig. 5.4

Humidity–temperature chart (see also the Appendix).

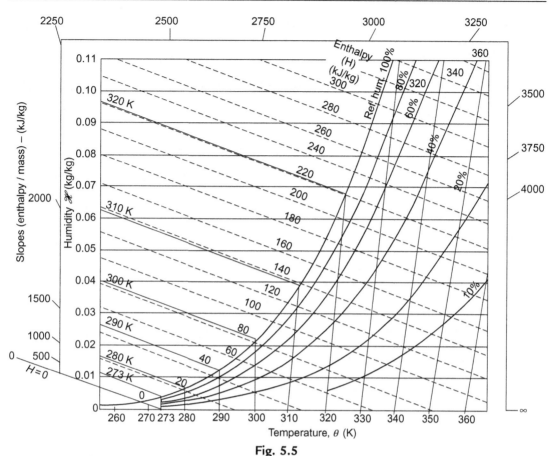

Fig. 5.5

Humidity–enthalpy diagram for air–water vapour system at atmospheric pressure.

temperature as well and, as previously mentioned, enable the change in composition of a gas to be followed as it is humidified by contact with water at the adiabatic saturation temperature of the gas.

Example 5.4

Air containing 0.005 kg water vapour per kg of dry air is heated to 325 K in a dryer and passed to the lower shelves. It leaves these shelves at 60% humidity and is reheated to 325 K and passed over another set of shelves, again leaving at 60% humidity. This is again repeated for the third and fourth sets of shelves, after which the air leaves the dryer. On the assumption that the material on each shelf has reached the wet-bulb temperature and that heat losses from the dryer may be neglected, determine.

(a) the temperature of the material on each tray,
(b) the amount of water removed in kg/s if 5 m^3/s moist air leaves the dryer,
(c) the temperature to which the inlet air would have to be raised to carry out the drying in a single stage.

Solution

For each of the four sets of shelves, the condition of the air is changed to 60% humidity along an adiabatic cooling line.

Initial condition of air: $\theta = 325$ K, $H = 0.005$ kg/kg

On humidifying to 60% humidity,

$$\theta = 301\text{K}, \quad H = 0.015\text{kg/kg and } \theta_w = 296\text{K}$$

At the end of the second pass,

$$\theta = 308\text{K}, \quad H = 0.022\text{kg/kg and } \theta_w = 301\text{K}$$

At the end of the third pass,

$$\theta = 312\text{K}, \quad H = 0.027\text{kg/kg and } \theta_w = 305\text{K}$$

At the end of the fourth pass,

$$\theta = 315\text{K}, \quad H = 0.032\text{kg/kg and } \theta_w = 307\text{K}$$

Thus, the temperatures of the material on each of the trays are

$$\underline{\underline{296\text{K, } 301\text{K, } 305\text{K, and } 307\text{K}}}$$

Total increase in humidity

$$= (0.032 - 0.005) = 0.027\,\text{kg/kg}$$

The air leaving the system is at 315 K and 60% humidity.

From Fig. 5.4, specific volume of dry air $= 0.893$ m^3/kg.

Specific volume of saturated air (*saturated volume*) $= 0.968$ m^3/kg.

Therefore, by interpolation, the humid volume of air of 60% humidity $= 0.937$ m^3/kg.

$$\text{Mass of air passing through the dryer} = \left(\frac{5}{0.937}\right) = 5.34\,\text{kg/s.}$$

$$\text{Mass of water evaporated} = (5.34 \times 0.027) = \underline{\underline{0.144\,\text{kg/s.}}}$$

If the material is to be dried by air in a single pass, the air must be heated before entering the dryer such that its wet-bulb temperature is 307 K.

For air with a humidity of 0.005 kg/kg, this corresponds to a dry-bulb temperature of $\underline{370\text{K}}$.

The various steps in this calculation are shown in Fig. 5.6.

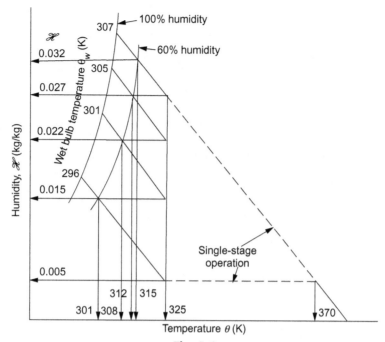

Fig. 5.6
Humidification stages for Example 5.4 (schematic).

5.3.2 Enthalpy–Humidity Chart

In the calculation of enthalpies, it is necessary to define some standard reference state at which the enthalpy is taken as zero. It is most convenient to take the melting point of the material constituting the vapour as the reference temperature and the liquid state of the material as its standard state.

If H is the enthalpy of the humid gas per unit mass of dry gas, H_a the enthalpy of the dry gas per unit mass, H_w the enthalpy of the vapour per unit mass, C_a the specific heat of the gas at constant pressure, C_w the specific heat of the vapour at constant pressure, θ the temperature of the humid gas, θ_0 the reference temperature, λ the latent heat of vaporisation of the liquid at θ_0, and H the humidity of the gas, then, for an unsaturated gas,

$$H = H_a + H_w \mathscr{H} \tag{5.13}$$

where

$$H_a = C_a(\theta - \theta_0) \tag{5.14}$$

and

$$H_w = C_w(\theta - \theta_0) + \lambda \qquad (5.15)$$

Thus, in Eq. (5.13),

$$H = (C_a + HC_w)(\theta - \theta_0) + H\lambda$$
$$= (\theta - \theta_0)(s + H\lambda) \qquad (5.16)$$

If the gas contains more liquid or vapour than is required to saturate it at the temperature in question, either the gas will be supersaturated or the excess material will be present in the form of liquid or solid according to whether the temperature θ is greater or less than the reference temperature θ_0. The supersaturated condition is unstable and will not be considered further.

If the temperature θ is greater than θ_0 and if the humidity H is greater than the humidity H_0 of saturated gas, the enthalpy H per unit mass of dry gas is given by

$$H = C_a(\theta - \theta_0) + H_0[C_w(\theta - \theta_0) + \lambda] + C_L(H - H_0)(\theta - \theta_0) \qquad (5.17)$$

where C_L is the specific heat of the liquid.

If the temperature θ is less than θ_0, the corresponding enthalpy H is given by

$$H = C_a(\theta - \theta_0) + H_0[C_w(\theta - \theta_0) + \lambda] + (H - H_0)[C_s(\theta - \theta_0) + \lambda_f] \qquad (5.18)$$

where C_s is the specific heat of the solid and λ_f is the latent heat of freezing of the liquid, a negative quantity.

Eqs (5.16)–(5.18) give the enthalpy in terms of the temperature and humidity of the humid gas for the three conditions: $\theta = \theta_0$, $\theta > \theta_0$, and $\theta < \theta_0$, respectively. Thus, given the percentage humidity and the temperature, the humidity may be obtained from Fig. 5.4, the enthalpy calculated from Eqs (5.16), (5.17), or (5.18) and plotted against the humidity, usually with enthalpy as the abscissa. Such a plot is shown in Fig. 5.7 for the air–water system, which includes the curves for 100% humidity and for some lower value, say $Z\%$.

Considering the nature of the isothermals for the three conditions dealt with previously, at constant temperature θ, the relation between enthalpy and humidity for an unsaturated gas is

$$H = \text{constant} + [C_w(\theta - \theta_0) + \lambda]H \qquad (5.19)$$

Thus, the isothermal is a straight line of slope $(C_w(\theta - \theta_0) + \lambda)$ with respect to the humidity axis. At the reference temperature θ_0, the slope is λ; at higher temperatures, the slope is greater than λ; and at lower temperatures, it is less than λ. Because the latent heat is normally large compared with the sensible heat, the slope of the isothermals remains positive down to very low temperatures. Since the humidity is plotted as the ordinate, the slope of the isothermal relative to the X-axis decreases with increase in temperature. When $\theta > \theta_0$ and

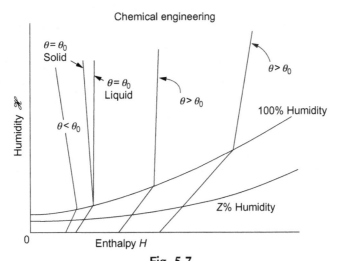

Fig. 5.7

Humidity-enthalpy diagram for air–water system—rectangular axes.

$H > H_0$ the saturation humidity, the vapour phase consists of a saturated gas with liquid droplets in suspension. The relation between enthalpy and humidity at constant temperature θ is

$$H = \text{constant} + C_L(\theta - \theta_0)H \qquad (5.20)$$

The isothermal is therefore a straight line of slope $C_L(\theta - \theta_0)$. At the reference temperature θ_0, the slope is zero, and the isothermal is parallel to the humidity axis. At higher temperatures, the slope has a small positive value. When $\theta < \theta_0$ and $H > H_0$, solid particles are formed, and the equation of the isothermal is

$$H = \text{constant} + \left[C_s(\theta - \theta_0) + \lambda_f\right]H \qquad (5.21)$$

This represents a straight line of slope $(C_s(\theta - \theta_0) + \lambda_f)$. Both $C_s(\theta - \theta_0)$ and λ_f are negative, and therefore, the slopes of all these isothermals are negative. When $\theta = \theta_0$, the slope is λ_f. In the supersaturated region, therefore, there are two distinct isothermals at temperature θ_0: one corresponds to the condition where the excess vapour is present in the form of liquid droplets and the other to the condition where it is present as solid particles. The region between these isothermals represents conditions where a mixture of liquid and solid is present in the saturated gas at the temperature θ_0.

The shape of the humidity-enthalpy line for saturated air is such that the proportion of the total area of the diagram representing saturated, as opposed to supersaturated, air is small when rectangular axes are used. In order to enable greater accuracy to be obtained in the use of the diagram, oblique axes are normally used, as in Fig. 5.5, so that the isothermal for unsaturated gas at the reference temperature θ_0 is parallel to the humidity axis.

It should be noted that the curves of humidity plotted against either temperature or enthalpy have a discontinuity at the point corresponding to the freezing point of the humidifying material. Above the temperature θ_0, the lines are determined by the vapour–liquid equilibrium and below it by the vapour–solid equilibrium.

Two cases may be considered to illustrate the use of enthalpy–humidity charts. These are the mixing of two streams of humid gas and the addition of liquid or vapour to a gas.

Mixing of two streams of humid gas

Consider the mixing of two gases of humidities H_1 and H_2, at temperatures θ_1 and θ_2, and with enthalpies H_1 and H_2 to give a mixed gas of temperature θ, enthalpy H, and humidity H. If the masses of dry gas concerned are m_1, m_2, and m, respectively, then taking a balance on the dry gas, vapour, and enthalpy,

$$m_1 + m_2 = m \tag{5.22}$$
$$m_1 H_1 + m_2 H_2 = mH \tag{5.23}$$

and

$$m_1 H_1 + m_2 H_2 = mH \tag{5.24}$$

Elimination of m gives

$$m_1(H - H_1) = m_2(H_2 - H) \tag{5.25}$$

and

$$m_1(H - H_1) = m_2(H_2 - H)$$

Dividing these two equations,

$$\frac{(H - H_1)}{H - H_1} = \frac{(H - H_2)}{H - H_2} \tag{5.26}$$

The condition of the resultant gas is therefore represented by a point on the straight line joining (H_1 and H_1) and (H_2 and H_2). The humidity H is given, from Eq. (5.25), by

$$\frac{(H - H_1)}{(H_2 - H)} = \frac{m_2}{m_1} \tag{5.27}$$

The gas formed by mixing two unsaturated gases may be either unsaturated, saturated, or supersaturated. The possibility of producing supersaturated gas arises because the 100% humidity line on the humidity–enthalpy diagram is concave towards the humidity axis.

Example 5.5

In an air-conditioning system, 1 kg/s air at 350 K and 10% humidity is mixed with 5 kg/s air at 300 K and 30% humidity. What is the enthalpy, humidity, and temperature of the resultant stream?

Solution
From Fig. 5.4,

at $\theta_1 = 350$ K and humidity $= 10\%$; $H_1 = 0.043$ kg/kg.

at $\theta_2 = 300$ K and humidity $= 30\%$; $H_2 = 0.0065$ kg/kg.

Thus, in Eq. (5.23),

$$(1 \times 0.043) + (5 \times 0.0065) = (1 + 5)H$$

and

$$H = 0.0125 \, \text{kg/kg}$$

From Fig. 5.5,

at $\theta_1 = 350$ K and $H_1 = 0.043$ kg/kg; $H_1 = 192$ kJ/kg.

at $\theta_2 = 300$ K and $H_2 = 0.0065$ kg/kg; $H_2 = 42$ kJ/kg.

Thus, in Eq. (5.25),

$$1(H - 192) = 5(42 - H)$$

and

$$H = 67 \, \text{kJ/kg}$$

From Fig. 5.5,

at $H = 67$ kJ/kg and $H = 0.0125$ kg/kg

$$\theta = 309 \, \text{K}$$

The data used in this example are shown in Fig. 5.8.

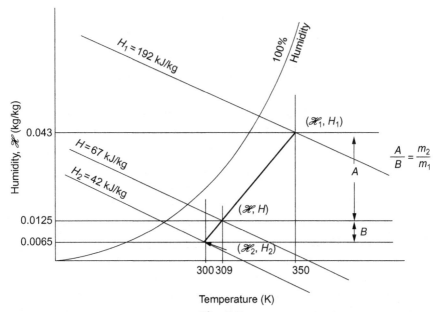

Fig. 5.8

Data used in Example 5.5.

Addition of liquid or vapour to a gas

If a mass m_3 of liquid or vapour of enthalpy H_3 is added to a gas of humidity H_1 and enthalpy H_1 and containing a mass m_1 of dry gas, then

$$m_1(H - H_1) = m_3 \tag{5.28}$$

$$m_1(H - H_1) = m_3 H_3 \tag{5.29}$$

Thus,

$$\frac{H - H_1}{H - H_1} = H_3 \tag{5.30}$$

where H and H are the humidity and enthalpy of the gas produced on mixing.

The composition and properties of the mixed stream are therefore represented by a point on the straight line of slope H_3, relative to the humidity axis, which passes through the point (H_1 and H_1). In Fig. 5.5, the edges of the plot are marked with points that, when joined to the origin, give a straight line of the slope indicated. Thus, in using the chart, a line of slope H_3 is drawn through the origin and a parallel line drawn through the point (H_1 and H_1). The point representing the final gas stream is then given from Eq. (5.28):

$$(H - H_1) = \frac{m_3}{m_1}$$

It can be seen from Fig. 5.5 that for the air–water system, a straight line, of slope equal to the enthalpy of dry saturated steam (2675 kJ/kg), is almost parallel to the isothermals, so that the addition of live steam has only a small effect on the temperature of the gas. The addition of water spray, even if the water is considerably above the temperature of the gas, results in a lowering of the temperature after the water has evaporated. This arises because the latent heat of vaporisation of the liquid constitutes the major part of the enthalpy of the vapour. Thus, when steam is added, it gives up a small amount of sensible heat to the gas, whereas when hot liquid is added, a small amount of sensible heat is given up, and a very much larger amount of latent heat is absorbed from the gas.

Example 5.6

0.15 kg/s steam at atmospheric pressure and superheated to 400 K is bled into an airstream at 320 K and 20% relative humidity. What is the temperature, enthalpy, and relative humidity of the mixed stream if the air is flowing at 5 kg/s? How much steam would be required to provide an exit temperature of 330 K and what would be the humidity of this mixture?

Solution

Steam at atmospheric pressure is saturated at 373 K at which the latent heat

$$= 2258 \, kJ/kg$$

Taking the specific heat of superheated steam as 2.0 kJ/kg K,

$$\text{Enthalpy of the steam}: H_3 = 4.18(373 - 273) + 2258 + 2.0(400 - 373)$$
$$= 2730 \text{kJ/kg}$$

From Fig. 5.5,

at $\theta_1 = 320$ K and 20% relative humidity, $H_1 = 0.013$ kg/kg and $H_1 = 83$ kJ/kg.

The line joining the axis and slope $H_3 = 2730$ kJ/kg at the edge of the chart is now drawn in, and a parallel line is drawn through (H_1, H_1).

Thus,

$$(H - H_1) = \frac{m_3}{m_1} = \left(\frac{0.15}{5}\right) = 0.03 \text{ kg/kg}$$

and

$$H = (0.03 + 0.013) = 0.043 \text{ kg/kg}$$

At the intersection of $H = 0.043$ kg/kg and the line through $(H_1$ and $H_1)$,

$$\underline{\underline{H = 165 \text{kJ/kg and} \theta = 324 \text{K}}}$$

When $\theta = 330$ K, the intersection of this isotherm and the line through H gives an outlet stream in which $H = 0.094$ kg/kg (83% relative humidity) and $H = 300$ kJ/kg.

Thus, in Eq. (5.28),

$$m_3 = 5(0.094 - 0.013) = \underline{\underline{0.41 \text{ kg/s}}}$$

The data used in this example are shown in Fig. 5.9.

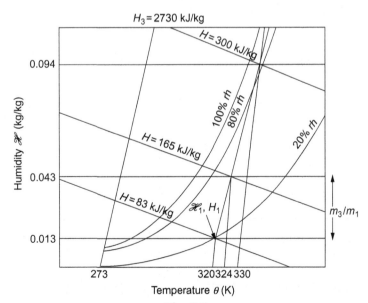

Fig. 5.9

Data used in Example 5.6 (schematic).

5.4 Determination of Humidity

The most important methods for determining humidity are as follows:

(1) *Chemical methods.* A known volume of the gas is passed over a suitable absorbent, the increase in mass of which is measured. The efficiency of the process can be checked by arranging a number of vessels containing absorbent in series and ascertaining that the increase in mass in the last of these is negligible. The method is very accurate but is laborious. Satisfactory absorbents for water vapour are phosphorus pentoxide dispersed in pumice and concentrated sulphuric acid.

(2) *Determination of the wet-bulb temperature.* Eq. (5.8) gives the humidity of a gas in terms of its temperature, its wet-bulb temperature, and various physical properties of the gas and vapour. The wet-bulb temperature is normally determined as the temperature attained by the bulb of a thermometer, which is covered with a piece of material that is maintained saturated with the liquid. The gas should be passed over the surface of the wet bulb at a high enough velocity (>5 m/s) (a) for the condition of the gas stream not to be affected appreciably by the evaporation of liquid, (b) for the heat transfer by convection to be large compared with that by radiation and conduction from the surroundings, and (c) for the ratio of the coefficients of heat and mass transfer to have reached a constant value. The gas should be passed long enough for equilibrium to be attained, and for accurate work, the liquid should be cooled nearly to the wet-bulb temperature before it is applied to the material.

The stream of gas over the liquid surface may be produced by a small fan or other similar means (Fig. 5.10A). The crude forms of wet-bulb thermometer, which make no provision for the rapid passage of gas, cannot be used for accurate determinations of humidity.

(3) *Determination of the dew point.* The dew point is determined by cooling a highly polished surface in the gas and observing the highest temperature at which condensation takes place (Fig. 5.10B). The humidity of the gas is equal to the humidity of saturated gas at the dew point. The instrument illustrated in Fig. 5.10C incorporates a polished gold mirror that is cooled using a thermoelectric module that utilises the *Peltier effect.*

(4) *Measurement of the change in length of a hair or fibre.* The length of a hair or fibre is influenced by the humidity of the surrounding atmosphere. Many forms of apparatus for automatic recording of humidity depend on this property. The method has the disadvantage that the apparatus needs frequent calibration because the zero tends to shift. This difficulty is most serious when the instrument is used over a wide range of humidities. A typical hair hygrometer is shown in Fig. 5.10D.

(5) *Measurement of conductivity of a fibre.* If a fibre is impregnated with an electrolyte, such as lithium chloride, its electrical resistance will be governed by its moisture content, which in turn depends on the humidity of the atmosphere in which it is situated.

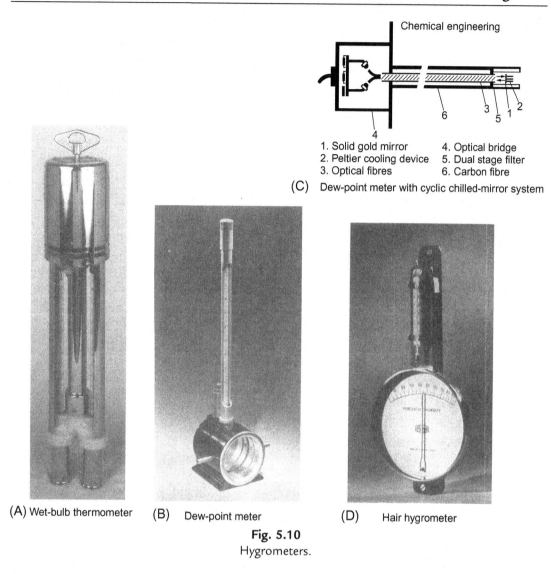

Chemical engineering

1. Solid gold mirror   4. Optical bridge
2. Peltier cooling device   5. Dual stage filter
3. Optical fibres   6. Carbon fibre

(C)   Dew-point meter with cyclic chilled-mirror system

(A) Wet-bulb thermometer   (B)   Dew-point meter   (D)   Hair hygrometer

Fig. 5.10
Hygrometers.

In a lithium chloride cell, a skein of very fine fibres is wound on a plastic frame carrying the electrodes, and the current flowing at a constant applied voltage gives a direct measure of the relative humidity.

(6) *Measurement of heat of absorption on to a surface.*

(7) *Electrolytic hygrometry* in which the quantity of electricity required to electrolyse water absorbed from the atmosphere on to a thin film of desiccant is measured.

(8) *Piezoelectric hygrometry* employing a quartz crystal with a hygroscopic coating in which moisture is alternately absorbed from a wet gas and desorbed in a dry gas stream, the dynamics is a function of the gas humidity.

(9) *Capacitance meters* in which the electric capacitance is a function of the degree of deposition of moisture from the atmosphere.

(10) *Observation of colour changes* in active ingredients, such as cobaltous chloride.

Further details of instruments for the measurement of humidity are given in Volume 3. Reference should also be made to standard works on psychrometry.[4,5,6]

5.5 Humidification and Dehumidification

5.5.1 Methods of Increasing Humidity

The following methods may be used for increasing the humidity of a gas:

1. Live steam may be added directly in the required quantity. It has been shown that this produces only a slight increase in the temperature, but the method is not generally favoured because any impurities that are present in the steam may be added at the same time.

2. Water may be sprayed into the gas at such a rate that, on complete vaporisation, it gives the required humidity. In this case, the temperature of the gas will fall as the latent heat of vaporisation must be supplied from the sensible heat of the gas and liquid.

3. The gas may be mixed with a stream of gas of higher humidity. This method is frequently used in laboratory work when the humidity of a gas supplied to an apparatus is controlled by varying the proportions in which two gas streams are mixed.

4. The gas may be brought into contact with water in such a way that only part of the liquid is evaporated. This is perhaps the most common method and will now be considered in more detail.

In order to obtain a high rate of humidification, the area of contact between the air and the water is made as large as possible by supplying the water in the form of a fine spray; alternatively, the interfacial area is increased by using a packed column. Evaporation occurs if the humidity at the surface is greater than that in the bulk of the air, that is, if the temperature of the water is above the dew point of the air.

When humidification is carried out in a packed column, the water that is not evaporated can be recirculated so as to reduce the requirements of fresh water. As a result of continued recirculation, the temperature of the water will approach the adiabatic saturation temperature of the air, and the air leaving the column will be cooled—in some cases to within $1^\circ K$ of the temperature of the water. If the temperature of the air is to be maintained constant, or raised, the water must be heated.

Two methods of changing the humidity and temperature of a gas from A (θ_1 and H_1) to B (θ_2 and H_2) may be traced on the humidity chart as shown in Fig. 5.11. The first method

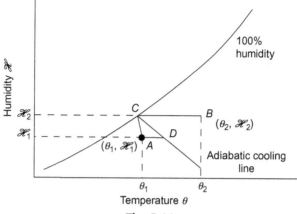

Fig. 5.11

Two methods of changing conditions of gas from (θ_1 and H_1) to (θ_2 and H_2).

consists of saturating the air by water artificially maintained at the dew point of air of humidity H_2 (line AC) and then heating at constant humidity to θ_2 (line CB). In the second method, the air is heated (line AD) so that its adiabatic saturation temperature corresponds with the dew point of air of humidity H_2. It is then saturated by water at the adiabatic saturation temperature (line DC) and heated at constant humidity to θ_2 (line CB). In this second method, an additional operation—the preliminary heating—is carried out on the air, but the water temperature automatically adjusts itself to the required value.

Since complete humidification is not always attained, an allowance must be made when designing air humidification cycles. For example, if only 95% saturation is attained, the adiabatic cooling line should be followed only to the point corresponding to that degree of saturation, and therefore, the gas must be heated to a slightly higher temperature before adiabatic cooling is commenced.

Example 5.7

Air at 300 K and 20% humidity is to be heated in two stages with intermediate saturation with water to 90% humidity so that the final stream is at 320 K and 20% humidity. What is the humidity of the exit stream and the conditions at the end of each stage?

Solution

At $\theta_1 = 300$ K and 20% humidity: $H_1 = 0.0045$ kg/kg, from Fig. 13.4, and
at $\theta_2 = 320$ K and 20% humidity: $H_2 = \underline{\underline{0.0140\,\text{kg/kg}}}$

When $H_2 = 0.0140$ kg/kg, air is saturated at 292 K and has a humidity of 90% at 293 K.

The adiabatic cooling line corresponding to 293 K intersects with $H = 0.0045$ kg/kg at a temperature $\theta = 318$ K.

Thus, the stages are.

(i)   heat the air at $H = 0.0045$ kg/kg from 300 to 318 K,
(ii) saturate with water at an adiabatic saturation temperature of 293 K until 90% humidity is attained. At the end of this stage,

$$H = 0.0140 \, \text{kg/kg and } \theta = 294.5 \, \text{K}$$

(iii) heat the saturated air at $H = 0.0140$ kg/kg from 294.5 to 320 K

5.5.2 Dehumidification

Dehumidification of air can be effected by bringing it into contact with a cold surface, either liquid or solid. If the temperature of the surface is lower than the dew point of the gas, condensation takes place, and the temperature of the gas falls. The temperature of the surface tends to rise because of the transfer of latent and sensible heat from the air. It would be expected that the air would cool at constant humidity until the dew point was reached, and that subsequent cooling would be accompanied by condensation. It is found in practice that this occurs only when the air is well mixed. Normally, the temperature and humidity are reduced simultaneously throughout the whole of the process. The air in contact with the surface is cooled below its dew point, and condensation of vapour therefore occurs before the more distant air has time to cool. Where the gas stream is cooled by cold water, countercurrent flow should be employed because the temperature of the water and air are changing in opposite directions.

The humidity can be reduced by compressing air, allowing it to cool again to its original temperature, and draining off the water that has condensed. During compression, the partial pressure of the vapour is increased, and condensation takes place as soon as it reaches the saturation value. Thus, if air is compressed to a high pressure, it becomes saturated with vapour, but the partial pressure is a small proportion of the total pressure. Compressed air from a cylinder therefore has a low humidity. Gas is frequently compressed before it is circulated so as to prevent condensation in the mains.

Many large air-conditioning plants incorporate automatic control of the humidity and temperature of the issuing air. Temperature control is effected with the aid of a thermocouple or resistance thermometer and humidity control by means of a thermocouple recording the difference between the wet- and dry-bulb temperatures.

5.6 Water Cooling

5.6.1 Cooling Towers

Cooling of water can be carried out on a small scale either by allowing it to stand in an open pond or by the spray pond technique in which it is dispersed in spray form and then collected in a large, open pond. Cooling takes place both by the transference of sensible heat and by evaporative cooling as a result of which sensible heat in the water provides the latent heat of vaporisation.

On the large scale, air and water are brought into countercurrent contact in a cooling tower that may employ either natural draught or mechanical draught. The water flows down over a series of wooden slats that give a large interfacial area and promote turbulence in the liquid. The air is humidified and heated as it rises, while the water is cooled mainly by evaporation.

The natural-draught cooling tower depends on the chimney effect produced by the presence in the tower of air and vapour of higher temperature and therefore of lower density than the surrounding atmosphere. Thus, atmospheric conditions and the temperature and quantity of the water will exert a very important effect on the operation of the tower. Not only will these factors influence the quantity of air drawn through the tower, but also they will affect the velocities and flow patterns and hence the transfer coefficients between gas and liquid. One of the prime considerations in design therefore is to construct a tower in such a way that the resistance to airflow is low. Hence, the packings and distributors must be arranged in open formation. The draught of a cooling tower at full load is usually only about 50 N/m^{27} and the air velocity in the region of 1.2–1.5 m/s, so that under the atmospheric conditions prevailing in the United Kingdom, the air usually leaves the tower in a saturated condition. The density of the airstream at outlet is therefore determined by its temperature. Calculation of conditions within the tower is carried out in the manner described in the following pages. It is, however, necessary to work with a number of assumed airflow rates and to select the one which fits both the transfer conditions and the relationship between air rate and pressure difference in the tower.

The *natural-draught cooling tower* consists of an empty shell, constructed either of timber or ferroconcrete, where the upper portion is empty and merely serves to increase the draught. The lower portion, amounting to about 10%–12% of the total height, is usually fitted with grids on to which the water is fed by means of distributors or sprays as shown in Fig. 5.12. The shells of cooling towers are now generally constructed in ferroconcrete in a shape corresponding approximately to a hyperboloid of revolution. The shape is chosen mainly for constructional reasons, but it does take account of the fact that the entering air will have a radial velocity component; the increase in cross-section towards the top causes a reduction in the outlet velocity, and there is a small recovery of kinetic energy into pressure energy.

The *mechanical-draught cooling tower* may employ forced draught with the fan at the bottom, or induced draught with the fan driving the moist air out at the top. The air velocity can be increased appreciably above that in the natural-draught tower, and a greater depth of packing can be used. The tower will extend only to the top of the packing unless atmospheric conditions are such that a chimney must be provided in order to prevent recirculation of the moist air. The danger of recirculation is considerably less with the induced-draught type because the air is expelled with a higher velocity. Mechanical-draught towers are generally confined to small installations and to conditions where the water must be cooled to as low a temperature as possible. In some cases, it is possible to cool the water to within 1°K of the wet-bulb temperature of the air. Although the initial cost of the tower is less, maintenance and operating

Fig. 5.12
Water-cooling tower. View of spray distribution system.

costs are of course higher than in natural-draught towers that are now used for all large installations. A typical steel-framed mechanical-draught cooling tower is shown in Fig. 5.13.

The operation of the conventional water-cooling tower is often characterised by the discharge of a plume consisting of a suspension of minute droplets of water in air. This is formed when the hot humid air issuing from the top of the tower mixes with the ambient atmosphere, and precipitation takes place as described earlier (Section 5.3.2). In the *hybrid* (or wet/dry) cooling tower,[8] mist formation is avoided by cooling *part* of the water in a finned-tube exchanger bundle that thus generates a supply of warm dry air that is then blended with the air issuing from the evaporative section. By adjusting the proportion of the water fed to the heat exchanger, the plume can be completely eliminated.

In the cooling tower, the temperature of the liquid falls and the temperature and humidity of the air rise, and its action is thus similar to that of an air humidifier. The limiting temperature to which the water can be cooled is the wet-bulb temperature corresponding to the condition of the air at inlet. The enthalpy of the airstream does not remain constant since the temperature of the liquid changes rapidly in the upper portion of the tower. Towards the

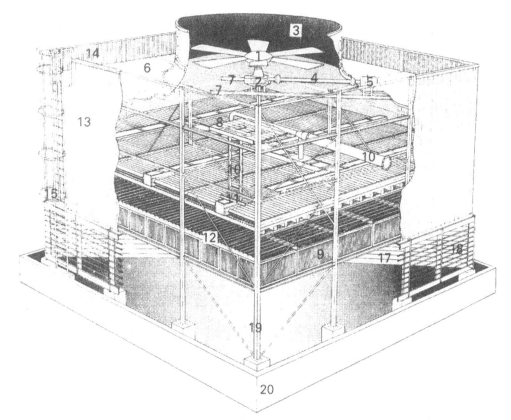

Fig. 5.13

Visco 2000 series steel-framed, mechanical-draught, water-cooling tower. (1) fan assembly; (2) gearbox; (3) fan stack; (4) drive shaft assembly; (5) motor; (6) fan deck; (7) mechanical equipment supports; (8) drift eliminators (PVC or timber—timber shown); (9) cooling tower packing (plastic plate or wooden lath); (10) inlet water distribution pipe; (11) open-type distribution system; (12) timber laths for even water distribution; (13) cladding; (14) cladding extended to form handrail; (15) access ladder; (16) internal access ladder to distribution system and drift eliminators; (17) diagonal wind baffles; (18) air inlet louvres; (19) steel structures with horizontal and diagonal ties; (20) cold water sump. *Some structural members have been omitted for clarity.*

bottom, however, the temperature of the liquid changes less rapidly because the temperature differences are smaller. At the top of the tower, the temperature falls from the bulk of the liquid to the interface and then again from the interface to the bulk of the gas. Thus, the liquid is cooled by transfer of sensible heat and by evaporation at the surface. At the bottom of a tall tower, however, the temperature gradient in the liquid is in the same direction, though smaller, but the temperature gradient in the gas is in the opposite direction. Transfer of sensible heat to the interface therefore takes place from the bulk of the liquid and from the bulk of the gas, and all the cooling is caused by the evaporation at the interface. In most cases, about 80% of the heat loss from the water is accounted for by evaporative cooling.

Fig. 5.14
Natural-draught water-cooling towers.

5.6.2 Design of Natural-Draught Towers

The airflow through a natural-draught or hyperbolic-type tower (Fig. 5.14) is due largely to the difference in density between the warm air in the tower and the external ambient air; thus, a draught is created in the stack by a chimney effect that eliminates the need for mechanical fans. It has been noted by McKelvey and Brooke[9] that natural-draught towers commonly operate at a pressure difference of some 50 N/m^2 under full load, and above the packing, the mean air velocity is typically 1–2 m/s. The performance of a natural-draught tower differs from that of a mechanical-draught installation in that the cooling achieved depends upon the relative humidity and the wet-bulb temperature. It is important therefore, at the design stage, to determine correctly and to specify the density of the inlet and exit airstreams in addition to the usual tower design conditions of water temperature range, how closely the water temperature should approach the wet-bulb temperature of the air, and the quantity of water to be handled. Because the performance depends to a large extent on atmospheric humidity, the outlet water temperature is difficult to control with natural-draught towers.

In the design of natural-draught towers, a ratio of height-to-base diameter of 3:2 is normally used, and a design method has been proposed by Chilton.[10] Chilton has shown that the duty coefficient D of a tower is approximately constant over the normal range of operation and is related to tower size by an efficiency factor or performance coefficient C_t given by

$$D_t = \frac{19.50 A_b z_t^{0.5}}{C_t^{1.5}} \tag{5.31}$$

where for water loadings in excess of 1 kg/m^2 s, C_t is usually about 5.2 though lower values are obtained with new packings that are being developed.

The duty coefficient is given by the following equation (in which SI units must be used as it is not dimensionally consistent):

$$\frac{W_L}{D_t} = 0.00369 \frac{\Delta H'}{\Delta T} (\Delta T' + 0.0752 \Delta H')^{0.5} \tag{5.32}$$

where W_L (kg/s) is the water load in the tower, $\Delta H'$ (kJ/kg) the change in enthalpy of the air passing through the tower, ΔT (°K) the change in water temperature in passing through the tower, and $\Delta T'$ (°K) the difference between the temperature of the air leaving the packing and the dry-bulb temperature of the inlet air. The air leaving the packing inside the tower is assumed to be saturated at the mean of the inlet and outlet water temperatures. Any divergence between theory and practice of a few degrees in this respect does not significantly affect the results as the draught component depends on the ratio of the change of density to change in enthalpy and not on change in temperature alone.[11] The use of Eqs (5.31) and (5.32) is illustrated in the following example.

Example 5.8

What are the diameter and height of a hyperbolic natural-draught cooling tower handling 4810 kg/s of water with the following temperature conditions:

$$\text{Water entering the tower} = 301 \text{ K}$$
$$\text{Water leaving the tower} = 294 \text{ K}$$
$$\text{Air : dry bulb} = 287 \text{ K}$$
$$\text{Wet bulb} = 284 \text{ K}$$

Solution

Temperature range for the water, $\Delta T = (301 - 294) = 7°\text{K}$.

At a mean water temperature of 0.5(301 + 294) = 297.5 K, the enthalpy = 92.6 kJ/kg.

At a dry-bulb temperature of 287 K, the enthalpy = 49.5 kJ/kg.

$$\therefore \ \Delta T' = (297.5 - 287) = 10.5°\text{K}$$

and

$$\Delta H' = (92.6 - 49.5) = 43.1 \text{ kJ/kg}$$

In Eq. (5.32),

$$\frac{4810}{D_t} = 0.00369 \left(\frac{43.1}{7}\right) [10.5 + (0.0752 \times 43.1)]^{0.5}$$

and

$$D_t = 57,110$$

Taking C_t as 5.0 and assuming as a first estimate a tower height of 100 m, then in Eq. (5.31),

$$57,110 = 19.50 A_b \frac{100^{0.5}}{5.0^{1.5}}$$

and

$$A_b = 3274\,\text{m}^2$$

Thus, the internal diameter of the column at sill level $= \left(\frac{3274 \times 4}{\pi}\right)^{0.5}$

$$= \underline{64.6\,\text{m}}$$

Since this gives a height/diameter ratio of $(100\!:\!64.6) \approx 3\!:\!2$, the design is acceptable.

5.6.3 Height of Packing for Both Natural and Mechanical Draught Towers

The height of a water-cooling tower can be determined[12] by setting up a material balance on the water, an enthalpy balance, and rate equations for the transfer of heat in the liquid and gas and for mass transfer in the gas phase. There is no concentration gradient in the liquid, and therefore, there is no resistance to mass transfer in the liquid phase.

Considering the countercurrent flow of water and air in a tower of height z (Fig. 5.15), the mass rate of flow of air per unit cross-section G' is constant throughout the whole height of

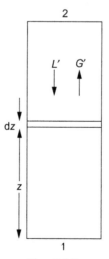

Fig. 5.15
Flow in water-cooling tower.

the tower, and because only a small proportion of the total supply of water is normally evaporated (1%–5%), the liquid rate per unit area L' can be taken as constant. The temperature, enthalpy, and humidity will be denoted by the symbols θ, H, and H, respectively; suffixes G, L, 1, 2, and f are being used to denote conditions in the gas and liquid, at the bottom and top of the column, and of the air in contact with the water.

The five basic equations for an incremental height of column, dz., are the following:

(1) Water balance:

$$dL' = G' \, dH \tag{5.33}$$

(2) Enthalpy balance:

$$G' \, dH_G = L' \, dH_L \tag{5.34}$$

since only a small proportion of the liquid is evaporated.

Now,

$$H_G = s(\theta_G - \theta_0) + \lambda H \tag{5.35}$$

and

$$H_L = C_L(\theta_L - \theta_0) \tag{5.36}$$

Thus,

$$G' \, dH_G = L' C_L \, d\theta_L \tag{5.37}$$

and

$$dH_G = s \, d\theta_G + \lambda \, dH \tag{5.38}$$

Integration of this expression over the whole height of the column, on the assumption that the physical properties of the materials do not change appreciably, gives

$$G'(H_{G2} - H_{G1}) = L' C_L(\theta_{L2} - \theta_{L1}) \tag{5.39}$$

(3) Heat transfer from the body of the liquid to the interface:

$$h_L a \, dz (\theta_L - \theta_f) = L' C_L \, d\theta_L \tag{5.40}$$

where h_L is the heat-transfer coefficient in the liquid phase and a is the interfacial area per unit volume of column. It will be assumed that the area for heat transfer is equal to that available for mass transfer, though it may be somewhat greater if the packing is not completely wetted.

Rearranging Eq. (5.40):

$$\frac{d\theta_L}{(\theta_L - \theta_f)} = \frac{h_L a}{L'C_L} dz \tag{5.41}$$

(4)   Heat transfer from the interface to the bulk of the gas:

$$h_G a dz(\theta_f - \theta_G) = G's d\theta_G \tag{5.42}$$

where h_G is the heat-transfer coefficient in the gas phase.

Rearranging,

$$\frac{d\theta_G}{(\theta_f - \theta_G)} = \frac{h_G a}{G's} dz \tag{5.43}$$

(5)   Mass transfer from the interface to the gas:

$$h_D \rho a dz(H_f - H) = G' dH \tag{5.44}$$

where h_D is the mass-transfer coefficient for the gas and ρ is the mean density of the air (see Eq. 5.6).

Rearranging,

$$\frac{dH}{H_f - H} = \frac{h_D a \rho}{G'} dz \tag{5.45}$$

These equations cannot be integrated directly since the conditions at the interface are not necessarily constant nor can they be expressed directly in terms of the corresponding property in the bulk of the gas or liquid.

If the Lewis relation (Eq. 5.11) is applied, it is possible to obtain workable equations in terms of enthalpy instead of temperature and humidity. Thus, writing h_G as $h_D \rho s$, from Eq. (5.42),

$$G's d\theta_G = h_D \rho a dz(s\theta_f - s\theta_G) \tag{5.46}$$

and from Eq. (5.44),

$$G'\lambda dH = h_D \rho a dz(\lambda H_f - \lambda H) \tag{5.47}$$

Adding these two equations gives

$$G'(s\,\mathrm{d}\theta_G + \lambda\,\mathrm{d}H) = h_D\rho a\,\mathrm{d}z\left[\left(s\theta_f + \lambda H_f\right) - \left(s\theta_G + \lambda H\right)\right]$$

$$G'\mathrm{d}H_G = h_D\rho a\,\mathrm{d}z\left(H_f - H_G\right) \quad \text{(from Eq. 13.35)} \tag{5.48}$$

or

$$\frac{\mathrm{d}H_G}{\left(H_f - H_G\right)} = \frac{h_D a\rho}{G'}\mathrm{d}z \tag{5.49}$$

The use of an enthalpy driving force, as in Eq. (5.48), was first suggested by Merkel,[13] and the following development of the treatment was proposed by Mickley.[12]

Combining Eqs (5.37), (5.40), and (5.48) gives

$$\frac{\left(H_G - H_f\right)}{\left(\theta_L - \theta_f\right)} = -\frac{h_L}{h_D\rho} \tag{5.50}$$

From Eqs (5.46) and (5.48),

$$\frac{\left(H_G - H_f\right)}{\left(\theta_G - \theta_f\right)} = \frac{\mathrm{d}H_G}{\mathrm{d}\theta_G} \tag{5.51}$$

and from Eqs (5.46) and (5.44),

$$\frac{\left(H - H_f\right)}{\theta_G - \theta_f} = \frac{\mathrm{d}H}{\mathrm{d}\theta_G} \tag{5.52}$$

These equations are now employed in the determination of the required height of a cooling tower for a given duty. The method consists of the graphical evaluation of the relation between the enthalpy of the body of gas and the enthalpy of the gas at the interface with the liquid. The required height of the tower is then obtained by integration of Eq. (5.49).

It is supposed that water is to be cooled at a mass rate L' per unit area from a temperature θ_{L2} to θ_{L1}. The air will be assumed to have a temperature θ_{G1}, a humidity H_1, and an enthalpy H_{G1} (which can be calculated from the temperature and humidity), at the inlet point at the bottom of the tower, and its mass flow per unit area will be taken as G'. The change in the condition of the liquid and gas phases will now be followed on an enthalpy–temperature diagram (Fig. 5.16). The enthalpy–temperature curve PQ for saturated air is plotted either using calculated data or from the humidity chart (Fig. 5.4). The region below this line relates to unsaturated air and the region above it to supersaturated air. If it is assumed that the air in contact with the liquid surface is saturated with water vapour, this curve represents the relation between air enthalpy H_f and temperature θ_f at the interface.

The curve connecting air enthalpy and water temperature is now drawn using Eq. (5.39). This is known as the operating line and is a straight line of slope $(L'C_L/G')$, passing through the

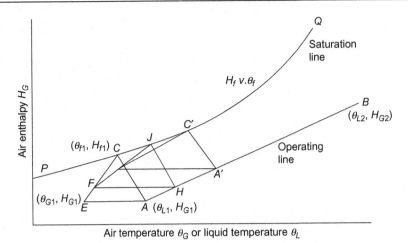

Fig. 5.16
Construction for determining the height of water-cooling tower.

points $A(\theta_{L1}$ and $H_{G1})$ and $B(\theta_{L2}$ and $H_{G2})$. Since $(\theta_{L1}$ and $H_{G1})$ are specified, the procedure is to draw a line through $(\theta_{L1}$ and $HG_1)$ of slope $(L'C_L/G')$ and to produce it to a point whose abscissa is equal to θ_{L2}. This point B then corresponds to conditions at the top of the tower, and the ordinate gives the enthalpy of the air leaving the column.

Eq. (5.50) gives the relation between liquid temperature, air enthalpy, and conditions at the interface, for any position in the tower, and is represented by a family of straight lines of slope $-(h_L/h_D\rho)$. The line for the bottom of the column passes through the point $A(\theta_{L1}, H_{G1})$ and cuts the enthalpy–temperature curve for saturated air at the point C, representing conditions at the interface. The difference in ordinates of points A and C is the difference in the enthalpy of the air at the interface and that of the bulk air at the bottom of the column.

Similarly, line $A'C'$, parallel to AC, enables the difference in the enthalpies of the bulk air and the air at the interface to be determined at some other point in the column. The procedure can be repeated for a number of points and the value of $(H_f - H_G)$ obtained as a function of H_G for the whole tower.

Now,

$$\frac{\mathrm{d}H_G}{(H_f - H_G)} = \frac{h_D a\rho}{G'}\mathrm{d}z \quad (\text{Eq.}13.49)$$

On integration,

$$z = \int_1^2 \mathrm{d}z = \frac{G'}{h_D a\rho}\int_1^2 \frac{\mathrm{d}H_G}{(H_f - H_G)} \tag{5.53}$$

assuming h_D to remain approximately constant.

Since $(H_f - H_G)$ is now known as a function of H_G, $1/(H_f - H_G)$ can be plotted against H_G and the integral evaluated between the required limits. The height of the tower is thus determined.

The integral in Eq. (5.53) cannot be evaluated by taking a logarithmic mean driving force because the saturation line PQ is far from linear. Carey and Williamson[14] have given a useful approximate method of evaluating the integral. They assume that the enthalpy difference $(H_f - H_G) = \Delta H$ varies in a parabolic manner. The three fixed points taken to define the parabola are at the bottom and top of the column (ΔH_1 and ΔH_2, respectively) and ΔH_m, the value at the mean water temperature in the column. The effective mean driving force is $f \Delta H_m$, where f is a factor for converting the driving force at the mean water temperature to the effective value. In Fig. 5.17, $(\Delta H_m/\Delta H_1)$ is plotted against $(\Delta H_m/\Delta H_2)$, and contours representing constant values of f are included.

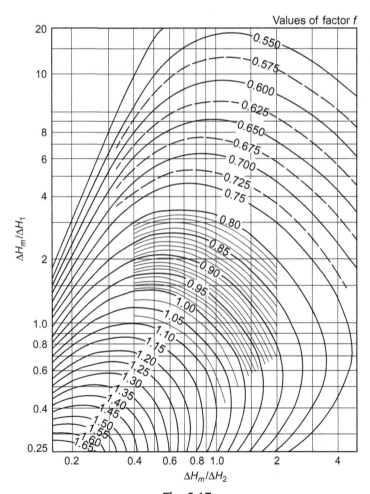

Fig. 5.17

Correction factor f for obtaining the mean effective driving force in column.

Using the mean driving force, integration of Eq. (5.53) gives

$$\frac{(H_{G2} - H_{G1})}{f\Delta H_m} = \frac{h_D a \rho}{G'} z \qquad (5.54)$$

or

$$z = \frac{G'}{h_D a \rho} \frac{(H_{G2} - H_{G1})}{f\Delta H_m}$$

5.6.4 Change in Air Condition

The change in the humidity and temperature of the air is now obtained. The enthalpy and temperature of the air are known only at the bottom of the tower, where fresh air is admitted. Here, the condition of the air may be represented by a point E with coordinates (H_{G1} and θG_1). Thus, the line AE (Fig. 5.16) is parallel to the temperature axis.

Since

$$\frac{H_G - H_f}{\theta_G - \theta_f} = \frac{dH_G}{d\theta_G} \qquad (5.51)$$

the slope of the line EC is ($dH_G/d\theta_G$) and represents the rate of change of air enthalpy with air temperature at the bottom of the column. If the gradient ($dH_G/d\theta_G$) is taken as constant over a small section, the point F, on EC, will represent the condition of the gas at a small distance from the bottom. The corresponding liquid temperature is found by drawing through F a line parallel to the temperature axis. This cuts the operating line at some point H, which indicates the liquid temperature. The corresponding value of the temperature and enthalpy of the gas at the interface is then obtained by drawing a line through H, parallel to AC. This line then cuts the curve for saturated air at a point J, which represents the conditions of the gas at the interface. The rate of change of enthalpy with temperature for the gas is then given by the slope of the line FJ. Again, this slope can be considered to remain constant over a small height of the column, and the condition of the gas is thus determined for the next point in the tower. The procedure is then repeated until the curve representing the condition of the gas has been extended to a point whose ordinate is equal to the enthalpy of the gas at the top of the column. This point is obtained by drawing a straight line through B, parallel to the temperature axis. The final point on the line then represents the condition of the air that leaves the top of the water-cooling tower.

The size of the individual increments of height that are considered must be decided for the particular problem under consideration and will depend, primarily, on the rate of change of the gradient $(dH_G/d\theta_G)$. It should be noted that, for the gas to remain unsaturated throughout the whole of the tower, the line representing the condition of the gas must be below the curve for saturated gas. If at any point in the column, the air has become saturated, it is liable to become supersaturated as it passes further up the column and comes into contact with hotter liquid. It is difficult to define precisely what happens beyond this point as partial condensation may occur, giving rise to a mist. Under these conditions, the preceding equations will no longer be applicable. However, an approximate solution is obtained by assuming that once the airstream becomes saturated, it remains so during its subsequent contact with the water through the column.

5.6.5 Temperature and Humidity Gradients in a Water Cooling Tower

In a water-cooling tower, the temperature profiles depend on whether the air is cooler or hotter than the surface of the water. Near the top, hot water makes contact with the exit air that is at a lower temperature, and sensible heat is therefore transferred both from the water to the interface and from the interface to the air. The air in contact with the water is saturated at the interface temperature, and humidity therefore falls from the interface to the air. Evaporation followed by mass transfer of water vapour therefore takes place, and latent heat is carried away from the interface in the vapour. The sensible heat removed from the water is then equal to the sum of the latent and sensible heats transferred to the air. Temperature and humidity gradients are then as shown in Fig. 5.18A.

If the tower is sufficiently tall, the interface temperature can fall below the dry-bulb temperature of the air (but not below its wet-bulb temperature), and sensible heat will then be transferred from both the air and the water to the interface. The corresponding temperature and humidity profiles are given in Fig. 5.18B. In this part of the tower, therefore, the sensible heat removed from the water will be that transferred as latent heat *less* the sensible heat transferred from the air.

5.6.6 Evaluation of Heat and Mass Transfer Coefficients

In general, coefficients of heat and mass transfer in the gas phase and the heat-transfer coefficient for the liquid phase are not known. They may be determined, however, by carrying out tests in the laboratory or pilot scale using the same packing. If, for the air–water system, a small column is operated at steady water and air rates and the temperature of the water at the top and bottom and the initial and final temperatures and humidities of the airstream are noted, the operating line for the system is obtained. Assuming a value of the ratio $-(h_L/h_D\rho)$, for the slope of the tie-line AC, the graphical construction is carried out similar to the construction discussed in Fig. 5.16, starting with the conditions at the bottom of the tower. The condition of the gas at the top of the tower is thus calculated and compared with the measured value. If the difference is significant, another value of $-(h_L/h_D\rho)$ is assumed and the procedure repeated. Now that the slope of the tie line is known, the value of the integral of

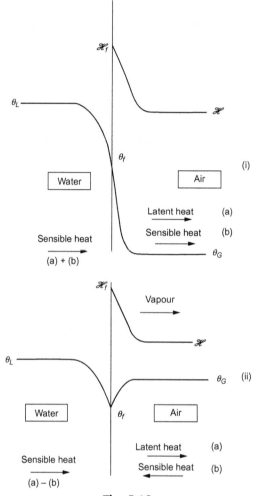

Fig. 5.18

Temperature and humidity gradients in a water-cooling tower, (i) upper sections of tower and (ii) bottom of tower.

$dH_G/(H_f - H_G)$ over the whole column can be calculated. Since the height of the column is known, the product $h_D a$ is found by solution of Eq. (5.49). $h_G a$ may then be calculated using the Lewis relation. The values of the three transfer coefficients are therefore obtained at any given flow rates from a single experimental run. The effect of liquid and gas rate may be found by means of a series of similar experiments.

Several workers have measured heat- and mass-transfer coefficients in water-cooling towers and in humidifying towers. Thomas and Houston,[15] using a tower 2 m high and 0.3 m square in cross-section, fitted with wooden slats, give the following equations for heat- and mass-transfer coefficients for packed heights >75 mm:

$$h_G a = 3.0 L'^{0.26} G'^{0.72} \tag{5.55}$$

$$h_L a = 1.04 \times 10^4 L'^{0.51} G'^{1.00} \tag{5.56}$$

$$h_D a = 2.95 L'^{0.26} G'^{0.72} \tag{5.57}$$

In these equations, L' and G' are expressed in kg/m^2 s, s in J/kg K, $h_G a$ and $h_L a$ in W/m^3 K, and $h_D a$ in s^{-1}. A comparison of the gas- and liquid-film coefficients may then be made for a number of gas and liquid rates. Taking the humid heat s as 1.17×10^3 J/kg K,

	$L' = G' = 0.5$ kg/m^2 s	$L' = G' = 1.0$ kg/m^2 s	$L' = G' = 2.0$ kg/m^2 s
$h_G a$	1780	3510	6915
$h_L a$	3650	10, 400	29, 600
$h_L a / h_G a$	2.05	2.96	4.28

Cribb[16] quotes values of the ratio h_L / h_G ranging from 2.4 to 8.5.

It is seen that the liquid-film coefficient is generally considerably higher than the gas-film coefficient but that it is not always safe to ignore the resistance to transfer in the liquid phase.

Lowe and Christie[17] used a 1.3 m square experimental column fitted with a number of different types of packing and measured heat- and mass-transfer coefficients and pressure drops. They showed that in most cases,

$$h_D a \alpha L'^{1-n} G'^n \tag{5.58}$$

The index n was found to vary from about 0.4 to 0.8 according to the type of packing. It will be noted that when $n \approx 0.75$, there is close agreement with the results given by Eq. (5.57).

The heat-transfer coefficient for the liquid is often large compared with that for the gas phase. As a first approximation, therefore, it can be assumed that the whole of the resistance to heat transfer lies within the gas phase and that the temperature at the water-air interface is equal to the temperature of the bulk of the liquid. Thus, everywhere in the tower, $\theta_f = \theta_L$. This simplifies the calculations, since the lines AC, HJ, and so on have a slope of $-\infty$, that is, they become parallel to the enthalpy axis.

Some workers have attempted to base the design of humidifiers on the overall heat-transfer coefficient between the liquid and gas phases. This treatment is not satisfactory since the quantities of heat transferred through the liquid and through the gas are not the same, as some of the heat is utilised in effecting evaporation at the interface. In fact, at the bottom of a tall tower, the transfer of heat in both the liquid and the gas phases may be towards the interface, as already indicated. A further objection to the use of overall coefficients is that the Lewis relation may be applied only to the heat- and mass-transfer coefficients in the gas phase.

In the design of commercial units, nomographs[18,19] are available, which give a performance characteristic (KaV/L'), where K is a mass-transfer coefficient (kg water/m^2 s) and V is the active cooling volume (m^3/m^2 plan area), as a function of θ, θ_w, and (L'/G'). For a given duty, (KaV/L') is calculated from

$$\frac{KaV}{C_L L'} = \int_{\theta_1}^{\theta_2} \frac{d\theta}{(H_f - H_G)} \tag{5.59}$$

and then, a suitable tower with this value of (KaV/L') is sought from performance curves.[20,21] In normal applications, the performance characteristic varies between 0.5 and 2.5.

Example 5.9

Water is to be cooled from 328 to 293 K by means of a countercurrent airstream entering at 293 K with a relative humidity of 20%. The flow of air is 0.68 m^3/m^2 s, and the water throughput is 0.26 kg/m^2 s. The whole of the resistance to heat and mass transfer may be assumed to be in the gas phase, and the product, $(h_D a)$, may be taken as 0.2 (m/s)(m^2/m^3), that is, 0.2 s^{-1}.

What is the required height of packing and the condition of the exit airstream?

Solution

Assuming the latent heat of water at 273 K = 2495 kJ/kg,

$$\text{specific heat of air} = 1.003 \text{ kJ/kg K.}$$

$$\text{and specific heat of water vapour} = 2.006 \text{ kJ/kg K.}$$

The enthalpy of the inlet airstream:

$$H_{G1} = 1.003(293 - 273) + H[2495 + 2.006(293 - 273)]$$

From Fig. 5.4,

$$\text{at } \theta = 293 \text{K and 20\% RH, } H = 0.003 \text{ kg/kg, and hence}$$
$$H_{G1} = (1.003 \times 20) + 0.003[2495 + (2.006 \times 20)]$$
$$= 27.67 \text{kJ/kg}$$

In the inlet air, water vapour = 0.003 kg/kg dry air.or

$$\frac{(0.003/18)}{(1/29)} = 0.005 \text{kmol/kmol dry air}$$

$$\text{Thus, flow of dry air} = (1 - 0.005)0.68 = 0.677 \text{m}^3/\text{m}^2 \text{s}$$

$$\text{Density of air at 293 K} = \left(\frac{29}{22.4}\right)\left(\frac{273}{293}\right) = 1.206 \text{kg/m}^3$$

and

$$\text{Mass flow of dry air} = (1.206 \times 0.677) = 0.817 \text{kg/m}^2 \text{s}$$

Slope of operating line:

$$(L'C_L/G') = \frac{(0.26 \times 4.18)}{0.817} = 1.33$$

The coordinates of the bottom of the operating line are

$$\theta_{L1} = 293\,\text{K}, \quad H_{G1} = 27.67\,\text{kJ/kg}$$

Hence, on an enthalpy-temperature diagram, the operating line of slope 1.33 is drawn through the point (293 and 27.67) = (θ_{L1} and H_{G1}).

The top point of the operating line is given by $\theta_{L2} = 328$ K, and H_{G2} is found to be 76.5 kJ/kg (Fig. 5.19).

From Figs 5.4 and 5.5, the curve representing the enthalpy of saturated air as a function of temperature is obtained and drawn in. Alternatively, this plot may be calculated from

$$H_F = C_a(\theta_f - 273) + H_0\left[C_w(\theta_f - 273) + \lambda\right]\text{kJ/kg}$$

The curve represents the relation between enthalpy and temperature at the interface, that is, H_f as a function of θ_f.

It now remains to evaluate the integral $\int \text{d}H_G/(H_f - H_G)$ between the limits, $H_{G1} = 27.7$ kJ/kg and $H_{G2} = 76.5$ kJ/kg. Various values of H_G between these limits are selected and the value of 9 obtained from the operating line. At this value of θ_1, now θ_f, the corresponding value of H_f is obtained from the curve for saturated air. The working is as follows:

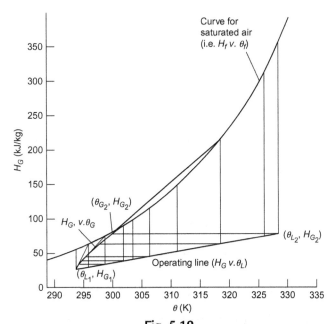

Fig. 5.19
Calculation of the height of a water-cooling tower.

H_G	$\theta = \theta_f$	H_f	$(H_f - H_G)$	$1/(H_f - H_G)$
27.7	293	57.7	30	0.0330
30	294.5	65	35	0.0285
40	302	98	58	0.0172
50	309	137	87	0.0114
60	316	190	130	0.0076
70	323	265	195	0.0051
76.5	328	355	279	0.0035

A plot of $1/(H_f - H_G)$ and H_G is now made as shown in Fig. 5.20 from which the area under the curve $= 0.65$. This value may be checked using the approximate solution of Carey and Williamson.[14]

At the bottom of the column,

$$H_{G1} = 27.7\,kJ/kg, \quad H_{f1} = 57.7\,kJ/kg \quad \therefore \Delta H_1 = 30\,kJ/kg$$

At the top of the column,

$$H_{G2} = 76.5\,kJ/kg, \quad H_{f2} = 355\,kJ/kg \quad \therefore \Delta H_2 = 279\,kJ/kg$$

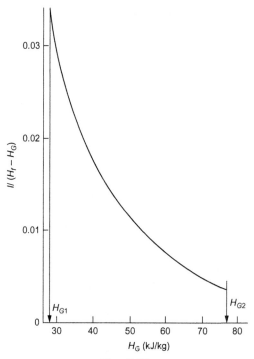

Fig. 5.20

Evaluation of the integral of $dH_G/(H_f - H_G)$.

At the mean water temperature of $0.5(328 + 293) = 310.5$ K,

$$H_{Gm} = 52\,kJ/kg, \quad H_f = 145\,kJ/kg \quad \therefore \Delta H_m = 93\,kJ/kg$$

$$\frac{\Delta H_m}{\Delta H_1} = 3.10, \quad \frac{\Delta H_m}{\Delta H_2} = 0.333,$$

and from Fig. 5.16, $f = 0.79$.

Thus,

$$\frac{(H_{G2} - H_{G1})}{f \Delta H_m} = \frac{(76.5 - 27.7)}{(0.79 \times 93)} = 0.66$$

which agrees well with the value (0.65) obtained by graphical integration.

Thus, in Eq. (5.53),

$$\text{Height of packing, } z = \int_{H_{G1}}^{H_{G2}} \frac{dH_G}{(H_f - H_G)} \frac{G'}{h_D a \rho}$$

$$= \frac{(0.65 \times 0.817)}{(0.2 \times 1.206)}$$

$$= 2.20\,\underline{m}$$

Assuming that the resistance to mass transfer lies entirely within the gas phase, the lines connecting θ_L and θ_f are parallel with the enthalpy axis.

In Fig. 5.18, a plot of H_G and θ_G is obtained using the construction given in Section 5.6.4 and shown in Fig. 5.15. From this curve, the value of θ_{G2} corresponding to $H_{G2} = 76.5$ kJ/kg is 300 K. From Fig. 5.5, under these conditions, the exit air has a humidity of 0.019 kg/kg that from Fig. 5.4 corresponds to a relative humidity of 83%.

5.6.7 Humidifying Towers

If the main function of the tower is to produce a stream of humidified air, the final temperature of the liquid will not be specified, and the humidity of the gas leaving the top of the tower will be given instead. It is therefore not possible to fix any point on the operating line, though its slope can be calculated from the liquid and gas rates. In designing a humidifier, therefore, it is necessary to calculate the temperature and enthalpy and hence the humidity of the gas leaving the tower for a number of assumed water-outlet temperatures and thereby determine the outlet water temperature resulting in the air leaving the tower with the required humidity. The operating line for this water-outlet temperature is then used in the calculation of the height of the tower required to effect this degree of humidification. The calculation of the dimensions of a humidifier is therefore rather more tedious than that for the water-cooling tower.

In a humidifier in which the make-up liquid is only a small proportion of the total liquid circulating, its temperature approaches the adiabatic saturation temperature θ_s, and remains constant, so that there is no temperature gradient in the liquid. The gas in contact with the liquid surface is approximately saturated and has a humidity H_s.

Thus,

$$d\theta_L = 0$$

and

$$\theta_{L1} = \theta_{L2} = \theta_L = \theta_f = \theta_s.$$

Hence,

$$-G's d\theta_G = h_G a dz(\theta_G - \theta_s) \quad \text{(from Eq.13.42)}$$

and

$$-G' dH = h_D \rho a dz(H - H_s) \quad \text{(from Eq.13.44)}$$

Integration of these equations gives

$$\ln \frac{(\theta_{G1} - \theta_s)}{(\theta_{G2} - \theta_s)} = \frac{h_G a}{G's} z \tag{5.60}$$

and

$$\ln \frac{(H_s - H_1)}{(H_s - H_2)} = \frac{h_D a \rho}{G'} z \tag{5.61}$$

assuming h_G, h_D, and s remain approximately constant.

From these equations, the temperature θ_{G2} and the humidity H_2 of the gas leaving the humidifier may be calculated in terms of the height of the tower. Rearrangement of Eq. (5.61) gives

$$\ln\left(1 + \frac{H_1 + H_2}{H_s - H_1}\right) = -\frac{h_D a \rho}{G'} z$$

or

$$\frac{(H_2 - H_1)}{(H_s - H_1)} = 1 - e^{-h_D a \rho z / G'} \tag{5.62}$$

Thus, the ratio of the actual increase in humidity produced in the saturator to the maximum possible increase in humidity (i.e. the production of saturated gas) is equal to $(1 - e^{-h_D a \rho z / G'})$, and complete saturation of the gas is reached exponentially. A similar relation exists for the change in the temperature of the gas stream:

$$\frac{(\theta_{G1} - \theta_{G2})}{(\theta_{G2} - \theta_s)} = 1 - e^{-h_G az/G's} \tag{5.63}$$

Further, the relation between the temperature and the humidity of the gas at any stage in the adiabatic humidifier is given by.

$$\frac{dH}{d\theta_G} = \frac{(H - H_s)}{(\theta_G - \theta_s)} \quad \text{(from Eq.13.52)}$$

On integration,

$$\ln \frac{(H_s - H_2)}{(H_s - H_1)} = \ln \frac{(\theta_{G2} - \theta_s)}{(\theta_{G1} - \theta_s)} \tag{5.64}$$

or

$$\frac{(H_s - H_2)}{H_s - H_1} = \frac{(\theta_{G2} - \theta_s)}{(\theta_{G1} - \theta_s)} \tag{5.65}$$

5.7 Systems Other than Air–Water

Calculations involving to systems where the Lewis relation is not applicable are very much more complicated because the adiabatic saturation temperature and the wet-bulb temperature do not coincide. Thus, the significance of the adiabatic cooling lines on the psychrometric chart is very much restricted. They no longer represent the changes that take place in a gas as it is humidified by contact with liquid initially at the adiabatic saturation temperature of the gas but simply give the compositions of all gases with the same adiabatic saturation temperature.

Calculation of the change in the condition of the liquid and the gas in a humidification tower is rendered more difficult since Eq. (5.49), which was derived for the air–water system, is no longer applicable. Lewis and White[22] have developed a method of integration of these equations that gives calculation based on the use of a *modified enthalpy* in place of the true enthalpy of the system.

For the air–water system, from Eq. (5.11),

$$h_G = h_D \rho s \tag{5.66}$$

This relationship applies quite closely for the conditions normally encountered in practice. For other systems, the relation between the heat- and mass-transfer coefficients in the gas phase is given by

$$h_G = b h_D \rho s \tag{5.67}$$

where *b* is approximately constant and generally has a value greater than unity.

For these systems, Eq. (5.46) becomes

$$G's\,d\theta_G = bh_D\rho a\,dz\left(s\theta_f - s\theta_G\right) \tag{5.68}$$

Adding Eqs (5.68) and (5.47) to obtain the relationship corresponding to Eq. (5.48) gives

$$G'(s\,d\theta_G + \lambda\,dH) = h_D\rho a\,dz\left[\left(bs\theta_f + \lambda H_f\right) - \left(ds\theta_G + \lambda H\right)\right] \tag{5.69}$$

Lewis and White use a *modified latent heat of vaporisation* λ' defined by

$$b = \frac{\lambda}{\lambda'} \tag{5.70}$$

and a *modified enthalpy* per unit mass of dry gas defined by

$$H'_G = s(\theta_G - \theta_0) + \lambda'H \tag{5.71}$$

Substituting in Eq. (5.67), from Eqs (5.38), (5.70), and (5.71),

$$G'\,dH_G = bh_D\rho a\,dz\left(H'_f - H'_G\right) \tag{5.72}$$

and

$$\frac{dH_G}{\left(H'_f - H'_G\right)} = \frac{bh_D\rho a}{G'}\,dz \tag{5.73}$$

Combining Eqs (5.37), (5.40), and (5.72),

$$\frac{\left(H'_G - H'_f\right)}{\left(\theta_L - \theta_f\right)} = -\frac{h_L}{h_D\rho b} \quad \text{(cf. Eq. 13.50)} \tag{5.74}$$

From Eqs (5.66) and (5.72),

$$\frac{\left(H'_G - H'_f\right)}{\left(\theta_G - \theta_f\right)} = \frac{dH_G}{d\theta_G} \quad \text{(cf. Eq. 13.51)} \tag{5.75}$$

From Eqs (5.44) and (5.67),

$$\frac{\left(H - H_f\right)}{\left(\theta_G - \theta_f\right)} = b\frac{dH}{d\theta_G} \quad \text{(cf. Eq. 13.52)} \tag{5.76}$$

The calculation of conditions within a countercurrent column operating with a system other than air–water is carried out in a similar manner to that already described by applying Eqs (5.73), (5.74), and (5.75) in conjunction with Eq. (3.39) of Vol. 1A:

$$G'(H_{G2} - H_{G1}) = L'C_L(\theta_{L2} - \theta_{L1}) \quad \text{(Eq.13.39)}$$

On an enthalpy-temperature diagram (Fig. 5.20), the enthalpy of saturated gas is plotted against its temperature. If equilibrium between the liquid and gas exists at the interface, this curve PQ represents the relation between gas enthalpy and temperature at the interface (H_f vs θ_f). The modified enthalpy of saturated gas is then plotted against temperature (curve RS) to give the relation between H'_f and θ_f. Since b is greater than unity, RS will lie below PQ. By combining Eqs (5.35), (5.70), and (5.72), H'_G is obtained in terms of H_G:

$$H'_G = \frac{1}{b}[H_G + (b-1)s(\theta_G - \theta_0)] \tag{5.77}$$

H'_G may be conveniently plotted against H_G for a number of constant temperatures. If b and s are constant, a series of straight lines is obtained. The operating line AB given by Eq. (5.39) is drawn in Fig. 5.21. Point A has coordinates (θ_{L1} and H_{G1}) corresponding to the bottom of the column. Point a has coordinates (θ_{L1} and H'_{G1}), H'_{G1} being obtained from Eq. (5.77).

From Eq. (5.72), a line through a, of slope $-(h_L/h_D\rho b)$, will intersect curve RS at c, (θ_{f1} and H'_{f1}) to give the interface conditions at the bottom of the column. The corresponding air enthalpy is given by C, (θ_{f1} and H_{f1}). The difference between the ordinates of c and a then gives the driving force in terms of modified enthalpy at the bottom of the column ($H'_{f1} - H'_{G1}$). A similar construction at other points, such as A', enables the driving force to be calculated at any other point. Hence, ($H'_{f1} - H'_{G1}$) is obtained as a function of H_G throughout the column. The height

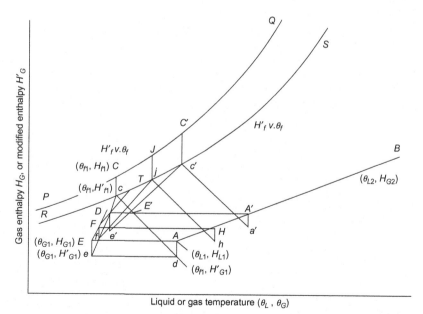

Fig. 5.21
Construction for height of a column for vapour other than water.

of column corresponding to a given change in air enthalpy can be obtained from Eq. (5.71) since the left-hand side can now be evaluated.

Thus,

$$\int_{H_{G1}}^{H_{G2}} \frac{dH_G}{\left(H_f' - H_G'\right)} = \frac{bh_D a\rho}{G'} z \qquad (5.78)$$

The change in the condition of the gas stream is obtained as follows: E, with coordinates (θ_{G1}, H_{G1}), represents the condition of the inlet gas. The modified enthalpy of this gas is given by $e(\theta_{G1}$ and $H'_{G1})$. From Eq. (5.75), it is seen that ec gives the rate of change of gas enthalpy with temperature $(dH_G/d\theta_G)$ at the bottom of the column. Thus, ED, parallel to ec, describes the way in which gas enthalpy changes at the bottom of the column. At some arbitrary small distance from the bottom, F represents the condition of the gas, and H gives the corresponding liquid temperature. In exactly the same way, the next small change is obtained by drawing a line hj through h parallel to ac. The slope of fj gives the new value of $(dH_G/d\theta_G)$, and therefore, the gas condition at a higher point in the column is obtained by drawing FT parallel to fj. In this way, the change in the condition of the gas through the column can be followed by continuing the procedure until the gas enthalpy reaches the value H_{G2} corresponding to the top of the column.

A detailed description of the method of construction of psychrometric charts is given by Shallcross and Low,[23] who illustrate their method by producing charts for three systems: air–water, air–benzene, and air–toluene at pressures of 1 and 2 bar.

Example 5.10

In a countercurrent packed column, n-butanol flows down at a rate of 0.25 kg/m^2 s and is cooled from 330 to 295 K. Air at 290 K, initially free of n-butanol vapour, is passed up the column at the rate of 0.7 m^3/m^2 s. Calculate the required height of tower and the condition of the exit air.

Data:

Mass-transfer coefficient per unit volume, $h_D a = 0.1$ s^{-1}.
Psychrometric ratio, $h_G/(h_D \rho_A s) = b = 2.34$.
Heat-transfer coefficients, $h_L = 3h_G$.
Latent heat of vaporisation of n-butanol, $\lambda = 590$ kJ/kg.
Specific heat of liquid n-butanol, $C_L = 2.5$ kJ/kg K.
Humid heat of gas, $s = 1.05$ kJ/kg K.

Temperature (K)	Vapour pressure of butanol (kN/m^2)
295	0.59
300	0.86
305	1.27
310	1.75
315	2.48

320	3.32
325	4.49
330	5.99
335	7.89
340	10.36
345	14.97
350	17.50

Solution

The first stage is to calculate the enthalpy of the saturated gas by way of the saturated humidity, H_0 given by

$$H_0 = \frac{P_{w0}}{P - P_{w0}} \frac{M_w}{M_A} = \frac{P_{w0}}{(101.3 - P_{w0})} \left(\frac{74}{29}\right)$$

The enthalpy is then

$$H_f = \frac{1}{(1 + H_0)} \times 1.001 \left(\theta_f - 273\right) + H_0 \left[2.5\left(\theta_f - 273\right) + 590\right] \text{kJ/kg}$$

where 1.001 kJ/kg K is the specific heat of dry air.

Thus,

$$H_f = \frac{1.001\theta_f - 273.27}{(1 + H_0)} + H_0 \left(2.5\theta_f - 92.5\right) \text{kJ/kg moist air}$$

The results of this calculation are presented in the following table, and H_f is plotted against θ_f in Fig. 5.21.

The modified enthalpy at saturation H'_f is given by

$$H'_f = \frac{\left(1.001\theta_f - 273.27\right)}{(1 + H_0)} + H_0 \left[2.5\left(\theta_f - 273\right) + \lambda'\right]$$

where from Eq. (5.70), $\lambda' = \lambda/b = (590/2.34)$ or 252 kJ/kg.

$$\therefore \ H'_f = \frac{\left(1.001\theta_f - 273.27\right)}{(1 + H_0)} + H_0 \left(2.5\theta_f - 430.5\right) \text{kJ/kg moist air}$$

These results are also given in the following table and plotted as H'_f against θ_f in Fig. 5.21:

θ_f (K)	P_{w0} (kN/m^2)	H_0 (kg/kg)	$(1.001\theta_f - 273.27)/(1 + H_0)$ (kJ/kg)	$H_0(2.5\theta_f - 92.5)$ (kJ/kg)	H_f (kJ/kg)	$H_0(2.5\theta_f - 430.5)$ (kJ/kg)	H'_f (kJ/kg)
295	0.59	0.0149	21.70	9.61	31.31	4.57	26.28
300	0.86	0.0218	24.45	14.33	40.78	6.97	33.42
305	1.27	0.0324	31.03	21.71	52.74	10.76	41.79

310	1.75	0.0448	35.45	30.58	66.03	15.43	50.88
315	2.48	0.0640	39.52	44.48	84.00	22.85	62.37
320	3.32	0.0864	43.31	61.13	104.44	31.92	75.23
325	4.49	0.1183	46.55	85.18	131.73	45.19	91.74
330	5.99	0.1603	49.18	117.42	166.60	63.23	112.41
335	7.89	0.2154	51.07	160.47	211.54	87.67	138.73
340	10.36	0.2905	51.97	220.05	272.02	121.87	173.83
345	14.97	0.4422	49.98	340.49	390.47	191.03	241.01
350	17.50	0.5325	50.30	416.68	466.98	236.70	287.00

The bottom of the operating line (point a) has coordinates $\theta_{L1} = 295$ K and H_{G1}, where

$$H_{G1} = 1.05(290 - 273) = 17.9 \, \text{kJ/kg}.$$

At a mean temperature of, say, 310 K, the density of air is

$$\left(\frac{29}{22.4}\right)\left(\frac{273}{310}\right) = 1.140 \, \text{kg/m}^3$$

and

$$G' = (0.70 \times 1.140) = 0.798 \, \text{kg/m}^2 \text{s}$$

Thus, the slope of the operating line becomes

$$\frac{L'C_L}{G'} = \frac{(0.25 \times 2.5)}{0.798} = 0.783 \, \text{kJ/kgK}$$

and this is drawn in as AB in Fig. 5.22, and at $\theta_{L2} = 330$ K, $H_{G2} = 46$ kJ/kg.

From Eq. (5.77), $H'_G = (H_G + (b - 1)s(\theta_G - \theta_0))/b$

$$\therefore H'_{G1} = \frac{17.9 + (2.34 - 1)1.05(290 - 273)}{2.34} = 17.87 \, \text{kJ/kg}$$

Point a coincides with the bottom of the column.

A line is drawn through a of slope $-\dfrac{h_L}{h_D \rho b} = -\left(\dfrac{3h_G}{h_D \rho b}\right) \cdot \left(\dfrac{h_D \rho s}{h_G}\right)$.

$$= -3s = -3.15 \, \text{kJ/kgK}$$

This line meets curve RS at c (θ_{f1} and H'_{f1}) to give the interface conditions at the bottom of the column. The corresponding air enthalpy is given by point C whose coordinates are

$$\theta_{f1} = 293 \text{K} \quad H_{f1} = 29.0 \, \text{kJ/kg}$$

The difference between the ordinates of c and a gives the driving force in terms of the modified enthalpy at the bottom of the column or

$$\left(H'_{f1} - H'_{G1}\right) = (23.9 - 17.9) = 6.0 \, \text{kJ/kg}$$

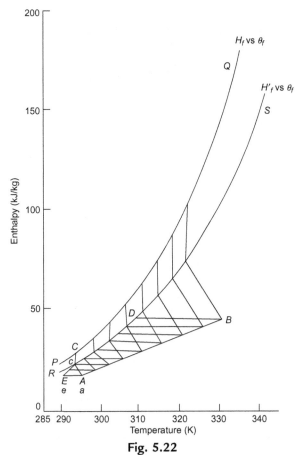

Fig. 5.22

Graphical construction for Example 5.10.

A similar construction is made at other points along the operating line with the results shown in the following table.

θ_f (K)	H_G (kJ/ kg)	H'_G (kJ/ kg)	H'_f (kJ/ kg)	$(H'_f - H'_G)$ (kJ/kg)	$1/(H'_f - H'_G)$ (kg/kJ)	Mean value in interval	Interval	Value of integral over interval
295	17.9	17.9	23.9	6.0	0.167	0.155	4.1	0.636
300	22.0	22.0	29.0	7.0	0.143	0.126	4.0	0.504
305	26.0	26.0	35.3	9.3	0.108	0.096	4.0	0.384
310	30.0	30.0	42.1	12.1	0.083	0.073	4.0	0.292
315	34.0	34.0	50.0	16.0	0.063	0.057	4.1	0.234
320	38.1	38.1	57.9	19.8	0.051	0.046	3.9	0.179
325	42.0	42.0	66.7	24.7	0.041	0.0375	4.0	0.150
330	46.0	46.0	75.8	29.8	0.034	Value of integral = 2.379		

from which

$$\int_{H_{G1}}^{H_{G2}} \frac{dH_G}{H_f' - H_G'} = 2.379$$

Substituting in Eq. (5.78),

$$\frac{bh_D \rho az}{G'} = 2.379$$

and

$$z = \frac{(2.379 \times 0.798)}{(2.34 \times 0.1)} = \underline{\underline{8.1\,m}}$$

It remains to evaluate the change in gas conditions.

Point e, ($\theta_{G1} = 290$ K and $H_{G1} = 17.9$ kJ/kg) represents the condition of the inlet gas. ec is now drawn in, and from Eq. (5.75), this represents $dH_G/d\theta_G$. As for the air–water system, this construction is continued until the gas enthalpy reaches H_{G2}. The final point is given by D at which $\underline{\theta_{G2} = 308\,K}$.

It is fortuitous that, in this problem, $H'_G = H_G$. This is not always the case, and reference should be made to Section 5.7 for elaboration of this point.

5.8 Nomenclature

		Units in SI system	Dimensions in M, N, L, T, θ
A	Interfacial area	m^2	L^2
A_b	Base area of hyperbolic tower	m^2	L^2
a	Interfacial area per unit volume of column	m^2/m^3	L^{-1}
b	Psychrometric ratio ($h/h_D \rho_A s$)	–	L^2 T$^{-2}\theta^{-1}$
C_a	Specific heat of gas at constant pressure	J/kg K	
C_L	Specific heat of liquid	J/kg K	L^2 T$^{-2}\theta^{-1}$
C_p	Specific heat of gas and vapour mixture at constant pressure	J/kg K	L^2 T$^{-2}\theta^{-1}$
C_s	Specific heat of solid	J/kg K	L^2 T$^{-2}\theta^{-1}$
C_t	Performance coefficient or efficiency factor	–	–
C_w	Specific heat of vapour at constant pressure	J/kg K	L^2 T$^{-2}\theta^{-1}$

c	Mass concentration of vapour	kg/m^3	$\mathbf{ML^{-3}}$
C_0	Mass concentration of vapour in saturated gas	kg/m^3	$\mathbf{ML^{-3}}$
D_t	Duty coefficient of tower (Eq. 5.31)	–	–
f	Correction factor for mean driving force	–	–
G'	Mass rate of flow of gas per unit area	kg/m^2 s	$\mathbf{ML^{-2}T^{-1}}$
H	Enthalpy of humid gas per unit mass of dry gas	J/kg	$\mathbf{L^2T^{-2}}$
H_a	Enthalpy per unit mass, of dry gas	J/kg	$\mathbf{L^2T^{-2}}$
H_w	Enthalpy per unit mass, of vapour	J/kg	$\mathbf{L^2T^{-2}}$
H_1	Enthalpy of stream of gas, per unit mass of dry gas	J/kg	$\mathbf{L^2T^{-2}}$
H_2	Enthalpy of another stream of gas, per unit mass of dry gas	J/kg	$\mathbf{L^2T^{-2}}$
H_3	Enthalpy per unit mass of liquid or vapour	J/kg	$\mathbf{L^2T^{-2}}$
H'	Modified enthalpy of humid gas defined by (5.69)	J/kg	$\mathbf{L^2T^{-2}}$
ΔH	Enthalpy driving force $(H_f - H_G)$	J/kg	$\mathbf{L^2T^{-2}}$
$\Delta H'$	Change in air enthalpy on passing through tower	J/kg	$\mathbf{L^2T^{-2}}$
h	Heat-transfer coefficient	W/m^2 K	$\mathbf{MT^{-3}\boldsymbol{\theta}^{-1}}$
h_D	Mass-transfer coefficient	kmol/ (kmol/m^3) m^2 s	$\mathbf{LT^{-1}}$
h_G	Heat-transfer coefficient for gas phase	W/m^2 K	$\mathbf{MT^{-3}\boldsymbol{\theta}^{-1}}$
h_L	Heat-transfer coefficient for liquid phase	W/m^2 K	$\mathbf{MT^{-3}\boldsymbol{\theta}^{-1}}$
H	Humidity	kg/kg	–
H_s	Humidity of gas saturated at the adiabatic saturation temperature	kg/kg	–
H_w	Humidity of gas saturated at the wet-bulb temperature	kg/kg	–
H_0	Humidity of saturated gas	kg/kg	–
H_1	Humidity of a gas stream	kg/kg	–
H_2	Humidity of second gas stream	kg/kg	–
L'	Mass rate of flow of liquid per unit area	kg/m^2 s	$\mathbf{ML^{-2}T^{-1}}$

M_A	Molecular weight of gas	kg/kmol	$\mathbf{MN^{-1}}$
M_w	Molecular weight of vapour	kg/kmol	$\mathbf{MN^{-1}}$
m, m_1, m_2	Masses of dry gas	kg	$\mathbf{M}$
m_3	Mass of liquid or vapour	kg	$\mathbf{M}$
P	Total pressure	N/m^2	$\mathbf{ML^{-1}T^{-2}}$
P_A	Mean partial pressure of gas	N/m^2	$\mathbf{ML^{-1}T^{-2}}$
P_w	Partial pressure of vapour	N/m^2	$\mathbf{ML^{-1}T^{-2}}$
P_{w0}	Partial pressure of vapour in saturated gas	N/m^2	$\mathbf{ML^{-1}T^{-2}}$
Q	Rate of transfer of heat to liquid surface	W	$\mathbf{ML^{-1}T^{-2}}$
$\mathbf{R}$	Universal gas constant	8314 J/ kmol K	$\mathbf{MN^{-1}L^{-2}T^{-2}\theta^{-1}}$
S	Humid heat of gas	J/kg K	$\mathbf{L^2T^{-2}\theta^{-1}}$
T	Absolute temperature	K	$\boldsymbol{\theta}$
ΔT	Change in water temperature in passing through the tower	K	$\boldsymbol{\theta}$
$\Delta T'$	(Temperature of air leaving packing—ambient dry-bulb temperature)	K	$\boldsymbol{\theta}$
V	Active volume per plan area of column	m^3/m^2	$\mathbf{L}$
W_L	Water loading on tower	kg/s	$\mathbf{MT^{-1}}$
w	Rate of evaporation	kg/s	$\mathbf{MT^{-1}}$
Z	Percentage humidity	–	–
z	Height from bottom of tower	m	$\mathbf{L}$
Z_t	Height of cooling tower	m	$\mathbf{L}$
θ	Temperature of gas stream	K	$\boldsymbol{\theta}$
θ_0	Reference temperature, taken as the melting point of the material	K	$\boldsymbol{\theta}$
θ_s	Adiabatic saturation temperature	K	$\boldsymbol{\theta}$
θ_w	Wet-bulb temperature	K	$\boldsymbol{\theta}$
13.8.1.1.1. λ	Latent heat of vaporisation per unit mass, at datum temperature	J/kg	$\mathbf{L^2T^{-2}}$
λ_f	Latent heat of freezing per unit mass, at datum temperature	J/kg	$\mathbf{L^2T^{-2}}$
λ'	Modified latent heat of vaporisation per unit mass defined by (5.68)	J/kg	$\mathbf{L^2T^{-2}}$
ρ	Mean density of gas and vapour	kg/m^3	$\mathbf{ML^{-3}}$

ρ_A	Mean density of gas at partial pressure P_A	kg/m^3	$\mathbf{ML}^{-3}$
Pr	Prandtl number	–	–
Sc	Schmidt number	–	–

Suffixes 1, 2, f, L, G denote conditions at the bottom of the tower, the top of the tower, the interface, the liquid, and the gas, respectively

Suffix m refers to the mean water temperature

References

1. BS 1339:1965. *British Standard 1339 Definitions, formulae and constants relating to the humidity of the air.* London: British Standards Institution; 1981.
2. Sherwood TK, Comings EW. An experimental study of the wet bulb hygrometer. *Trans Am Inst Chem Eng* 1932;**28**:88.
3. Grosvenor MM. Calculations for dryer design. *Trans Am Inst Chem Eng* 1908;**1**:184.
4. Wexler A. In: Ruskin RE, editor. *Humidity and moisture. Measurements and control in science and industry. Principles and methods of humidity measurement in gases*, vol. 1. New York: Reinhold; 1965.
5. Hickman MJ. *Measurement of humidity.* 4th ed. *National physical laboratory. Notes on applied science no. 4*, London: HMSO; 1970.
6. Meadowcroft DB. Chapter 6. Chemical analysis—moisture measurement. In: Noltingk BE, editor. *Instrumentation reference book.* London: Butterworth; 1988.
7. Wood B, Betts P. A contribution to the theory of natural draught cooling towers. *Proc Inst Mech Eng (Steam Group)* 1950;**163**:54.
8. Clark R. Cutting the fog. *The chemical engineer (London)* 1992;**529**:22.
9. McKelvey KK, Brooke M. *The industrial cooling tower.* New York: Elsevier; 1959.
10. Chilton CH. Performance of natural-draught cooling towers. *Proc Inst Elec Eng* 1952;**99**:440.
11. Perry RH, Green DW, editors. *Perry's chemical engineers' handbook.* 6th ed. New York: McGraw-Hill; 1984.
12. Mickley HS. Design of forced draught air conditioning equipment. *Chem Eng Prog* 1949;**45**:739.
13. Merkel, F. Verdunstungs-Kühlung. *Ver Deut Ing Forschungsarb* 1925;275.
14. Carey WF, Williamson GJ. Gas cooling and humidification: design of packed towers from small scale tests. *Proc Inst Mech Eng (Steam Group)* 1950;**163**:41.
15. Thomas WJ, Houston P. Simultaneous heat and mass transfer in cooling towers. *Br Chem Eng* 1959;**4**:160. 217.
16. Cribb G. Liquid phase resistance in water cooling. *Br Chem Eng* 1959;**4**:264.
17. Lowe HJ, Christie DG. *Heat transfer and pressure drop data on cooling tower packings, and model studies of the resistance of natural-draught towers to airflowInst Mech Eng Symposium on Heat Transfer*; 1962. Paper 113, 933.
18. Wood B, Betts P. A total heat–temperature diagram for cooling tower calculations. *Engineer* 1950;**189** (4912):337. (4913) 349.
19. Zivi SM, Brand BB. Analysis of cross-flow cooling towers. *Refrig Eng* 1956;**64**(8):31. 90.
20. *Counter-flow cooling tower performance.* Kansas City: J. F. Pritchard & Co.; 1957
21. Cooling Tower Institute. *Performance curves.* Houston; 1967.
22. Lewis JG, White RR. Simplified humidification calculations. *Ind Eng Chem* 1953;**45**:486.
23. Shallcross DC, Low SL. Construction of psychrometric charts for systems other than water vapour in air. *Chem Eng Res Design* 1994;**72**:763. Errata: *Chem Eng Res Design* 1995;73:865.

Further Reading

1. Backhurst JR, Harker JH, Porter JE. *Problems in heat and mass transfer.* London: Edward Arnold; 1974.
2. Burger R. Cooling tower drift elimination. *Chem Eng Progr* 1975;**71**(7):73.
3. DeMonbrun JR. Factors to consider in selecting a cooling tower. *Chem Eng* 1968;**75**(19):106.
4. Donohue JM, Nathan CC. Unusual problems in cooling water treatment. *Chem Eng Progr* 1975;**71**(7):88.
5. Eckert ERG, Drake RM. *Analysis of heat and mass transfer.* New York: McGraw-Hill; 1972.
6. Elgawhary AW. Spray cooling system design. *Chem Eng Progr* 1975;**71**(7):83.
7. Friar F. Cooling-tower basin design. *Chem Eng* 1974;**81**(15):122.
8. Guthrie KM. Capital cost estimating. *Chem Eng* 1969;**76**(6):114.
9. Hall WA. Cooling tower plume abatement. *Chem Eng Progr* 1971;**67**(7):52.
10. Hansen EP, Parker JJ. Status of big cooling towers. *Power Eng* 1967;**71**(5):38.
11. Holzhauer R. Industrial cooling towers. *Plant Eng* 1975;**29**(15):60.
12. Incropera FP, De Witt DP. *Fundamentals of heat and mass transfer.* 4th ed. New York: Wiley; 1996.
13. Industrial Water Society. *Guide to mechanical draught evaporative cooling towers: selection, operation and maintenance.* London: Industrial Water Society; 1987.
14. Jackson J. *Cooling towers.* London: Butterworths; 1951.
15. Jordan DR, Bearden MD, McIlhenny WF. Blowdown concentration by electrodialysis. *Chem Eng Progr* 1975;**71**:77.
16. Juong JF. How to estimate cooling tower costs. *Hydrocarbon Process* 1969;**48**(7):200.
17. Kelly GM. *Cooling tower design and evaluation parameters.* ASME paper, *75-IPWR-9*; 1975.
18. Kolflat TD. Cooling tower practices. *Power Eng* 1974;**78**(1):32.
19. Maze RW. Practical tips on cooling tower sizing. *Hydrocarbon Process* 1967;**46**(2):123.
20. Maze RW. Air cooler or water tower—which for heat disposal? *Chem Eng* 1975;**83**(1):106.
21. McCabe WL, Smith JC, Harriott P. *Unit operations of chemical engineering.* New York: McGraw-Hill; 1985.
22. McKelvey KK, Brooke M. *The industrial cooling tower.* New York: Elsevier; 1959.
23. Meytsar J. Estimate cooling tower requirements easily. *Hydrocarbon Process* 1978;**57**(11):238.
24. Nelson WL. What is cost of cooling towers? *Oil Gas J* 1967;**65**(47):182.
25. Norman WS. *Absorption, distillation and cooling towers.* London: Longmans; 1961.
26. Olds FC. Cooling towers. *Power Eng* 1972;**76**(12):30.
27. Paige PM. Costlier cooling towers require a new approach to water-systems design. *Chem Eng* 1967;**74**(14):93.
28. Park JE, Vance JM. Computer model of crossflow towers. *Chem Eng Progr* 1971;**67**:55.
29. Picciotti M. Design quench water towers. *Hydrocarbon Process* 1977;**56**(6):163.
30. Picciotti M. Optimize quench water systems. *Hydrocarbon Process* 1977;**56**(9):179.
31. Rabb A. Are dry cooling towers economical? *Hydrocarbon Process* 1968;**47**(2):122.
32. Uchiyama T. Cooling tower estimates made easy. *Hydrocarbon Process* 1976;**55**(12):93.
33. Walker R. *Water supply, treatment and distribution.* New York: Prentice-Hall; 1978.
34. Wrinkle RB. Performance of counterflow cooling tower cells. *Chem Eng Progr* 1971;**67**:45.

CHAPTER 6

Transport Processes in Microfluidic Applications

6.1 Introduction

Traditional chemical engineering unit operations involve the design and operation of equipments and processes at the macroscale such as distillation columns, absorption and stripping towers, agitators, and heat exchangers. In the past two decades, there have been substantial advances in the use of microscale devices for carrying out various operations such as chemical- and bio-molecular detection and sensing, cell capture and counting, generation of mono-disperse drops of radii in the range of $O(10)$ μm, etc.[1] These developments have been driven by the need for rapid analyses of DNA, high-throughput screening, disease diagnosis for patients in remote locations, and drug discovery via combinatorial synthesis. Chemical engineers have contributed substantially to these recent developments, and the basic principles of fluid flow and transport phenomena still remain relevant even in the design of such microscale devices, although additional physicochemical effects due to surface tension, electrostatic interactions, and possible slip at the wall start to play an important role in such applications. The science and engineering of systems and processes that involve the transport and manipulation of fluids in devices and flow configurations, which have their smallest dimensions in the range 1–100 μm is now commonly referred to as 'microfluidics'.[2] Owing to the extremely small dimensions involved in microfluidic devices, the volumes processed are very small, typically in the range of 10^{-9}–10^{-15} L. The first applications of microfluidic devices arguably were in the area of analysis and sensing of chemical and biological molecules, eventually leading to 'lab-on-a-chip' (or micrototal analysis systems, abbreviated sometimes as μ-TAS) devices, which entail the design and operation of an entire chemical or biochemical laboratory on the surface of silicon or polymer-based 'chips'. The objective of such a device is to take a liquid sample, transport, mix (with reagents), characterise the sample, all in a single portable device. Many fabrication technologies from the electronics industry were modified and adapted in the manufacture of such devices. Further, 'soft fabrication' using elastomeric templates[3–5] were also developed for the fabrication of microfluidic devices. Some of the main advantages of microfluidic devices include: (1) the use of very small quantities of samples and reagents, (2) the ability to carry out separation and detection (sensing) at a very high resolution and sensitivity, (3) relatively low cost, and the short

Coulson and Richardson's Chemical Engineering. https://doi.org/10.1016/B978-0-08-102550-5.00006-7

processing time involved in the operations, and (4) portability of the device which makes the possibility to use bio/chemical analysis at remote locations. The major applications of microfluidic devices have been in the areas of DNA sequencing and genome analysis, synthesis of nanoparticles and nanowires, sensing of biological molecules, and in drug discovery. One of the most important commercial application of a microfluidic process is in the design of ink-jet printers, wherein very tiny drops of ink of uniform diameter (~50 μm) need to be produced rapidly, which is achieved by a precise design of the orifice of the printer. Since the subject of microfluidics is very pertinent to modern chemical engineering, the objective of this chapter is to provide an introduction to the basic physical principles underlying microfluidic applications, and the challenges involved in the design and development of such devices.

One interesting and important contrast in the engineering design and analysis of microfluidic systems vis-a-vis conventional chemical engineering unit operations is that in microfluidic devices, we often need to 'scale down' (as opposed to the conventional 'scale-up') of physical processes from the laboratory scale experiments and analysis. It is often assumed in traditional engineering design that, in the exercise of scale-up, the physical processes at play remain unchanged across the scales. However, in microfluidic applications, it is possible to have fundamental physical phenomena to change drastically whilst scaling down. For instance, new physical effects that were not relevant (e.g. surface or van der Waals forces) at macroscales could start becoming important in microscale devices. Hence scaling-down in microfluidic devices must be done with some caution.

6.2 Fluid Flow in Microchannels

An essential component in the analysis and design of microfluidic processes and devices is the understanding of fluid flow behaviour in microchannels and tubes. The fluids that are handled in microfluidic systems are often liquids, and hence can be treated as mechanically incompressible to a good accuracy. However, there may be some cases, such as microfluidic gas sensors, where gases may be involved in the flow. Even there, the Mach number of the flow is usually smaller than unity, thus allowing us to treat the flows as incompressible. Despite the small length scales involved, it is often accurate to invoke the continuum hypothesis, and treat the fluids within the realm of continuum fluid mechanics. However, in some instances of gas flow, it is possible for the gas flow to be in the 'Knudsen regime' where the mean-free path for collisions between the gas molecules could be comparable or even larger than the channel width, thus rendering the continuum hypothesis invalid. In such cases, more sophisticated kinetic theories must be employed to describe the flows accurately. In this chapter, we will assume that the fluids (i.e. both liquids and gases) flowing in microchannels to be incompressible, and that the processes can be analysed within the realm of the continuum hypothesis. Thus, whilst microfluidic devices are small compared to macroscale engineering devices, their dimensions are not so small that the continuum approximation breaks down and the inherently molecular nature of the fluids becomes important. Even for gases in microchannels, many billions of molecules occupy a characteristic volume element of the flow, and the fluid behaves like a continuum. Hence, in most microfluidic applications involving Newtonian fluids, the Navier-Stokes equations can be safely assumed to provide an accurate description of the flow.[4]

The next important aspect of flow in microfluidic devices is the geometry: most applications involve flows in rectangular microchannels, which are long compared to their width. Further, the shortest dimension in the cross-section available for flow is of $O(100)$ μm. In order to estimate the relative importance of inertial and viscous forces in microscale flows, it is useful to consider the well-known Reynolds number $N_{Re} \equiv DU\rho/\mu$, where D is the smallest dimension in the cross-section, U is the average velocity of the flow, ρ is the density of the fluid, and μ is the viscosity of the fluid. Here, it is being assumed for simplicity that the fluid flowing in the microchannel is Newtonian, although, it is often the case that there are dissolved macromolecules (such as DNA) which can impart a non-Newtonian rheology to the fluid. We next estimate a typical Reynolds number in a microfluidic flow: assuming the flow of water $(\rho \sim 10^3 \text{ kg/m}^3, \eta \sim 10^{-3} \text{ N s/m}^2)$ in a channel with width $\sim 10^{-4}$ m with velocity ~ 1 m/s, we have $Re \sim 100$. Even with such a large estimate for flow velocity, the Reynolds number is ~ 100. For such Reynolds numbers, the flow in channels remains laminar. For rectangular channels of large aspect ratio, the onset of transition from laminar flow to turbulence happens at $Re \sim 1200$. More realistic velocities in microfluidic devices are in the range 1 μm/s to 1 cm/s, and the corresponding Reynolds numbers range between 10^{-6} and 10, thus indicating that viscous forces dominate inertial forces in microscale flows. This is one of the most important features of microfluidic flows: owing to the small dimensions involved, the flow inside a rigid channel is largely dominated by viscous effects. At sufficiently low Reynolds numbers, the nonlinear inertial terms of the Navier-Stokes equations can often be neglected thus resulting in a linear system of differential equations which are somewhat more tractable in terms of their solution. However, additional physical effects such as surface tension and electrical forces dominate over pressure and inertial forces. Thus, unlike traditional fluid mechanics, where the governing equations (for Newtonian fluids) are well developed, in microfluidics, the possibility of new physical phenomena at play often makes the modelling of such phenomena more challenging, and hence must be addressed on a case-to-case basis.

For laminar flows in channels with infinite aspect ratio (i.e. for large ratios of width to depth of the channel), the governing Navier-Stokes equations allow for an analytical solution for the laminar flow velocity profile in the channel, which is the well-known parabolic velocity distribution (Fig. 6.1):

$$u_x(z) = \frac{\triangle P B^2}{2\mu L}\left[1 - \left(\frac{z}{B}\right)^2\right]. \tag{6.1}$$

Here, x is the flow direction and z is the direction perpendicular to the flow. The gap-width of the channel in which fluid is flowing is $2B$. The extent of the channel in the y-direction is

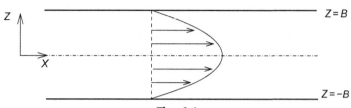

Fig. 6.1

Schematic representation of the steady, fully developed laminar flow velocity profile in a rectangular channel of half-width B.

considered large enough so that variations in that direction are negligible except in regions very close to the side walls. The previous equation is derived using the usual assumptions of steady, fully developed flow, and $\triangle P$ is the pressure drop across the length L of the channel. The average fluid velocity in the channel is obtained by integrating this equation with respect to z over the channel gap-width $2B$ to give

$$U \equiv \langle u_x \rangle = \frac{1}{3} \frac{\triangle PB^2}{\mu L}, \tag{6.2}$$

and the volumetric flow rate Q is the product of the above average velocity and the cross-sectional area for flow $2BW$:

$$Q \equiv 2BW \langle u_x \rangle = \frac{2}{3} \frac{\triangle PB^3 W}{\mu L}, \tag{6.3}$$

where W is the width of the channel in the y-direction, and the above result is valid for $W \gg B$. One important consequence of the previous equation for microfluidic flows is the cost of pumping the fluid when the geometry is scaled down: when the gap-width is decreased by a factor of 2 (keeping all other conditions constant), the pressure gradient required to achieve the same volumetric flow rate Q increases by eight times. This simple calculation illustrates the difficulty encountered in scaling down flow operations in the microscale. Consequently, enormous pressure gradients are required to pump the fluid at reasonable flow rates. To circumvent this problem, often other forms of driving the fluid are considered, such as electroosmotic flows, wherein the driving force is an applied electric field across the ends of the channel. For rectangular channels with height B, width W, where the width is not necessarily much larger than the height, there will be corrections to Eq. (6.2), and the first correction due to the finite aspect ratio (W/B) is of $O(W/B)$.[6]

Another important assumption in the derivation of the above velocity profile in a channel is the 'no-slip' boundary condition, which states that the tangential velocity (here the u_x component) should go to zero at stationary rigid surfaces. Whilst the validity of the no-slip condition has been debated extensively, it has been widely accepted as the appropriate boundary condition for macroscale flows. In microfluidic flows, it is possible that owing to the small volumes of the flow, and the concomitant large surface area to volume ratios, additional surface effects can play an important role leading to the violation of the no-slip boundary condition valid for macroscale flows. However, experiments in microchannels and tubes have repeatedly shown that the phenomenological no-slip condition still remains a valid boundary condition even for microscale flows, when the liquid wets the boundaries.[6] For partially wetting liquids, there is some experimental evidence that the results are in agreement with a slip boundary condition, and this is attributed to the presence of gas or vapour cavities along the solid surface.

One important aspect of fluid flows in channels and tubes is the transition from laminar to turbulent flows. It is well known since the experiments of Reynolds in 1883 that the laminar flow in a circular tube undergoes a transition to turbulent flow at Reynolds number above 2000. This transition from an orderly and steady laminar flow to a more chaotic and unsteady turbulent flow has enormous consequences for transport processes occurring in these geometries. For instance, in a laminar flow, the streamlines are parallel, and hence any transport of species or heat across streamlines must happen purely by diffusion or conduction. By contrast, in turbulent flows, there are fluctuating turbulent eddies, which move randomly in all directions and tend to enhance the transport even in directions perpendicular to the mean flow. The question then arises whether the nature of the laminar-turbulent transition is the same in microchannels or whether it is influenced by the smallness of the geometry. Experiments again have unambiguously shown that even in tubes of diameter ~ 100 μm, the laminar-turbulent transition of a Newtonian fluid is very similar to macroscale pipe flows, and the transition occurs at a Reynolds number of above 2000. This experimental result unambiguously places most microfluidic flows in the laminar regime, and this has important implications in the analysis of transport in such microscale flows.

6.3 Dimensionless Groups in Microfluidics

In engineering applications involving scale-up (or scale-down), it is often useful to represent the ratios of various physical forces in terms of dimensionless groups. A dimensionless group or a dimensionless number is a ratio of two competing physical effects, and its magnitude is an immediate indicator of which of the two forces are dominant in a given context. Here, we provide a brief discussion of the dimensionless groups relevant to microfluidic processes. A more detailed discussion on the relevance and physical implications of dimensionless groups in the microfluidic context can be found in Squires and Quake.[7] Chapter 1 of Vol. 1A of this book has a general discussion on dimensional analysis and dimensionless groups relevant to fluid mechanics and unit operations.

6.3.1 Reynolds Number

The most important dimensionless number to characterise fluid flow is the Reynolds number $Re = DU\rho/\mu$, where D is the length scale relevant to the flow, U is the velocity scale relevant to the flow, and μ and ρ are, respectively, the viscosity and density of the fluid. The Reynolds number is the ratio of inertial to viscous force per unit volume of the fluid. For a Newtonian fluid, the inertial force per unit volume can be estimated to be $\rho U^2/D$, and the viscous force per unit volume can be estimated to be $\mu U/D^2$, and $Re = (\rho U^2/D)/(\mu U/D^2)$. When Re is small compared to unity, then usually inertial forces can be considered to be negligible. The major mathematical consequence of this is that the nonlinearities in the Navier-Stokes equations can be neglected, which will result in the *linear* Stokes equations. There are many analytical solution methods that can be used to solve for Stokes flows, which are discussed in the classic book by Happel and Brenner[8] and in the monograph by Kim and Karilla.[9] When Re becomes of the order of unity, then fluid inertia starts to become as important as the viscous stresses. Instead of a straight circular tube, if one considers the flow in a slightly curved circular tube with radius of curvature R_c large compared with the tube diameter D ($R_c \gg D$), due to finite inertia, centrifugal forces become important and drive a secondary flow famously referred to as the 'Dean flow'. In microfluidic applications, such secondary flows can improve mixing in the direction perpendicular to the main flow, and this has been exploited in 'serpentine' micromixers, wherein secondary flows in twisted tubes have exploited to enhance mixing.

In the earlier discussion, we considered only the convective nonlinearity when we estimated the inertial forces in the Navier-Stokes equations. In addition to the steady convective nonlinear inertial term, there is also the unsteady, linear inertial term proportional to $\rho \frac{\partial u}{\partial t}$. The inertial force density arising from this term can be estimated as $\rho U/\tau$, where τ is the timescale required for the flow to reach a new steady state from an existing steady state. For instance, if there is a sudden increase in the pressure drop across a microchannel, it is relevant to estimate the timescale required to achieve the new steady state. This can be estimated by balancing the unsteady inertial force per unit volume $\rho U/\tau$ with the viscous force per unit volume $\mu U/D^2$ resulting in $\tau \sim D^2/\nu$, where $\nu = \mu/\rho$ is the kinematic viscosity of the fluid. This balance indicates that the timescale τ required to reach a new steady state is nothing but the momentum diffusion time scale. For microchannels of $D \sim 100$ μm, we can estimate this time scale for water as $\tau \sim 10$ ms. Thus, the time required to reach a steady state in microscale flows is typically small owing to the small dimensions involved. Thus, whilst the traditional inertial nonlinearities in the Navier-Stokes equations (for a Newtonian fluid) are perhaps not very important in microscale flows, it is possible that new physical effects such as viscoelasticity or surface tension can give rise to a new type of nonlinearity in such flows. The relevant dimensionless groups that will quantify these additional physical effects will be addressed in the following sections.

6.3.2 Weissenberg Number

Many microfluidic devices handle liquids in which there are small amounts of dissolved polymers. The addition of polymers to an otherwise Newtonian solvent imparts elasticity to the solution, and such polymer solutions are *viscoelastic*, and exhibit flow behaviour that is intermediate between that of a Newtonian fluid and an elastic solid. The presence of elasticity or partial 'memory' in a viscoelastic fluid is characterised by a relaxation time λ. For a Newtonian fluid (e.g. air or water), the relaxation time $\lambda = 0$. In other words, if there is flow of a Newtonian fluid in a tube under conditions that inertial effects are negligible, then if the applied pressure drop is set to zero suddenly, the flow in the tube will come to a stop instantaneously in the absence of inertia. However, for the flow of polymer solutions which are viscoelastic, even when inertia is negligible, the flow in the tube will undergo an 'elastic recoil' under the sudden removal of the applied pressure gradient. This is somewhat similar to the recoil of an elastic rubber band under the removal of tension. However, unlike an elastic rubber band, the recoil in a viscoelastic fluid is partial.

The presence of a nonzero relaxation time λ in a viscoelastic fluid introduces the possibility of two dimensionless numbers. First, it is possible to make λ dimensionless using the convective timescale for the flow D/U, thus yielding a dimensionless group called Weissenberg (sometimes also called Deborah) number $Wi = \lambda U/D$. The Newtonian fluid limit is approached when $Wi \to 0$, and increase in Wi signifies the importance of elasticity in a given flow. When Wi is small, the polymer chains in the solution relax rather quickly, but when Wi is $O(1)$ or larger, the polymer chains are substantially distorted from their equilibrium conformations. Second, it is possible to make λ dimensionless using the momentum diffusion timescale R^2/ν thus yielding another dimensionless group called the 'Elasticity number' $El = \lambda\nu/R^2$. Whilst the Weissenberg number is a flow-dependent dimensionless number, the elasticity number is independent of flow parameters and is a function only of material and geometric parameters. Owing to the small dimensions involved in microfluidic processes, it is possible to achieve large values of elasticity number for the same polymer solution. To achieve comparable values of elasticity number in macroscale applications, it is necessary to have a highly viscous solvent and a polymer chain with a very long relaxation time. Thus, microfluidic applications involving viscoelastic solutions can potentially operate in parametric regimes of high El, which are often not accessible in macroscale flows of similar solutions.

6.3.3 Capillary Number

When there are two coflowing immiscible liquid layers in a microfluidic device, the interface between the two immiscible liquid layers is characterised by an interfacial tension γ, which affects the dynamics of the free surface. The presence of surface tension in an interface implies that it costs energy to deform a flat interface, and γ is the energy per unit area required to create new interface. There are many applications involving the formation of droplets with radii of $O(100)$ μm, and here again interfacial tension plays a very important role in the breakup of a liquid thread into tiny droplets. Typically stresses induced by interfacial deformation are estimated as γ/R, where R is the radius of the droplet. If the viscous stresses around the drop are estimated as $\mu U/h$, where h is the gap width of the microfluidic channel, then a balance of capillary and viscous forces yields $\gamma/R \sim \mu U/h$, thus yielding $R/h \sim \gamma/(\mu U)$, where the dimensionless group 'Capillary number' is defined as $Ca = \mu U/\gamma$. When $Ca \ll 1$, then surface tension dominates viscous stresses, and hence a drop can be expected to remain spherical in shape, whilst for $Ca \gg 1$, then the drop shape is easily deformed by the viscous stresses of the surrounding liquid.

6.3.4 Knudsen Number

There are many technological applications involving the flow of gases in microscale devices. In that context, the question arises when the continuum approximation holds good, and when it can be expected to be doubtful. Unlike liquids where there are frequent molecular collisions, in dilute gases, the molecules largely move in a ballistic manner which is interrupted by intermolecular collisions which are quite rare in dilute gases. The average distance between molecular collisions is called the mean-free path, which can be estimated using the kinetic theory of gases to be $\lambda_f \sim 1/(na^2)$, where n is the number density of molecules and a is the molecular radius. An ideal gas at 20°C and 1 atm would have $\lambda_f \sim 100$ nm. The Knudsen number $Kn = \lambda_f/L$ is the ratio of the mean-free path to the length scale L characterising the microfluidic device. As $Kn \to 0$, the continuum hypothesis is valid, but for $Kn \sim O(1)$, noncontinuum physics becomes more relevant. The validity of the no-slip boundary condition becomes questionable under such circumstances. Often, a 'Maxwell slip condition'[10] is employed:

$$u_x = \beta \frac{du_x}{dy},$$ (6.4)

where β is called the slip length and is of the same order as the mean-free path λ_f. Another consequence in gas flows is that density of the gas is more strongly dependent on its temperature than that for a liquid. In such cases, the assumption of constant density (i.e. mechanically incompressible fluid) may not be accurate, and compressibility effects can become important in microfluidic geometries.

6.3.5 Peclet Number

The Peclet number is the heat (or) mass transfer analogue of the Reynolds number in that it measures the relative importance of convective flux to diffusive flux. If we consider the flow of a species of concentration C_0 in a channel of width h and velocity U_0, then the convective flux of the species is estimated as $U_0 C_0$, whilst the diffusive flux of the species over a length scale of h is estimated as $\mathcal{D}C_0/h$, where $\mathcal{D}$ is the species diffusivity. The ratio of these two fluxes is the Peclet number $Pe = U_0 C_0/(\mathcal{D}C_0/h) - U_0 h/\mathcal{D}$. When $Pe \ll 1$, then diffusive effects dominate over convective transport, whilst the opposite is generally true for $Pe \gg 1$. As mentioned earlier, the flow in microfluidic processes is often laminar with parallel streamlines, and hence

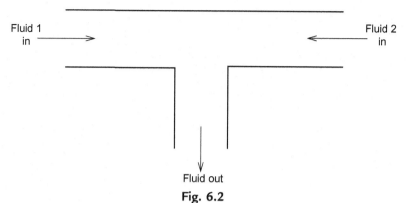

Fig. 6.2

Schematic representation mixing of two fluid streams in a 'T'-shaped microchannel.

diffusion often dominates the transport in the direction perpendicular to the flow streamlines. Convective transport, however, dominates the transport along the flow direction when $Pe \sim O(1)$ or larger. This balance between convection along the flow direction and diffusion in the cross-stream direction is a very important feature in microfluidic processes. In the microfluidic context, the physical implication of Peclet number is best illustrated[7] by this example. Consider a 'T'-shaped junction wherein two fluids are injected to flow alongside in the straight section of the 'T' channel, as shown in Fig. 6.2. A relevant question in any microfluidic process is the distance required along the channel such that the two streams are homogenised. The diffusive timescale for species to diffuse a length of h is $h^2/\mathcal{D}$. In this diffusive time, the fluid would have moved a distance $Z \sim U_0 h^2/\mathcal{D}$ along the length of the channel. Thus, the ratio of the length along the channel for homogenisation to the width h of the channel is $Z/h \sim U_0 h/\mathcal{D} \sim Pe$. Thus, the magnitude of Peclet number will dictate the homogenisation of two streams in a microfluidic channel, and this has important implications in the design of such channels.

6.4 Alternative Ways of Driving Microscale Flows

As discussed earlier in this chapter, a major problem with driving laminar microfluidic flows with the aid of applied pressure gradients is that whilst scaling down the operation, the pressure difference is inversely proportional to the cube of the shortest cross-sectional dimension for the same flow rate. This means that to achieve the same volumetric throughput, one must have very high performance pumping mechanisms, which may not be feasible always. To overcome this difficulty, other driving forces such as applied electric fields and gradients in interfacial tension are used sometimes. Here, we will provide a brief overview of driving forces other than pressure gradients used for driving microscale flows.

Electric fields can be applied to liquids to drive flows provided there are mobile ionic charges in the liquids which can couple to the applied electric field. It is well known, for instance, that at the boundary between a polar liquid like water and a solid surface, there are dissociated mobile charges in the liquid which have a sign that is opposite to the relatively immobile charges in the solid surface. This configuration is commonly referred to as the 'electrical double layer'. The solid surface may also get charged due to specific adsorption of ions. Due to a net nonzero charge density in the liquid close to the solid surface, there is a nonzero electrical body force in a unit volume of fluid next to the solid boundary under the influence of an externally applied electric field. The bulk of the fluid is electrically neutral, and so there is no net electrical force acting in the bulk. A balance of this electrical body force and viscous stresses in the fluid gives rise to 'electro-osmotic' flow, where the fluid moves relative to a stationary, charged solid surface. One important feature of electro-osmotic flows is that the nature of the velocity profile in the microfluidic channel is very different from the parabolic velocity profile caused by applied pressure gradients. The velocity profile is very uniform (almost like a plug flow) except in thin regions close to the solid boundaries, wherein the velocity sharply changes to satisfy the no-slip boundary condition at the solid surfaces. The rather uniform nature of the velocity profile has important consequences on transport of heat and mass in such systems, and in the processing of thermally sensitive materials.

Another way to drive flows in the microscale, when a gas-liquid or liquid-liquid interface is present, is by controlling spatial variations of surface tension (Marangoni flows) created with thermal, chemical, electrical, or light gradients. To realise these driving forces, it is often necessary to pattern or modify the geometrical, chemical and mechanical features of the microchannel.[6] Thus, both electrical and interfacial tension (or 'capillary')-based driving forces are potential alternatives to pressure gradients to drive fluid flow in microfluidic devices. However, one caveat with these driving forces is that the system becomes very sensitive to contamination of solid surfaces.

6.5 Transport Processes in Microscale Flows

A common theme that connects various microfluidic applications is the transport, mixing, reaction, and separation of chemical species in microfluidic devices. Whilst the fluid flows can often be treated as laminar, and hence convective transport of momentum (i.e. inertial stresses) does not play an important role, the same cannot be said in terms of convective transport of heat and mass in microscale flows. For instance, whilst the Reynolds number $Re = U_0 R/\nu$ can be small compared with unity, $Pe = U_0 R/D$ can still be $O(1)$ or large. This can be achieved by realising that $Pe = ReSc$, where the Schmidt number $Sc = \nu/D$, and even if $Re \ll 1$, the Schmidt number can be such that $Sc \gg 1$ yielding $Pe \sim O(1)$ or larger. When Pe is large, it is often tempting to conclude that convective transport would dominate diffusive effects. Whilst this is largely true in the bulk of the flow, near the solid boundaries, diffusive processes are dominant thus giving rise to concentration boundary layers. Hence, a clear understanding and analysis of the two competing effects (diffusion vs. convection) is very important in the rational design of microfluidic devices.

One important application in which the competition between convection and diffusion is best explained is surface-based biosensors, in which analytes in a solution react with receptors on the surface.[11,12] In DNA microarrays, such surface-based sensors are used to capture specific genetic sequences. The design of such sensors involves issues like the time taken for a target molecule to bind with the receptor on the surface, the number of molecules that successfully bind at the surface, etc. When a liquid sample containing target molecules flows in a microfluidic device, the target molecules are convected by the flow, whilst diffusion is the only mechanism for the target molecule to reach the surface, in the direction perpendicular to the flow. The issues involved here are somewhat similar to a solid catalyst surface in which reactions happen, and the fluid flows adjacent to the solid catalyst.

The equation governing the evolution of the concentration of a species is described by the convective-diffusion equation

$$\frac{\partial c}{\partial t} + \mathbf{v} \cdot \nabla c = D \nabla^2 c. \tag{6.5}$$

When the fluid velocity field $\mathbf{v}$ is known by a solution of the Navier-Stokes equation, then the previous equation is a linear partial differential equation for the concentration c. When we nondimensionalise the lengths with the channel half-width B, time with B/U_0 where U_0 is the mean velocity of the flow, we arrive at the following nondimensional convective-diffusion equation for c:

$$\frac{\partial c}{\partial t^*} + \mathbf{v}^* \cdot \nabla^* c = \frac{1}{Pe} \nabla^{*2} c. \tag{6.6}$$

Since the equation is homogeneous in c, it is not necessary to render c nondimensional, although it is possible to do so using available scales in the problem. The previous equation shows that the diffusive term in the right-hand side is multiplied by the term $1/Pe$, and for $Pe \gg 1$, then it is tempting to neglect the diffusion term completely. However, a complete neglect of the diffusion term will lead to unphysical results, since very close to the solid surface, the fluid velocity field obeys the no-slip condition, and hence the strength of convection is small near solid boundaries, and the only mechanism for transport across the solid-fluid interface is diffusion. Thus, even at high Peclet numbers, there will be thin regions called 'concentration boundary layers' where diffusion remains important. Outside the boundary layer, however, convection remains the only dominant mode of transport.

In the absence of any flow, let us consider a channel in which one half of the channel (between the wall and the centreline) has a species with concentration c_∞, and the other half has the concentration of the species to be 0. This is a simple one-dimensional unsteady diffusion problem in the absence of flow, and in the limit when the walls are taken to infinity, there is an analytical solution for the unsteady diffusion equation

$$\frac{\partial c}{\partial t} = D \frac{\partial^2 c}{\partial x^2}, \tag{6.7}$$

where x is the direction perpendicular to the walls. The solution for this unsteady diffusion equation, when the walls are placed at infinity, is achieved by the use of similarity transforms, which yields

$$c(x,t) = \frac{1}{2} c_\infty \mathrm{erfc}\left(\frac{x}{2\sqrt{Dt}}\right), \tag{6.8}$$

where 'erfc' is the complementary error function. For the case where the walls are at infinity, as $t \to \infty$, the concentration takes the value $c = c_\infty/2$ everywhere in the domain. The earlier solution also conveys an important (time-dependent) length scale $l_D = \sqrt{Dt}$, which is the length over which diffusion has taken place at time t, and hence l_D is referred to the 'diffusion length'. Now, when the walls are not at infinite distance, but separated by a width of $2B$, then we can estimate the time it takes for the species to diffuse over the entire cross-section of the channel: $t_D \sim B^2/D$. Despite the small dimensions, the diffusion time t_D can be very large in microfluidic applications due to the very small diffusion coefficients D of some macromolecules.

Let us next consider a microfluidic channel in which two different fluids are brought into contact, and diffusion happens transverse to the flow, whilst convection transports species in the flow direction. When the channels are shallow relative to their width, then the transverse transport of the species is similar to the unsteady one-dimensional diffusion example discussed earlier, except for the fact that the flow direction acts like a 'time', and the earlier solution can be reinterpreted with the residence time (i.e. the time a fluid particle has stayed in the channel) at a given location y/U_0 being considered as 'time' in Eq. (6.8). Thus, the diffusion time for equalisation of concentration in the transverse direction is still proportional to B^2/D. The residence time for the species in a channel of length L is L/U_0, and the ratio of the diffusion time to residence time is $(BU_0/D)(B/L) = PeB/L$. If we want complete homogenisation of the concentration, then the residence time must be comparable to the diffusion time, which yields $L/B \sim Pe$. For $Pe \gg 1$, this implies that the length of the channel must be $O(Pe)B$ for homogenisation to occur. If the length of the channel is not large, then the residence time in the channel is not sufficient for homogenisation of concentration to occur. In applications where complete homogenisation is desired, then this analysis will serve as a guide to estimate the length of the microchannel in a specific application. In some cases, where it is necessary to minimise mixing, then this geometry is ideal. To enhance mixing, it is often necessary to have (convective) flow patterns in the system that aid in reducing the lengths over which diffusion must occur, which will lead to reducing the diffusion time from B^2/D. Whilst this is possible in macroscale applications by external stirring and agitation, in microscale flows, the channel walls are patterned such that fluid motion results in convective transport that tends to reduce the lengths over which diffusion must occur.[12]

Before closing this discussion, it is pertinent to discuss another phenomenon that could be relevant in certain types of microfluidic transport. In some applications, it may be required to introduce a reagent in the form of a finite 'pulse' which is localised in space and time, and the consequent evolution of this 'pulse' input in both space and time is relevant to its detection/sensing. This phenomenon was first addressed by G.I. Taylor, and later improved upon by R. Aris, and hence is referred to as the 'Taylor-Aris dispersion', who showed the coupled effects of axial convection, radial diffusion, and axial diffusion in this problem. To understand this phenomenon, let us first consider the hypothetical case where only convection is present, and both axial and radial diffusions are 'turned off'. Here, due to the parabolic nature of the velocity profile, the pulse will be elongated differentially at various radial positions, and the region near the centreline will keep moving with the maximum velocity, whilst the region immediately adjacent to the wall will remain stationary at the same location where the pulse was originally introduced. Thus, the pulse will be 'stretched' spatially in the form of a paraboloid, and the spatial extent of the paraboloid will increase linearly with time due to convective motion. Now, let us 'switch on' diffusion, and in general, diffusion will happen in directions where there is a concentration gradient. In this problem, where we have a stretched pulse, there will be concentration gradients both in the flow and radial directions. Diffusion is normally thought of as a 'spreading' mechanism. Axial diffusion will still tend to cause axial spreading. However, radial diffusion tends to spread the pulse near the nose of the paraboloid from the fast moving central region to the slower moving region near the wall. Thus, the species front end of the paraboloid cannot move as rapidly as it would if there was no radial diffusion. At the rear end of the paraboloid, there is diffusion from the wall region to the central region, thereby assisting the species at the rear end to 'catch up' with the front. Thus, diffusion in the radial direction inhibits the axial spread of the pulse. Taylor and Aris showed that this process can be described, in the frame of centre-of-mass of the pulse, by an unsteady diffusion equation for the effective concentration (averaged over the radial direction), with an effective diffusion coefficient[12]

$$D_{\text{eff}} = D(1 + Pe^2/A), \tag{6.9}$$

where A is a constant that depends on the geometry, which takes the value of 48 for Hagen-Poiseuille flow in a tube. Thus, at long times, the pulse spreads diffusively in the axial direction with the spread growing as $\sqrt{D_{\text{eff}}t}$, rather than as a linear function of time, if only convection is dominant.

6.6 Analysis of a Model Surface-Based Sensor

In this section, we use the principles of flow and transport and analyse the phenomena that occur in a biosensor, which will aid in the design of the sensor. The presentation in this section closely follows the review of Squires et al.[11] In many biosensors, a solution containing target molecules is made to flow in a microfluidic device. A rectangular area in the bottom wall of the channel is functionalised with receptors, with which the target molecules should chemically undergo 'binding' at a specified rate. Examples include microarrays for proteins and DNA for capturing specific sequences of proteins or DNA in biotechnological and diagnostic applications. Microfluidic devices are ideal for quantifying the binding on the surface even with very small sample volumes. Whilst much focus has been paid to the chemical binding and associated kinetics, the transport of the target molecules to the rectangular patch plays an equally important role, and affects the performance of the sensor. Let us assume that the target solution with concentration of target molecules c_0 flows with a volumetric flow rate Q through a channel of height H and width W_c, as shown in Fig. 6.3. The bottom wall of the channel (say) has a rectangular 'patch' of length L (in the flow direction) and width W_s (along the width of the channel, along the wall surface and in the direction perpendicular to the flow direction). This rectangular patch is the sensing area, wherein it has been functionalised with suitable receptors, with an area concentration of b_m per unit area. We further assume that the target molecules can bind only with the receptors in the rectangular sensing patch, and not with adjacent area in the surface wall. The binding constants for the target molecules to bind with the receptors are assumed to be k_{on} and k_{off}. The channel has width W_c, height H perpendicular to the wall, and $W_c \gg H$. First, let us analyse the phenomenon in the absence of any convective transport, and then examine the role of convection on this process. As target molecules diffuse and reach the sensor, assuming that the binding rates are fast, the target molecules are instantaneously bound at the sensor surface, thus creating a zone of low concentration of target molecules near the sensor surface. From basic principles of transient diffusion, this 'depletion zone' will grow as $\delta \propto \sqrt{\mathcal{D}t}$, where $\mathcal{D}$ is the diffusivity of the target molecules. At very early times, the depletion zone will be initially flat, until a time of $L^2/\mathcal{D}$, whereupon, the depletion zone will move radially until a timescale of $H^2/\mathcal{D}$, as shown in Fig. 6.4. This radially growing depletion zone will span the entire channel width after times $H^2/\mathcal{D}$, and then the depletion zone grows in the

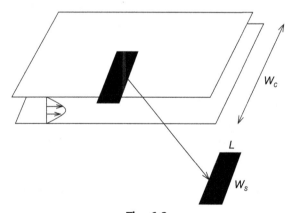

Fig. 6.3
Schematic representation of the 3D configuration of the surface-based sensor. The length of the sensor 'strip' is L and its width is W_s. The channel width is W_c, whilst the height of the channel (i.e. the distance between the channel walls) is H.

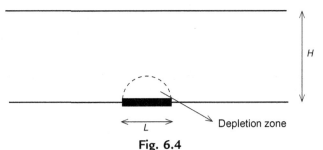

Fig. 6.4
Spreading of the depletion zone in the sensor.

direction parallel to the channel walls, and the spatial extent of this zone is proportional to $\sqrt{\mathcal{D}t}$ which grows indefinitely, until it reaches the two ends of the channel.

It is of importance to estimate how many target molecules are bound with time. This can be estimated from Fick's law for diffusion $j_D = -\mathcal{D}\nabla C$. The gradient of concentration can be estimated as $(c_0 - 0)/\delta$, where the surface concentration of the target molecules is taken to be zero. Thus, $j_D \propto \mathcal{D}c_0/\delta$. This flux must be multiplied by a suitable area to obtain the estimate for the number of molecules that are bound on the sensor per unit time. Assuming that W_s is the width of the sensor in the third direction (perpendicular to its length and height), $J_D = \mathcal{D}c_0/\delta(LW_s)$, which becomes (after using $\delta \propto \sqrt{\mathcal{D}t}$), $J_D = LW_s c_0 \sqrt{\mathcal{D}/t}$. For times $t \gg H^2/\mathcal{D}$, the depletion zone spreads into the channel, and the number of molecules that bind per unit time is estimated as $J_D \propto HW_c c_0 \sqrt{\mathcal{D}/t}$, which is independent of the size of the sensor L. There is no steady state that is reached in this purely diffusive system, as the depletion zone continues to grow into the channel, unless the length of the channel is reached. As time increases, the rate at which target molecules binds with the sensor decreases as $1/\sqrt{t}$.

Next, we examine the role of convective transport (i.e. flow) in the above phenomena. The presence of convective transport arrests the growth of depletion layer by supplying target molecules at a fixed rate $J_C \propto c_0 Q/(W_c H)$, with Q being the volumetric flow rate of the fluid, and W_c being the width of the channel (which may not be equal to the sensor width W_s). Thus, upstream of the sensor, by balancing the diffusive and convective fluxes, one obtains $c_0 Q/(W_c H) \sim \mathcal{D}c_0/\delta_s$, yielding $\delta_s \propto \mathcal{D}W_c H/Q$. Thus, when there is convective transport, if the extent of the depletion zone $\delta(t)$ is smaller than δ_s (its steady value, obtained under the balance of convection and diffusion along the channel length), then convection is not strong enough to supply the target molecules. If $\delta(t) > \delta_s$, then convective transport dominates diffusion, and the depletion zone is compressed. At sufficiently slow flow rates, all the target molecules will be captured by the sensor, and the rate of binding of target molecules per time is estimated as $J_D \propto C_0 Q$. For diffusion to be dominant, the steady extent of the depletion zone $\delta_s \propto \mathcal{D}HW_c/Q$ must be greater than the channel height H, yielding $Q \ll \mathcal{D}W_c$. This can also be seen differently as the ratio of diffusive time for the target molecule across the channel height $(H^2/\mathcal{D})$ to the convective time to be transported a distance H given by $Q/(H^2 W_c)$. This ratio is nothing but a Peclet number $Pe_H = (H^2/D)/(H^2 W_c/Q) = Q/(DW_c)$, where the subscript H denotes the length scale used in the Peclet number. For $Pe_H \ll 1$, diffusion dominates, and $\delta_s \sim H/Pe_H \gg 1$, that is, the depletion zone is much longer than the channel height H. When $Pe_H \gg 1$, convective transport dominates, wherein molecules are swept downstream before they can diffuse, and only the target molecules that are very close to the sensor patch have a chance of binding with the sensor. Very close to the sensor surface, the target molecules residing in that zone do not explore the full parabolic velocity profile across the height of the channel. Instead, they see a locally linear flow $u = \dot{\gamma} z$ at a height z above the sensor, where $\dot{\gamma} = 6Q/(H^2 W_c)$ is the shear rate of the parabolic flow. The thickness of the depletion zone δ_s can be estimated in this limit as follows. The depletion layer zone δ_s is that length scale wherein the time for convection past the sensor (of length L) $L/(\dot{\gamma}\delta_s)$ is comparable to timescale for diffusion across the length δ_s, which is δ_s^2/D. Balancing the two times, one obtains $\delta_s/L \sim Pe_s^{-1/3}$, where $Pe_s = \dot{\gamma}L^2/D$ is the Peclet number based on local shear flow near the sensor surface. The number of target molecules captured in this limit can then be estimated as $J_D \sim Dc_0/\delta_s W_s L \sim Dc_0 W_s Pe_s^{1/3}$. This relation suggests that the capture rate of the target molecules has only a weak dependence on the flow rate (i.e. Pe_s), since the exponent in Pe_s is only 1/3. To obtain a tenfold increase in capture rate of the target molecules, the volumetric flow rate must be increased thousandfold. The two Peclet numbers Pe_H and Pe_s signify different physical phenomena: If $Pe_H \gg 1$, the extent of the depletion zone is thin compared with channel height H, whilst if $Pe_H \ll 1$, then the extent of the depletion zone extends upstream and is large compared with channel height. When $Pe_H \gg 1$, then one needs to estimate Pe_s, which denotes whether the size of the depletion zone is large or small compared with the sensor length L. The kinetics of binding of the target molecule can give rise to additional dynamics, if that is a slow process, and the reader is referred to the review article of Squires et al.[11] for more discussion.

6.7 Mixing in Microfluidic Devices

One of the major challenges in microfluidic devices is the fact that most often the flows are laminar, and hence the transport rates in microscale flows are dominated by diffusion. This is in contrast to macroscale flows, where the flow velocities are high enough to ensure turbulent flows, where the transport rates are substantially increased by turbulent eddies. This mechanism is not realisable in microscale flows. In macroscale fluid flow operations, it is often possible to improve mixing by inducing a secondary (transverse) flows, but these secondary flows are driven by fluid inertia. When scaled-down to smaller dimensions that are more typical of microfluidic applications, the inertially-driven secondary flows will become less important, as the driving inertial forces diminish as Reynolds number decreases due to scaling down. Thus, any mechanism that relies on inertial forces to drive secondary flows is bound to be irrelevant to microfluidic applications, and hence strategies to improve mixing should be such that they work even for Stokes flows.

There is another way to realise chaotic flow trajectories, even though the velocity field itself is not chaotic, and this methodology is referred to as 'chaotic mixing'. This is very different from turbulent flows, where the velocity fields are themselves chaotic, and this leads to particle trajectories being chaotic as a result. In chaotic mixing at low Reynolds numbers, the mixing occurs by a dramatic decrease of characteristic length scales over which diffusion is important. Many approaches have been proposed to enhance mixing in microscale flows, but here we will describe a rather successful design called the 'herringbone micromixer'[13] which is a passive chaotic mixer, where there is no external activation of the mixing. The mixer design has a channel whose walls are structure in the form of herringbone-shaped grooves. When the fluid flows past this structure, the 'troughs' in the grooves promote circulation of the fluid, resulting in a helically shaped movement of the fluid. Further, the herringbone patterns are changed significantly after a few herringbones, and this results in a displacement of the centre of the fluid helices. This change in the fluid helices is equivalent to a time-dependent modification of the flow, and this engenders a chaotic flow in the herringbone mixer. It has been shown[13] that the mixing length in a herringbone micromixer is proportional to $\log(Pe)$ instead of being proportional Pe itself (as in the case of straight microchannels, as discussed earlier), thus leading to a significant decrease in mixing lengths in such geometries.

6.8 Further Reading in Microfluidics

Whilst the present chapter has attempted to provide a brief overview of the subject of microfluidics to a chemical engineering audience, interested readers are directed here to more detailed textbooks and review articles on the subject. There are several textbooks that have appeared in the last decade or so, which introduce the reader to the principles and challenges involved in microfluidics. We provide here a few such references, with the disclaimer that we have not attempted to be comprehensive, but only representative. The textbooks by Kirby,[12] Bruus,[14] and Tabeling[15] provide very good introductions and a comprehensive overview of the fundamental phenomena relevant to microfluidic applications, covering all the basic principles behind the various applications. The review articles by Stone et al.,[6] Stone and Kim,[2] and Squires and Quake[7] are a good place to start for a chemical engineer to understand the phenomena and challenges in microfluidic applications, and how fundamental principles of fluid flow and transport phenomena play a crucial role in the analysis and design of microfluidic processes. A general survey on the broad subject of microfluidics is given by Whitesides.[1] Jensen[16] provides a critical overview of reaction engineering at the microscale. The review by Lauga and Stone[10] discusses in detail the relevance of no-slip boundary condition in the microfluidic context.

References

1. Whitesides GM. The origins and the future of microfluidics. *Nature* 2006;**442**:05058.
2. Stone HA, Kim S. Microfluidics: basic issues, applications and challenges. *AIChE J* 2001;**47**:1250–4.

3. Quake SR, Scherer A. From micro- to nanofabrication with soft materials. *Science* 2000;**290**:1536–40.

4. Whitesides GM, Stroock AD. Flexible methods for microfluidics. *Phys Today* 2001;**54**:42–8.

5. McDonald JC, Whitesides GM. Poly(dimethylsiloxane) as a material for fabricating microfluidic devices. *Acc Chem Res* 2002;**35**:491–9.

6. Stone HA, Stroock AD, Ajdrai A. Engineering flows in small devices: microfluidics toward a lab-on-a-chip. *Ann Rev Fluid Mech* 2004;**36**:381–12411.

7. Squires TM, Quake SR. Microfluidics: fluid physics at the nanoliter scale. *Rev Mod Phys* 2005;**77**:977–121026.

8. Happel J, Brenner H. *Low Reynolds number hydrodynamics*. Dordrecht: Springer; 1983.

9. Kim S, Karilla SJ. *Microhydrodynamics: principles and selected applications*. New York: Dover; 2005.

10. Lauga E, Stone HA. Microfluidics: the no-slip boundary condition. In: Tropea C, Yarin A, Foss JF, editors. *Handbook of experimental fluid dynamics*. New York: Springer; 2007. p. 1219–40 [chapter 19].

11. Squires TM, Messinger RJ, Manalis SR. Making it stick: convection, reaction, and diffusion in surface-based biosensors. *Nat Biotech* 2008;**26**:417–26.

12. Kirby B. *Micro- and nanoscale fluid mechanics: transport in microfluidic devices*. Cambridge: Cambridge University Press; 2010.

13. Stroock AD, Dertinger SKW, Ajdari A, Mezic I, Stone HA, Whitesides GM. Chaotic mixers for microchannels. *Science* 2002;**295**:647–51.

14. Bruus H. *Theoretical microfluidics*. New York: Oxford University Press; 2008.

15. Tabeling P. *Introduction to microfluidics*. New York: Oxford University Press; 2005.

16. Jensen KF. Microreaction engineering—is small better? *Chem Eng Sci* 2001;**56**(2):293–303 16th International conference on chemical reactor engineering.

Appendix

A.1 Tables of Physical Properties

Table A1 Thermal conductivities of liquids[a]

Liquid	k (W/m K)	(K)	K (Btu/h ft. °F)	Liquid	K (W/m K)	(K)	K (Btu/h ft. °F)
Acetic acid, 100%	0.171	293	0.099	Hexane (n-)	0.138	303	0.080
50%	0.35	293	0.20		0.135	333	0.078
Acetone	0.177	303	0.102	Heptyl alcohol (n-)	0.163	303	0.094
	0.164	348	0.095		0.157	348	0.091
Allyl alcohol	0.180	298–303	0.104	Hexyl alcohol (n-)	0.161	303	0.093
Ammonia	0.50	258–303	0.29		0.156	348	0.090
Ammonia, aqueous	0.45	293	0.261				
	0.50	333	0.29	Kerosene	0.149	293	0.086
Amyl acetate	0.144	283	0.083		0.140	348	0.081
Amyl alcohol (n-)	0.163	303	0.094				
	0.154	373	0.089	Mercury	8.36	301	4.83
Amyl alcohol (iso-)	0.152	303	0.088	Methyl alcohol, 100%	0.215	293	0.124
	0.151	348	0.087	80%	0.267	293	0.154
Aniline	0.173	273–293	0.100	60%	0.329	293	0.190
				40%	0.405	293	0.234
Benzene	0.159	303	0.092	20%	0.492	293	0.284
	0.151	333	0.087	100%	0.197	323	0.114
Bromobenzene	0.128	303	0.074	Methyl chloride	0.192	258	0.111
	0.121	373	0.070		0.154	303	0.089
Butyl acetate (n-)	0.147	298–303	0.085				
Butyl alcohol (n-)	0.168	303	0.097	Nitrobenzene	0.164	303	0.095
	0.164	348	0.095		0.152	373	0.088
Butyl alcohol (iso-)	0.157	283	0.091	Nitromethane	0.216	303	0.125
					0.208	333	0.120
Calcium chloride brine							
30%	0.55	303	0.32	Nonane (n-)	0.145	303	0.084
15%	0.59	303	0.34		0.142	333	0.082
Carbon disulphide	0.161	303	0.093	Octane (n-)	0.144	303	0.083
	0.152	348	0.088		0.140	333	0.081
Carbon tetrachloride	0.185	273	0.107	Oils, petroleum	0.138–0.156	273	0.08–0.09
	0.163	341	0.094	Oil, castor	0.180	293	0.104
Chlorobenzene	0.144	283	0.083		0.173	373	0.100

Continued

547

Table A1 Thermal conductivities of liquids—cont'd

Liquid	k (W/m K)	(K)	K (Btu/h ft. °F)	Liquid	K (W/m K)	(K)	K (Btu/h ft. °F)
Chloroform	0.138	303	0.080		0.168	293	0.097
Cymene (para)	0.135	303	0.078	Oil, olive	0.164	373	0.095
	0.137	333	0.079				
Decane (n-)	0.147	303	0.085	Paraldehyde	0.145	303	0.084
	0.144	333	0.083		0.135	373	0.078
Dichlorodifluoromethane	0.099	266	0.057	Pentane (n-)	0.135	303	0.078
	0.092	289	0.053		0.128	348	0.074
	0.083	311	0.048	Perchloroethylene	0.159	323	0.092
	0.074	333	0.043	Petroleum ether	0.130	303	0.075
	0.066	355	0.038		0.126	348	0.073
Dichloroethane	0.142	323	0.082	Propyl alcohol (n-)	0.171	303	0.099
Dichloromethane	0.192	258	0.111		0.164	348	0.095
	0.166	303	0.096	Propyl alcohol (iso-)	0.157	303	0.091
Ethyl acetate	0.175	293	0.101		0.155	333	0.090
Ethyl alcohol, 100%	0.182	293	0.105	Sodium	0.85	373	49
80%	0.237	293	0.137		0.80	483	46
60%	0.305	293	0.176	Sodium chloride brine, 25.0%	0.57	303	0.33
40%	0.388	293	0.224	12.5%	0.59	303	0.34
20%	0.486	293	0.281	Sulphuric acid 90%	0.36	303	0.21
100%	0.151	323	0.087	60%	0.43	303	0.25
Ethyl benzene	0.149	303	0.086	30%	0.52	303	0.30
	0.142	333	0.082	Sulphur dioxide	0.22	258	0.128
Ethyl bromide	0.121	293	0.070		0.192	303	0.111
Ethyl ether	0.138	303	0.080				
	0.135	348	0.078	Toluene	0.149	303	0.086
Ethyl iodide	0.111	313	0.064		0.145	348	0.084
	0.109	348	0.063	β-trichloroethane	0.133	323	0.077
Ethylene glycol	0.265	273	0.153	Trichloroethylene	0.138	323	0.080
Gasoline	0.135	303	0.078	Turpentine	0.128	288	0.074
Glycerol, 100%	0.284	293	0.164	Vaseline	0.184	288	0.106
80%	0.327	293	0.189				
60%	0.381	293	0.220	Water	0.57	273	0.330
40%	0.448	293	0.259		0.615	303	0.356
20%	0.481	293	0.278		0.658	333	0.381
100%	0.284	373	0.164		0.688	353	0.398
Heptane (n-)	0.140	303	0.081	Xylene (ortho-)	0.155	293	0.090
	0.137	333	0.079	(meta-)	0.155	293	0.090

A linear variation with temperature may be assumed. The extreme values given constitute also the temperature limits over which the data are recommended.
[a]By permission from *Heat Transmission*, by W. H. McAdams, copyright 1942, McGraw-Hill.

Table A2 Latent heats of vaporisation[a]

No.	Compound	Range $\theta_c - \theta$ (°F)	θ_c (°F)	Range $\theta_c - \theta$ (K)	θ_c (K)
18	Acetic acid	180–405	610	100–225	594
22	Acetone	216–378	455	120–210	508
29	Ammonia	90–360	271	50–200	406
13	Benzene	18–720	552	10–00	562
16	Butane	162–360	307	90–200	426
21	Carbon dioxide	18–180	88	10–100	304
4	Carbon disulphide	252–495	523	140–275	546

Table A2 Latent heats of vaporisation—cont'd

No.	Compound	Range $\theta_c - \theta$ (°F)	θ_c (°F)	Range $\theta_c - \theta$ (K)	θ_c (K)
2	Carbon tetrachloride	54–450	541	30–250	556
7	Chloroform	252–495	505	140–275	536
8	Dichloromethane	270–450	421	150–250	489
3	Diphenyl	315–720	981	175–400	800
25	Ethane	45–270	90	25–150	305
26	Ethyl alcohol	36–252	469	20–140	516
28	Ethyl alcohol	252–540	469	140–300	516
17	Ethyl chloride	180–450	369	100–250	460
13	Ethyl ether	18–720	381	10–00	467
2	Freon-11 (CCl_3F)	126–450	389	70–250	471
2	Freon-12 (CCl_2F_2)	72–360	232	40–200	384
5	Freon-21 ($CHCl_2F$)	126–450	354	70–250	451
6	Freon-22 ($CHClF_2$)	90–306	205	50–170	369
1	Freon-113 (CCl_2F-$CClF_2$)	162–450	417	90–250	487
10	Heptane	36–540	512	20–300	540
11	Hexane	90–450	455	50–225	508
15	Isobutane	144–360	273	80–200	407
27	Methanol	72–450	464	40–250	513
20	Methyl chloride	126–450	289	70–250	416
19	Nitrous oxide	45–270	97	25–150	309
9	Octane	54–540	565	30–300	569
12	Pentane	36–360	387	20–200	470
23	Propane	72–360	205	40–200	369
24	Propyl alcohol	36–360	507	20–200	537
14	Sulphur dioxide	162–288	314	90–160	430
30	Water	180–900	705	100–500	647

Example: for water at 373 K, $\theta_c - \theta = (647 - 373) = 274$ K, and the latent heat of vaporisation is 2257 kJ/kg.
[a]By permission from *Heat Transmission,* by W. H. McAdams, Copyright 1942, McGraw-Hill.

Table A3 Specific heats of liquids[a]

No.	Liquid	Range (K)
29	Acetic acid, 100%	273–353
32	Acetone	293–323
52	Ammonia	203–323
37	Amyl alcohol	223–298
26	Amyl acetate	273–373
30	Aniline	273–403
23	Benzene	283–353
27	Benzyl alcohol	253–303
10	Benzyl chloride	243–303
49	Brine, 25% $CaCl_2$	233–293
51	Brine, 25% NaCl	233–293
44	Butyl alcohol	273–373
2	Carbon disulphide	173–298
3	Carbon tetrachloride	283–333
8	Chlorobenzene	273–373
4	Chloroform	273–323
21	Decane	193–298

Continued

Table A3   Specific heats of liquids—cont'd

No.	Liquid	Range (K)
6A	Dichloroethane	243–333
5	Dichloromethane	233–323
15	Diphenyl	353–393
22	Diphenylmethane	303–373
16	Diphenyl oxide	273–473
16	Dowtherm A	273–473
24	Ethyl acetate	223–298
42	Ethyl alcohol, 100%	303–353
46	Ethyl alcohol, 95%	293–353
50	Ethyl alcohol, 50%	293–353
25	Ethyl benzene	273–373
1	Ethyl bromide	278–298
13	Ethyl chloride	243–313
36	Ethyl ether	173–298
7	Ethyl iodide	273–373
39	Ethylene glycol	233–473
2A	Freon-11 (CCl_3F)	253–343
6	Freon-12 (CCl_2F_2)	233–288
4A	Freon-21 ($CHCl_2F$)	253–343
7A	Freon-22 ($CHClF_2$)	253–333
3A	Freon-113 ($CCl_2F\text{-}CClF_2$)	253–343
38	Glycerol	233–293
28	Heptane	273–333
35	Hexane	193–293
48	Hydrochloric acid, 30%	293–373
41	Isoamyl alcohol	283–373
43	Isobutyl alcohol	273–373
47	Isopropyl alcohol	253–323
31	Isopropyl ether	193–293
40	Methyl alcohol	233–293
13A	Methyl chloride	193–293
14	Naphthalene	363–473
12	Nitrobenzene	273–373
34	Nonane	223–298
33	Octane	223–298
3	Perchloroethylene	243–413
45	Propyl alcohol	253–373
20	Pyridine	223–298
9	Sulphuric acid, 98%	283–318
11	Sulphur dioxide	253–373
23	Toluene	273–333
53	Water	283–473
19	Xylene (*ortho*)	273–373
18	Xylene (*meta*)	273–373
17	Xylene (*para*)	273–373

Table A4 Specific heats at constant pressure of gases and vapours at 101.3 kN/m^{2a}

No.	Gas	Range (K)	No.	Gas	Range (K)
10	Acetylene	273–473	1	Hydrogen	273–873
15	Acetylene	473–673	2	Hydrogen	873–1673
16	Acetylene	673–1673	35	Hydrogen bromide	273–1673
27	Air	273–1673	30	Hydrogen chloride	273–1673
12	Ammonia	273–873	20	Hydrogen fluoride	273–1673
14	Ammonia	873–1673	36	Hydrogen iodide	273–1673
18	Carbon dioxide	273–673	19	Hydrogen sulphide	273–973
24	Carbon dioxide	673–1673	21	Hydrogen sulphide	973–1673
26	Carbon monoxide	273–1673	5	Methane	273–573
32	Chlorine	273–473	6	Methane	573–973
34	Chlorine	473–1673	7	Methane	973–1673
3	Ethane	273–473	25	Nitric oxide	273–973
9	Ethane	473–873	28	Nitric oxide	973–1673
8	Ethane	873–1673	26	Nitrogen	273–1673
4	Ethylene	273–473	23	Oxygen	273–773
11	Ethylene	473–873	29	Oxygen	773–1673
13	Ethylene	873–1673	33	Sulphur	573–1673
17B	Freon-11 (CCl_3F)	273–423	22	Sulphur dioxide	273–673
17C	Freon-21 ($CHCl_2F$)	273–423	31	Sulphur dioxide	673–1673
17A	Freon-22 ($CHClF_2$)	273–423	17	Water	273–1673
17D	Freon-113 (CCl_2F-$CClF_2$)	273–423			

[a]By permission from *Heat Transmission*, by W. H. McAdams, Copyright, 1942, McGraw-Hill.

Table A5 Viscosity of water[a]

Temperature (θ) (K)	Viscosity (μ) (mN s/m^2)	Temperature (θ) (K)	Viscosity (μ) (mN s/m^2)	Temperature (θ) (K)	Viscosity (μ) (mN s/m^2)
273	1.7921	306	0.7523	340	0.4233
274	1.7313	307	0.7371	341	0.4174
275	1.6728	308	0.7225	342	0.4117
276	1.6191	309	0.7085	343	0.4061
277	1.5674	310	0.6947	344	0.4006
278	1.5188	311	0.6814	345	0.3952
279	1.4728	312	0.6685	346	0.3900
280	1.4284	313	0.6560	347	0.3849

Continued

Table A5 Viscosity of water—cont'd

Temperature (θ) (K)	Viscosity (μ) (mN s/m^2)	Temperature (θ) (K)	Viscosity (μ) (mN s/m^2)	Temperature (θ) (K)	Viscosity (μ) (mN s/m^2)
281	1.3860	314	0.6439	348	0.3799
282	1.3462	315	0.6321	349	0.3750
283	1.3077	316	0.6207	350	0.3702
284	1.2713	317	0.6097	351	0.3655
285	1.2363	318	0.5988	352	0.3610
286	1.2028	319	0.5883	353	0.3565
287	1.1709	320	0.5782	354	0.3521
288	1.1404	321	0.5683	355	0.3478
289	1.1111	322	0.5588	356	0.3436
290	1.0828	323	0.5494	357	0.3395
291	1.0559	324	0.5404	358	0.3355
292	1.0299	325	0.5315	359	0.3315
293	1.0050	326	0.5229	360	0.3276
293.2	1.0000	327	0.5146	361	0.3239
294	0.9810	328	0.5064	362	0.3202
295	0.9579	329	0.4985	363	0.3165
296	0.9358	330	0.4907	364	0.3130
297	0.9142	331	0.4832	365	0.3095
298	0.8937	332	0.4759	366	0.3060
299	0.8737	333	0.4688	367	0.3027
300	0.8545	334	0.4618	368	0.2994
301	0.8360	335	0.4550	369	0.2962
302	0.8180	336	0.4483	370	0.2930
303	0.8007	337	0.4418	371	0.2899
304	0.7840	338	0.4355	372	0.2868
305	0.7679	339	0.4293	373	0.2838

[a]Calculated from the formula

$$1/\mu = 21.482\left[(\theta - 281.435) + \sqrt{(8078.4 + (\theta - 281.435)^2)}\right] - 1200 (\mu \text{ in Ns/m}^2)$$

By permission from *Fluidity and Plasticity*, by E. C. Bingham. Copyright 1922, McGraw-Hill Book Company Inc.

Table A6 Thermal conductivities of gases and vapours

Substance	k (W/m K)	(K)	k (Btu/h ft. °F)	Substance	k (W/m K)	(K)	k (Btu/h ft. °F)
Acetone	0.0098	273	0.0057	Chlorine	0.0074	273	0.0043
	0.0128	319	0.0074	Chloroform	0.0066	273	0.0038
	0.0171	373	0.0099		0.0080	319	0.0046
	0.0254	457	0.0147		0.0100	373	0.0058
Acetylene	0.0118	198	0.0068		0.0133	457	0.0077
	0.0187	273	0.0108	Cyclohexane	0.0164	375	0.0095
	0.0242	323	0.0140				
	0.0298	373	0.0172	Dichloro-difluoro-methane	0.0083	273	0.0048
Air	0.0164	173	0.0095		0.0111	323	0.0064
	0.0242	273	0.0140		0.0139	373	0.0080
	0.0317	373	0.0183		0.0168	423	0.0097
	0.0391	473	0.0226				
	0.0459	573	0.0265	Ethane	0.0114	203	0.0066

Table A6 Thermal conductivities of gases and vapours—cont'd

Substance	k (W/m K)	(K)	k (Btu/h ft. °F)	Substance	k (W/m K)	(K)	k (Btu/h ft. °F)
Ammonia	0.0164	213	0.0095		0.0149	239	0.0086
	0.0222	273	0.0128		0.0183	273	0.0106
	0.0272	323	0.0157		0.0303	373	0.0175
	0.0320	373	0.0185	Ethyl acetate	0.0125	319	0.0072
Benzene	0.0090	273	0.0052		0.0166	373	0.0096
	0.0126	319	0.0073		0.0244	457	0.0141
	0.0178	373	0.0103	Alcohol	0.0154	293	0.0089
	0.0263	457	0.0152		0.0215	373	0.0124
	0.0305	485	0.0176	Chloride	0.0095	273	0.0055
Butane (*n*-)	0.0135	273	0.0078		0.0164	373	0.0095
	0.0234	373	0.0135		0.0234	457	0.0135
(iso-)	0.0138	273	0.0080		0.0263	485	0.0152
	0.0241	373	0.0139	Ether	0.0133	273	0.0077
Carbon dioxide	0.0118	223	0.0068		0.0171	319	0.0099
	0.0147	273	0.0085		0.0227	373	0.0131
	0.0230	373	0.0133		0.0327	457	0.0189
	0.0313	473	0.0181		0.0362	485	0.0209
	0.0396	573	0.0228	Ethylene	0.0111	202	0.0064
Disulphide	0.0069	273	0.0040		0.0175	273	0.0101
	0.0073	280	0.0042		0.0267	323	0.0131
Monoxide	0.0071	84	0.0041		0.0279	373	0.0161
	0.0080	94	0.0046	Heptane (*n*-)	0.0194	473	0.0112
	0.0234	213	0.0135		0.0178	373	0.0103
Tetrachloride	0.0071	319	0.0041	Hexane (*n*-)	0.0125	273	0.0072
	0.0090	373	0.0052		0.0138	293	0.0080
	0.0112	457	0.0065	Hexene	0.0106	273	0.0061
Hydrogen	0.0113	173	0.065		0.0109	373	0.0189
	0.0144	223	0.083		0.0225	457	0.0130
	0.0173	273	0.100		0.0256	485	0.0148
	0.0199	323	0.115	Methylene chloride	0.0067	273	0.0039
	0.0223	373	0.129		0.0085	319	0.0049
	0.0308	573	0.178		0.0109	373	0.0063
Hydrogen and carbon dioxide		273			0.0164	485	0.0095
0% H$_2$	0.0144		0.0083	Nitric oxide	0.0178	203	0.0103
20%	0.0286		0.0165		0.0239	273	0.0138
40%	0.0467		0.0270	Nitrogen	0.0164	173	0.0095
60%	0.0709		0.0410		0.0242	273	0.0140
80%	0.1070		0.0620		0.0277	323	0.0160
100%	0.173		0.10		0.0312	373	0.0180
Hydrogen and nitrogen		273		Nitrous oxide	0.0116	201	0.0067
0% H$_2$	0.0230		0.0133		0.0157	273	0.0087
20%	0.0367		0.0212		0.0222	373	0.0128
40%	0.0542		0.0313				
60%	0.0758		0.0438	Oxygen	0.0164	173	0.0095
80%	0.1098		0.0635		0.0206	223	0.0119
Hydrogen and nitrous oxide		273			0.0246	273	0.0142
0% H$_2$	0.0159		0.0092		0.0284	323	0.0164
20%	0.0294		0.0170		0.0321	373	0.0185
40%	0.0467		0.0270				
60%	0.0709		0.0410	Pentane (n-)	0.0128	273	0.0074
80%	0.112		0.0650		0.0144	293	0.0083
Hydrogen sulphide	0.0132	273	0.0076	(iso-)	0.0125	273	0.0072
					0.0220	373	0.0127
Mercury	0.0341	473	0.0197	Propane	0.0151	273	0.0087
Methane	0.0173	173	0.0100		0.0261	373	0.0151
	0.0251	223	0.0145				
	0.0302	273	0.0175	Sulphur dioxide	0.0087	273	0.0050
	0.0372	323	0.0215		0.0119	373	0.0069
Methyl alcohol	0.0144	273	0.0083				
	0.0222	373	0.0128	Water vapour	0.0208	319	0.0120

Continued

Table A6   Thermal conductivities of gases and vapours—cont'd

Substance	k (W/m K)	(K)	k (Btu/h ft. °F)	Substance	k (W/m K)	(K)	k (Btu/h ft. °F)
Acetate	0.0102	273	0.0059		0.0237	373	0.0137
	0.0118	293	0.0068		0.0324	473	0.0187
Chloride	0.0092	273	0.0053		0.0429	573	0.0248
	0.0125	319	0.0072		0.0545	673	0.0315
	0.0163	373	0.0094		0.0763	773	0.0441

The extreme temperature values given constitute the experimental range. For extrapolation to other temperatures, it is suggested that the data given be plotted as log k versus log T or that use be made of the assumption that the ratio $C_p\mu/k$ is practically independent of temperature (and of pressure, within moderate limits)

By permission from *Heat Transmission*, by W. H. McAdams, Copyright 1942, McGraw-Hill.

Table A7 Viscosities of gases[a] coordinates for use with graph on facing page

No.	Gas	X	Y
1	Acetic acid	7.7	14.3
2	Acetone	8.9	13.0
3	Acetylene	9.8	14.9
4	Air	11.0	20.0
5	Ammonia	8.4	16.0
6	Argon	10.5	22.4
7	Benzene	8.5	13.2
8	Bromine	8.9	19.2
9	Butene	9.2	13.7
10	Butylene	8.9	13.0
11	Carbon dioxide	9.5	18.7
12	Carbon disulphide	8.0	16.0
13	Carbon monoxide	11.0	20.0
14	Chlorine	9.0	18.4
15	Chloroform	8.9	15.7
16	Cyanogen	9.2	15.2
17	Cyclohexane	9.2	12.0
18	Ethane	9.1	14.5
19	Ethyl acetate	8.5	13.2
20	Ethyl alcohol	9.2	14.2
21	Ethyl chloride	8.5	15.6
22	Ethyl ether	8.9	13.0
23	Ethylene	9.5	15.1
24	Fluorine	7.3	23.8
25	Freon-11 (CCl_8F)	10.6	15.1
26	Freon-12 (CCl_2F_2)	11.1	16.0
27	Freon-21 ($CHCl_2F$)	10.8	15.3
28	Freon-22 ($CHClF_2$)	10.1	17.0
29	Freon-113 ($CCl_2F-CClF_3$)	11.3	14.0
30	Helium	10.9	20.5
31	Hexane	8.6	11.8
32	Hydrogen	11.2	12.4

Table A7 Viscosities of gases coordinates for use with graph on facing page—cont'd

No.	Gas	X	Y
33	$3H_2 + 1 N_2$	11.2	17.2
34	Hydrogen bromide	8.8	20.9
35	Hydrogen chloride	8.8	18.7
36	Hydrogen cyanide	9.8	14.9
37	Hydrogen iodide	9.0	21.3
38	Hydrogen sulphide	8.6	18.0
39	Iodine	9.0	18.4
40	Mercury	5.3	22.9
41	Methane	9.9	15.5
42	Methyl alcohol	8.5	15.6
43	Nitric oxide	10.9	20.5
44	Nitrogen	10.6	20.0
45	Nitrosyl chloride	8.0	17.6
46	Nitrous oxide	8.8	19.0
47	Oxygen	11.0	21.3
48	Pentane	7.0	12.8
49	Propane	9.7	12.9
50	Propyl alcohol	8.4	13.4
51	Propylene	9.0	13.8
52	Sulphur dioxide	9.6	17.0
53	Toluene	8.6	12.4
54	2, 3, 3-trimethylbutane	9.5	10.5
55	Water	8.0	16.0
56	Xenon	9.3	23.0

To convert to lb/ft.-h multiply by 2.42.

[a]By permission from *Perry's Chemical Engineers' Handbook,* by Perry, R. H. and Green, D. W. (eds) 6th edn. Copyright 1984, McGraw-Hill Book Company Inc.

Table A8 Viscosities and densities of liquids[a] coordinates for graph on following page

No.	Liquid	X	Y	Density at 293 K (kg/m^3)
1	Acetaldehyde	15.2	4.8	783 (291 K)
2	Acetic acid, 100%	12.1	14.2	1049
3	Acetic acid, 70%	9.5	17.0	1069
4	Acetic anhydride	12.7	12.8	1083
5	Acetone, 100%	14.5	7.2	792
6	Acetone, 35%	7.9	15.0	948
7	Allyl alcohol	10.2	14.3	854
8	Ammonia, 100%	12.6	2.0	817 (194 K)
9	Ammonia, 26%	10.1	13.9	904
10	Amyl acetate	11.8	12.5	879
11	Amyl alcohol	7.5	18.4	817
12	Aniline	8.1	18.7	1022

Continued

Table A8 **Viscosities and densities of liquids coordinates for graph on following page—cont'd**

No.	Liquid	X	Y	Density at 293 K (kg/m³)
13	Anisole	12.3	13.5	990
14	Arsenic trichloride	13.9	14.5	2163
15	Benzene	12.5	10.9	880
16	Brine, $CaCl_2$, 25%	6.6	15.9	1228
17	Brine, NaCl, 25%	10.2	16.6	1186 (298 K)
18	Bromine	14.2	13.2	3119
19	Bromotoluene	20.0	15.9	1410
20	Butyl acetate	12.3	11.0	882
21	Butyl alcohol	8.6	17.2	810
22	Butyric acid	12.1	15.3	964
23	Carbon dioxide	11.6	0.3	1101 (236 K)
24	Carbon disulphide	16.1	7.5	1263
25	Carbon tetrachloride	12.7	13.1	1595
26	Chlorobenzene	12.3	12.4	1107
27	Chloroform	14.4	10.2	1489
28	Chlorosulphonic acid	11.2	18.1	1787 (298 K)
29	Chlorotoluene, *ortho*	13.0	13.3	1082
30	Chlorotoluene, *meta*	13.3	12.5	1072
31	Chloroluene, *para*	13.3	12.5	1070
32	Cresol, *meta*	2.5	20.8	1034
33	Cyclohexanol	2.9	24.3	962
34	Dibromoethane	12.7	15.8	2495
35	Dichloroethane	13.2	12.2	1256
36	Dichloromethane	14.6	8.9	1336
37	Diethyl oxalate	11.0	16.4	1079
38	Dimethyl oxalate	12.3	15.8	1148 (327 K)
39	Diphenyl	12.0	18.3	992 (346 K)
40	Dipropyl oxalate	10.3	17.7	1038 (273 K)
41	Ethyl acetate	13.7	9.1	901
42	Ethyl alcohol, 100%	10.5	13.8	789
43	Ethyl alcohol, 95%	9.8	14.3	804
44	Ethyl alcohol, 40%	6.5	16.6	935
45	Ethyl benzene	13.2	11.5	867
46	Ethyl bromide	14.5	8.1	1431
47	Ethyl chloride	14.8	6.0	917 (279 K)
48	Ethyl ether	14.5	5.3	708 (298 K)
49	Ethyl formate	14.2	8.4	923
50	Ethyl iodide	14.7	10.3	1933
51	Ethylene glycol	6.0	23.6	1113
52	Formic acid	10.7	15.8	1220
53	Freon-11 (CCl_3F)	14.4	9.0	1494 (290 K)
54	Freon-12 (CCl_2F_2)	16.8	5.6	1486 (293 K)

Table A8 Viscosities and densities of liquids coordinates for graph on following page—cont'd

No.	Liquid	X	Y	Density at 293 K (kg/m^3)
55	Freon-21 (CHCl$_2$F)	15.7	7.5	1426 (273 K)
56	Freon-22 (CHClF$_2$)	17.2	4.7	3870 (273 K)
57	Freon-113 (CCl$_2$F-CClF$_2$)	12.5	11.4	1576
58	Glycerol, 100%	2.0	30.0	1261
59	Glycerol, 50%	6.9	19.6	1126
60	Heptane	14.1	8.4	684
61	Hexane	14.7	7.0	659
62	Hydrochloric acid, 31.5%	13.0	16.6	1157
63	Isobutyl alcohol	7.1	18.0	779 (299 K)
64	Isobutyric acid	12.2	14.4	949
65	Isopropyl alcohol	8.2	16.0	789
66	Kerosene	10.2	16.9	780–820
67	Linseed oil, raw	7.5	27.2	934 ± 4 (288 K)
68	Mercury	18.4	16.4	13,546
69	Methanol, 100%	12.4	10.5	792
70	Methanol, 90%	12.3.	11.8	820
71	Methanol, 40%	7.8	15.5	935
72	Methyl acetate	14.2	8.2	924
73	Methyl chloride	15.0	3.8	952 (273 K)
74	Methyl ethyl ketone	13.9	8.6	805
75	Naphthalene	7.9	18.1	1145
76	Nitric acid, 95%	12.8	13.8	1493
77	Nitric acid, 60%	10.8	17.0	1367
78	Nitrobenzene	10.6	16.2	1205 (291 K)
79	Nitrotoluene	11.0	17.0	1160
80	Octane	13.7	10.0	703
81	Octyl alcohol	6.6	21.1	827
82	Pentachloroethane	10.9	17.3	1671 (298 K)
83	Pentane	14.9	5.2	630 (291 K)
84	Phenol	6.9	20.8	1071 (298 K)
85	Phosphorus tribromide	13.8	16.7	2852 (288 K)
86	Phosphorus trichloride	16.2	10.9	1574
87	Propionic acid	12.8	13.8	992
88	Propyl alcohol	9.1	16.5	804
89	Propyl bromide	14.5	9.6	1353
90	Propyl chloride	14.4	7.5	890
91	Propyl iodide	14.1	11.6	1749
92	Sodium	16.4	13.9	970
93	Sodium hydroxide, 50%	3.2	25.8	1525
94	Stannic chloride	13.5	12.8	2226
95	Sulphur dioxide	15.2	7.1	1434 (273 K)

Continued

Table A8   Viscosities and densities of liquids coordinates for graph on following page—cont'd

No.	Liquid	X	Y	Density at 293 K (kg/m³)
96	Sulphuric acid, 110%	7.2	27.4	1980
97	Sulphuric acid, 98%	7.0	24.8	1836
98	Sulphuric acid, 60%	10.2	21.3	1498
99	Sulphuryl chloride	15.2	12.4	1667
100	Tetrachloroethane	11.9	15.7	1600
101	Tetrachloroethylene	14.2	12.7	1624 (288 K)
102	Titanum tetrachloride	14.4	12.3	1726
103	Toluene	13.7	10.4	866
104	Trichloroethylene	14.8	10.5	1466
105	Turpentine	11.5	14.9	861–867
106	Vinyl acetate	14.0	8.8	932
107	Water	10.2	13.0	998
108	Xylene, *ortho*	13.5	12.1	881
109	Xylene, *meta*	13.9	10.6	867
110	Xylene, *para*	13.9	10.9	861

[a]By permission from *Perry's Chemical Engineers' Handbook*, by Perry, R. H. and Green, D. W. (eds), 6th edn. Copyright 1984, McGraw-Hill.

Table A9   Critical constants of gases[a]

	Critical Temperature T_c (K)	Critical Pressure P_c (MN/m²)	Compressibility Constant in Critical State Z_c
Paraffins			
Methane	191	4.64	0.290
Ethane	306	4.88	0.284
Propane	370	4.25	0.276
n-Butane	425	3.80	0.274
Isobutane	408	3.65	0.282
n-Pentane	470	3.37	0.268
Isopentane	461	3.33	0.268
Neopentane	434	3.20	0.268
n-Hexane	508	3.03	0.264
n-Heptane	540	2.74	0.260
n-Octane	569	2.49	0.258
Mono-olefins			
Ethylene	282	5.07	0.268
Propylene	365	4.62	0.276
1-Butene	420	4.02	0.276
1-Pentene	474	4.05	

Table A9 Critical constants of gases—cont'd

	Critical Temperature T_c (K)	Critical Pressure P_c (MN/m^2)	Compressibility Constant in Critical State Z_c
Miscellaneous organic compounds			
Acetic acid	595	5.78	0.200
Acetone	509	4.72	0.237
Acetylene	309	6.24	0.274
Benzene	562	4.92	0.274
1, 3-Butadiene	425	4.33	0.270
Cyclohexane	553	4.05	0.271
Dichlorodifluoromethane (Freon-12)	385	4.01	0.273
Diethyl ether	467	3.61	0.261
Ethyl alcohol	516	6.38	0.249
Ethylene oxide	468	7.19	0.25
Methyl alcohol	513	7.95	0.220
Methyl chloride	416	6.68	0.276
Methyl ethyl ketone	533	4.00	0.26
Toluene	594	4.21	0.27
Trichlorofluoromethane (Freon-11)	471	4.38	0.277
Trichlorotrifluoroethane (Freon-113)	487	3.41	0.274
Elementary gases			
Bromine	584	10.33	0.307
Chlorine	417	7.71	0.276
Helium	5.3	0.23	0.300
Hydrogen	33.3	1.30	0.304
Neon	44.5	2.72	0.307
Nitrogen	126	3.39	0.291
Oxygen	155	5.08	0.29
Miscellaneous inorganic compounds			
Ammonia	406	11.24	0.242
Carbon dioxide	304	7.39	0.276
Carbon monoxide	133	3.50	0.294
Hydrogen chloride	325	8.26	0.266
Hydrogen sulphide	374	9.01	0.284
Nitric oxide (NO)	180	6.48	0.25
Nitrous oxide (N_2O)	310	7.26	0.271
Sulphur	1313	11.75	
Sulphur dioxide	431	7.88	0.268
Sulphur trioxide	491	8.49	0.262
Water	647	22.1	0.23

[a]Selected values from K. A. Kobe and R. E. Lynn, Jr., *Chem. Rev.*, 52, 117 (1953), with permission.

Table A10 Emissivities of surfaces[a]

Surface	T (K)	Emissivity
A. Metals and metallic oxides		
Aluminium		
Highly polished plate	500–850	0.039–0.057
Polished plate	296	0.040
Rough plate	299	0.055
Plate oxidised at 872 K	472–872	0.11–0.19
Aluminium-surfaced roofing	311	0.216
Brass		
Hard-rolled, polished	294	0.038
Polished	311–589	0.096
Rolled plate, natural surface	295	0.06
Rubbed with coarse emery	295	0.20
Dull plate	322–622	0.22
Oxidised	472–872	0.61–0.59
Chromium—see copper-nickel alloys		
Polished electrolytic	353	0.018
Commercial, emeried, and polished	292	0.030
Commercial, scraped shiny	295	0.072
Polished	390	0.023
Plate, covered with thick oxide	498	0.78
Plate heated to 872 K	472–872	0.57–0.57
Cuprous oxide	1072–1372	0.66–0.54
Molten copper	1350–1550	0.16–0.13
Gold		
Pure, highly polished	500–900	0.018–0.35
Iron and steel		
Electrolytic iron, highly polished	450–500	0.052–0.064
Polished iron	700–1300	0.144–0.377
Freshly emeried iron	293	0.242
Polished cast iron	473	0.21
Wrought iron, highly polished	311–522	0.28
Cast iron, newly turned	295	0.435
Steel casting, polished	1044–1311	0.52–0.56
Ground sheet steel	1211–1372	0.55–0.61
Smooth sheet iron	1172–1311	0.55–0.60
Cast iron, turned	1155–1261	0.60–0.70
Oxidised surfaces		
Iron plate, completely rusted	293	0.685
Sheet steel, rolled, and oxidised	295	0.657
Iron	373	0.736
Cast iron, oxidised at 872 K	472–872	0.64–0.78
Steel, oxidised at 872 K	472–872	0.79–0.79
Smooth electrolytic iron	500–800	0.78–0.82
Iron oxide	772–1472	0.85–0.89
Ingot iron, rough	1200–1390	0.87–0.95

Table A10   Emissivities of surfaces—cont'd

Surface	T (K)	Emissivity
Sheet steel with rough oxide layer	297	0.80
Cast iron, strongly oxidised	311–522	0.95
Wrought iron, dull oxidised	294–633	0.94
Steel plate, rough	311–644	0.94–0.97
Molten metal		
Cast iron	1572–1672	0.29–0.29
Mild steel	1872–2070	0.28–0.28
Lead		
Pure, unoxidised	400–500	0.057–0.075
Grey, oxidised	297	0.281
Oxidised at 472 K	472	0.03
Mercury	273–373	0.09–0.12
Molybdenum		
Filament	1000–2866	0.096–0.292
Monel		
Metal oxidised at 872 K	472–872	0.41–0.46
Nickel		
Electroplated on polished iron and polished	296	0.045
Technically pure, polished	500–600	0.07–0.087
Electroplated on pickled iron, unpolished	293	0.11
Wire	460–1280	0.096–0.186
Plate, oxidised by heating to 872 K	472–872	0.37–0.48
Nickel oxide	922–1527	0.59–0.86
Nickel alloys		
Chromonickel	325–1308	0.64–0.76
Nickelin, grey oxidised	294	0.262
KA-28 alloy, rough brown, after heating	489–763	0.44–0.36
KA-28 alloy, after heating at 800 K	489–800	0.62–0.73
NCT 3 alloy, oxidised from service	489–800	0.90–0.97
NCT 6 alloy, oxidised from service	544–836	0.89–0.82
Platinum		
Pure, polished plate	500–900	0.054–0.104
Strip	1200–1900	0.12–0.17
Filament	300–1600	0.036–0.192
Wire	500–1600	0.073–0.182
Silver		
Polished, pure	500–900	0.0198–0.0324
Polished	310–644	0.0221–0.0312

Continued

Table A10 Emissivities of surfaces—cont'd

Surface	T (K)	Emissivity
Steel—see iron tantalum		
Filament	1600–3272	0.194–0.31
Tin		
Bright tinned iron sheet	298	0.043 and 0.064
Tungsten		
Filament, aged	300–3588	0.032–0.35
Filament	3588	0.39
Zinc		
Commercially pure, polished	500–600	0.045–0.053
Oxidised by heating to 672 K	672	0.11
Galvanised sheet iron, fairly bright	301	0.228
Galvanised sheet iron, grey oxidised	297	0.276
B. *Refractories, building materials, paints, etc.*		
Asbestos		
Board	297	0.96
Paper	311–644	0.93–0.945
Brick		
Red, rough	294	0.93
Silica, unglazed	1275	0.80
Silica, glazed, rough	1475	0.85
Grog, glazed	1475	0.75
Carbon		
T-carbon	400–900	0.81–0.79
Filament	1311–1677	0.526
Candle soot	372–544	0.952
Lampblack-water-glass coating	372–456	0.957–0.952
Thin layer on iron plate	294	0.927
Thick coat	293	0.967
Lampblack, 0.08 mm or thicker	311–644	0.945
Enamel		
White fused on iron	292	0.897
Glass		
Smooth	295	0.937
Gypsum		
0.5 mm thick on blackened plate	294	0.903
Marble		
Light grey, polished	295	0.931
Oak		
Planed	294	0.895

Table A10 Emissivities of surfaces—cont'd

Surface	T (K)	Emissivity
Oil layers		
On polished nickel		
Polished surface alone		0.045
0.025 mm oil		0.27
0.050 mm oil		0.46
0.125 mm oil		0.72
Thick oil layer		0.82
On aluminium foil		
Aluminium foil alone	373	0.087
1 coat of oil	373	0.561
2 coats of oil	373	0.574
Paints, lacquers, varnishes		
Snow white enamel on rough iron plate	296	0.906
Black, shiny lacquer sprayed on iron	298	0.875
Black, shiny shellac on tinned iron sheet	294	0.821
Black matt shellac	350–420	0.91
Black lacquer	311–366	0.80–0.95
Matt black lacquer	311–366	0.96–0.98
White lacquer	311–366	0.80–0.95
Oil paints	373	0.92–0.96
Aluminium paint	373	0.27–0.67
After heating to 600 K	422–622	0.35
Aluminium lacquer	294	0.39
Paper, thin		
Pasted on tinned iron plate	292	0.924
Pasted on rough iron plate	292	0.929
Pasted on black lacquered plate	292	0.944
Roofing	294	0.91
Plaster, lime, rough	283–361	0.91
Porcelain, glazed	295	0.924
Quartz, rough, fused	294	0.932
Refractory materials		
Poor radiators	872–1272	0.65–0.75
Good radiators	872–1272	0.80–0.90
Rubber		
Hard, glossy plate	296	0.945
Soft, grey, rough	298	0.859
Serpentine, polished	296	0.900
Water	273–373	0.95–0.963

[a]From Hottel, H. C. and Sarofim, A. F.: Radiation Heat Transfer (McGraw-Hill, New York, 1967).

A.2 Steam Tables

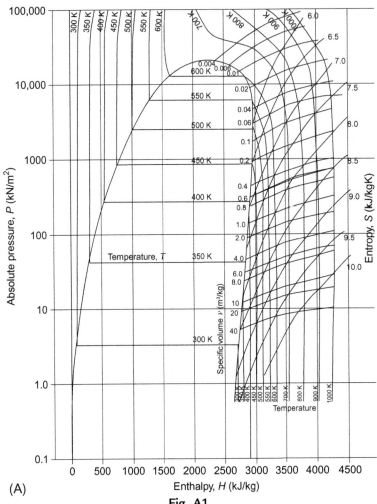

(A)

Fig. A1

Pressure-enthalpy diagram for water and steam.

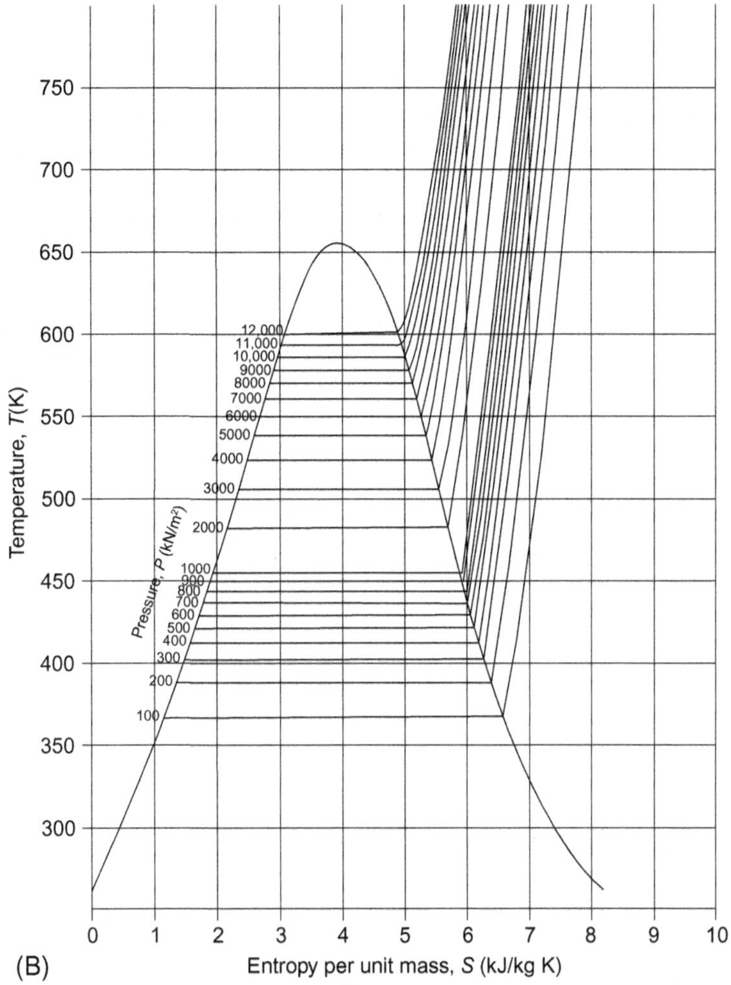

Fig. A2

Temperature-entropy diagram for water and steam.

Table A11 Properties of saturated steam (SI units)

Absolute Pressure (kN/m²)	Temperature (°C) θ_s	Temperature (K) T_s	Enthalpy per Unit Mass (H_s) (kJ/kg) Water	Latent	Steam	Entropy per Unit Mass (S_s) (kJ/kg K) Water	Latent	Steam	Specific Volume (v) (m³/kg) Water	Steam
			Datum, triple point of water							
0.611	0.01	273.16	0.0	2501.6	2501.6	0	9.1575	9.1575	0.0010002	206.16
1.0	6.98	280.13	29.3	2485.0	2514.4	0.1060	8.8706	8.9767	0.001000	129.21
2.0	17.51	290.66	73.5	2460.2	2533.6	0.2606	8.4640	8.7246	0.001001	67.01
3.0	24.10	297.25	101.0	2444.6	2545.6	0.3543	8.2242	8.5785	0.001003	45.67
4.0	28.98	302.13	121.4	2433.1	2554.5	0.4225	8.0530	8.4755	0.001004	34.80
5.0	32.90	306.05	137.8	2423.8	2561.6	0.4763	7.9197	8.3960	0.001005	28.19
6.0	36.18	309.33	151.5	2416.0	2567.5	0.5209	7.8103	8.3312	0.001006	23.74
7.0	39.03	312.18	163.4	2409.2	2572.6	0.5591	7.7176	8.2767	0.001007	20.53
8.0	41.54	314.69	173.9	2403.2	2577.1	0.5926	7.6370	8.2295	0.001008	18.10
9.0	43.79	316.94	183.3	2397.9	2581.1	0.6224	7.5657	8.1881	0.001009	16.20
10.0	45.83	318.98	191.8	2392.9	2584.8	0.6493	7.5018	8.1511	0.001010	14.67
12.0	49.45	322.60	206.9	2384.2	2591.2	0.6964	7.3908	8.0872	0.001012	12.36
14.0	52.58	325.73	220.0	2376.7	2596.7	0.7367	7.2966	8.0333	0.001013	10.69
16.0	55.34	328.49	231.6	2370.0	2601.6	0.7721	7.2148	7.9868	0.001015	9.43
18.0	57.83	330.98	242.0	2363.9	2605.9	0.8036	7.1423	7.9459	0.001016	8.45
20.0	60.09	333.24	251.5	2358.4	2609.9	0.8321	7.0773	7.9094	0.001017	7.65
25.0	64.99	338.14	272.0	2346.4	2618.3	0.8933	6.9390	7.8323	0.001020	6.20
30.0	69.13	342.28	289.3	2336.1	2625.4	0.9441	6.8254	7.7695	0.001022	5.23
35.0	72.71	345.86	304.3	2327.2	2631.5	0.9878	6.7288	7.7166	0.001025	4.53
40.0	75.89	349.04	317.7	2319.2	2636.9	1.0261	6.6448	7.6709	0.001027	3.99
45.0	78.74	351.89	329.6	2312.0	2641.7	1.0603	6.5703	7.6306	0.001028	3.58
50.0	81.35	354.50	340.6	2305.4	2646.0	1.0912	6.5035	7.5947	0.001030	3.24
60.0	85.95	359.10	359.9	2293.6	2653.6	1.1455	6.3872	7.5327	0.001033	2.73
70.0	89.96	363.11	376.8	2283.3	2660.1	1.1921	6.2883	7.4804	0.001036	2.37
80.0	93.51	366.66	391.7	2274.0	2665.8	1.2330	6.2022	7.4352	0.001039	2.09
90.0	96.71	369.86	405.2	2265.6	2670.9	1.2696	6.1258	7.3954	0.001041	1.87
100.0	99.63	372.78	417.5	2257.9	2675.4	1.3027	6.0571	7.3598	0.001043	1.69
101.325	100.00	373.15	419.1	2256.9	2676.0	1.3069	6.0485	7.3554	0.0010437	1.6730
105	101.00	374.15	423.3	2254.3	2677.6	1.3182	6.0252	7.3434	0.001045	1.618
110	102.32	375.47	428.8	2250.8	2679.6	1.3330	5.9947	7.3277	0.001046	1.549
115	103.59	376.74	434.2	2247.4	2681.6	1.3472	5.9655	7.3127	0.001047	1.486
120	104.81	377.96	439.4	2244.1	2683.4	1.3609	5.9375	7.2984	0.001048	1.428
125	105.99	379.14	444.4	2240.9	2685.2	1.3741	5.9106	7.2846	0.001049	1.375
130	107.13	380.28	449.2	2237.8	2687.0	1.3868	5.8847	7.2715	0.001050	1.325
135	108.24	381.39	453.9	2234.8	2688.7	1.3991	5.8597	7.2588	0.001050	1.279
140	109.32	382.47	458.4	2231.9	2690.3	1.4109	5.8356	7.2465	0.001051	1.236
145	110.36	383.51	462.8	2229.0	2691.8	1.4225	5.8123	7.2347	0.001052	1.196
150	111.37	384.52	467.1	2226.2	2693.4	1.4336	5.7897	7.2234	0.001053	1.159
155	112.36	385.51	471.3	2223.5	2694.8	1.4445	5.7679	7.2123	0.001054	1.124
160	113.32	386.47	475.4	2220.9	2696.2	1.4550	5.7467	7.2017	0.001055	1.091
165	114.26	387.41	479.4	2218.3	2697.6	1.4652	5.7261	7.1913	0.001056	1.060
170	115.17	388.32	483.2	2215.7	2699.0	1.4752	5.7061	7.1813	0.001056	1.031
175	116.06	389.21	487.0	2213.3	2700.3	1.4849	5.6867	7.1716	0.001057	1.003
180	116.93	390.08	490.7	2210.8	2701.5	1.4944	5.6677	7.1622	0.001058	0.977
185	117.79	390.94	494.3	2208.5	2702.8	1.5036	5.6493	7.1530	0.001059	0.952
190	118.62	391.77	497.9	2206.1	2704.0	1.5127	5.6313	7.1440	0.001059	0.929
195	119.43	392.58	501.3	2203.8	2705.1	1.5215	5.6138	7.1353	0.001060	0.907
200	120.23	393.38	504.7	2201.6	2706.3	1.5301	5.5967	7.1268	0.001061	0.885
210	121.78	394.93	511.3	2197.2	2708.5	1.5468	5.5637	7.1105	0.001062	0.846
220	123.27	396.42	517.6	2193.0	2710.6	1.5628	5.5321	7.0949	0.001064	0.810
230	124.71	397.86	523.7	2188.9	2712.6	1.5781	5.5018	7.0800	0.001065	0.777
240	126.09	399.24	529.6	2184.9	2714.5	1.5929	5.4728	7.0657	0.001066	0.746
250	127.43	400.58	535.4	2181.0	2716.4	1.6072	5.4448	7.0520	0.001068	0.718
260	128.73	401.88	540.9	2177.3	2718.2	1.6209	5.4179	7.0389	0.001069	0.692
270	129.99	403.14	546.2	2173.6	2719.9	1.6342	5.3920	7.0262	0.001070	0.668
280	131.21	404.36	551.5	2170.1	2721.5	1.6471	5.3669	7.0140	0.001071	0.646
290	132.39	405.54	556.5	2166.6	2723.1	1.6596	5.3427	7.0022	0.001072	0.625
300	133.54	406.69	561.4	2163.2	2724.7	1.6717	5.3192	6.9909	0.001074	0.606
320	135.76	408.91	570.9	2156.7	2727.6	1.6948	5.2744	6.9692	0.001076	0.570
340	137.86	411.01	579.9	2150.4	2730.3	1.7168	5.2321	6.9489	0.001078	0.538
360	139.87	413.02	588.5	2144.4	2732.9	1.7376	5.1921	6.9297	0.001080	0.510
380	141.79	414.94	596.8	2138.6	2735.3	1.7575	5.1541	6.9115	0.001082	0.485
400	143.63	416.78	604.7	2132.9	2737.6	1.7764	5.1179	6.8943	0.001084	0.462

Table A11   Properties of saturated steam (SI units)—cont'd

Absolute Pressure (kN/m²)	Temperature (°C) θ_s	Temperature (K) T_s	Enthalpy per Unit Mass (H_s) (kJ/kg) Water	Latent	Steam	Entropy per Unit Mass (S_s) (kJ/kg K) Water	Latent	Steam	Specific Volume (v) (m³/kg) Water	Steam
420	145.39	418.54	612.3	2127.5	2739.8	1.7946	5.0833	6.8779	0.001086	0.442
440	147.09	420.24	619.6	2122.3	2741.9	1.8120	5.0503	6.8622	0.001088	0.423
460	148.73	421.88	626.7	2117.2	2743.9	1.8287	5.0186	6.8473	0.001089	0.405
480	150.31	423.46	633.5	2112.2	2745.7	1.8448	4.9881	6.8329	0.001091	0.389
500	151.85	425.00	640.1	2107.4	2747.5	1.8604	4.9588	6.8192	0.001093	0.375
520	153.33	426.48	646.5	2102.7	2749.3	1.8754	4.9305	6.8059	0.001095	0.361
540	154.77	427.92	652.8	2098.1	2750.9	1.8899	4.9033	6.7932	0.001096	0.348
560	156.16	429.31	658.8	2093.7	2752.5	1.9040	4.8769	6.7809	0.001098	0.337
580	157.52	430.67	664.7	2089.3	2754.0	1.9176	4.8514	6.7690	0.001100	0.326
600	158.84	431.99	670.4	2085.0	2755.5	1.9308	4.8267	6.7575	0.001101	0.316
620	160.12	433.27	676.0	2080.8	2756.9	1.9437	4.8027	6.7464	0.001102	0.306
640	161.38	434.53	681.5	2076.7	2758.2	1.9562	4.7794	6.7356	0.001104	0.297
660	162.60	435.75	686.8	2072.7	2759.5	1.9684	4.7568	6.7252	0.001105	0.288
680	163.79	436.94	692.0	2068.8	2760.8	1.9803	4.7348	6.7150	0.001107	0.280
700	164.96	438.11	697.1	2064.9	2762.0	1.9918	4.7134	6.7052	0.001108	0.272
720	166.10	439.25	702.0	2061.1	2763.2	2.0031	4.6925	6.6956	0.001109	0.266
740	167.21	440.36	706.9	2057.4	2764.3	2.0141	4.6721	6.6862	0.001110	0.258
760	168.30	441.45	711.7	2053.7	2765.4	2.0249	4.6522	6.6771	0.001112	0.252
780	169.37	442.52	716.3	2050.1	2766.4	2.0354	4.6328	6.6683	0.001114	0.246
800	170.41	443.56	720.9	2046.6	2767.5	2.0457	4.6139	6.6596	0.001115	0.240
820	171.44	444.59	725.4	2043.0	2768.5	2.0558	4.5953	6.6511	0.001116	0.235
840	172.45	445.60	729.9	2039.6	2769.4	2.0657	4.5772	6.6429	0.001118	0.229
860	173.43	446.58	734.2	2036.2	2770.4	2.0753	4.5595	6.6348	0.001119	0.224
880	174.40	447.55	738.5	2032.8	2771.3	2.0848	4.5421	6.6269	0.001120	0.220
900	175.36	448.51	742.6	2029.5	2772.1	2.0941	4.5251	6.6192	0.001121	0.215
920	176.29	449.44	746.8	2026.2	2773.0	2.1033	4.5084	6.6116	0.001123	0.210
940	177.21	450.36	750.8	2023.0	2773.8	2.1122	4.4920	6.6042	0.001124	0.206
960	178.12	451.27	754.8	2019.8	2774.6	2.1210	4.4759	6.5969	0.001125	0.202
980	179.01	452.16	758.7	2016.7	2775.4	2.1297	4.4602	6.5898	0.001126	0.198
1000	179.88	453.03	762.6	2013.6	2776.2	2.1382	4.4447	6.5828	0.001127	0.194
1100	184.06	457.21	781.1	1998.6	2779.7	2.1786	4.3712	6.5498	0.001133	0.177
1200	187.96	461.11	798.4	1984.3	2782.7	2.2160	4.3034	6.5194	0.001139	0.163
1300	191.60	464.75	814.7	1970.7	2785.4	2.2509	4.2404	6.4913	0.001144	0.151
1400	195.04	468.19	830.1	1957.7	2787.8	2.2836	4.1815	6.4651	0.001149	0.141
1500	198.28	471.43	844.6	1945.3	2789.9	2.3144	4.1262	6.4406	0.001154	0.132
1600	201.37	474.52	858.5	1933.2	2791.7	2.3436	4.0740	6.4176	0.001159	0.124
1700	204.30	477.45	871.8	1921.6	2793.4	2.3712	4.0246	6.3958	0.001163	0.117
1800	207.11	480.26	884.5	1910.3	2794.8	2.3976	3.9776	6.3751	0.001168	0.110
1900	209.79	482.94	896.8	1899.3	2796.1	2.4227	3.9327	6.3555	0.001172	0.105
2000	212.37	485.52	908.6	1888.7	2797.2	2.4468	3.8899	6.3367	0.001177	0.0996
2200	217.24	490.39	930.9	1868.1	2799.1	2.4921	3.8094	6.3015	0.001185	0.0907
2400	221.78	494.93	951.9	1848.5	2800.4	2.5342	3.7348	6.2690	0.001193	0.0832
2600	226.03	499.18	971.7	1829.7	2801.4	2.5736	3.6652	6.2388	0.001201	0.0769
3000	233.84	506.99	1008.3	1794.0	2802.3	2.6455	3.5383	6.1838	0.001216	0.0666
3500	242.54	515.69	1049.7	1752.2	2802.0	2.7252	3.3976	6.1229	0.001235	0.0570
4000	250.33	523.48	1087.4	1712.9	2800.3	2.7965	3.2720	6.0685	0.001252	0.0498
4500	257.41	530.56	1122.1	1675.6	2797.7	2.8612	3.1579	6.0191	0.001269	0.0440
5000	263.92	537.07	1154.5	1639.7	2794.2	2.9207	3.0528	5.9735	0.001286	0.0394
6000	275.56	548.71	1213.7	1571.3	2785.0	3.0274	2.8633	5.8907	0.001319	0.0324
7000	285.80	558.95	1267.5	1506.0	2773.4	3.1220	2.6541	5.8161	0.001351	0.0274
8000	294.98	568.13	1317.2	1442.7	2759.9	3.2077	2.5393	5.7470	0.001384	0.0235
9000	303.31	576.46	1363.8	1380.8	2744.6	3.2867	2.3952	5.6820	0.001418	0.0205
10,000	310.96	584.11	1408.1	1319.7	2727.7	3.3606	2.2592	5.6198	0.001453	0.0180
11,000	318.04	591.19	1450.6	1258.8	2709.3	3.4304	2.1292	5.5596	0.001489	0.0160
12,000	324.64	597.79	1491.7	1197.5	2698.2	3.4971	2.0032	5.5003	0.001527	0.0143
14,000	336.63	609.78	1571.5	1070.9	2642.4	3.6241	1.7564	5.3804	0.0016105	0.01150
16,000	347.32	620.47	1650.4	934.5	2584.9	3.7470	1.5063	5.2533	0.0017102	0.00931
18,000	356.96	630.11	1734.8	779.0	2513.9	3.8766	1.2362	5.1127	0.0018399	0.007497
20,000	365.71	638.86	1826.6	591.6	2418.2	4.0151	0.9259	4.9410	0.0020374	0.005875
22,000	373.68	646.83	2010.3	186.3	2196.6	4.2934	0.2881	4.5814	0.0026675	0.003735
22,120	374.15	647.30	2107.4	0	2107.4	4.4429	0	4.4429	0.0031700	0.003170

Adapted from the Abridged Callendar Steam Tables by permission of Messrs. Edward Arnold (Publishers) Ltd.

Table A12 Properties of saturated steam (Centigrade and Fahrenheit units)

Pressure		Temperature		Enthalpy per Unit Mass						Entropy (Btu/lb.°F)		Specific
				Centigrade Units (kcal/kg)			Fahrenheit Units (Btu/lb)					Volume (ft³/lb)
Absolute (lb/in.²)	Vacuum (in. Hg)	(°C)	(°F)	Water	Latent	Steam	Water	Latent	Steam	Water	Steam	Steam
0.5	28.99	26.42	79.6	26.45	582.50	608.95	47.6	1048.5	1096.1	0.0924	2.0367	643.0
0.6	28.79	29.57	85.3	29.58	580.76	610.34	53.2	1045.4	1098.6	0.1028	2.0214	540.6
0.7	28.58	32.28	90.1	32.28	579.27	611.55	58.1	1042.7	1100.8	0.1117	2.0082	466.6
0.8	28.38	34.67	94.4	34.66	577.95	612.61	62.4	1040.3	1102.7	0.1196	1.9970	411.7
0.9	28.17	36.80	98.2	36.80	576.74	613.54	66.2	1038.1	1104.3	0.1264	1.9871	368.7
1.0	27.97	38.74	101.7	38.74	575.60	614.34	69.7	1036.1	1105.8	0.1326	1.9783	334.0
1.1	27.76	40.52	104.9	40.52	574.57	615.09	72.9	1034.3	1107.2	0.1381	1.9702	305.2
1.2	27.56	42.17	107.9	42.17	573.63	615.80	75.9	1032.5	1108.4	0.1433	1.9630	281.1
1.3	27.35	43.70	110.7	43.70	572.75	616.45	78.7	1030.9	1109.6	0.1484	1.9563	260.5
1.4	27.15	45.14	113.3	45.12	571.94	617.06	81.3	1029.5	1110.8	0.1527	1.9501	243.0
1.5	26.95	46.49	115.7	46.45	571.16	617.61	83.7	1028.1	1111.8	0.1569	1.9442	228.0
1.6	26.74	47.77	118.0	47.73	570.41	618.14	86.0	1026.8	1112.8	0.1609	1.9387	214.3
1.7	26.54	48.98	120.2	48.94	569.71	618.65	88.2	1025.5	1113.7	0.1646	1.9336	202.5
1.8	26.33	50.13	122.2	50.08	569.06	619.14	90.2	1024.4	1114.6	0.1681	1.9288	191.8
1.9	26.13	51.22	124.2	51.16	568.47	619.63	92.1	1023.3	1115.4	0.1715	1.9243	182.3
2.0	25.92	52.27	126.1	52.22	567.89	620.11	94.0	1022.2	1116.2	0.1749	1.9200	173.7
3.0	23.88	60.83	141.5	60.78	562.89	623.67	109.4	1013.2	1122.6	0.2008	1.8869	118.7
4.0	21.84	67.23	153.0	67.20	559.29	626.49	121.0	1006.7	1127.7	0.2199	1.8632	90.63
5.0	19.80	72.38	162.3	72.36	556.24	628.60	130.2	1001.6	1131.8	0.2348	1.8449	73.52
6.0	17.76	76.72	170.1	76.71	553.62	630.33	138.1	996.6	1134.7	0.2473	1.8299	61.98
7.0	15.71	80.49	176.9	80.52	551.20	631.72	144.9	992.2	1137.1	0.2582	1.8176	53.64
8.0	13.67	83.84	182.9	83.89	549.16	633.05	151.0	988.5	1139.5	0.2676	1.8065	47.35
9.0	11.63	86.84	188.3	86.88	547.42	634.30	156.5	985.2	1141.7	0.2762	1.7968	42.40
10.0	9.59	89.58	193.2	89.61	545.82	635.43	161.3	982.5	1143.8	0.2836	1.7884	38.42
11.0	7.55	92.10	197.8	92.15	544.26	636.41	165.9	979.6	1145.5	0.2906	1.7807	35.14
12.0	5.50	94.44	202.0	94.50	542.75	637.25	170.1	976.9	1147.0	0.2970	1.7735	32.40
13.0	3.46	96.62	205.9	96.69	541.34	638.03	173.9	974.6	1148.5	0.3029	1.7672	30.05
14.0	1.42	98.65	209.6	98.73	540.06	638.79	177.7	972.2	1149.9	0.3086	1.7613	28.03
14.696	Gauge (lb/in.²)	100.00	212.0	100.06	539.22	639.28	180.1	970.6	1150.7	0.3122	1.7574	26.80
15	0.3	100.57	213.0	100.65	538.9	639.5	181.2	970.0	1151.2	0.3137	1.7556	26.28
16	1.3	102.40	216.3	102.51	537.7	640.2	184.5	967.9	1152.4	0.3187	1.7505	24.74
17	2.3	104.13	219.5	104.27	536.5	640.8	187.6	965.9	1153.5	0.3231	1.7456	23.38
18	3.3	105.78	222.4	105.94	535.5	641.4	190.6	964.0	1154.6	0.3276	1.7411	22.17
19	4.3	107.36	225.2	107.53	534.5	642.0	193.5	962.2	1155.7	0.3319	1.7368	21.07
20	5.3	108.87	228.0	109.05	533.6	642.6	196.3	960.4	1156.7	0.3358	1.7327	20.09
21	6.3	110.32	230.6	110.53	532.6	643.1	198.9	958.8	1157.7	0.3396	1.7287	19.19
22	7.3	111.71	233.1	111.94	531.7	643.6	201.4	957.2	1158.6	0.3433	1.7250	18.38
23	8.3	113.05	235.5	113.30	530.8	644.1	203.9	955.6	1159.5	0.3468	1.7215	17.63
24	9.3	114.34	237.8	114.61	530.0	644.6	206.3	954.0	1160.3	0.3502	1.7181	16.94
25	10.3	115.59	240.1	115.87	529.2	645.1	208.6	952.5	1161.1	0.3534	1.7148	16.30
26	11.3	116.80	242.2	117.11	528.4	645.5	210.8	951.1	1161.9	0.3565	1.7118	15.72
27	12.3	117.97	244.4	118.31	527.6	645.9	212.9	949.7	1162.6	0.3595	1.7089	15.17
28	13.3	119.11	246.4	119.47	526.8	646.3	215.0	948.3	1163.3	0.3625	1.7060	14.67
29	14.3	120.21	248.4	120.58	526.1	646.7	217.0	947.0	1164.0	0.3654	1.7032	14.19
30	15.3	121.3	250.3	121.7	525.4	647.1	219.0	945.6	1164.6	0.3682	1.7004	13.73
32	17.3	123.3	254.0	123.8	524.1	647.9	222.7	943.1	1165.8	0.3735	1.6952	12.93
34	19.3	125.3	257.6	125.8	522.8	648.6	226.3	940.7	1167.0	0.3785	1.6905	12.21
36	21.3	127.2	260.9	127.7	521.5	649.2	229.7	938.5	1168.2	0.3833	1.6860	11.58
38	23.3	128.9	264.1	129.5	520.3	649.8	233.0	936.4	1169.4	0.3879	1.6817	11.02
40	25.3	130.7	267.2	131.2	519.2	650.4	236.1	934.4	1170.5	0.3923	1.6776	10.50
42	27.3	132.3	270.3	132.9	518.0	650.9	239.1	932.3	1171.4	0.3964	1.6737	10.30
44	29.3	133.9	273.1	134.5	516.9	651.4	242.0	930.3	1172.3	0.4003	1.6700	9.600
46	31.3	135.4	275.8	136.0	515.9	651.9	244.9	928.3	1173.2	0.4041	1.6664	9.209
48	33.3	136.9	278.5	137.5	514.8	652.3	247.6	926.4	1174.0	0.4077	1.6630	8.848
50	35.3	138.3	281.0	139.0	513.8	652.8	250.2	924.6	1174.8	0.4112	1.6597	8.516
52	37.3	139.7	283.5	140.4	512.8	653.2	252.7	922.9	1175.6	0.4146	1.6566	8.208
54	39.3	141.0	285.9	141.8	511.8	653.6	255.2	921.1	1176.3	0.4179	1.6536	7.922
56	41.3	142.3	288.3	143.1	510.9	654.0	257.6	919.4	1177.0	0.4211	1.6507	7.656
58	43.3	143.6	290.5	144.4	510.0	654.4	259.9	917.8	1177.7	0.4242	1.6478	7.407
60	45.3	144.9	292.7	145.6	509.2	654.8	262.2	916.2	1178.4	0.4272	1.6450	7.175
62	47.3	146.1	294.9	146.8	508.4	655.2	264.4	914.6	1179.0	0.4302	1.6423	6.957

Table A12   Properties of saturated steam (Centigrade and Fahrenheit units)—cont'd

Pressure		Temperature		Enthalpy per Unit Mass						Entropy (Btu/lb.°F)		Specific Volume (ft³/lb) Steam
				Centigrade Units (kcal/kg)			Fahrenheit Units (Btu/lb)					
Absolute (lb/in.²)	Vacuum (in. Hg)	(°C)	(°F)	Water	Latent	Steam	Water	Latent	Steam	Water	Steam	
64	49.3	147.3	296.9	148.0	507.6	655.6	266.5	913.1	1179.6	0.4331	1.6398	6.752
66	51.3	148.4	299.0	149.2	506.7	655.9	268.6	911.6	1180.2	0.4359	1.6374	6.560
68	53.3	149.5	301.0	150.3	505.9	656.2	270.7	910.1	1180.8	0.4386	1.6350	6.378
70	55.3	150.6	302.9	151.5	505.0	656.5	272.7	908.7	1181.4	0.4412	1.6327	6.206
72	57.3	151.6	304.8	152.6	504.2	656.8	274.6	907.4	1182.0	0.4437	1.6304	6.044
74	59.3	152.6	306.7	153.6	503.4	657.0	276.5	906.0	1182.5	0.4462	1.6282	5.890
76	61.3	153.6	308.5	154.7	502.6	657.3	278.4	904.6	1183.0	0.4486	1.6261	5.743
78	63.3	154.6	310.3	155.7	501.8	657.5	280.3	903.2	1183.5	0.4510	1.6240	5.604
80	65.3	155.6	312.0	156.7	501.1	657.8	282.1	901.9	1184.0	0.4533	1.6219	5.472
82	67.3	156.5	313.7	157.7	500.3	658.0	283.9	900.6	1184.5	0.4556	1.6199	5.346
84	69.3	157.5	315.4	158.6	499.6	658.2	285.6	899.4	1185.0	0.4579	1.6180	5.226
86	71.3	158.4	317.1	159.6	498.9	658.5	287.3	898.1	1185.4	0.4601	1.6161	5.110
88	73.3	159.4	318.7	160.5	498.3	658.8	289.0	896.8	1185.8	0.4622	1.6142	5.000
90	75.3	160.3	320.3	161.5	497.6	659.1	290.7	895.5	1186.2	0.4643	1.6124	4.896
92	77.3	161.2	321.9	162.4	496.9	659.3	292.3	894.3	1186.6	0.4664	1.6106	4.796
94	79.3	162.0	323.3	163.3	496.3	659.6	293.9	893.1	1187.0	0.4684	1.6088	4.699
96	81.3	162.8	324.8	164.1	495.7	659.8	295.5	891.9	1187.4	0.4704	1.6071	4.607
98	83.3	163.6	326.6	165.0	495.0	660.0	297.0	890.8	1187.8	0.4723	1.6054	4.519
100	85.3	164.4	327.8	165.8	494.3	660.1	298.5	889.7	1188.2	0.4742	1.6038	4.434
105	90.3	166.4	331.3	167.9	492.7	660.6	302.2	886.9	1189.1	0.4789	1.6000	4.230
110	95.3	168.2	334.8	169.8	491.2	661.0	305.7	884.2	1189.9	0.4833	1.5963	4.046
115	100.3	170.0	338.1	171.7	489.8	661.5	309.2	881.5	1190.7	0.4876	1.5927	3.880
120	105.3	171.8	341.3	173.6	488.3	661.9	312.5	878.9	1191.4	0.4918	1.5891	3.729
125	110.3	173.5	344.4	175.4	486.9	662.3	315.7	876.4	1192.1	0.4958	1.5856	3.587
130	115.3	175.2	347.3	177.1	485.6	662.7	318.8	874.0	1192.8	0.4997	1.5823	3.456
135	120.3	176.8	350.2	178.8	484.2	663.0	321.9	871.5	1193.4	0.5035	1.5792	3.335
140	125.3	178.3	353.0	180.5	482.9	663.4	324.9	869.1	1194.0	0.5071	1.5763	3.222
145	130.3	179.8	355.8	182.1	481.6	663.7	327.8	866.8	1194.6	0.5106	1.5733	3.116
150	135.3	181.3	358.4	183.7	480.3	664.0	330.6	864.5	1195.1	0.5140	1.5705	3.015

Adapted from the Abridged Callendar Steam Tables by permission of Messrs. Edward Arnold (Publishers) Ltd.

Table A13 Enthalpy of superheated steam, H (kJ/kg)

Pressure P (kN/m²)	Saturation		Temperature, θ (°C)	100	200	300	400	500	600	700	800
	T_s (K)	S_s (kJ/kg)	Temperature, T (K)	373.15	473.15	573.15	673.15	773.15	873.15	973.15	1073.15
100	372.78	2675.4		2676.0	2875.4	3074.6	3278.0	3488.0	3705.0	3928.0	4159.0
200	393.38	2706.3			2870.4	3072.0	3276.4	3487.0	3704.0	3927.0	4158.0
300	406.69	2724.7			2866.0	3069.7*	3275.0	3486.0	3703.1	3927.0	4158.0
400	416.78	2737.6			2861.3	3067.0	3273.5	3485.0	3702.0	3926.0	4157.0
500	425.00	2747.5			2856.0	3064.8	3272.1	3484.0	3701.2	3926.0	4156.8
600	431.99	2755.5			2850.7	3062.0	3270.0	3483.0	3701.0	3925.0	4156.2
700	438.11	2762.0			2845.5	3059.5	3269.0	3482.6	3700.2	3924.0	4156.0
800	443.56	2767.5			2839.7	3057.0	3266.8	3480.4	3699.0	3923.8	4155.0
900	448.56	2772.1			2834.0	3055.0	3266.2	3479.5	3698.6	3923.0	4155.0
1000	453.03	2776.2			2828.7	3051.7	3264.3	3478.0	3697.5	3922.8	4154.0
2000	485.59	2797.2				3024.8	3248.0	3467.0	3690.0	3916.0	4150.0
3000	506.98	2802.3				2994.8	3231.7	3456.0	3681.6	3910.4	4145.0
4000	523.49	2800.3				2962.0	3214.8	3445.0	3673.4	3904.0	4139.6
5000	537.09	2794.2				2926.0	3196.9	3433.8	3665.4	3898.0	4135.5
6000	548.71	2785.0				2886.0	3178.0	3421.7	3657.0	3891.8	4130.0
7000	558.95	2773.4				2840.0	3159.1	3410.0	3648.8	3886.0	4124.8
8000	568.13	2759.9				2785.0	3139.5	3398.0	3640.4	3880.8	4121.0
9000	576.46	2744.6					3119.0	3385.5	3632.0	3873.6	4116.0
10,000	584.11	2727.7					3097.7	3373.6	3624.0	3867.2	4110.8
11,000	591.19	2709.3					3075.6	3361.0	3615.5	3862.0	4106.0
12,000	597.79	2698.2					3052.9	3349.0	3607.0	3855.3	4101.2

Adapted from the Abridged Callendar Steam Tables by permission of Messrs. Edward Arnold (Publishers) Ltd.

Table A14 Entropy of superheated steam, S (kJ/kg K)

Pressure P (kN/m²)	Saturation		Temperature, θ (°C) / Temperature, T (K)							
	T_s (K)	H_s (kJ/kg)	100 / 373.15	200 / 473.15	300 / 573.15	400 / 673.15	500 / 773.15	600 / 873.15	700 / 973.15	800 / 1073.15
100	372.78	7.3598	7.362	7.834	8.216	8.544	8.834	9.100	9.344	9.565
200	393.38	7.1268		7.507	7.892	8.222	8.513	8.778	9.020	9.246
300	406.69	6.9909		7.312	7.702	8.033	8.325	8.591	8.833	9.057
400	416.78	6.8943		7.172	7.566	7.898	8.191	8.455	8.700	8.925
500	425.00	6.8192		7.060	7.460	7.794	8.087	8.352	8.596	8.820
600	431.99	6.7575		6.968	7.373	7.708	8.002	8.268	8.510	8.738
700	438.11	6.7052		6.888	7.298	7.635	7.930	8.195	8.438	8.665
800	443.56	6.6596		6.817	7.234	7.572	7.867	8.133	8.375	8.602
900	448.56	6.6192		6.753	7.176	7.515	7.812	8.077	8.321	8.550
1000	453.03	6.5828		6.695	7.124	7.465	7.762	8.028	8.272	8.502
2000	485.59	6.3367			6.768	7.128	7.431	7.702	7.950	8.176
3000	506.98	6.1838			6.541	6.922	7.233	7.508	7.756	7.985
4000	523.49	6.0685			6.364	6.770	7.090	7.368	7.620	7.850
5000	537.07	5.9735			6.211	6.647	6.977	7.258	7.510	7.744
6000	548.71	5.8907			6.060	6.542	6.880	7.166	7.422	7.655
7000	558.95	5.8161			5.933	6.450	6.798	7.088	7.345	7.581
8000	568.13	5.7470			5.792	6.365	6.724	7.020	7.280	7.515
9000	576.46	5.6820				6.288	6.659	6.958	7.220	7.457
10,000	584.11	5.6198				6.215	6.598	6.902	7.166	7.405
11,000	591.19	5.5596				6.145	6.540	6.850	7.117	7.357
12,000	597.79	5.5003				6.077	6.488	6.802	7.072	7.312

Adapted from the Abridged Callendar Steam Tables by permission of Messrs. Edward Arnold (Publishers) Ltd.

A.3 Mathematical Tables

Table A15 Laplace transforms[a]

No.	Transform $\overline{f}(p) = \int_0^\infty e^{-pt} f(t)\,dt$	Function $f(t)$
1	$\frac{1}{p}$	1
2	$\frac{1}{p^2}$	t
3	$\frac{1}{p^n}$ $n = 1, 2, 3, \ldots$	$\frac{t^{n-1}}{(n-1)!}$
4	$\frac{1}{\sqrt{p}}$	$\frac{1}{\sqrt{(\pi t)}}$
5	$\frac{1}{p^{3/2}}$	$2\sqrt{\frac{t}{\pi}}$
6	$\frac{1}{p^{n+\frac{1}{2}}}$ $n = 1, 2, 3, \ldots$	$\frac{2^n t^{n-\frac{1}{2}}}{[1.3.5\ldots(2n-1)]\sqrt{\pi}}$
7	$\frac{\Gamma(k)}{p^k}$ $k > 0$	t^{k-1}
8	$\frac{1}{p-a}$	e^{at}
9	$\frac{1}{(p-a)^2}$	te^{at}
10	$\frac{1}{(p-a)^n}$ $n = 1, 2, 3, \ldots$	$\frac{1}{(n-1)!} t^{n-1} e^{at}$
11	$\frac{\Gamma(k)}{(p-a)^k}$ $k > 0$	$t^{k-1} e^{at}$
12	$\frac{1}{(p-a)(p-b)}$ $a \neq b$	$\frac{1}{a-b}\left(e^{at} - e^{bt}\right)$
13	$\frac{p}{(p-a)(p-b)}$ $a \neq b$	$\frac{1}{a-b}\left(ae^{at} - be^{bt}\right)$
14	$\frac{1}{(p-a)(p-b)(p-c)}$ $a \neq b \neq c$	$-\frac{(b-c)e^{at} + (c-a)e^{at} + (a-b)e^{ct}}{(a-b)(b-c)(c-a)}$
15	$\frac{1}{p^2+a^2}$	$\frac{1}{a}\sin at$
16	$\frac{p}{p^2+a^2}$	$\cos at$
17	$\frac{1}{p^2-a^2}$	$\frac{1}{a}\sinh at$
18	$\frac{p}{p^2-a^2}$	$\cosh at$
19	$\frac{1}{p(p^2+a^2)}$	$\frac{1}{a^2}(1 - \cos at)$
20	$\frac{1}{p^2(p^2+a^2)}$	$\frac{1}{a^3}(at - \sin at)$
21	$\frac{1}{(p^2+a^2)^2}$	$\frac{1}{2a^3}(\sin at - at\cos at)$
22	$\frac{p}{(p^2+a^2)^2}$	$\frac{t}{2a}\sin at$
23	$\frac{p^2}{(p^2+a^2)^2}$	$\frac{1}{2a}(\sin at + at\cos at)$
24	$\frac{p^2-a^2}{(p^2+a^2)^2}$	$t\cos at$
25	$\frac{p}{(p^2+a^2)(p^2+b^2)}$ $a^2 \neq b^2$	$\frac{\cos at - \cos bt}{b^2-a^2}$
26	$\frac{1}{(p-a)^2+b^2}$	$\frac{1}{b} e^{at}\sin bt$
27	$\frac{p-a}{(p-a)^2+b^2}$	$e^{at}\cos bt$
28	$\frac{3a^2}{p^3+a^3}$	$e^{-at} - e^{at/2}\left(\cos\frac{at\sqrt{3}}{2} - \sqrt{3}\sin\frac{at\sqrt{3}}{2}\right)$
29	$\frac{4a^3}{p^4+4a^4}$	$\sin at\cosh at - \cos at\sinh at$
30	$\frac{p}{p^4+4a^4}$	$\frac{1}{2a^2}\sin at\sinh at$
31	$\frac{1}{p^4-a^4}$	$\frac{1}{2a^3}(\sinh at - \sin at)$
32	$\frac{p}{p^4-a^4}$	$\frac{1}{2a^2}(\cosh at - \cos at)$

Table A15   Laplace transforms—cont'd

No.	Transform $\bar{f}(p) = \int_0^\infty e^{-pt} f(t)\,dt$	Function $f(t)$
33	$\dfrac{8a^3 p^2}{(p^2+a^2)^3}$	$(1+a^2 t^2)\sin at - at\cos at$
34	$\dfrac{1}{p}\left(\dfrac{p-1}{p}\right)^n$	$\dfrac{e^t}{n!}\dfrac{d^n}{dt^n}(t^n e - t) = $ Laguerre polynomial of degree n
35	$\dfrac{p}{(p-a)^{3/2}}$	$\dfrac{1}{\sqrt{(\pi t)}} - e^{at}(1+2at)$
36	$\sqrt{(p-a)} - \sqrt{(p-b)}$	$\dfrac{1}{2\sqrt{(\pi t^3)}}\left(e^{bt} - e^{at}\right)$
37	$\dfrac{i}{\sqrt{p}+a}$	$\dfrac{i}{\sqrt{(\pi t)}} - ae^{a^2 t}\,\mathrm{erfc}\left(a\sqrt{t}\right)$
38	$\dfrac{\sqrt{p}}{p-a^2}$	$\dfrac{1}{\sqrt{(\pi t)}} + ae^{a^2 t}\,\mathrm{erf}\left(a\sqrt{t}\right)$
39	$\dfrac{\sqrt{p}}{p+a^2}$	$\dfrac{1}{\sqrt{(\pi t)}} - \dfrac{2ae^{-a^2 t}}{\sqrt{\pi}}\int_0^{a\sqrt{t}} e^{\lambda^2}\,d\lambda$
40	$\dfrac{1}{\sqrt{p}(p-a^2)}$	$\dfrac{1}{a}e^{a^2 t}\,\mathrm{erf}\left(a\sqrt{t}\right)$
41	$\dfrac{1}{\sqrt{p}(p+a^2)}$	$\dfrac{2e^{-a^2 t}}{a\sqrt{\pi}}\int_0^{a\sqrt{t}} e^{\lambda^2}\,d\lambda$
42	$\dfrac{b^2-a^2}{(p-a^2)(b+\sqrt{p})}$	$e^{a^2 t}\left[b - a\,\mathrm{erf}\left(a\sqrt{t}\right)\right] - be^{b^2 t}\,\mathrm{erfc}\left(b\sqrt{t}\right)$
43	$\dfrac{1}{\sqrt{p}(\sqrt{p}+a)}$	$e^{a^2 t}\,\mathrm{erfc}\left(a\sqrt{t}\right)$
44	$\dfrac{1}{(p+a)\sqrt{(p+b)}}$	$\dfrac{1}{\sqrt{(b-a)}}e^{-at}\,\mathrm{erf}\left[\sqrt{(b-a)}\sqrt{t}\right]$
45	$\dfrac{b^2-a^2}{\sqrt{p}(p-a^2)(\sqrt{p}+b)}$	$e^{a^2 t}\left[\dfrac{b}{a}\,\mathrm{erf}\left(a\sqrt{t}\right) - 1\right] + e^{b^2 t}\,\mathrm{erfc}\left(b\sqrt{t}\right)$
46	$\dfrac{(1-p)^n}{p^{n+1/2}}$	$\dfrac{n!}{(2n)!\sqrt{(\pi t)}}H_{2n}\left(\sqrt{t}\right)$ where $H_n(t) = e^{t^2}\dfrac{d^n}{dt^n}e^{-t^2}$ is the Hermite polynomial
47	$\dfrac{(1-p)^n}{p^{n+3/2}}$	$-\dfrac{n!}{\sqrt{\pi(2n+1)!}}H_{2n+1}\left(\sqrt{t}\right)$
48	$\dfrac{\sqrt{(p+2a)}}{\sqrt{p}} - 1$	$ae^{-at}[I_1(at) + I_0(at)]$
49	$\dfrac{1}{\sqrt{(p+a)}\sqrt{(p+b)}}$	$e^{-\frac{1}{2}(a+b)t}I_0\left(\dfrac{a-b}{2}t\right)$
50	$\dfrac{\Gamma(k)}{(p+a)^k(p+b)^k}\,k>0$	$\sqrt{\pi}\left(\dfrac{t}{a-b}\right)^{k-\frac{1}{2}}e^{-\frac{1}{2}(a+b)t}I_{k-\frac{1}{2}}\left(\dfrac{a-b}{2}t\right)$
51	$\dfrac{1}{\sqrt{(p+a)}(p+b)^{3/2}}$	$te^{-\frac{1}{2}(a+b)t}\left[I_0\left(\dfrac{a-b}{2}t\right) + I_1\left(\dfrac{a-b}{2}t\right)\right]$
52	$\dfrac{\sqrt{(p+2a)} - \sqrt{p}}{\sqrt{(p+2a)} + \sqrt{p}}$	$\dfrac{1}{t}e^{-at}I_1(at)$
53	$\dfrac{(a-b)^k}{\left[\sqrt{(p+a)} + \sqrt{(p+b)}\right]^{2k}}\,k>0$	$\dfrac{k}{t}e^{-\frac{1}{2}(a+b)t}I_k\left(\dfrac{a-b}{2}t\right)$
54	$\dfrac{\left[\sqrt{(p+a)} + \sqrt{p}\right]^{-2j}}{\sqrt{p}\sqrt{(p+a)}}\,j>-1$	$\dfrac{1}{a^j}e^{-\frac{1}{2}at}I_j\left(\dfrac{1}{2}at\right)$
55	$\dfrac{1}{\sqrt{(p^2+a^2)}}$	$J_0(at)$
56	$\dfrac{\left[\sqrt{(p^2+a^2)} - p\right]^j}{\sqrt{(p^2+a^2)}}\,j>1$	$a^j J_j(at)$
57	$\dfrac{1}{(p^2+a^2)^k}\,k>0$	$\dfrac{\sqrt{\pi}}{\Gamma(k)}\left(\dfrac{t}{2a}\right)^{k-\frac{1}{2}}J_{k-\frac{1}{2}}(at)$
58	$\left[\sqrt{(p^2+a^2)} - p\right]^k\,k>0$	$\dfrac{ka^k}{t}J_k(at)$

Continued

Table A15 Laplace transforms—cont'd

No.	Transform $\overline{f}(p) = \int_0^\infty e^{-pt} f(t)\,dt$	Function $f(t)$
59	$\left[\dfrac{p-\sqrt{(p^2-a^2)}\,]^j}{\sqrt{(p^2-a^2)}}\right]\quad j>-1$	$d^j I_j(at)$
60	$\dfrac{1}{(p^2-a^2)^k}\quad k>0$	$\dfrac{\sqrt{\pi}}{\Gamma(k)}\left(\dfrac{t}{2a}\right)^{k-\frac{1}{2}} I_{k-\frac{1}{2}}(at)$
61	$\dfrac{e^{-kp}}{p}$	$S_k(t) = \begin{cases} 0 & \text{when } 0<t<k \\ 1 & \text{when } t>k \end{cases}$
62	$\dfrac{e^{-kp}}{p^2}$	$\begin{cases} 0 & \text{when } 0<t<k \\ t-k & \text{when } t>k \end{cases}$
63	$\dfrac{e^{-kp}}{p^j}\quad j>0$	$\begin{cases} 0 & \text{when } 0<t<k \\ \dfrac{(t-k)^{j-1}}{\Gamma(j)} & \text{when } t>k \end{cases}$
64	$\dfrac{1-e^{-kp}}{p}$	$\begin{cases} 1 & \text{when } 0<t<k \\ 0 & \text{when } t>k \end{cases}$
65	$\dfrac{1}{p(1-e^{-kp})} = \dfrac{1+\coth\frac{1}{2}kp}{2p}$	$S(k,t)=n \text{ when } (n-1)k<t<nk \ \ n=1,2,3,\dots$
66	$\dfrac{1}{p(e^{kp}-a)}$	$\begin{cases} 0 & \text{when } 0<t<k \\ 1+a+a^2+\cdots+a^{n-1} & \\ \quad\text{when } nk<t<(n+1)k \ \ n=1,2,3,\dots \end{cases}$
67	$\dfrac{1}{p}\tanh kp$	$M(2k,t)=(-1)^{n-1}$ $\text{when } 2k(n-1)<t<2kn \ n=1,2,3,\dots$
68	$\dfrac{1}{p(1+e^{-kp})}$	$\dfrac{1}{2}M(k,t)+\dfrac{1}{2}=\dfrac{1-(-1)n}{2}$ $\text{when } (n-1)k<t<nk$
69	$\dfrac{1}{p^2}\tanh kp$	$H(2k,t)=\begin{cases} t & \text{when } 0<t<2k \\ 4k-t & \text{when } 2k<t<4k \end{cases}$
70	$\dfrac{1}{p\sinh kp}$	$2S(2k,t+k)-2=2(n-1)$ $\text{when } (2n-3)k<t<(2n-1)k \ \ t>0$
71	$\dfrac{1}{p\cosh kp}$	$M(2k,t+3k)+1=1+(-1)^n$ $\text{when } (2n-3)k<t<(2n-1)k \ \ t>0$
72	$\dfrac{1}{p}\coth kp$	$2S(2k,t)-1=2n-1$ $\text{when } 2k(n-1)<t<2kn$
73	$\dfrac{k}{p^2+k^2}\coth\dfrac{\pi p}{2k}$	$\lvert\sin kt\rvert$
74	$\dfrac{1}{(p^2+1)(1-e^{-\pi p})}$	$\begin{cases} \sin t & \text{when } (2n-2)\pi<t<(2n-1)\pi \\ 0 & \text{when } (2n-1)\pi<t<2n\pi \end{cases}$
75	$\dfrac{1}{p}e^{-k/p}$	$J_0\left[2\sqrt{(kt)}\right]$
76	$\dfrac{1}{\sqrt{p}}e^{-k/p}$	$\dfrac{1}{\sqrt{(\pi t)}}\cos 2\sqrt{(kt)}$
77	$\dfrac{1}{\sqrt{p}}e^{k/p}$	$\dfrac{1}{\sqrt{(\pi t)}}\cosh 2\sqrt{(kt)}$
78	$\dfrac{1}{p^{3/2}}e^{-k/p}$	$\dfrac{1}{\sqrt{(\pi k)}}\sin 2\sqrt{(kt)}$
79	$\dfrac{1}{p^{3/2}}e^{k/p}$	$\dfrac{1}{\sqrt{(\pi k)}}\sinh 2\sqrt{(kt)}$
80	$\dfrac{1}{p^j}e^{k/p}\quad j>0$	$\left(\dfrac{t}{k}\right)^{(j-1)/2} J_{j-1}\left[2\sqrt{(kt)}\right]$
81	$\dfrac{1}{p^j}e^{k/p}\quad j>0$	$\left(\dfrac{t}{k}\right)^{(j-1)/2} I_{j-1}\left[2\sqrt{(kt)}\right]$

Table A15   Laplace transforms—cont'd

No.	Transform $\overline{f}(p) = \int_0^\infty e^{-pt} f(t) \mathrm{d}t$	Function $f(t)$
82	$e^{-k\sqrt{p}} \quad k > 0$	$\dfrac{k}{2\sqrt{(\pi t^3)}} \exp\left(-\dfrac{k^2}{4t}\right)$
83	$\dfrac{1}{p} e^{-k\sqrt{p}} \quad k \geq 0$	$\mathrm{erfc}\left(\dfrac{k}{2\sqrt{t}}\right)$
84	$\dfrac{1}{\sqrt{p}} e^{-k\sqrt{p}} \quad k \geq 0$	$\dfrac{1}{\sqrt{(\pi t)}} \exp\left(-\dfrac{k^2}{4t}\right)$
85	$p^{-3/2} e^{-k\sqrt{p}} \quad k \geq 0$	$2\sqrt{\dfrac{t}{\pi}} \left[\exp\left(-\dfrac{k^2}{4t}\right)\right] - k\,\mathrm{erfc}\left(\dfrac{k}{2\sqrt{t}}\right)$
86	$\dfrac{a e^{-k\sqrt{p}}}{p(a + \sqrt{p})} \quad k \geq 0$	$-\exp(ak)\exp(a^2 t)\,\mathrm{erfc}\left(a\sqrt{t} + \dfrac{k}{2\sqrt{t}}\right) + \mathrm{erfc}\left(\dfrac{k}{2\sqrt{t}}\right)$
87	$\dfrac{e^{-k\sqrt{p}}}{\sqrt{p}(a + \sqrt{p})} \quad k \geq 0$	$\exp(ak)\exp(a^2 t)\,\mathrm{erfc}\left(a\sqrt{t} + \dfrac{k}{2\sqrt{t}}\right)$
88	$\dfrac{e^{-k\sqrt{[p(p+a)]}}}{\sqrt{[p(p+a)]}}$	$\begin{cases} 0 & \text{when } 0 < t < k \\ \exp\left(-\dfrac{1}{2}at\right) I_0\left[\dfrac{1}{2}a\sqrt{(t^2 - k^2)}\right] & \text{when } t > k \end{cases}$
89	$\dfrac{e^{-k\sqrt{(p^2 + a^2)}}}{\sqrt{(p^2 + a^2)}}$	$\begin{cases} 0 & \text{when } 0 < t < k \\ J_0\left[a\sqrt{(t^2 - k^2)}\right] & \text{when } t > k \end{cases}$
90	$\dfrac{e^{-k\sqrt{(p^2 - a^2)}}}{\sqrt{(p^2 - a^2)}}$	$\begin{cases} 0 & \text{when } 0 < t < k \\ I_0\left[a\sqrt{(t^2 + k^2)}\right] & \text{when } t > k \end{cases}$
91	$\dfrac{e^{-k[\sqrt{(p^2 + a^2)} - p]}}{\sqrt{(p^2 + a^2)}} \quad k \geq 0$	$J_0\left[a\sqrt{(t^2 + 2kt)}\right]$
92	$e^{-kp} - e^{-k\sqrt{(p^2 + a^2)}}$	$\begin{cases} 0 & \text{when } 0 < t < k \\ \dfrac{ak}{\sqrt{(t^2 - k^2)}} J_1\left[a\sqrt{(t^2 - k^2)}\right] \\ \quad \text{when } t > k \end{cases}$
93	$e^{-k\sqrt{(p^2 - a^2)}} - e^{-kp}$	$\begin{cases} 0 & \text{when } 0 < t < k \\ \dfrac{ak}{\sqrt{(t^2 - k^2)}} J_1\left[a\sqrt{(t^2 - k^2)}\right] \\ \quad \text{when } t > k \end{cases}$
94	$\dfrac{a^j e^{-k\sqrt{(p^2 + a^2)}}}{\sqrt{(p^2 + a^2)}\left[\sqrt{(p^2 + a^2)} + p\right]^j} \quad j > -1$	$\begin{cases} 0 & \text{when } 0 < t < k \\ \left(\dfrac{t - k}{t + k}\right)^{(1/2)j} J_j\left[a\sqrt{(t^2 - k^2)}\right] \\ \quad \text{when } t > k \end{cases}$
95	$\dfrac{1}{p}\ln p$	$\lambda - \ln t \quad \lambda = -0.5772\ldots$
96	$\dfrac{1}{p^k}\ln p \quad k > 0$	$t^{k-1}\left\{\dfrac{\lambda}{[\Gamma(k)]^2} - \dfrac{\ln t}{\Gamma(k)}\right\}$
97[b]	$\dfrac{\ln p}{p - a} \quad a > 0$	$(\exp at)[\ln a - Ei(-at)]$
98[c]	$\dfrac{1}{p}\ln(1 + kp) \quad k > 0$	$\cos t Si(t) - \sin t Ci(t)$
99[c]	$\dfrac{p \ln p}{p^2 + 1}$	$-\sin t Si(0 - \cos t Ci(t)$
100[b]	$\dfrac{1}{p}\ln(1 + kp) \quad k > 0$	$-Ei\left(-\dfrac{t}{k}\right)$
101	$\ln\dfrac{p - a}{p - b}$	$\dfrac{1}{t}\left(e^{bt} - e^{at}\right)$
102[c]	$\dfrac{1}{p}\ln(1 + k^2 p^2)$	$-2Ci\left(\dfrac{t}{k}\right)$
103[c]	$\dfrac{1}{p}\ln(p^2 + a^2) \quad a > 0$	$2\ln a - 2Ci(at)$
104[c]	$\dfrac{1}{p^2}\ln(p^2 + a^2) \quad a > 0$	$\dfrac{2}{a}[at \ln a + \sin at - at Ci(at)]$

Continued

Table A15   Laplace transforms—cont'd

No.	Transform $\bar{f}(p) = \int_0^\infty e^{-pt} f(t) dt$	Function $f(t)$
105	$\ln \frac{p^2 + a^2}{p^2}$	$\frac{2}{t}(1 - \cos at)$
106	$\ln \frac{p^2 - a^2}{p^2}$	$\frac{2}{t}(1 - \cosh at)$
107	$\tan^{-1} \frac{k}{p}$	$\frac{1}{t} \sin kt$
108[c]	$\frac{1}{p} \tan^{-1} \frac{k}{p}$	$\mathrm{Si}(kt)$
109	$\exp(k^2 p^2)\, \mathrm{erfc}\,(kp) \;\; k > 0$	$\frac{1}{k\sqrt{\pi}} \exp\left(-\frac{t^2}{4k^2}\right)$
110	$\frac{1}{p} \exp(k^2 p^2)\, \mathrm{erfc}\,(kp) \;\; k > 0$	$\mathrm{erf}\left(\frac{t}{2k}\right)$
111	$\exp(kp)\, \mathrm{erfc}\,\left[\sqrt{(kp)}\right] \;\; k > 0$	$\frac{\sqrt{k}}{\pi \sqrt{t(t+k)}}$
112	$\frac{1}{\sqrt{p}}\, \mathrm{erfc}\,\left[\sqrt{(kp)}\right]$	$\begin{cases} 0 & \text{when } 0 < t < k \\ (\pi t)^{-\frac{1}{2}} & \text{when } t > k \end{cases}$
113	$\frac{1}{\sqrt{p}} \exp(kp)\, \mathrm{erfc}\,\left[\sqrt{(kp)}\right] \; k > 0$	$\frac{1}{\sqrt{[\pi(t+k)]}}$
114	$\mathrm{erf}\left(\frac{k}{\sqrt{p}}\right)$	$\frac{1}{\pi t} \sin\left(2k\sqrt{t}\right)$
115	$\frac{1}{\sqrt{p}} \exp\left(\frac{k^2}{p}\right) \mathrm{erfc}\left(\frac{k}{\sqrt{p}}\right)$	$\frac{1}{\sqrt{(\pi t)}} \exp\left(-2k\sqrt{t}\right)$
116[d]	$K_0(kp)$	$\begin{cases} 0 & \text{when } 0 < t < k \\ (t^2 - k^2)^{-\frac{1}{2}} & \text{when } t > k \end{cases}$
117[d]	$K_0\left(k\sqrt{p}\right)$	$\frac{1}{2t} \exp\left(-\frac{k^2}{4t}\right)$
118[d]	$\frac{1}{p} \exp(kp) K_1(kp)$	$\frac{1}{k} \sqrt{[t(t+2k)]}$
119[d]	$\frac{1}{\sqrt{p}} K_1\left(k\sqrt{p}\right)$	$\frac{1}{k} \exp\left(-\frac{k^2}{4t}\right)$
120[d]	$\frac{1}{\sqrt{p}} \exp\left(\frac{k}{p}\right) K_0\left(\frac{k}{p}\right)$	$\frac{2}{\sqrt{(\pi t)}} K_0\left[2\sqrt{(2kt)}\right]$
121[e]	$\pi \exp(-kp) I_0(kp)$	$\begin{cases} [t(2k-t)]^{-\frac{1}{2}} & \text{when } 0 < t < 2k \\ 0 & \text{when } t > 2k \end{cases}$
122[e]	$\exp(-kp) I_1(kp)$	$\begin{cases} \dfrac{k-t}{\pi k \sqrt{[t(2k-t)]}} & \text{when } 0 < t < 2k \\ 0 & \text{when } t > 2k \end{cases}$
123	Unity	Unit impulse

These functions are tabulated in M. Abramowitz and I. A. Stegun, *Handbook of Mathematical Functions*, Dover Publications, New York, 1965.

[a] By permission from *Operational Mathematics* by R. V. Churchill, McGraw-Hill 1958.
[b] $Ei(-t) = -\int_t^\infty \frac{e^{-x}}{x} dx \, (\text{for } t > 0) = $ exponential integral function.
[c] $Si(t) = \int_0^t \frac{\sin x}{x} dx = $ sine integral function and $Ci(t) = -\int_t^\infty \frac{\cos x}{x} dx = $ cosine integral function.
[d] $K_n(x)$ denotes the Bessel function of the second kind for the imaginary argument.
[e] $I_n(x)$ denotes the Bessel function of the first kind for the imaginary argument.

Table A16 Error function and its derivative

x	$\frac{2}{\sqrt{\pi}}e^{-x^2}$	Erfx	x	$\frac{2}{\sqrt{\pi}}e^{-x^2}$	Erfx
0.00	1.12837	0.00000	0.50	0.87878	0.52049
0.01	1.12826	0.01128	0.51	0.86995	0.52924
0.02	1.12792	0.02256	0.52	0.86103	0.53789
0.03	1.12736	0.03384	0.53	0.85204	0.54646
0.04	1.12657	0.04511	0.54	0.84297	0.55493
0.05	1.12556	0.05637	0.55	0.83383	0.56332
0.06	1.12432	0.06762	0.56	0.82463	0.57161
0.07	1.12286	0.07885	0.57	0.81536	0.57981
0.08	1.12118	0.09007	0.58	0.80604	0.58792
0.09	1.11927	0.10128	0.59	0.79666	0.59593
0.10	1.11715	0.11246	0.60	0.78724	0.60385
0.11	1.11480	0.12362	0.61	0.77777	0.61168
0.12	1.11224	0.13475	0.62	0.76826	0.61941
0.13	1.10946	0.14586	0.63	0.75872	0.62704
0.14	1.10647	0.15694	0.64	0.74914	0.63458
0.15	1.10327	0.16799	0.65	0.73954	0.64202
0.16	1.09985	0.17901	0.66	0.72992	0.64937
0.17	1.09623	0.18999	0.67	0.72027	0.65662
0.18	1.09240	0.20093	0.68	0.71061	0.66378
0.19	1.08837	0.21183	0.69	0.70095	0.67084
0.20	1.08413	0.22270	0.70	0.69127	0.67780
0.21	1.07969	0.23352	0.71	0.68159	0.68466
0.22	1.07506	0.24429	0.72	0.67191	0.69143
0.23	1.07023	0.25502	0.73	0.66224	0.69810
0.24	1.06522	0.26570	0.74	0.65258	0.70467
0.25	1.06001	0.27632	0.75	0.64293	0.71115
0.26	1.05462	0.28689	0.76	0.63329	0.71753
0.27	1.04904	0.29741	0.77	0.62368	0.72382
0.28	1.04329	0.30788	0.78	0.61408	0.73001
0.29	1.03736	0.31828	0.79	0.60452	0.73610
0.30	1.03126	0.32862	0.80	0.59498	0.74210
0.31	1.02498	0.33890	0.81	0.58548	0.74800
0.32	1.01855	0.34912	0.82	0.57601	0.75381
0.33	1.01195	0.35927	0.83	0.56659	0.75952
0.34	1.00519	0.36936	0.84	0.55720	0.76514
0.35	0.99828	0.37938	0.85	0.54786	0.77066
0.36	0.99122	0.38932	0.86	0.53858	0.77610
0.37	0.98401	0.39920	0.87	0.52934	0.78143
0.38	0.97665	0.40900	0.88	0.52016	0.78668
0.39	0.96916	0.41873	0.89	0.51103	0.79184
0.40	0.96154	0.42839	0.90	0.50196	0.79690
0.41	0.95378	0.43796	0.91	0.49296	0.80188
0.42	0.94590	0.44746	0.92	0.48402	0.80676
0.43	0.93789	0.45688	0.93	0.47515	0.81156
0.44	0.92977	0.46622	0.94	0.46635	0.81627
0.45	0.92153	0.47548	0.95	0.45761	0.82089
0.46	0.91318	0.48465	0.96	0.44896	0.82542
0.47	0.90473	0.49374	0.97	0.44037	0.82987
0.48	0.89617	0.50274	0.98	0.43187	0.83423
0.49	0.88752	0.51166	0.99	0.42345	0.83850

Continued

Table A16   Error function and its derivative—cont'd

x	$\frac{2}{\sqrt{\pi}}e^{-x^2}$	Erfx	x	$\frac{2}{\sqrt{\pi}}e^{-x^2}$	Erfx
1.00	0.41510	0.84270	1.50	0.11893	0.96610
1.01	0.40684	0.84681	1.51	0.11540	0.96727
1.02	0.39867	0.85083	1.52	0.11195	0.96841
1.03	0.39058	0.85478	1.53	0.10859	0.96951
1.04	0.38257	0.85864	1.54	0.10531	0.97058
1.05	0.37466	0.86243	1.55	0.10210	0.97162
1.06	0.36684	0.86614	1.56	0.09898	0.97262
1.07	0.35911	0.86977	1.57	0.09593	0.97360
1.08	0.35147	0.87332	1.58	0.09295	0.97454
1.09	0.34392	0.87680	1.59	0.09005	0.97546
1.10	0.33647	0.88020	1.60	0.08722	0.97634
1.11	0.32912	0.88353	1.61	0.08447	0.97720
1.12	0.32186	0.88678	1.62	0.08178	0.97803
1.13	0.31470	0.88997	1.63	0.07917	0.97884
1.14	0.30764	0.89308	1.64	0.07662	0.97962
1.15	0.30067	0.89612	1.65	0.07414	0.98037
1.16	0.29381	0.89909	1.66	0.07173	0.98110
1.17	0.28704	0.90200	1.67	0.06938	0.98181
1.18	0.28037	0.90483	1.68	0.06709	0.98249
1.19	0.27381	0.90760	1.69	0.06487	0.98315
1.20	0.26734	0.91031	1.70	0.06271	0.98379
1.21	0.26097	0.91295	1.71	0.06060	0.98440
1.22	0.25471	0.91553	1.72	0.05856	0.98500
1.23	0.24854	0.91805	1.73	0.05657	0.98557
1.24	0.24248	0.92050	1.74	0.05464	0.98613
1.25	0.23652	0.92290	1.75	0.05277	0.98667
1.26	0.23065	0.92523	1.76	0.05095	0.98719
1.27	0.22489	0.92751	1.77	0.04918	0.98769
1.28	0.21923	0.92973	1.78	0.04747	0.98817
1.29	0.21367	0.93189	1.79	0.04580	0.98864
1.30	0.20820	0.93400	1.80	0.04419	0.98909
1.31	0.20284	0.93606	1.81	0.04262	0.98952
1.32	0.19757	0.93806	1.82	0.04110	0.98994
1.33	0.19241	0.94001	1.83	0.03963	0.99034
1.34	0.18734	0.94191	1.84	0.03820	0.99073
1.35	0.18236	0.94376	1.85	0.03681	0.99111
1.36	0.17749	0.94556	1.86	0.03547	0.99147
1.37	0.17271	0.94731	1.87	0.03417	0.99182
1.38	0.16802	0.94901	1.88	0.03292	0.99215
1.39	0.16343	0.95067	1.89	0.03170	0.99247
1.40	0.15894	0.95228	1.90	0.03052	0.99279
1.41	0.15453	0.95385	1.91	0.02938	0.99308
1.42	0.15022	0.95537	1.92	0.02827	0.99337
1.43	0.14600	0.95685	1.93	0.02721	0.99365
1.44	0.14187	0.95829	1.94	0.02617	0.99392
1.45	0.13783	0.95969	1.95	0.02517	0.99417
1.46	0.13387	0.96105	1.96	0.02421	0.99442
1.47	0.13001	0.96237	1.97	0.02328	0.99466
1.48	0.12623	0.96365	1.98	0.02237	0.99489
1.49	0.12254	0.96489	1.99	0.02150	0.99511
			2.00	0.02066	0.99532

Problems

A solutions manual is available for the problems in Volume 1 of *Chemical Engineering* from booksellers and from

Heinemann Customer Services
Halley Court
Jordan Hill
Oxford OX2 8YW
The United Kingdom

Tel: 01865 888180
E-mail: bhuk.orders@repp.co.uk

1.1. Calculate the time taken for the distant face of a brick wall, of thermal diffusivity, $D_H = 0.0042$ cm^2/s and thickness $l = 0.45$ m, initially at 290 K, to rise to 470 K if the near face is suddenly raised to a temperature of $\theta' = 870$ K and maintained at that temperature. Assume that all the heat flow is perpendicular to the faces of the wall and that the distant face is perfectly insulated.

1.2. Calculate the time for the distant face to reach 470 K under the same conditions as Problem 9.1, except that the distant face is not perfectly lagged but a very large thickness of material of the same thermal properties as the brickwork is stacked against it.

1.3. Benzene vapour, at atmospheric pressure, condenses on a plane surface 2 m long and 1 m wide, maintained at 300 K and inclined at an angle of 45 degrees to the horizontal. Plot the thickness of the condensate film and the point heat-transfer coefficient against the distance from the top of the surface.

1.4. It is desired to warm 0.9 kg/s of air from 283 to 366 K by passing it through the pipes of a bank consisting of 20 rows with 20 pipes in each row. The arrangement is in line with centre-to-centre spacing, in both directions, equal to twice the pipe diameter. Flue gas, entering at 700 K and leaving at 366 K with a free flow mass velocity of 10 kg/m^{2}s, is passed across the outside of the pipes.

   Neglecting gas radiation, how long should the pipes be?

   For simplicity, the outer and inner pipe diameters may be taken as 12 mm.

   Values of k and μ, which may be used for both air and flue gases, are given below. The specific heat capacity of air and flue gases is 1.0 kJ/kg K.

Temperature (K)	Thermal Conductivity k (W/m K)	Viscosity μ (mN s/m^2)
250	0.022	0.0165
500	0.040	0.0276
800	0.055	0.0367

1.5. A cooling coil, consisting of a single length of tubing through which water is circulated, is provided in a reaction vessel, the contents of which are kept uniformly at 360 K by means of a stirrer. The inlet and outlet temperatures of the cooling water are 280 and 320 K, respectively. What would the outlet water temperature become if the length of the cooling coil were increased five times? Assume the overall heat-transfer coefficient to be constant over the length of the tube and independent of the water temperature.

1.6. In an oil cooler, 216 kg/h of hot oil enters a thin metal pipe of diameter 25 mm. An equal mass flow of cooling water passes through the annular space between the pipe and a larger concentric pipe with the oil and water moving in opposite directions. The oil enters at 420 K and is to be cooled to 320 K. If the water enters at 290 K, what length of pipe will be required? Take coefficients of 1.6 kW/m^2 K on the oil side and 3.6 kW/m^2 K on the water side and 2.0 kJ/kg K for the specific heat of the oil.

1.7. The walls of a furnace are built up to 150 mm thickness of a refractory of thermal conductivity 1.5 W/m K. The surface temperatures of the inner and outer faces of the refractory are 1400 and 540 K, respectively.

If a layer of insulating material 25 mm thick, of thermal conductivity 0.3 W/m K, is added, what temperatures will its surfaces attain assuming the inner surface of the furnace to remain at 1400 K? The coefficient of heat transfer from the outer surface of the insulation to the surroundings, which are at 290 K, may be taken as 4.2, 5.0, 6.1, and 7.1 W/m K, for surface temperatures of 370, 420, 470, and 520 K, respectively. What will be the reduction in heat loss?

1.8. A pipe of outer diameter 50 mm, maintained at 1100 K, is covered with 50 mm of insulation of thermal conductivity 0.17 W/m K.

Would it be feasible to use a magnesia insulation that will not stand temperatures above 615 K and has a thermal conductivity of 0.09 W/m K for an additional layer thick enough to reduce the outer surface temperature to 370 K in the surroundings at 280 K? Take the surface coefficient of heat transfer by radiation and convection as 10 W/m^2 K.

1.9. In order to warm 0.5 kg/s of a heavy oil from 311 to 327 K, it is passed through tubes of inside diameter 19 mm and length 1.5 m, forming a bank, on the outside of which steam is condensing at 373 K. How many tubes will be needed?

In calculating *Nu*, *Pr*, and *Re*, the thermal conductivity of the oil may be taken as 0.14 W/m K and the specific heat as 2.1 kJ/kg K, irrespective of temperature. The viscosity is to be taken at the mean oil temperature. Viscosity of the oil at 319 and 373 K is 154 and 19.2 mN s/m^2, respectively.

1.10. A metal pipe of 12 mm outer diameter is maintained at 420 K. Calculate the rate of heat loss per metre run in the surroundings uniformly at 290 K, (a) when the pipe is covered with 12 mm thickness of a material of thermal conductivity 0.35 W/m K and surface emissivity 0.95 and (b) when the thickness of the covering material is reduced to 6 mm but the outer surface is treated so as to reduce its emissivity to 0.10.

The coefficients of radiation from a perfectly black surface in the surroundings at 290 K are 6.25, 8.18, and 10.68 W/m² K at 310, 370, and 420 K, respectively.

The coefficients of convection may be taken as $1.22(\theta/d)^{0.25}$ W/m² K, where θ (K) is the temperature difference between the surface and the surrounding air and d (m) is the outer diameter.

1.11. A condenser consists of 30 rows of parallel pipes of outer diameter 230 mm and thickness 1.3 mm, with 40 pipes, each 2 m long, per row. Water, inlet temperature 283 K, flows through the pipes at 1 m/s, and steam at 372 K condenses on the outside of the pipes. There is a layer of scale 0.25 mm thick, of thermal conductivity 2.1 W/m K, on the inside of the pipes.

Taking the coefficients of heat transfer on the water side as 4.0 and on the steam side as 8.5 kW/m² K, calculate the outlet water temperature and the total mass of steam condensed per second. The latent heat of steam at 372 K is 2250 kJ/kg. The density of water is 1000 kg/m³.

1.12. In an oil cooler, water flows at the rate of 360 kg/h per tube through metal tubes of outer diameter 19 mm and thickness 1.3 mm, along the outside of which oil flows in the opposite direction at the rate of 75 g/s per tube.

If the tubes are 2 m long and the inlet temperatures of the oil and water are, respectively, 370 and 280 K, what will be the outlet oil temperature? The coefficient of heat transfer on the oil side is 1.7 and on the water side 2.5 kW/m² K, and the specific heat of the oil is 1.9 kJ/kg K.

1.13. Waste gases flowing across the outside of a bank of pipes are being used to warm air that flows through the pipes. The bank consists of 12 rows of pipes with 20 pipes, each 0.7 m long, per row. They are arranged in line, with centre-to-centre spacing equal in both directions to one-and-a-half times the pipe diameter. Both inner and outer diameter may be taken as 12 mm. Air, mass velocity of 8 kg/m²s, enters the pipes at 290 K. The initial gas temperature is 480 K, and the total mass of the gases crossing the pipes per second is the same as the total mass of the air flowing through them.

Temperature (K)	Thermal Conductivity (W/m K)	Viscosity (mN/s/m²)
250	0.022	0.0165
310	0.027	0.0189
370	0.030	0.0214
420	0.033	0.0239
480	0.037	0.0260
Specific heat=1.00 kJ/kg K		

Neglecting gas radiation, estimate the outlet temperature of the air. The physical constants for the waste gases that may be assumed the same as for air are as follows:

1.14. Oil is to be warmed from 300 to 344 K by passing it at 1 m/s through the pipes of a shell-and-tube heat exchanger. Steam at 377 K condenses on the outside of the pipes, which have outer and inner diameters of 48 and 41 mm, respectively. Due to fouling, the inside diameter has been reduced to 38 mm, and the resistance to heat transfer of the pipe wall and dirt together, based on this diameter, is 0.0009 m² K/W.

It is known from previous measurements under similar conditions that the oil-side coefficients of heat transfer for a velocity of 1 m/s, based on a diameter of 38 mm, vary with the temperature of the oil as follows:

Oil temperature (K)	300	311	322	333	344
Oil-side coefficient of heat transfer (W/m² K)	74	80	97	136	244

The specific heat and density of the oil may be assumed constant at 1.9 kJ/kg K and 900 kg/m³, respectively, and any resistance to heat transfer on the steam side neglected.

Find the length of tube bundle required.

1.15. It is proposed to construct a heat exchanger to condense 7.5 kg/s of *n*-hexane at a pressure of 150 kN/m², involving a heat load of 4.5 MW. The hexane is to reach the condenser from the top of a fractionating column at its condensing temperature of 356 K.

From experience, it is anticipated that the overall heat-transfer coefficient will be 450 W/m² K. Cooling water is available at 289 K.

Outline the proposals that you would make for the type and size of the exchanger and explain the details of the mechanical construction that you consider require special attention.

1.16. A heat exchanger is to be mounted at the top of a fractionating column about 15 m high to condense 4 kg/s of *n*-pentane at 205 kN/m², corresponding to a condensing temperature of 333 K. Give an outline of the calculations you would make to obtain an approximate idea of the size and construction of the exchanger required.

For purposes of standardisation, 19 mm outer diameter tubes of 1.65 mm wall thickness will be used, and these may be 2.5, 3.6, or 5 m in length. The film coefficient for condensing pentane on the outside of a horizontal tube bundle may be taken as 1.1 kW/m² K. The condensation is effected by pumping water through the tubes, the initial water temperature being 288 K.

The latent heat of condensation of pentane is 335 kJ/kg.

For these 19 mm tubes, a water velocity of 1 m/s corresponds to a flowrate of 200 g/s of water.

1.17. An organic liquid is boiling at 340 K on the inside of a metal surface of thermal conductivity 42 W/m K and thickness 3 mm. The outside of the surface is heated by condensing steam. Assuming that the heat-transfer coefficient from steam to the outer metal surface is constant at 11 kW/m² K, irrespective of the steam temperature, find the value of the steam temperature to give a maximum rate of evaporation.

The coefficients of heat transfer from the inner metal surface to the boiling liquid that depend upon the temperature difference are as follows:

Temperature Difference Metal Surface to Boiling Liquid (K)	Heat-Transfer Coefficient Metal Surface to Boiling Liquid (kW/m² K)
22.2	4.43
27.8	5.91
33.3	7.38
36.1	7.30
38.9	6.81
41.7	6.36
44.4	5.73
50.0	4.54

1.18. It is desired to warm an oil of specific heat 2.0 kJ/kg K from 300 to 325 K by passing it through a tubular heat exchanger with metal tubes of inner diameter 10 mm. Along the outside of the tubes flows water, inlet temperature of 372 K and outlet temperature of 361 K.

The overall heat-transfer coefficient from water to oil, based on the inside area of the tubes, may be assumed constant at 230 W/m² K, and 75 g/s of oil is to be passed through each tube.

The oil is to make two passes through the heater. The water makes one pass along the outside of the tubes. Calculate the length of the tubes required.

1.19. A condenser consists of a number of metal pipes of outer diameter 25 mm and thickness 2.5 mm. Water, flowing at 0.6 m/s, enters the pipes at 290 K, and it is not permissible that it should be discharged at a temperature in excess of 310 K.

If 1.25 kg/s of a hydrocarbon vapour is to be condensed at 345 K on the outside of the pipes, how long should each pipe be, and how many pipes should be needed?

Take the coefficient of heat transfer on the water side as 2.5 and on the vapour side as 0.8 kW/m² K and assume that the overall coefficient of heat transfer from vapour to water, based upon these figures, is reduced by 20% by the effects of the pipe walls, dirt, and scale.

The latent heat of the hydrocarbon vapour at 345 K is 315 kJ/kg.

1.20. An organic vapour is being condensed at 350 K on the outside of a bundle of pipes through which water flows at 0.6 m/s, its inlet temperature being 290 K. The outer and inner diameters of the pipes are 19 mm and 15 mm, respectively, but a layer of scale 0.25 mm thick and thermal conductivity 2.0 W/m K has formed on the inside of the pipes.

If the coefficients of heat transfer on the vapour and water sides, respectively, are 1.7 and 3.2 kW/m² K and it is required to condense 25 g/s of vapour on each of the pipes, how long should these be, and what will be the outlet temperature of the water?

The latent heat of condensation is 330 kJ/kg.

Neglect any resistance to heat transfer in the pipe walls.

1.21. A heat exchanger is required to cool continuously 20 kg/s of warm water from 360 to 335 K by means of 25 kg/s of cold water, inlet temperature of 300 K.

Assuming that the water velocities are such as to give an overall coefficient of heat transfer of 2 kW/m² K, assumed constant, calculate the total area of surface required (a) in a counterflow heat exchanger, i.e., one in which the hot and cold fluids flow in opposite directions, and (b) in a multipass heat exchanger, with the cold water making two passes through the tubes and the hot water making one pass along the outside of the tubes. In case (b) assume that the hot water flows in the same direction as the inlet cold water and that its temperature over any cross-section is uniform.

1.22. Find the heat loss per unit area of surface through a brick wall 0.5 m thick when the inner surface is at 400 K and the outside at 310 K. The thermal conductivity of the brick may be taken as 0.7 W/m K.

1.23. A furnace is constructed with 225 mm of firebrick, 120 mm of insulating brick, and 225 mm of building brick. The inside temperature is 1200 K and the outside temperature 330 K. If the thermal conductivities are 1.4, 0.2, and 0.7 W/m K, find the heat loss per unit area and the temperature at the junction of the firebrick and insulating brick.

1.24. Calculate the total heat loss by radiation and convection from an unlagged horizontal steam pipe of 50 mm outside diameter at 415 K to air at 290 K.

1.25. Toluene is continuously nitrated to mononitrotoluene in a cast-iron vessel of 1 m diameter fitted with a propeller agitator of 0.3 m diameter driven at 2 Hz. The temperature is maintained at 310 K by circulating cooling water at 0.5 kg/s through a stainless steel coil of 25 mm outside diameter and 22 mm inside diameter wound in the form of a helix of 0.81 m diameter. The conditions are such that the reacting material may be considered to have the same physical properties as 75% sulphuric acid. If the mean water temperature is 290 K, what is the overall heat-transfer coefficient?

1.26. 7.5 kg/s of pure isobutane is to be condensed at a temperature of 331.7 K in a horizontal tubular exchanger using a water inlet temperature of 301 K. It is proposed to use 19 mm outside diameter tubes of 1.6 mm wall arranged on a 25 mm triangular pitch. Under these conditions, the resistance of the scale may be taken as 0.0005 m² K/W. Determine the number and arrangement of the tubes in the shell.

1.27. 37.5 kg/s of crude oil is to be heated from 295 to 330 K by heat exchange with the bottom product from a distillation column. The bottom product, flowing at 29.6 kg/s, is to be cooled from 420 to 380 K. There is available tubular exchanger with an inside shell diameter of 0.60 m having one pass on the shell side and two passes on the tube side. It has 324 tubes, 19 mm outside diameter with 2.1 mm wall and 3.65 m long, arranged on a 25 mm square pitch and supported by baffles with a 25% cut, spaced at 230 mm intervals. Would this exchanger be suitable?

1.28. A 150 mm internal diameter steam pipe is carrying steam at 444 K and is lagged with 50 mm of 85% magnesia. What will be the heat loss to the air at 294 K?

1.29. A refractory material that has an emissivity of 0.40 at 1500 K and 0.43 at 1420 K is at a temperature of 1420 K and is exposed to black furnace walls at a temperature of 1500 K. What is the rate of gain of heat by radiation per unit area?

1.30. The total emissivity of clean chromium as a function of surface temperature T (K) is given approximately by

$$e = 0.38 \left(1 - \frac{263}{T}\right).$$

Obtain an expression for the absorptivity of solar radiation as a function of surface temperature and compare the absorptivity and emissivity at 300, 400, and 1000 K.

Assume that the sun behaves as a black body at 5500 K.

1.31. Repeat Problem 9.30 for the case of aluminium, assuming the emissivity to be 1.25 times that for chromium.

1.32. Calculate the heat transferred by solar radiation on the flat concrete roof of a building, 8×9 m, if the surface temperature of the roof is 330 K. What would be the effect of covering the roof with a highly reflecting surface such as polished aluminium separated from the concrete by an efficient layer of insulation?

The total emissivity of concrete at 330 K is 0.89, whilst the total absorptivity of solar radiation (sun temperature $= 5500$ K) at this temperature is 0.60. Use the data from Problem 9.31 that should be solved first for aluminium.

1.33. A rectangular iron ingot of $15 \times 15 \times 30$ cm is supported at the centre of a reheating furnace. The furnace has walls of silica brick at 1400 K, and the initial temperature of the ingot is 290 K. How long will it take to heat the ingot to 600 K?

It may be assumed that the furnace is large compared with the ingot size and that the ingot remains at uniform temperature throughout its volume. Convection effects are negligible.

The total emissivity of the oxidised iron surface is 0.78, and both emissivity and absorptivity are independent of the surface temperature:

$$\text{Density of iron} = 7.2 \, \text{Mg/m}^3$$
$$\text{Specific heat capacity of iron} = 0.50 \, \text{kJ/kg K}$$

1.34. A wall is made of brick, of thermal conductivity 1.0 W/m K, 230 mm thick, lined on the inner face with plaster of thermal conductivity 0.4 W/m K and of thickness 10 mm. If a temperature difference of 30 K is maintained between the two outer faces, what is the heat flow per unit area of wall?

1.35. A 50 mm diameter pipe of circular cross-section and with walls 3 mm thick is covered with two concentric layers of lagging, the inner layer having a thickness of 25 mm and a thermal conductivity of 0.08 W/m K, and the outer layer has a thickness of 40 mm and a thermal conductivity of 0.04 W/m K. What is the rate of heat loss per metre length of pipe if the temperature inside the pipe is 550 K and the outside surface temperature is 330 K?

1.36. The temperature of oil leaving a cocurrent flow cooler is to be reduced from 370 to 350 K by lengthening the cooler. The oil and water flowrates and inlet temperatures and the other dimensions of the cooler will remain constant. The water enters at 285 K and the oil at 420 K. The water leaves the original cooler at 310 K. If the original length is 1 m, what must be the new length?

1.37. In a countercurrent-flow heat exchanger, 1.25 kg/s of benzene (specific heat of 1.9 kJ/kg K and density of 880 kg/m^3) is to be cooled from 350 to 300 K with water at 290 K. In the heat exchanger, tubes of 25 mm external and 22 mm internal diameter are employed, and the water passes through the tubes. If the film coefficients for the water and benzene are 0.85 and 1.70 kW/m^2 K, respectively, and the scale resistance can be neglected, what total length of tube will be required if the minimum quantity of water is to be used and its temperature is not to be allowed to rise above 320 K?

1.38. Calculate the rate of loss of heat from a 6 m long horizontal steam pipe of 50 mm internal diameter and 60 mm external diameter when carrying steam at 800 kN/m^2. The temperature of the surroundings is 290 K.

What would be the cost of steam saved by coating the pipe with a 50 mm thickness of 85% magnesia lagging of thermal conductivity 0.07 W/m K, if steam costs £0.5 per 100 kg? The emissivity of the surface of the bare pipe and of the lagging may be taken as 0.85, and the coefficient h for heat loss by natural convection is given by

$$h = 1.65(\Delta T)^{0.25} \text{ W/m}^2 \text{ K},$$

where ΔT is the temperature difference in K.

Take the Stefan-Boltzmann constant as 5.67×10^{-3} W/m^2 K^4.

1.39. A stirred reactor contains a batch of 700 kg reactants of specific heat 3.8 kJ/kg K initially at 290 K, which is heated by dry saturated steam at 170 kN/m^2 fed to a helical coil. During the heating period, the steam supply rate is constant at 0.1 kg/s, and condensate leaves at the temperature of the steam. If heat losses are neglected, calculate the true temperature of the reactants when a thermometer immersed in the material reads 360 K. The bulb of the thermometer is approximately cylindrical and is 100 mm long by 10 mm diameter with a water equivalent of 15 g, and the overall heat-transfer coefficient to the thermometer is 300 W/m^2 K. What would a thermometer with a similar bulb of half the length and half the heat capacity indicate under these conditions?

1.40. How long will it take to heat 0.18 m^3 of liquid of density 900 kg/m^3 and specific heat 2.1 kJ/kg K from 293 to 377 K in a tank fitted with a coil of area 1 m^2? The coil is fed with steam at 383 K, and the overall heat-transfer coefficient can be taken as constant at 0.5 kW/m² K. The vessel has an external surface of 2.5 m², and the coefficient for heat transfer to the surroundings at 293 K is 5 W/m² K.

The batch system of heating is to be replaced by a continuous countercurrent heat exchanger in which the heating medium is a liquid entering at 388 K and leaving at 333 K. If the heat-transfer coefficient is 250 W/m² K, what heat exchange area is required? Heat losses may be neglected.

1.41. The radiation received by the earth's surface on a clear day with the sun overhead is 1 kW/m², and an additional of 0.3 kW/m² is absorbed by the earth's atmosphere. Calculate approximately the temperature of the sun, assuming its radius to be 700,000 km and the distance between the sun and the earth to be 150,000,000 km. The sun may be assumed to behave as a black body.

1.42. A thermometer is immersed in a liquid that is heated at the rate of 0.05 K/s. If the thermometer and the liquid are both initially at 290 K, what rate of passage of liquid over the bulb of the thermometer is required if the error in the thermometer reading after 600 s is to be no more than 1 K? Take the water equivalent of the thermometer as 30 g and the heat-transfer coefficient to the bulb to be given by $U = 735\,u^{0.8}$ W/m² K. The area of the bulb is 0.01 m², where u is the velocity in m/s.

1.43. In a shell-and-tube heat exchanger with horizontal tubes 25 mm external diameter and 22 mm internal diameter, benzene is condensed on the outside by means of water flowing through the tubes at the rate of 0.03 m³/s. If the water enters at 290 K and leaves at 300 K and the heat-transfer coefficient on the water side is 850 W/m² K, what total length of tubing will be required?

1.44. In a contact sulphuric acid plant, the gases leaving the first converter are to be cooled from 845 to 675 K by means of the air required for the combustion of the sulphur. The air enters the heat exchanger at 495 K. If the flow of each of the streams is 2 m³/s at NTP, suggest a suitable design for a shell-and-tube heat exchanger employing the tubes of 25 mm internal diameter:
(a) Assume parallel cocurrent flow of the gas streams.
(b) Assume parallel countercurrent flow.
(c) Assume that the heat exchanger is fitted with baffles giving crossflow outside the tubes.

1.45. A large block of material of thermal diffusivity $D_H = 0.0042$ cm²/s is initially at a uniform temperature of 290 K, and one face is raised suddenly to 875 K and maintained at that temperature. Calculate the time taken for the material at a depth of 0.45 m to reach a temperature of 475 K on the assumption of unidirectional heat transfer and that the material can be considered to be infinite in extent in the direction of transfer.

1.46. A 50% glycerol-water mixture is flowing at a Reynolds number of 1500 through a 25 mm diameter pipe. Plot the mean value of the heat-transfer coefficient as a function of pipe length assuming that

$$Nu = 1.62 \left(RePr\frac{d}{l} \right)^{0.33}.$$

Indicate the conditions under which this is consistent with the predicted value $Nu = 4.1$ for fully developed flow.

1.47. A liquid is boiled at a temperature of 360 K using steam fed at a temperature of 380 K to a coil heater. Initially, the heat-transfer surfaces are clean, and an evaporation rate of 0.08 kg/s is obtained from each square metre of heating surface. After a period, a layer of scale of resistance 0.0003 m^2 K/W is deposited by the boiling liquid on the heat-transfer surface. On the assumption that the coefficient on the steam side remains unaltered and that the coefficient for the boiling liquid is proportional to its temperature difference raised to the power of 2.5, calculate the new rate of boiling.

1.48. A batch of reactants of specific heat 3.8 kJ/kg K and of mass 1000 kg is heated by means of a submerged steam coil of area 1 m^2 fed with steam at 390 K. If the overall heat-transfer coefficient is 600 W/m^2 K, calculate the time taken to heat the material from 290 to 360 K, if heat losses to the surroundings are neglected.

If the external area of the vessel is 10 m^2 and the heat-transfer coefficient to the surroundings at 290 K is 8.5 W/m^2 K, what will be the time taken to heat the reactants over the same temperature range, and what is the maximum temperature to which the reactants can be raised?

What methods would you suggest for improving the rate of heat transfer?

1.49. What do you understand by the terms 'black body' and 'grey body' when applied to radiant heat transfer?

Two large, parallel plates with grey surfaces are situated 75 mm apart; one has an emissivity of 0.8 and is at a temperature of 350 K, and the other has an emissivity of 0.4 and is at a temperature of 300 K. Calculate the net rate of heat exchange by radiation per unit area taking the Stefan–Boltzmann constant as 5.67×10^{-8} W/m^2 K^4. Any formula (other than Stefan's law) that you use must be proved.

1.50. A longitudinal fin on the outside of a circular pipe is 75 mm deep and 3 mm thick. If the pipe surface is at 400 K, calculate the heat dissipated per metre length from the fin to the atmosphere at 290 K if the coefficient of heat transfer from its surface may be assumed constant at 5 W/m^2 K. The thermal conductivity of the material of the fin is 50 W/m K, and the heat loss from the extreme edge of the fin may be neglected. It should be assumed that the temperature is uniformly 400 K at the base of the fin.

1.51. Liquid oxygen is distributed by road in large spherical insulated vessels, 2 m internal diameter, well lagged on the outside. What thickness of magnesia lagging, of thermal conductivity 0.07 W/m K, must be used so that not more than 1% of the liquid oxygen evaporates during a journey of 10 ks if the vessel is initially 80% full?

$$\text{Latent heat of vaporisation of oxygen} = 215\,\text{kJ/kg}$$
$$\text{Boiling point of oxygen} = 90\,\text{K}$$
$$\text{Density of liquid oxygen} = 1140\,\text{kg/m}^3$$
$$\text{Atmospheric temperature} = 288\,\text{K}$$

Heat-transfer coefficient from the outside

surface of the lagging to atmosphere $= 4.5\,\text{W/m}^2\text{K}$.

1.52. Benzene is to be condensed at the rate of 1.25 kg/s in a vertical shell-and-tube type of heat exchanger fitted with the tubes of 25 mm outside diameter and 2.5 m long. The vapour condenses on the outside of the tubes, and the cooling water enters at 295 K and passes through the tubes at 1.05 m/s. Calculate the number of tubes required if the heat exchanger is arranged for a single pass of the cooling water. The tube wall thickness is 1.6 mm.

1.53. One end of a metal bar 25 mm in diameter and 0.3 m long is maintained at 375 K, and heat is dissipated from the whole length of the bar to the surroundings at 295 K. If the coefficient of heat transfer from the surface is 10 W/m^2 K, what is the rate of loss of heat? Take the thermal conductivity of the metal as 85 W/m K.

1.54. A shell-and-tube heat exchanger consists of 120 tubes of internal diameter 22 mm and length 2.5 m. It is operated as a single-pass condenser with benzene condensing at a temperature of 350 K on the outside of the tubes and water of inlet temperature 290 K passing through the tubes. Initially, there is no scale on the walls, and a rate of condensation of 4 kg/s is obtained with a water velocity of 0.7 m/s through the tubes. After prolonged operation, a scale of resistance 0.0002 m^2 K/W is formed on the inner surface of the tubes. To what value must the water velocity be increased in order to maintain the same rate of condensation on the assumption that the transfer coefficient on the water side is proportional to the velocity raised to the 0.8 power and that the coefficient for the condensing vapour is 2.25 kW/m^2 K based on the inside area? The latent heat of vaporisation of benzene is 400 kJ/kg.

1.55. Derive an expression for the radiant heat-transfer rate per unit area between two large parallel planes of emissivities e_1 and e_2 and at absolute temperatures T_1 and T_2, respectively.

Two such planes are situated 2.5 mm apart in air: One has an emissivity of 0.1 and is at 350 K, and the other has an emissivity of 0.05 and is at 300 K. Calculate the percentage change in the total heat-transfer rate by coating the first surface so as to reduce its emissivity to 0.025:

$$\text{Stefan-Boltzmann constant} = 5.67 \times 10^{-8}\,\text{W/m}^2\text{K}^4$$
$$\text{Thermal conductivity of air} = 0.026\,\text{W/mK}$$

1.56. Water flows at 2 m/s through a 2.5 m length of a 25 mm diameter tube. If the tube is at 320 K and the water enters and leaves at 293 and 295 K, respectively, what is the value of the heat-transfer coefficient? How would the outlet temperature change if the velocity were increased by 50%?

1.57. A liquid hydrocarbon is fed at 295 K to a heat exchanger consisting of a 25 mm diameter tube heated on the outside by condensing steam at atmospheric pressure. The flowrate of the hydrocarbon is measured by means of a 19 mm orifice fitted to the 25 mm feed pipe. The reading on a differential manometer containing the hydrocarbon over water is 450 mm, and the coefficient of discharge of the metre is 0.6.

Calculate the initial rate of rise of temperature (K/s) of the hydrocarbon as it enters the heat exchanger. The outside film coefficient $=6.0$ kW/m^2 K. The inside film coefficient h is given by

$$\frac{hd}{k}=0.023\left(\frac{ud\rho}{\mu}\right)^{0.8}\left(\frac{C_p\mu}{k}\right)^{0.4}.$$

where

$u=$ linear velocity of hydrocarbon (m/s),
$d=$ tube diameter (m),
$\rho=$ liquid density (800 kg/m^3),
$\mu=$ liquid viscosity (9×10^{-4} Ns/m^2),
$C_p=$ specific heat of liquid (1.7×10^3 J/kg K),
$K=$ thermal conductivity of liquid (0.17 W/m K).

1.58. Water passes at 1.2 m/s through a series of 25 mm diameter tubes 5 m long maintained at 320 K. If the inlet temperature is 290 K, at what temperature would you expect it to leave?

1.59. Heat is transferred from one fluid stream to a second fluid across a heat-transfer surface. If the film coefficients for the two fluids are, respectively, 1.0 and 1.5 kW/m^2 K, the metal is 6 mm thick (thermal conductivity of 20 W/m K), and the scale coefficient is equivalent to 850 W/m^2 K, what is the overall heat-transfer coefficient?

1.60. A pipe of outer diameter 50 mm carries hot fluid at 1100 K. It is covered with a 50 mm layer of insulation of thermal conductivity 0.17 W/m K. Would it be feasible to use magnesia insulation, which will not stand temperatures above 615 K and has a thermal conductivity of 0.09 W/m K for an additional layer thick enough to reduce the outer surface temperature to 370 K in the surroundings at 280 K? Take the surface coefficient of transfer by radiation and convection as 10 W/m^2 K.

1.61. A jacketed reaction vessel containing 0.25 m^3 of liquid of specific gravity 0.9 and specific heat 3.3 kJ/kg K is heated by means of steam fed to a jacket on the walls. The contents of the tank are agitated by a stirrer rotating at 3 Hz. The heat-transfer area is 2.5 m^2, and the steam temperature is 380 K. The outside film heat-transfer coefficient is 1.7 kW/m^2 K, and the 10 mm thick wall of the tank has a thermal conductivity of 6.0 W/m K. The inside film coefficient was found to be 1.1 kW/m^2 K for a stirrer speed of 1.5 Hz and to be proportional to the two-thirds power of the speed of rotation.

Neglecting heat losses and the heat capacity of the tank, how long will it take to raise the temperature of the liquid from 295 to 375 K?

1.62. By dimensional analysis, derive a relationship for the heat-transfer coefficient h for natural convection between a surface and a fluid on the assumption that the coefficient is a function of the following variables:

k = thermal conductivity of the fluid
C_p = specific heat of the fluid
ρ = density of the fluid
μ = viscosity of the fluid
β_g = the product of the coefficient of cubical expansion of the fluid and the acceleration due to gravity
l = a characteristic dimension of the surface
ΔT = the temperature difference between the fluid and the surface

Indicate why each of these quantities would be expected to influence the heat-transfer coefficient and explain how the orientation of the surface affects the process.

Under what conditions is heat transfer by convection important in chemical engineering?

1.63. A shell-and-tube heat exchanger is used for preheating the feed to an evaporator. The liquid of specific heat 4.0 kJ/kg K and specific gravity 1.1 passes through the inside of the tubes and is heated by steam condensing at 395 K on the outside. The exchanger heats liquid at 295 K to an outlet temperature of 375 K when the flowrate is 175 cm^3/s and to 370 K when the flowrate is 325 cm^3/s. What are the heat-transfer area and the value of the overall heat-transfer coefficient when the flow is 175 cm^3/s?

Assume that the film heat-transfer coefficient for the liquid in the tubes is proportional to the 0.8 power of the velocity, the transfer coefficient for the condensing steam remains constant at 3.4 kW/m^2 K and that the resistance of the tube wall and scale can be neglected.

1.64. 0.1 m^3 of liquid of specific heat capacity 3 kJ/kg K and density 950 kg/m^3 is heated in an agitated tank fitted with a coil, of heat-transfer area 1 m^2, supplied with steam at 383 K. How long will it take to heat the liquid from 293 to 368 K, if the tank, of external area 20 m^2, is losing heat to the surroundings at 293 K? To what temperature will the system fall in 1800 s if the steam is turned off?

$$\text{Overall heat transfer coefficient in coil} = 2000 \, \text{W}/\text{m}^2\text{K}$$

$$\text{Heat transfer coefficient to surroundings} = 10 \, \text{W}/\text{m}^2\text{K}$$

1.65. The contents of a reaction vessel are heated by means of steam at 393 K supplied to a heating coil that is totally immersed in the liquid. When the vessel has a layer of lagging 50 mm thick on its outer surfaces, it takes 1 h to heat the liquid from 293 to 373 K. How long will it take if the thickness of lagging is doubled?

$$\text{Outside temperature} = 293 \text{ K}$$

$$\text{Thermal conductivity of lagging} = 0.05 \text{ W/mK}$$

Coefficient for heat loss by radiation and convection from outside surface of vessel $= 10 \text{ W/m}^2 \text{ K}$:

$$\text{Outside area of vessel} = 8 \text{ m}^2$$

$$\text{Coil area} = 0.2 \text{ m}^2$$

Overall, heat-transfer coefficient for steam coil $= 300 \text{ W/m}^2 \text{ K}$.

1.66. A smooth tube in a condenser that is 25 mm in internal diameter and 10 m long is carrying cooling water, and the pressure drop over the length of the tube is $2 \times 10^4 \text{ N/m}^2$. If vapour at a temperature of 353 K is condensing on the outside of the tube and the temperature of the cooling water rises from 293 K at inlet to 333 K at outlet, what is the value of the overall heat-transfer coefficient based on the inside area of the tube? If the coefficient for the condensing vapour is 15,000 W/m^2 K, what is the film coefficient for the water? If the latent heat of vaporisation is 800 kJ/kg, what is the rate of condensation of vapour?

1.67. A chemical reactor, 1 m in diameter and 5 m long, operates at a temperature of 1073 K. It is covered with a 500 mm thickness of lagging of thermal conductivity 0.1 W/m K. The heat loss from the cylindrical surface to the surroundings is 3.5 kW. What is the heat-transfer coefficient from the surface of the lagging to the surroundings at a temperature of 293 K? How would the heat loss be altered if the coefficient were halved?

1.68. An open cylindrical tank 500 mm in diameter and 1 m deep is three-quarters filled with a liquid of density 980 kg/m^3 and of specific heat capacity 3 kJ/kg K. If the heat-transfer coefficient from the cylindrical walls and the base of the tank is 10 W/m^2 K and from the surface is 20 W/m^2 K, what area of heating coil, fed with steam at 383 K, is required to heat the contents from 288 to 368 K in a half hour? The overall heat-transfer coefficient for the coil may be taken as 100 W/m^2 K, the surroundings are at 288 K, and the heat capacity of the tank itself may be neglected.

1.69. Liquid oxygen is distributed by road in large spherical vessels, 1.82 m in internal diameter. If the vessels were unlagged and the coefficients for heat transfer from the outside of the vessel to the atmosphere were 5 W/m^2 K, what proportion of the contents would evaporate during a journey lasting an hour? Initially, the vessels are 80% full.

What thickness of lagging would be required to reduce the losses to one-tenth?

$$\text{Atmospheric temperature} = 288 \text{ K}$$

$$\text{Boiling point of oxygen} = 90 \text{ K}$$

$$\text{Density of oxygen} = 1140 \text{ kg/m}^3$$

$$\text{Latent heat of vaporisation of oxygen} = 214 \text{ kJ/kg}$$

$$\text{Thermal conductivity of lagging} = 0.07 \text{ W/mK}$$

1.70. Water at 293 K is heated by passing through a 6.1 m coil of 25 mm internal diameter pipe. The thermal conductivity of the pipe wall is 20 W/m K, and the wall thickness is 3.2 mm. The coil is heated by condensing steam at 373 K for which the film coefficient is 8 kW/m^2 K. When the water velocity in the pipe is 1 m/s, its outlet temperature is 309 K. What will the outlet temperature be if the velocity is increased to 1.3 m/s and if the coefficient of heat transfer to the water in the tube is proportional to the velocity raised to the 0.8 power?

1.71. Liquid is heated in a vessel by means of steam that is supplied to an internal coil in the vessel. When the vessel contains 1000 kg of liquid, it takes half an hour to heat the contents from 293 to 368 K if the coil is supplied with steam at 373 K. The process is modified so that liquid at 293 K is continuously fed to the vessel at the rate of 0.28 kg/s. The total contents of the vessel are always being maintained at 1000 kg. What is the equilibrium temperature that the contents of the vessel will reach, if heat losses to the surroundings are neglected and the overall heat-transfer coefficient remains constant?

1.72. The heat loss through a firebrick furnace wall 0.2 m thick is to be reduced by the addition of a layer of insulating brick to the outside. What is the thickness of insulating brick necessary to reduce the heat loss to 400 W/m^2? The inside furnace wall temperature is 1573 K, the ambient air adjacent to the furnace exterior is at 293 K, and the natural convection heat-transfer coefficient at the exterior surface is given by $h_0 = 3.0\Delta T^{0.25}$ W/m^2 K, where ΔT is the temperature difference between the surface and the ambient air:

$$\text{Thermal conductivity of firebrick} = 1.5\,\text{W}/\text{mK}$$
$$\text{Thermal conductivity of insulating brick} = 0.4\,\text{W}/\text{mK}$$

1.73. 2.8 kg/s of organic liquid of specific heat capacity 2.5 kJ/kg K is cooled in a heat exchanger from 363 to 313 K using water whose temperature rises from 293 to 318 K flowing countercurrently. After maintenance, the pipework is wrongly connected so that the two streams, flowing at the same rates as previously, are now in cocurrent flow. On the assumption that overall heat-transfer coefficient is unaffected, show that the new outlet temperatures of the organic liquid and the water will be 320.6 K and 314.5 K, respectively.

1.74. An organic liquid is cooled from 353 to 328 K in a single-pass heat exchanger. When the cooling water of initial temperature 288 K flows countercurrently, its outlet temperature is 333 K. With the water flowing cocurrently, its feed rate has to be increased in order to give the same outlet temperature for the organic liquid; the new outlet temperature of the water is 313 K. When the cooling water is flowing countercurrently, the film heat-transfer coefficient for the water is 600 W/m^2 K.

What is the coefficient when the exchanger is operating with cocurrent flow if its value is proportional to the 0.8 power of the water velocity?

Calculate the film coefficient from the organic liquid, on the assumptions that it remains unchanged and that heat-transfer resistances other than those attributable to the two liquids may be neglected.

1.75. A reaction vessel is heated by steam at 393 K supplied to a coil immersed in the liquid in the tank. It takes 1800 s to heat the contents from 293 to 373 K when the outside temperature is 293 K. When the outside and initial temperatures are only 278 K, it takes 2700 s to heat the contents to 373 K. The area of the steam coil is 2.5 m^2 and of the external surface is 40 m^2. If the overall heat-transfer coefficient from the coil to the liquid in the vessel is 400 W/m^2 K, show that the overall coefficient for transfer from the vessel to the surroundings is about 5 W/m^2 K.

1.76. Steam at 403 K is supplied through a pipe of 25 mm outside diameter. Calculate the heat loss per unit length to the surroundings at 293 K, on the assumption that there is a negligible drop in temperature through the wall of the pipe. The heat-transfer coefficient h from the outside of the pipe to the surroundings is given by

$$h = 1.22 \left(\frac{\Delta T}{d} \right)^{0.25} \text{W/m}^2 \text{ K}$$

where d is the outside diameter of the pipe (m) and ΔT is the temperature difference (K) between the surface and surroundings.

The pipe is then lagged with a 50 mm thickness of lagging of thermal conductivity 0.1 W/m K. If the outside heat-transfer coefficient is given by the same equation as for the bare pipe, by what factor is the heat loss reduced?

1.77. A vessel contains 1 tonne of liquid of specific heat capacity 4.0 kJ/kg K. It is heated by steam at 393 K that is fed to a coil immersed in the liquid, and heat is lost to the surroundings at 293 K from the outside of the vessel. How long does it take to heat the liquid from 293 to 353 K, and what is the maximum temperature to which the liquid can be heated? When the liquid temperature has reached 353 K, the steam supply is turned off for 2 h, and the vessel cools. How long will it take to reheat the material to 353 K?

Coil: area, 0.5 m^2. Overall heat-transfer coefficient to liquid, 600 W/m^2 K.
Outside of vessel area, 6 m^2. Heat-transfer coefficient to the surroundings, 10 W/m^2 K.

1.78. A bare thermocouple is used to measure the temperature of a gas flowing through a hot pipe. The heat-transfer coefficient between the gas and the thermocouple is proportional to the 0.8 power of the gas velocity, and the heat transfer by radiation from the walls to the thermocouple is proportional to the temperature difference.

When the gas is flowing at 5 m/s, the thermocouple reads 323 K. When it is flowing at 10 m/s, it reads 313 K, and when it is flowing at 15.0 m/s, it reads 309 K. Show that the gas temperature is about 298 K and calculate the approximate wall temperature. What temperature will the thermocouple indicate when the gas velocity is 20 m/s?

1.79. A hydrocarbon oil of density 950 kg/m^3 and specific heat capacity 2.5 kJ/kg K is cooled in a heat exchanger from 363 to 313 K by water flowing countercurrently. The temperature of the water rises from 293 to 323 K. If the flowrate of the hydrocarbon is 0.56 kg/s, what is the required flowrate of water?

After plant modifications, the heat exchanger is incorrectly connected so that the two streams are in cocurrent flow. What are the new outlet temperatures of hydrocarbon and water, if the overall heat-transfer coefficient is unchanged?

1.80. A reaction mixture is heated in a vessel fitted with an agitator and a steam coil of area 10 m^2 fed with steam at 393 K. The heat capacity of the system is equal to that of 500 kg of water. The overall coefficient of heat transfer from the vessel of area 5 m^2 is 10 W/m^2 K. It takes 1800 s to heat the contents from ambient temperature of 293 to 333 K. How long will it take to heat the system to 363 K, and what is the maximum temperature that can be reached?

$$\text{Specific heat capacity of water} = 4200 \, \text{J/kg K}.$$

1.81. A pipe, 50 mm outside diameter, is carrying steam at 413 K, and the coefficient of heat transfer from its outer surface to the surroundings at 288 K is 10 W/m^2 K. What is the heat loss per unit length?

It is desired to add lagging of thermal conductivity 0.03 W/m K as a thick layer to the outside of the pipe in order to cut heat losses by 90%. If the heat transfer from the outside surface of the lagging is 5 W/m^2 K, what thickness of lagging is required?

1.82. It takes 1800 s (0.5 h) to heat a tank of liquid from 293 to 333 K using steam supplied to an immersed coil when the steam temperature is 383 K. How long will it take when the steam temperature is raised to 393 K? The overall heat-transfer coefficient from the steam coil to the tank is 10 times the coefficient from the tank to the surroundings at a temperature of 293 K, and the area of the steam coil is equal to the outside area of the tank.

1.83. A thermometer is situated in a duct in an airstream that is at a constant temperature. The reading varies with the gas flowrate as follows:

Air Velocity (m/s)	Thermometer Reading (K)
6.1	553
7.6	543
12.2	533

The wall of the duct and the gas stream are at somewhat different temperatures. If the heat-transfer coefficient for radiant heat transfer from the wall to the thermometer remains constant and the heat-transfer coefficient between the gas stream and thermometer is proportional to the 0.8 power of the velocity, what is the true temperature of the airstream? Neglect any other forms of heat transfer.

2.1. Ammonia gas is diffusing at a constant rate through a layer of stagnant air 1 mm thick. Conditions are fixed so that the gas contains 50% by volume of ammonia at one boundary of the stagnant layer. The ammonia diffusing to the other boundary is quickly absorbed, and the concentration is negligible at that plane. The temperature is 295 K and the pressure is atmospheric, and under these conditions, the diffusivity of ammonia in air is 0.18 cm^2/s. Calculate the rate of diffusion of ammonia through the layer.

2.2. A simple rectifying column consists of a tube arranged vertically and supplied at the bottom with a mixture of benzene and toluene as vapour. At the top, a condenser returns some of the product as a reflux that flows in a thin film down the inner wall of the tube. The tube is insulated, and heat losses can be neglected. At one point in the column, the vapour contains 70 mol% benzene, and the adjacent liquid reflux contains 59 mol% benzene. The temperature at this point is 365 K. Assuming the diffusional resistance to vapour transfer to be equivalent to the diffusional resistance of a stagnant vapour layer of 0.2 mm thick, calculate the rate of interchange of benzene and toluene between vapour and liquid. The molar latent heats of the two materials can be taken as equal. The vapour pressure of toluene at 365 K is 54.0 kN/m^2, and the diffusivity of the vapours is 0.051 cm^2/s.

2.3. By what percentage would the rate of absorption be increased or decreased by increasing the total pressure from 100 to 200 kN/m^2 in the following cases?

(a)   The absorption of ammonia from a mixture of ammonia and air containing 10% of ammonia by volume, using pure water as solvent. Assume that all the resistance to mass transfer lies within the gas phase.

(b)   The same conditions as (a) but the absorbing solution exerts a partial vapour pressure of ammonia of 5 kN/m^2.

The diffusivity can be assumed to be inversely proportional to the absolute pressure.

2.4. In the Danckwerts model of mass transfer, it is assumed that the fractional rate of surface renewal s is constant and independent of surface age. Under such conditions, the expression for the surface age distribution function is se^{-st}. If the fractional rate of surface renewal was proportional to surface age (say $s = bt$, where b is a constant), show that the surface age distribution function would then assume the form

$$\left(\frac{2b}{\pi}\right)^{1/2} e^{-bt^2/2}.$$

2.5. By considering of the appropriate element of a sphere, show that the general equation for molecular diffusion in a stationary medium and in the absence of a chemical reaction is

$$\frac{\partial C}{\partial t} = D\left(\frac{\partial^2 C}{\partial r^2} + \frac{1}{r^2}\frac{\partial^2 C}{\partial \beta^2} + \frac{1}{r^2 \sin^2\beta}\frac{\partial^2 C}{\partial \varphi^2} + \frac{2}{r}\frac{\partial C}{\partial r} + \frac{\cot\beta}{r^2}\frac{\partial C}{\partial \beta}\right),$$

where C is the concentration of the diffusing substance, D is the molecular diffusivity, t is time, and r, β, and ϕ are spherical polar coordinates, β being the latitude angle.

2.6. Prove that for equimolecular counterdiffusion from a sphere to a surrounding stationary, infinite medium, the Sherwood number based on the diameter of the sphere is equal to 2.

2.7. Show that the concentration profile for unsteady-state diffusion into a bounded medium of thickness L, when the concentration at the interface is suddenly raised to a constant value C_i and kept constant at the initial value of C_0 at the other boundary, is

$$\frac{C - C_o}{C_i - C_o} = 1 - \frac{z}{L} = \frac{2}{\pi} \left[\sum_{n=1}^{n=\infty} \frac{1}{n} \exp\left(-\frac{n^2 \pi^2 Dt}{L^2}\right) \sin\frac{n z \pi}{L} \right].$$

Assume the solution to be the sum of the solution for infinite time (steady-state part) and the solution of a second unsteady-state part, which simplifies the boundary conditions for the second part.

2.8. Show that under the conditions specified in Problem 10.7 and assuming the Higbie model of surface renewal, the average mass flux at the interface is given by

$$(N_A)_t = (C_i - C_o)\frac{D}{L}\left\{1 + \frac{2L^2}{\pi^2 Dt} \sum_{n=1}^{n=\infty} \left[\frac{\pi^2}{6} - \frac{1}{n^2}\exp\left(-\frac{n^2 \pi^2 Dt}{L^2}\right)\right]\right\}.$$

Use the relation $\sum_{n=1}^{n=\infty} \frac{1}{n^2} = \frac{\pi^2}{6}$.

2.9. According to the simple penetration theory, the instantaneous mass flux, $(N_A)_t$, is

$$(N_A)_t = (C_i - C_o)\left(\frac{D}{\pi t}\right)^{0.5}$$

What is the equivalent expression for the instantaneous heat flux under analogous conditions?

Pure sulphur dioxide is absorbed at 295 K and atmospheric pressure into a laminar water jet. The solubility of SO_2, assumed constant over a small temperature range, is 1.54 kmol/m^3 under these conditions, and the heat of solution is 28 kJ/kmol.

Calculate the resulting jet surface temperature if the Lewis number is 90. Neglect heat transfer between the water and the gas.

2.10. In a packed column, operating at approximately atmospheric pressure and 295 K, a 10% ammonia–air mixture is scrubbed with water, and the concentration of ammonia is reduced to 0.1%. If the whole of the resistance to mass transfer may be regarded as lying within a thin laminar film on the gas side of the gas-liquid interface, derive from first principles an expression for the rate of absorption at any position in the column. At some intermediate point where the ammonia concentration in the gas phase has been reduced to 5%, the partial pressure of ammonia in equilibrium with the aqueous solution is 660 N/m^2, and the transfer rate is 10^{-3} kmol/m^{2}s. What is the thickness of the hypothetical gas film if the diffusivity of ammonia in air is 0.24 cm^2/s?

2.11. An open bowl, 0.3 m in diameter, contains water at 350 K evaporating into the atmosphere. If the air currents are sufficiently strong to remove the water vapour as it is formed and if the resistance to its mass transfer in air is equivalent to that of a 1 mm layer for conditions of molecular diffusion, what will be the rate of cooling due to evaporation? The water can be considered as well mixed, and the water equivalent of the system is equal to 10 kg. The diffusivity of water vapour in air may be taken as 0.20 cm²/s and the kilogram molecular volume at NTP as 22.4 m³.

2.12. Show by substitution that when a gas of solubility C^+ is absorbed into a stagnant liquid of infinite depth, the concentration at time t and depth x is

$$C^+ \operatorname{erfc} \frac{2}{2\sqrt{Dt}}.$$

Hence, on the basis of the simple penetration theory, show that the rate of absorption in a packed column will be proportional to the square root of the diffusivity.

2.13. Show that in steady-state diffusion through a film of liquid, accompanied by a first-order irreversible reaction, the concentration of solute in the film at depth z below the interface is given by

$$\frac{C}{C_i} = \frac{\sinh \sqrt{(k/D)}(zL - z)}{\sinh \sqrt{(k/D)}zL}$$

if $C = 0$ at $z = Z_L$ and $C = C_i$ at $z = 0$, corresponding to the interface. Hence, show that according to the 'film theory' of gas absorption, the rate of absorption per unit area of interface N_A is given by

$$N_A = K_L C_i \frac{\beta}{\tanh \beta},$$

where $\beta = \sqrt{(Dk)}/K_L$, D is the diffusivity of the solute, k the rate constant of the reaction, K_L the liquid film mass-transfer coefficient for physical absorption, C_i the concentration of solute at the interface, z the distance normal to the interface, and z_L the liquid film thickness.

2.14. The diffusivity of the vapour of a volatile liquid in air can be conveniently determined by Winkelmann's method, in which liquid is contained in a narrow diameter vertical tube maintained at a constant temperature, and an airstream is passed over the top of the tube sufficiently rapidly to ensure that the partial pressure of the vapour there remains approximately zero. On the assumption that the vapour is transferred from the surface of the liquid to the airstream by molecular diffusion, calculate the diffusivity of carbon tetrachloride vapour in air at 321 K and atmospheric pressure from the following experimentally obtained data:

Time From Commencement of Experiment (ks)	Liquid Level (cm)
0	0.00
1.6	0.25
11.1	1.29
27.4	2.32
80.2	4.39
117.5	5.47
168.6	6.70
199.7	7.38
289.3	9.03
383.1	10.48

The vapour pressure of carbon tetrachloride at 321 K is 37.6 kN/m^2, and the density of the liquid is 1540 kg/m^3. Take the kilogram molecular volume as 22.4 m^3.

2.15. Ammonia is absorbed in water from a mixture with air using a column operating at atmospheric pressure and 295 K. The resistance to transfer can be regarded as lying entirely within the gas phase. At a point in the column, the partial pressure of the ammonia is 6.6 kN/m^2. The back pressure at the water interface is negligible, and the resistance to transfer may be regarded as lying in a stationary gas film 1 mm thick. If the diffusivity of ammonia in air is 0.236 cm^2/s, what is the transfer rate per unit area at that point in the column? If the gas were compressed to 200 kN/m^2 pressure, how would the transfer rate be altered?

2.16. What are the general principles underlying the two-film, penetration and film-penetration theories for mass transfer across a phase boundary? Give the basic differential equations that have to be solved for these theories with the appropriate boundary conditions.

According to the penetration theory, the instantaneous rate of mass transfer per unit area (N_A) at some time t after the commencement of transfer is given by

$$(N_A)_t = \Delta C \sqrt{\frac{D}{\pi t}},$$

where ΔC is the concentration force and D is the diffusivity.

Obtain expressions for the average rates of transfer on the basis of the Higbie and Danckwerts assumptions.

2.17. A solute diffuses from a liquid surface at which its molar concentration is C_i into a liquid with which it reacts. The mass-transfer rate is given by Fick's law, and the reaction is first order with respect to the solute. In a steady-state process, the diffusion rate falls at a depth L to one-half the value at the interface. Obtain an expression for the concentration C of solute at a depth z from the surface in terms of the molecular diffusivity D and the reaction rate constant k. What is the molar flux at the surface?

2.18. 4 cm^3 of mixture formed by adding 2 cm^3 of acetone to 2 cm^3 of dibutyl phthalate is contained in a 6 mm diameter vertical glass tube immersed in a thermostat maintained at 315 K. A stream of air at 315 K and atmospheric pressure is passed over the open top of the tube to maintain a zero partial pressure of acetone vapour at that point. The liquid level is initially 11.5 mm below the top of the tube, and the acetone vapour is transferred to the airstream by molecular diffusion alone. The dibutyl phthalate can be regarded as completely non-volatile, and the partial pressure of acetone vapour may be calculated from Raoult's law on the assumption that the density of dibutyl phthalate is sufficiently greater than that of acetone for the liquid to be completely mixed.

Calculate the time taken for the liquid level to fall to 5 cm below the top of the tube, neglecting the effects of bulk flow in the vapour:

Kilogram molecular volume $= 22.4$ m^3.

Molecular weights of acetone and dibutyl phthalate $= 58$ and 278 kg/kmol, respectively.

Liquid densities of acetone and dibutyl phthalate $= 764$ and 1048 kg/m^3, respectively.

Vapour pressure of acetone at 315 K $= 60.5$ kN/m^2.

Diffusivity of acetone vapour in air at 315 K $= 1.23 \times 10^{-5}$ m^2/s.

2.19. A crystal is suspended in fresh solvent, and 5% of the crystal dissolves in 300 s. How long will it take before 10% of the crystal has dissolved? Assume that the solvent can be regarded as infinite in extent that the mass transfer in the solvent is governed by Fick's second law of diffusion and may be represented as a unidirectional process and that changes in the surface area of the crystal may be neglected. Start your derivations using Fick's second law.

2.20. In a continuous steady-state reactor, a slightly soluble gas is absorbed into a liquid in which it dissolves and reacts, the reaction being second order with respect to the dissolved gas. Calculate the reaction rate constant on the assumption that the liquid is semi-infinite in extent and that mass-transfer resistance in the gas phase is negligible. The diffusivity of the gas in the liquid is 10^{-8} m^2/s, the gas concentration in the liquid falls to one-half of its value in the liquid over a distance of 1 mm, and the rate of absorption at the interface is 4×10^{-6} kmol/m^2 s.

2.21. Experiments have been carried out on the mass transfer of acetone between air and a laminar water jet. Assuming that desorption produces random surface renewal with a constant fractional rate of surface renewal, s, but an upper limit on surface age equal to the life of the jet, τ, shows that the surface age frequency distribution function, $\phi(t)$, for this case is given by

$$\phi(t) = s \exp \frac{-st}{1 - \exp(-st)} \quad \text{for } 0 < t < \tau$$

$$\phi(t) = 0 \quad \text{for } t > \tau.$$

Hence, show that the enhancement, E, for the increase in value of the liquid-phase mass-transfer coefficient is

$$E = \frac{(\pi s\tau)^{1/2} \text{erf } (s\tau)^{1/2}}{2[1 - \exp(-st)]}$$

where E is defined as the ratio of the mass-transfer coefficient predicted by the conditions described above to the mass-transfer coefficient obtained from the penetration theory for a jet with an undisturbed surface. Assume that the interfacial concentration of acetone is practically constant.

2.22. Solute gas is diffusing into a stationary liquid, virtually free of solvent, and of sufficient depth for it to be regarded as semi-infinite in extent. In what depth of fluid below the surface will 90% of the material that has been transferred across the interface have accumulated in the first minute?

Diffusivity of gas in liquid $= 10^{-9} \text{m}^2/\text{s}$.

2.23. A chamber, of volume 1 m^3, contains air at a temperature of 293 K and a pressure of 101.3 kN/m^2, with a partial pressure of water vapour of 0.8 kN/m^2. A bowl of liquid with a free surface of 0.01 m^2 and maintained at a temperature of 303 K is introduced into the chamber. How long will it take for the air to become 90% saturated at 293 K, and how much water must be evaporated?

The diffusivity of water vapour in air is 2.4×10^{-5} m^2/s, and the mass-transfer resistance is equivalent to that of a stagnant gas film of thickness 0.25 mm. Neglect the effects of bulk flow.

Saturation vapour pressure of water $= 4.3$ kN/m^2 at 303 K and 2.3 kN/m^2 at 293 K.

2.24. A large deep bath contains molten steel, the surface of which is in contact with air. The oxygen concentration in the bulk of the molten steel is 0.03% by mass, and the rate of transfer of oxygen from the air is sufficiently high to maintain the surface layers saturated at a concentration of 0.16% by weight. The surface of the liquid is disrupted by gas bubbles rising to the surface at a frequency of 120 bubbles per m^2 of surface per second; each bubble disrupts and mixes about 15 cm^2 of the surface layer into the bulk.

On the assumption that the oxygen transfer can be represented by a surface renewal model, obtain the appropriate equation for mass transfer by starting with Fick's second law of diffusion and calculate the following:

(a) The mass-transfer coefficient
(b) The mean mass flux of oxygen at the surface
(c) The corresponding film thickness for a film model, giving the same mass-transfer rate

Diffusivity of oxygen in steel $= 1.2 \times 10^{-8}$ m^2/s
Density of molten steel $= 7100$ kg/m^3

2.25. Two large reservoirs of gas are connected by a pipe of length $2\,L$ with a full-bore valve at its midpoint. Initially, a gas A fills one reservoir and the pipe up to the valve, and gas B fills the other reservoir and the remainder of the pipe. The valve is opened rapidly, and the gases in the pipe mix by molecular diffusion.

Obtain an expression for the concentration of gas A in that half of the pipe in which it is increasing, as a function of distance y from the valve and time t after opening. The whole system is at a constant pressure, and the ideal gas law is applicable to both gases. It may be assumed that the rate of mixing in the vessels is high so that the gas concentrations at the two ends of the pipe do not change.

2.26. A pure gas is absorbed into a liquid with which it reacts. The concentration in the liquid is sufficiently low for the mass transfer to be governed by Fick's law, and the reaction is first order with respect to the solute gas. It may be assumed that the film theory may be applied to the liquid and that the concentration of solute gas falls from the saturation value to zero across the film. Obtain an expression for the mass-transfer rate across the gas-liquid interface in terms of the molecular diffusivity, D, the first-order reaction rate constant k, the film thickness L, and the concentration C_{AS} of solute in a saturated solution. The reaction is initially carried out at 293 K. By what factor will the mass-transfer rate across the interface change, if the temperature is raised to 313 K?

Reaction rate constant at 293 K	$= 2.5 \times 10^{-6}\,\text{s}^{-1}$
Energy of activation for reaction (in Arrhenius equation)	$= 26{,}430$ kJ/kmol
Universal gas constant $\mathbf{R}$	$= 8314$ J/kmol K
Molecular diffusivity D	$= 10^{-9}\,\text{m}^2/\text{s}$
Film thickness L	$= 10$ mm

Solubility of gas at 313 K is 80% of solubility at 293 K.

2.27. Using Maxwell's law of diffusion, obtain an expression for the effective diffusivity for a gas $\mathbf{A}$ in a binary mixture of $\mathbf{B}$ and $\mathbf{C}$, in terms of the diffusivities of $\mathbf{A}$ in the two pure components and the molar concentrations of $\mathbf{A}$, $\mathbf{B}$, and $\mathbf{C}$.

Carbon dioxide is absorbed in water from a 25% mixture in nitrogen. How will its absorption rate compare with that from a mixture containing 35% carbon dioxide, 40% hydrogen, and 25% nitrogen? It may be assumed that the gas-film resistance is controlling, that the partial pressure of carbon dioxide at the gas-liquid interface is negligible, and that the two-film theory is applicable, with the gas-film thickness the same in the two cases.

Diffusivity of CO_2, in hydrogen $3.5 \times 10^{-5}\,\text{m}^2/\text{s}$ and in nitrogen $1.6 \times 10^{-5}\,\text{m}^2/\text{s}$.

2.28. Given that, from the penetration theory for mass transfer across an interface, the instantaneous rate of mass transfer is inversely proportional to the square root of the time of exposure, obtain a relationship between exposure time in the Higbie model and surface renewal rate in the Danckwerts model that will give the same average mass-transfer rate. The age distribution function and average mass-transfer rate from the Danckwerts theory must be derived from first principles.

2.29. Ammonia is absorbed in a falling film of water in an absorption apparatus, and the film is disrupted and mixed at regular intervals as it flows down the column. The mass-transfer rate is calculated from the penetration theory on the assumption that all the relevant conditions apply. It is found from measurements that the mass-transfer rate immediately before mixing is only 16% of that calculated from the theory and the difference has been attributed to the existence of a surface film that remains intact and unaffected by the mixing process. If the liquid mixing process takes place every second, what thickness of surface film would account for the discrepancy?

$$\text{Diffusivity of ammonia in water} = 1.76 \times 10^{-9} \text{m}^2/\text{s}$$

2.30. A deep pool of ethanol is suddenly exposed to an atmosphere consisting of pure carbon dioxide, and unsteady-state mass transfer, governed by Fick's Law, takes place for 100 s. What proportion of the absorbed carbon dioxide will have accumulated in the 1 mm thick layer of ethanol closest to the surface?

$$\text{Diffusivity of carbon dioxide in ethanol} = 4 \times 10^{-9} \text{m}^2/\text{s}$$

2.31. A soluble gas is absorbed into a liquid with which it undergoes a second-order irreversible reaction. The process reaches a steady state with the surface concentration of reacting material remaining constant at C_{As} and the depth of penetration of the reactant being small compared with the depth of liquid that can be regarded as infinite in extent. Derive the basic differential equation for the process, and from this, derive an expression for the concentration and mass-transfer rate (moles per unit area and unit time) as a function of depth below the surface. Assume that mass transfer is by molecular diffusion.

If the surface concentration is maintained at 0.04 kmol/m³, the second-order rate constant k_2 is 9.5×10^3 m³/kmol s, and the liquid-phase diffusivity D is 1.8×10^{-9} m²/s, calculate the following:

(a) The concentration at a depth of 0.1 mm
(b) The molar rate of transfer at the surface (kmol/m²s)
(c) The molar rate of transfer at a depth of 0.1 mm
  It may be noted that if

$$\frac{dC_A}{dy} = q, \quad \text{then} \quad \frac{d^2C_A}{dy^2} = q\frac{dq}{dC_A}$$

2.32. In calculating the mass-transfer rate from the *penetration theory*, two models for the age distribution of the surface elements are commonly used—those due to Higbie and to

Danckwerts. Explain the difference between the two models and give examples of situations in which each of them would be appropriate.

In the Danckwerts model, it is assumed that the elements of the surface have an age distribution ranging from zero to infinity. Obtain the age distribution function for this model and apply it to obtain the average mass-transfer coefficient at the surface, given that from the penetration theory the mass-transfer coefficient for surface of age t is $\sqrt{D/(\pi t)}$, where D is the diffusivity.

If for unit area of surface the surface renewal rate is s, by how much will the mass-transfer coefficient be changed if no surface has an age exceeding $2/s$?

If the probability of surface renewal is linearly related to age, as opposed to being constant, obtain the corresponding form of the age distribution function.

It may be noted that

$$\int_0^\infty e^{-x^2}\,dx = \frac{\sqrt{\pi}}{2}$$

2.33. Explain the basis of the *penetration theory* for mass transfer across a phase boundary. What are the assumptions in the theory that lead to the result that the mass-transfer rate is inversely proportional to the square root of the time for which a surface element has been expressed? (Do *not* present a solution of the differential equation.) Obtain the age distribution function for the surface:

(a) On the basis of the Danckwerts' assumption that the probability of surface renewal is independent of its age

(b) On the assumption that the probability of surface renewal increases linearly with the age of the surface

Using the Danckwerts surface renewal model, estimate the following:

(a) At what age of a surface element is the mass-transfer rate equal to the mean value for the whole surface for a surface renewal rate (5) of 0.01 m^2/m^2s?

(b) For what proportion of the total mass transfer is surface of an age exceeding 10 s responsible?

2.34. At a particular location in a distillation column, where the temperature is 350 K and the pressure is 500 mm Hg, the mol fraction of the more volatile component in the vapour is 0.7 at the interface with the liquid and 0.5 in the bulk of the vapour. The molar latent heat of the more volatile component is 1.5 times that of the less volatile. Calculate the mass-transfer rates (kmol $m^{-2}s^{-1}$) of the two components. The resistance to mass transfer in the vapour may be considered to lie in a stagnant film of thickness 0.5 mm at the interface. The diffusivity in the vapour mixture is 2×10^{-5} m^2s^{-1}.

Calculate the mol fractions and concentration gradients of the two components at the midpoint of the film. Assume that the ideal gas law is applicable and that the universal gas constant $\mathbf{R} = 8314$ J/kmol K.

2.35. For the diffusion of carbon dioxide at atmospheric pressure and a temperature of 293 K, at what time will the concentration of solute 1 mm below the surface reach 1% of the value at the surface? At that time, what will the mass-transfer rate (kmol m^{-2}s^{-1}) be
(a)   at the free surface,
(b)   at the depth of 1 mm?
   The diffusivity of carbon dioxide in water may be taken as 1.5×10^{-9} m^{2}s^{-1}. In the literature, Henry's law constant **K** for carbon dioxide at 293 K is given as 1.08×10^6 where $K = P/X$, P being the partial pressure of carbon dioxide (mm Hg) and X the corresponding mol fraction in the water.

2.36. Experiments are carried out at atmospheric pressure on the absorption into water of ammonia from a mixture of hydrogen and nitrogen, both of which may be taken as insoluble in the water. For a constant mol fraction of 0.05 of ammonia, it is found that the absorption rate is 25% higher when the molar ratio of hydrogen to nitrogen is changed from 1:1 to 4:1. Is this result consistent with the assumption of a steady-state gas-film controlled process, and if not, what suggestions have you to make to account for the discrepancy?

> Neglect the partial pressure attributable to ammonia in the bulk solution.
> Diffusivity of ammonia in hydrogen $= 52 \times 10^{-6}$ m^{2}s^{-1}
> Diffusivity of ammonia in nitrogen $= 23 \times 10^{-6}$ m^{2}s^{-1}

2.37. Using a steady-state film model, obtain an expression for the mass-transfer rate across a laminar film of thickness L in the vapour phase for the more volatile component in a binary distillation process
(a)   where the molar latent heats of two components are equal,
(b)   where the molar latent heat of the less volatile component (LVC) is f times that of the more volatile component (MVC).
   For the case where the ratio of the molar latent heats f is 1.5, what is the ratio of the mass-transfer rate in case (b) to that in case (a) when the mole fraction of the MVC falls from 0.75 to 0.65 across the laminar film?

2.38. On the assumptions involved in the penetration theory of mass transfer across a phase boundary, the concentration C_A of a solute A at a depth y below the interface at a time t after the formation of the interface is given by

$$\frac{C_A}{C_{Ai}} = \mathrm{erfc}\left[\frac{y}{2\sqrt{(Dt)}}\right]$$

where C_{Ai} is the interface concentration, assumed constant, and D is the molecular diffusivity of the solute in the solvent. The solvent initially contains no dissolved solute. Obtain an expression for the molar rate of transfer of A per unit area at time t and depth y and at the free surface (at $y = 0$).

In a liquid-liquid extraction unit, spherical drops of solvent of uniform size are continuously fed to a continuous phase of lower density that is flowing vertically upwards and hence countercurrently with respect to the droplets. The resistance to mass transfer may be regarded as lying wholly within the drops, and the penetration theory may be applied. The upward velocity of the liquid, which may be taken as uniform over the cross-section of the vessel, is one-half of the terminal falling velocity of the droplets in the still liquid.

Occasionally, two droplets coalesce on formation giving rise to a single drop of twice the volume. What is the ratio of the mass-transfer rate (kmol/s) to a coalesced drop to that of a single droplet when each has fallen the same distance, which is to the bottom of the equipment?

The fluid resistance force acting on the droplet should be taken as that given by Stokes' law, which is $3\pi\mu du$ where μ is the viscosity of the continuous phase, d the drop diameter, and u its velocity relative to the continuous phase.

It may be noted that

$$\text{erfc}(x) = \frac{2}{\sqrt{\pi}} \int_x^\infty e^{-x^2} dx.$$

2.39. In a drop extractor, a dense organic solvent is introduced in the form of spherical droplets of diameter d and extracts a solute from an aqueous stream that flows upwards at a velocity u_o equal to half the terminal falling velocity of the droplets. On increasing the flowrate of the aqueous stream by 50% whilst maintaining the solvent rate constant, it is found that the average concentration of solute in the outlet stream of organic phase is decreased by 10%. By how much would the effective droplet size have had to change to account for this reduction in concentration? Assume that the penetration theory is applicable with the mass-transfer coefficient inversely proportional to the square root of the contact time between the phases and the continuous-phase resistance small compared with that in the droplets. The drag force F acting on the falling droplets may be calculated from Stokes' law, $F = 3\pi\mu du_o$, where μ is the viscosity of the aqueous phase. Clearly, state any assumptions made in your calculation.

2.40. According to the penetration theory for mass transfer across an interface, the ratio of the concentration C_A at a depth y and time t to the surface concentration C_{As} if the liquid is initially free of solute is given by

$$\frac{C_A}{C_{As}} = \text{erfc}\frac{y}{2\sqrt{(Dt)}}$$

where D is the diffusivity. Obtain a relation for the instantaneous rate of mass transfer at time t both at the surface ($y = 0$) and at a depth y.

What proportion of the total solute transferred into the liquid in the first 90 s of exposure will be retained in a 1 mm layer of liquid at the surface, and what proportion will be retained in the next 0.5 mm? Take the diffusivity as 2×10^{-9} m^2/s.

2.41. Obtain an expression for the effective diffusivity of component A in a gaseous mixture of A, B, and C, in terms of the binary diffusion coefficients D_{AB} for A in B and D_{AC} for A in C.

The gas-phase mass-transfer coefficient for the absorption of ammonia into water from a mixture of composition NH$_3$ 20%, N$_2$ 73%, and H$_2$ 7% is found experimentally to be 0.030 m/s. What would you expect the transfer coefficient to be for a mixture of composition NH$_3$ 5%, N$_2$ 60%, and H$_2$ 35%? All compositions are given on a molar basis. The total pressure and temperature are the same in both cases. The transfer coefficients are based on a steady-state film model, and the effective film thickness may be assumed constant. Neglect the solubility of N$_2$ and H$_2$ in water:

$$\text{Diffusivity of NH}_3 \text{ in N}_2 = 23 \times 10^{-6} \text{m}^2/\text{s}.$$
$$\text{Diffusivity of NH}_3 \text{ in H}_2 = 52 \times 10^{-6} \text{m}^2/\text{s}.$$

2.42. State the assumptions made in the penetration theory for the absorption of a pure gas into a liquid. The surface of an initially solute-free liquid is suddenly exposed to a soluble gas, and the liquid is sufficiently deep for no solute to have time to reach the bottom of the liquid. Starting with Fick's second law of diffusion, obtain an expression for (i) the concentration and (ii) the mass-transfer rate at a time t and a depth y below the surface.

After 50 s, at what depth y will the concentration have reached one-tenth the value at the surface? What is the mass-transfer rate (i) at the surface and (ii) at the depth y, if the surface concentration has a constant value of 0.1 kmol/m^3?

2.43. In a drop extractor, liquid droplets of approximate uniform size and spherical shape are formed at a series of nozzles and rise countercurrently through the continuous phase that is flowing downwards at a velocity equal to one-half of the terminal rising velocity of the droplets. The flowrates of both phases are then increased by 25%. Because of the greater shear rate at the nozzles, the mean diameter of the droplets is however only 90% of the original value. By what factor will the overall mass-transfer rate change?

It may be assumed that the penetration model may be used to represent the mass-transfer process. The depth of penetration is small compared with the radius of the droplets, and the effects of surface curvature may be neglected. From the penetration theory, the concentration C_A at a depth y below the surface at time t is given by

$$\frac{C_A}{C_{As}} = \text{erfc} \left[\frac{y}{2\sqrt{(Dt)}} \right] \quad \text{where} \quad \text{erfc } X = \frac{2}{\sqrt{\pi}} \int_0^\infty e^{-x^2} dx$$

where C_{As} is the surface concentration for the drops (assumed constant) and D is the diffusivity in the dispersed (droplet) phase. The droplets may be assumed to rise at their terminal velocities, and the drag force F on the droplet may be calculated from Stokes' law, $F = 3\pi\mu du$.

2.44. According to Maxwell's law, the partial pressure gradient in a gas that is diffusing in a two-component mixture is proportional to the product of the molar concentrations of the two components multiplied by its mass-transfer velocity relative to that of the second component. Show how this relationship can be adapted to apply to the absorption of a soluble gas from a multicomponent mixture in which the other gases are insoluble and obtain an effective diffusivity for the multicomponent system in terms of the binary diffusion coefficients.

Carbon dioxide is absorbed in alkaline water from a mixture consisting of 30% CO_2 and 70% N_2, and the mass-transfer rate is 0.1 kmol/s. The concentration of CO_2 in the gas in contact with the water is effectively zero. The gas is then mixed with an equal molar quantity of a second gas stream of molar composition 20% CO_2, 50% N_2, and 30% H_2. What will be the new mass-transfer rate if the surface area, temperature, and pressure remain unchanged? It may be assumed that a steady-state film model is applicable and that the film thickness is unchanged:

$$\text{Diffusivity of } CO_2 \text{ in } N_2 = 16 \times 10^{-6} \text{m}^2/\text{s}$$
$$\text{Diffusivity of } CO_2 \text{ in } H_2 = 35 \times 10^{-6} \text{m}^2/\text{s}$$

2.45. What is the penetration theory for mass transfer across a phase boundary? Give the details of the underlying assumptions.

From the penetration theory, the mass-transfer rate per unit area N_A is given in terms of the concentration difference ΔC_A between the interface and the bulk fluid, the molecular diffusivity D, and the age t of the surface element by

$$N_A = \sqrt{\frac{D}{\pi t}} \Delta C_A \text{ kmol/m}^2\text{s (in SI units)}$$

What is the mean rate of transfer if all the elements of the surface are exposed for the same time t_e before being remixed with the bulk?

Danckwerts assumed a random surface renewal process in which the probability of surface renewal is independent of its age. If s is the fraction of the total surface renewed per unit time, obtain the age distribution function for the surface and show that the mean mass-transfer rate N_A over the whole surface is

$$N_A = \sqrt{Ds} \Delta C_A \left(\text{kmol/m}^2\text{s, in SI units} \right)$$

In a particular application, it is found that the older surface is renewed more rapidly than the recently formed surface and that, after a time $1/s$, the surface renewal rate doubles, which increases from s to $2s$. Obtain the new age distribution function.

2.46. Derive the partial differential equation for unsteady-state unidirectional diffusion accompanied by an nth-order chemical reaction (rate constant k):

$$\frac{\partial C_A}{\partial t} = D\frac{\partial^2 C_A}{\partial y^2} - kC_A{}^n$$

where C_A is the molar concentration of reactant at position y at time t.

Explain why, when applying the equation to reaction in a porous catalyst particle, it is necessary to replace the molecular diffusivity D by an effective diffusivity D_e.

Solve the above equation for a first-order reaction under steady-state conditions, and obtain an expression for the mass-transfer rate per unit area at the surface of a catalyst particle that is in the form of a thin platelet of thickness $2L$.

Explain what is meant by the effectiveness factor η for a catalyst particle, and show that it is equal to $\frac{1}{\phi}\tanh\phi$ for the platelet referred to previously, where ϕ is the Thiele modulus, $L\sqrt{\frac{k}{D_e}}$.

For the case where there is a mass-transfer resistance in the fluid external to the particle (mass-transfer coefficient h_D), express the mass-transfer rate in terms of the bulk concentration C_{Ao}, rather than the concentration C_{As} at the surface of the particle.

For a bed of catalyst particles in the form of flat platelets, it is found that the mass-transfer rate is increased by a factor of 1.2 if the velocity of the external fluid is doubled. The mass-transfer coefficient h_D is proportional to the velocity raised to the power of 0.6. What is the value of h_D at the original velocity?

$$k = 1.6 \times 10^{-3}\,\text{s}^{-1},\ D_e = 10^{-8}\,\text{m}^2\text{s}^{-1}\ \text{catalyst thickness } (2L) = 10\,\text{mm}$$

2.47. Explain the basic concepts underlying the two-film theory for mass transfer across a phase boundary, and obtain an expression for film thickness.

Water evaporates from an open bowl at 349 K at the rate of 4.11×10^{-3} kg/m²s. What is the effective gas-film thickness?

The water is replaced by ethanol at 343 K. What will be its rate of evaporation (in kg/m²s) if the film thickness is unchanged?

What proportion of the total mass transfer will be attributable to bulk flow?

Data are as follows:

Vapour pressure of water at 349 $K = 301$ mm Hg.
Vapour pressure of ethanol at 343 $K = 541$ mm Hg.
Neglect the partial pressure of vapour in the surrounding atmosphere.
Diffusivity of water vapour in air $= 26 \times 10^{-6}$ m²/s.
Diffusivity of ethanol in air $= 12 \times 10^{-6}$ m²/s.
Density of mercury $= 13,600$ kg/m³.
Universal gas constant $R = 8314$ J/kmol K.

3.1. Calculate the thickness of the boundary layer at a distance of 75 mm from the leading edge of a plane surface over which water is flowing at the rate of 3 m/s. Assume that the flow in the boundary layer is streamline and that the velocity u of the fluid at a distance y from the surface may be represented by the relation $u = a + by + cy^2 + dy^3$, where the coefficients a, b, c, and d are independent of y. Take the viscosity of water as 1 mN s/m².

3.2. Water flows at a velocity of 1 m/s over a plane surface 0.6 m wide and 1 m long. Calculate the total drag force acting on the surface if the transition from streamline to turbulent flow in the boundary layer occurs when the Reynolds group $Re_x = 10^5$.

3.3. Calculate the thickness of the boundary layer at a distance of 150 mm from the leading edge of a surface over which oil, of viscosity 50 mN s/m² and density 990 kg/m³, flows with a velocity of 0.3 m/s. What is the displacement thickness of the boundary layer?

3.4. Calculate the thickness of the laminar sublayer when benzene flows through a pipe of 50 mm diameter at 0.003 m³/s. What is the velocity of the benzene at the edge of the laminar sublayer? Assume fully developed flow exists within the pipe.

3.5. Air is flowing at a velocity of 5 m/s over a plane surface. Derive an expression for the thickness of the laminar sublayer and calculate its value at a distance of 1 m from the leading edge of the surface.

Assume that within the boundary layer outside the laminar sublayer, the velocity of flow is proportional to the one-seventh power of the distance from the surface and that the shear stress R at the surface is given by

$$\frac{R}{\rho u_s^2} = 0.03 \left(\frac{u_s \rho x}{\mu} \right)^{-0.2},$$

where ρ is the density of the fluid (1.3 kg/m³ for air), μ the viscosity of the fluid (17×10^{-6} N s/m² for air), u_s the stream velocity (m/s), and x the distance from the leading edge (m).

3.6. Obtain the momentum equation for an element of boundary layer. If the velocity profile in the laminar region may be represented approximately by a sine function, calculate the boundary-layer thickness in terms of distance from the leading edge of the surface.

3.7. Explain the concepts of 'momentum thickness' and 'displacement thickness' for the boundary layer formed during flow over a plane surface. Develop a similar concept to displacement thickness in relation to heat flux across the surface for laminar flow and heat transfer by thermal conduction, for the case where the surface has a constant temperature and the thermal boundary layer is always thinner than the velocity boundary layer. Obtain an expression for this 'thermal thickness' in terms of the thicknesses of the velocity and temperature boundary layers.

Similar forms of cubic equations may be used to express velocity and temperature variations with distance from the surface.

For a Prandtl number, Pr, less than unity, the ratio of the temperature to the velocity boundary-layer thickness is equal to $Pr^{-1/3}$. Work out the 'thermal thickness' in terms of the thickness of the velocity boundary layer for a value of $Pr=0.7$.

3.8. Explain why it is necessary to use concepts, such as the displacement thickness and the momentum thickness, for a boundary layer in order to obtain a boundary-layer thickness that is largely independent of the approximation used for the velocity profile in the neighbourhood of the surface.

It is found that the velocity u at a distance y from the surface may be expressed as a simple power function ($u \propto y^n$) for the turbulent boundary layer at a plane surface. What is the value of n if the ratio of the momentum thickness to the displacement thickness is 1.78?

3.9. Derive the momentum equation for the flow of a fluid over a plane surface for conditions where the pressure gradient along the surface is negligible. By assuming a sine function for the variation of velocity with distance from the surface (within the boundary layer) for streamline flow, obtain an expression for the boundary-layer thickness as a function of distance from the leading edge of the surface.

3.10. Derive the momentum equation for the flow of a viscous fluid over a small plane surface.

Show that the velocity profile in the neighbourhood of the surface may be expressed as a sine function that satisfies the boundary conditions at the surface and at the outer edge of the boundary layer.

Obtain the boundary-layer thickness and its displacement thickness as a function of the distance from the leading edge of the surface, when the velocity profile is expressed as a sine function.

3.11. Derive the momentum equation for the flow of a fluid over a plane surface for conditions where the pressure gradient along the surface is negligible. By assuming a sine function for the variation of velocity with distance from the surface (within the boundary layer) for streamline flow, obtain an expression for the boundary-layer thickness as a function of distance from the leading edge of the surface.

3.12. Derive the momentum equation for the flow of a viscous fluid over a small plane surface. Show that the velocity profile in the neighbourhood of the surface may be expressed as a sine function that satisfies the boundary conditions at the surface and at the outer edge of the boundary layer.

Obtain the boundary-layer thickness and its displacement thickness as a function of the distance from the leading edge of the surface, when the velocity profile is expressed as a sine function.

4.1. If the temperature rise per metre length along a pipe carrying air at 12.2 m/s is 66 K, what will be the corresponding pressure drop for a pipe temperature of 420 K and an air temperature of 310 K?

The density of air at 310 K is 1.14 kg/m^3.

4.2. It is required to warm a quantity of air from 289 to 313 K by passing it through a number of parallel metal tubes of inner diameter 50 mm maintained at 373 K. The pressure drop must not exceed 250 N/m^2. How long should the individual tubes be?

The density of air at 301 K is 1.19 kg/m^3, and the coefficients of heat transfer by convection from tube to air are 45, 62, and 77 W/m^{2}K for velocities of 20, 24, and 30 m/s at 301 K, respectively.

4.3. Air at 330 K, flowing at 10 m/s, enters a pipe of inner diameter 25 mm, maintained at 415 K. The drop of static pressure along the pipe is 80 N/m^2 per metre length. Using the Reynolds analogy between heat transfer and friction, estimate the temperature of the air 0.6 m along the pipe.

4.4. Air flows at 12 m/s through a pipe of inside diameter 25 mm. The rate of heat transfer by convection between the pipe and the air is 60 W/m^{2}K. Neglecting the effects of temperature variation, estimate the pressure drop per metre length of pipe.

4.5. Air at 320 K and atmospheric pressure is flowing through a smooth pipe of 50 mm internal diameter, and the pressure drop over a 4 m length is found to be 1.5 kN/m^2. Using Reynolds analogy, by how much would you expect the air temperature to fall over the first metre of pipe length if the wall temperature there is kept constant at 295 K?

$$\text{Viscosity of air} = 0.018 \, \text{mN s/m}^2$$

$$\text{Specific heat of air} = 1.05 \, \text{kJ/kg K}$$

4.6. Obtain an expression for the simple Reynolds analogy between heat transfer and friction. Indicate the assumptions that are made in the derivation and the conditions under which you would expect the relation to be applicable.

The Reynolds number of a gas flowing at 2.5 kg/m^{2}s through a smooth pipe is 20,000. If the specific heat of the gas at constant pressure is 1.67 kJ/kg K, what will the heat-transfer coefficient be?

4.7. Explain Prandtl's concept of a 'mixing length'. What parallels can you draw between the mixing length and the mean free path of the molecules in a gas?

The ratio of the mixing length to the distance from the pipe wall has a constant value of 0.4 for the turbulent flow of a fluid in a pipe. What is the value of the pipe friction factor if the ratio of the mean velocity to the axial velocity is 0.8?

4.8. The velocity profile in the neighbourhood of a surface for a Newtonian fluid may be expressed in terms of a dimensionless velocity u^+ and a dimensionless distance y^+ from the surface. Obtain the relation between u^+ and y^+ in the laminar sublayer. Outside the laminar sublayer, the relation is

$$u^+ = 2.5 \ln y^+ + 5.5$$

At what value of y^+ does the transition from the laminar sublayer to the turbulent zone occur?

In the 'universal velocity profile', the laminar sublayer extends to values of $y^+=5$, the turbulent zone starts at $y^+=30$, and the range $5<y^+<30$, the buffer layer, is covered by a second linear relation between u^+ and $\ln y^+$. What is the *maximum* difference between the values of u^+, in the range $5<y^+<30$, using the two methods of representation of the velocity profile?

Definitions are as follows:

$$u^+ = \frac{u_x}{u^*}$$

$$y^+ = \frac{yu^*\rho}{\mu}$$

$$u^{*2} = \frac{R}{\rho}$$

where

u_x is the velocity at distance y from surface,

R is the wall shear stress,

p and μ are the density and the viscosity of fluid, respectively.

4.9. Calculate the rise in temperature of water passed at 4 m/s through a smooth 25 mm diameter pipe, 6 m long. The water enters at 300 K, and the temperature of the wall of the tube can be taken as approximately constant at 330 K. Use the following:
   (a) The simple Reynolds analogy
   (b) The Taylor–Prandtl modification
   (c) The buffer lays equation
   (d) $Nu=0.023Re^{0.8}Pr^{0.33}$
   Comment on the differences in the results so obtained.

4.10. Calculate the rise in temperature of a stream of air, entering at 290 K and passing at 4 m/s through the tube maintained at 350 K, other conditions remaining the same as detailed in Problem 12.9.

4.11. Air flows through a smooth circular duct of internal diameter 0.25 m at an average velocity of 15 m/s. Calculate the fluid velocity at points 50 and 5 mm from the wall. What will be the thickness of the laminar sublayer if this extends to $u^+=y^+=5$? The density of the air may be taken as 1.12 kg/m^3 and the viscosity as 0.02 mN s/m^2.

4.12. Obtain the Taylor–Prandtl modification of the Reynolds Analogy for momentum and heat transfer, and give the corresponding relation for mass transfer (no bulk flow).

An airstream at approximately atmospheric temperature and pressure and containing a low concentration of carbon disulphide vapour is flowing at 38 m/s through a series of

50 mm diameter tubes. The inside of the tubes is covered with a thin film of liquid, and both heat and mass transfer are taking place between the gas stream and the liquid film. The film heat-transfer coefficient is found to be 100 W/m^{2}K. Using a pipe friction chart and assuming the tubes to behave as smooth surfaces, calculate

(a) the film mass-transfer coefficient,

(b) the gas velocity at the interface between the laminar sublayer and the turbulent zone of the gas.

 Specific heat of air $= 1.0$ kJ/kg K

 Viscosity of air $= 0.02$ mN s/m^2

 Diffusivity of carbon disulphide vapour in air $= 1.1 \times 10^{-5}$ m^2/s

 Thermal conductivity of air $0.024 = $ W/m K

4.13. Obtain the Taylor–Prandtl modification of the Reynolds' analogy between momentum and heat transfer and write down the corresponding analogy for mass transfer. For a particular system, a mass-transfer coefficient of 8.71×10^{-8} m/s and a heat-transfer coefficient of 2730 W/m^2 K were measured for similar flow conditions. Calculate the ratio of the velocity in the fluid where the laminar sublayer terminates, to the stream velocity:

$$\text{Molecular diffusivity} = 1.5 \times 10^{-9} \text{m}^2/\text{s}$$

$$\text{Viscosity} = 1 \text{ mN s/m}^2$$

$$\text{Density} = 1000 \text{ kg/m}^3$$

$$\text{Thermal conductivity} = 0.48 \text{ W/m K}$$

$$\text{Specific heat capacity} = 4.0 \text{ kJ/kg K}$$

4.14. Heat and mass transfer are taking place simultaneously to a surface under conditions where the Reynolds analogy between momentum, heat, and mass transfer may be applied. The mass transfer is of a single component at a high concentration in a binary mixture and the other component of which undergoes no net transfer. Using the Reynolds analogy, obtain a relation between the coefficients for heat transfer and for mass transfer.

4.15. Derive the Taylor–Prandtl modification of the Reynolds Analogy between momentum and heat transfer.

 In a shell-and-tube condenser, water flows through the tubes that are 10 m long and 40 mm in diameter. The pressure drop across the tubes is 5.6 kN/m^2, and the effects of entry and exit losses may be neglected. The tube walls are smooth, and flow may be taken as fully developed. The ratio of the velocity at the edge of the laminar sublayer to the mean velocity of flow may be taken as $2 Re^{-0.125}$, where Re is the Reynolds number in the pipeline.

If the tube walls are at an approximately constant temperature of 393 K and the inlet temperature of the water is 293 K, estimate the outlet temperature.

Physical properties of water, density 1000 kg/m^3

Viscosity 1 mN s/m^2

Thermal conductivity 0.6 W/m K

Specific heat capacity 4.2 kJ/kg K

4.16. Explain the importance of the universal velocity profile and derive the relation between the dimensionless derivative of velocity u^+ and the dimensionless derivative of distance from the surface y^+, using the concept of Prandtl's mixing length λ_E.

It may be assumed that the fully turbulent portion of the boundary layer starts at $y^+ = 30$; that the ratio of the mixing length λ_E to the distance y from the surface, $\lambda_E/y = 0.4$; and that for a smooth surface $u^+ = 14$ at $y^+ = 30$.

If the laminar sublayer extends from $y^+ = 0$ to $y^+ = 5$, obtain the equation for the relation between u^+ and y^+ in the buffer zone, and show that the ratio of the eddy viscosity to the molecular viscosity increases linearly from 0 to 5 through this buffer zone.

4.17. Derive the Taylor–Prandtl modification of the Reynolds analogy between heat and momentum transfer and express it in a form in which it is applicable to pipe flow.

If the relationship between the Nusselt number Nu, Reynolds number Re, and Prandtl number Pr is

$$Nu = 0.023Re^{0.8}/Pr^{0.33}$$

calculate the ratio of the velocity at the edge of the laminar sublayer to the velocity at the pipe axis for water ($Pr = 10$) flowing at a Reynolds number (Re) of 10,000 in a smooth pipe. Use the pipe friction chart.

4.18. Obtain a dimensionless relation for the velocity profile in the neighbourhood of a surface for the turbulent flow of a liquid, using Prandtl's concept of a 'mixing length' (universal velocity profile). Neglect the existence of the buffer layer and assume that, outside the laminar sublayer, eddy transport mechanisms dominate. Assume that in the turbulent fluid, the mixing length λ_E is equal to 0.4 times the distance y from the surface and that the dimensionless velocity u^+ is equal to 5.5 when the dimensionless distance y^+ is unity.

Show that, if the Blasius relation is used for the shear stress R at the surface, the thickness of the laminar sublayer δ_b is approximately 1.07 times that calculated on the assumption that the velocity profile in the turbulent fluid is given by Prandtl's one-seventh power law.

Blasius equation is as follows:

$$\frac{\rho}{\rho u_s^2} = 0.028 \left(\frac{u_s \delta \rho}{\mu} \right)^{-0.25}$$

where

p and μ are the density and viscosity of the fluid,

u_s is the stream velocity,

δ is the total boundary-layer thickness.

4.19. Obtain the Taylor-Prandtl modification of the Reynolds analogy between momentum transfer and mass transfer (equimolecular counterdiffusion) for the turbulent flow of a fluid over a surface. Write down the corresponding analogy for heat transfer. State clearly the assumptions that are made. For turbulent flow over a surface, the film heat-transfer coefficient for the fluid is found to be 4 kW/m^2 K. What would the corresponding value of the mass-transfer coefficient be, given the following physical properties?

$$\text{Diffusivity } D = 5 \times 10^{-9} \, \text{m}^2/\text{s}$$
$$\text{Thermal conductivity } k = 0.6 \, \text{W}/\text{m K}$$
$$\text{Specific heat capacity } C_p = 4 \, \text{kJ}/\text{kg K}$$
$$\text{Density } \rho = 1000 \, \text{kg}/\text{m}^3$$
$$\text{Viscosity } \mu = 1 \, \text{mN s}/\text{m}^2$$

Assume that the ratio of the velocity at the edge of the laminar sublayer to the stream velocity is (a) 0.2 and (b) 0.6.

Comment on the difference in the two results.

4.20. By using the simple Reynolds analogy, obtain the relation between the heat-transfer coefficient and the mass-transfer coefficient for the gas phase for the absorption of a soluble component from a mixture of gases. If the heat-transfer coefficient is 100 W/m^2 K, what will the mass-transfer coefficient be for a gas of specific heat capacity C_p of 1.5 kJ/kg K and density 1.5 kg/m^3? The concentration of the gas is sufficiently low for bulk flow effects to be negligible.

4.21. The velocity profile in the neighbourhood of a surface for a Newtonian fluid may be expressed in terms of a dimensionless velocity u^+ and a dimensionless distance y^+ from the surface. Obtain the relation between u^+ and y^+ in the laminar sublayer. Outside the laminar sublayer, the relation takes the form

$$u^+ = 2.5 \, \ln y^+ + 5.5$$

At what value of y^+ does the transition from the laminar sublayer to the turbulent zone occur?

In the 'universal velocity profile', the laminar sublayer extends to values of $y^+ = 5$, the turbulent zone starts at $y^+ = 30$, and the range $5 < y^+ < 30$, the buffer layer, is covered by a second linear relation between u^+ and $\ln y^+$. What is the *maximum* difference between the values of u^+, in the range $5 < y^+ < 30$, using the two methods of representation of the velocity profile?

Definitions are as follows:

$$u^+ = \frac{u_x}{u^*}$$

$$y^+ = \frac{yu^*\rho}{\mu}$$

$$u^{*2} = R/\rho$$

where

u_x is velocity at distance y from surface,

R is wall shear stress,

ρ and μ are the density and viscosity of the fluid, respectively.

4.22. In the universal velocity profile, a 'dimensionless' velocity u^+ is plotted against ln y^+, where y^+ is a 'dimensionless' distance from the surface. For the region where eddy transport dominates (eddy kinematic viscosity), the ratio of the mixing length (λ_E) to the distance (y) from the surface may be taken as approximately constant and equal to 0.4. Obtain an expression for du^+/dy^+ in terms of y^+.

In the buffer zone, the ratio of du^+/dy^+ to y^+ is *twice* the value calculated above. Obtain an expression for the eddy kinematic viscosity E in terms of the kinematic viscosity (μ/ρ) and y^+. On the assumption that the eddy thermal diffusivity E_H and the eddy kinematic viscosity E are equal, calculate the value of the temperature gradient in a liquid flowing over the surface at $y^+ = 15$ (which lies within the buffer layer) for a surface heat flux of 1000 W/m² The liquid has a Prandtl number of 7 and a thermal conductivity of 0.62 W/m K.

4.23. Derive an expression relating the pressure drop for the turbulent flow of a fluid in a pipe to the heat-transfer coefficient at the walls on the basis of the simple Reynolds analogy. Indicate the assumptions that are made and the conditions under which to apply closely. Air at 320 K and atmospheric pressure is flowing through a smooth pipe of 50 mm internal diameter, and the pressure drop over a 4 m length is found to be 150 mm water gauge. By how much would be expected the air temperature to fall over the first metre if the wall temperature there is 290 K?

$$\text{Viscosity of air} = 0.018\,\text{mN s/m}^2$$

$$\text{Specific heat capacity }(C_p) = 1.05\,\text{kJ/kg K}$$

$$\text{Molecular volume} = 22.4\,\text{m}^3/\text{kmol at 1 bar and 273 K}$$

5.1. In a process in which benzene is used as a solvent, it is evaporated into dry nitrogen. The resulting mixture at a temperature of 297 K and a pressure of 101.3 kN/m² has a relative humidity of 60%. It is desired to recover 80% of the benzene present by cooling to 283 K and compressing to a suitable pressure. What must this pressure be?

Vapour pressures of benzene : at $297\,K = 12.2\,kN/m^2$; at $283\,K = 6.0\,kN/m^2$.

5.2. $0.6\,m^3/s$ of gas is to be dried from a dew point of 294 K to a dew point of 277.5 K. How much water must be removed, and what will be the volume of the gas after drying?

Vapour pressure of water at $294\,K = 2.5\,kN/m^2$

Vapour pressure of water at $277.5\,K = 0.85\,kN/m^2$

5.3. Wet material, containing 70% moisture, is to be dried at the rate of 0.15 kg/s in a countercurrent dryer to give a product containing 5% moisture (both on a wet basis). The drying medium consists of air heated to 373 K and containing water vapour equivalent to a partial pressure of $1.0\,kN/m^2$. The air leaves the dryer at 313 K and 70% saturated. Calculate how much air will be required to remove the moisture. The vapour pressure of water at 313 K may be taken as $7.4\,kN/m^2$.

5.4. $30,000\,m^3$ of coal gas (measured at 289 K and $101.3\,kN/m^2$ saturated with water vapour) is compressed to $340\,kN/m^2$ pressure and cooled to 289 K, and the condensed water is drained off. Subsequently, the pressure is reduced to $170\,kN/m^2$, and the gas is distributed at this pressure and at 289 K. What is the percentage humidity of the gas after this treatment?

The vapour pressure of water at 289 K is $1.8\,kN/m^2$.

5.5. A rotary countercurrent dryer is fed with ammonium nitrate containing 5% moisture at the rate of 1.5 kg/s and discharges the nitrate with 0.2% moisture. The air enters at 405 K and leaves at 355 K, the humidity of the entering air being 0.007 kg of moisture per kilogram of dry air. The nitrate enters at 294 K and leaves at 339 K.

Neglecting radiation losses, calculate the mass of dry air passing through the dryer and the humidity of the air leaving the dryer:

Latent heat of water at $294\,K = 2450\,kJ/kg$

Specific heat of ammonium nitrate $= 1.88\,kJ/kg\,K$

Specific heat of dry air $= 0.99\,kJ/kg\,K$

Specific heat of water vapour $= 2.01\,kJ/kg\,K$

5.6. Material is fed to a dryer at the rate of 0.3 kg/s, and the moisture removed is 35% of the wet charge. The stock enters and leaves the dryer at 324 K. The air temperature falls from 341 to 310 K, its humidity rising from 0.01 to 0.02 kg/kg.

Calculate the heat loss to the surroundings:

Latent heat of water at $324\,K = 2430\,kJ/kg$

Specific heat of dry air $= 0.99\,kJ/kg\,K$

Specific heat of water vapour $= 2.01\,kJ/kg\,K$

5.7. A rotary dryer is fed with sand at the rate of 1 kg/s. The feed is 50% wet, and the sand is discharged with 3% moisture. The air enters at 380 K with a solute humidity of 0.007 kg/kg. The wet sand enters at 294 K and leaves at 309 K, and the air leaves at 310 K.

Calculate the mass of air passing through the dryer and the humidity of the air leaving the dryer. Allow a radiation loss of 25 kJ/kg of dry air:

$$\text{Latent heat of water at } 294 \text{ K} = 2450 \text{ kJ/kg}$$
$$\text{Specific heat of sand} = 0.88 \text{ kJ/kg K}$$
$$\text{Specific heat of dry air} = 0.99 \text{ kJ/kg K}$$
$$\text{Specific heat of vapour} = 2.01 \text{ kJ/kg K}$$

5.8. Water is to be cooled in a packed tower from 330 to 295 K by means of air flowing countercurrently. The liquid flows at the rate of 275 cm^3/m^2 s and the air at 0.7 m^3/m^2 s. The entering air has a temperature of 295 K and a humidity of 20%. Calculate the required height of tower and the condition of the air leaving at the top.

The whole of the resistance to heat and mass transfer may be considered as being within the gas phase, and the product of the mass-transfer coefficient and the transfer surface per unit volume of column ($h_D a$) may be taken as 0.2 s^{-1}.

5.9. Water is to be cooled in a small packed column from 330 to 285 K by means of air flowing countercurrently. The rate of flow of liquid is 1400 cm^3/m^2 s, and the flowrate of the air, which enters at a temperature of 295 K and a humidity of 60%, is 3.0 m^3/m^2 s. Calculate the required height of tower if the whole of the resistance to heat and mass transfer may be considered as being in the gas phase, and the product of the mass-transfer coefficient and the transfer surface per unit volume of column is 2 s^{-1}.

What is the condition of the air that leaves at the top?

5.10. Air containing 0.005 kg of water vapour per kilogram of dry air is heated to 325 K in a dryer and passed to the lower shelves. It leaves these shelves at 60% humidity, reheated to 325 K, and passed over another set of shelves, again leaving at 60% humidity. This is again reheated for the third and fourth sets of shelves, after which the air leaves the dryer. On the assumption that the material in each shelf has reached the wet-bulb temperature and that heat losses from the dryer may be neglected, determine
   (a)   the temperature of the material on each tray,
   (b)   the rate of water removal if 5 m^3/s of moist air leaves the dryer,
   (c)   the temperature to which the inlet air would have to be raised to carry out the drying in a single stage.

5.11. 0.08 m^3/s of air at 305 K and 60% humidity is to be cooled to 275 K. Calculate, by use of a psychrometric chart, the amount of heat to be removed for each 10 K interval of the cooling process. What total mass of moisture will be deposited? What is the humid heat of the air at the beginning and end of the process?

5.12. A hydrogen stream at 300 K and atmospheric pressure has a dew point of 275 K. It is to be further humidified by adding to it (through a nozzle) saturated stream at 240 kN/m² at the rate of 1 kg steam, 30 kg of hydrogen feed. What will be the temperature and humidity of the resultant stream?

5.13. In a countercurrent packed column, n-butanol flows down at the rate of 0.25 kg/m² s and is cooled from 330 to 295 K. Air at 290 K, initially free of n-butanol vapour, is passed up the column at the rate of 0.7 m³/m² s. Calculate the required height of tower and the condition of the exit air.

Data are as follows:

$$\text{Mass transfer coefficient per unit volume}: h_D a = 0.1\,\text{s}^{-1}$$

$$\text{Psychrometric ratio}: \frac{h}{h_D \rho_A s} = 2.34$$

$$\text{Heat transfer coefficients}: h_L = 3 h_G$$

$$\text{Latent heat of vaporisation of n} - \text{butanol}, \lambda = 590\,\text{kJ/kg}$$

$$\text{Specific heat of liquid n} - \text{butanol}: C_L = 2.5\,\text{kJ/kg K}$$

$$\text{Humid heat of gas}: s = 1.05\,\text{kJ/kg K}$$

Temperature (K)	Vapour Pressure of n-Butanol (kN/m²)
295	0.59
300	0.86
305	1.27
310	1.75
315	2.48
320	3.32
325	4.49
330	5.99
335	7.89
340	10.36
345	14.97
350	17.50

5.14. Estimate the height and base diameter of a natural draught hyperbolic cooling tower that will handle a flow of 5000 kg/s water entering at 300 K and leaving at 294 K. The dry-bulb air temperature is 287 K, and the ambient wet-bulb temperature is 284 K.

Index

Note: Page numbers followed by *f* indicate figures and *t* indicate tables.

Printed in the United States
by Baker & Taylor Publisher Services